T0318354

CARBON CAPTURE AND STORAGE IN INTERNATIONAL ENERGY POLICY AND LAW

CARBON CAPTURE AND STORAGE IN INTERNATIONAL ENERGY POLICY AND LAW

Edited by

HIRDAN KATARINA DE MEDEIROS COSTA
Institute of Energy and Environment, University of São Paulo, São Paulo, Brazil

CAROLINA ARLOTA
Visiting Assistant Professor of Law at the University of Oklahoma, College of Law, Norman, OK, United States

ELSEVIER

Elsevier
Radarweg 29, PO Box 211, 1000 AE Amsterdam, Netherlands
The Boulevard, Langford Lane, Kidlington, Oxford OX5 1GB, United Kingdom
50 Hampshire Street, 5th Floor, Cambridge, MA 02139, United States

Notices
Knowledge and best practice in this field are constantly changing. As new research and experience broaden our understanding, changes in research methods, professional practices, or medical treatment may become necessary.

Practitioners and researchers must always rely on their own experience and knowledge in evaluating and using any information, methods, compounds, or experiments described herein. In using such information or methods they should be mindful of their own safety and the safety of others, including parties for whom they have a professional responsibility.

To the fullest extent of the law, neither the Publisher nor the authors, contributors, or editors, assume any liability for any injury and/or damage to persons or property as a matter of products liability, negligence or otherwise, or from any use or operation of any methods, products, instructions, or ideas contained in the material herein.

British Library Cataloguing-in-Publication Data
A catalogue record for this book is available from the British Library

Library of Congress Cataloging-in-Publication Data
A catalog record for this book is available from the Library of Congress

ISBN: 978-0-323-85250-0

For Information on all Elsevier publications
visit our website at https://www.elsevier.com/books-and-journals

Publisher: Graham Nisbet
Editorial Project Manager: Aleksandra Packowska
Production Project Manager: Kamesh Ramajogi
Cover Designer: Christian J. Bilbow

Typeset by MPS Limited, Chennai, India

Dedications

Hirdan's dedication: Luiza e Heloisa (grandmother and mother).

Carolina's dedication: à minha querida mamãe.

Contents

I

Conceptualizing international energy law and carbon capture and storage (CCS) in light of climate change

II

Case studies on CCS and related policies, and their consequences for climate change

4. The institutional approach of climate change at the multinational level: the new paradigm of the Brazilian legislative experience 61

Gustavo de Melo Ribeiro, Hirdan Katarina de Medeiros Costa, André Felipe Simões and Carolina Arlota

5. Carbon capture and storage technologies and efforts on climate change in Latin American and Caribbean countries 75

Romario de Carvalho Nunes and Hirdan Katarina de Medeiros Costa

6. Geologic CO_2 sequestration in the United States of America 107

Owen L. Anderson

7. The United Kingdom's experience in Carbon Capture and Storage projects: the current regulatory framework and related challenges   141

Isabela Morbach Machado e Silva and
Hirdan Katarina de Medeiros Costa

8. Regulatory framework carbon capture, utilization, and storage in Europe: a regulatory review and specific cases   155

Israel Araujo Lacerda de, Vitor Emanoel Siqueira Santos and
Hirdan Katarina de Medeiros Costa

9. Australian legislation on new mitigation technologies—the case of carbon capture and storage   167

Israel Araujo Lacerda de and Hirdan Katarina de Medeiros Costa

10. Carbon capture and storage: Intellectual property, innovation policy, and climate change   181

Matthew Rimmer

11. Negative-emission technologies and patent rights after COVID-19   205

Joshua D. Sarnoff

III

Comparative experiences around the world

IV

The current picture and future perspectives

List of contributors

Owen L. Anderson The University of Texas at Austin School of Law, Austin, TX, United States

Israel Araujo Lacerda de Institute of Energy and Environment, University of São Paulo, São Paulo, Brazil

Alexandre Sales Cabral Arlota Partner at Mattos Filho, Veiga Filho, Marrey Jr. e Quiroga Advogados, Rio de Janeiro, Brazil

Carolina Arlota Visiting Assistant Professor of Law at the University of Oklahoma, College of Law, Norman, OK, United States; Law and Economics Research Fellow, Capitalism and Rule of Law Project (Cap Law) at Antonin Scalia Law School, George Mason University, Fairfax, VA, United States

Hirdan Katarina de Medeiros Costa Institute of Energy and Environment, University of São Paulo, São Paulo, Brazil

Silvia Andrea Cupertino Institute of Energy and Environment, University of São Paulo, São Paulo, Brazil

Romario de Carvalho Nunes Institute of Energy and Environment, University of São Paulo, São Paulo, Brazil

Edmilson Moutinho dos Santos Institute of Energy and Environment, University of São Paulo, São Paulo, Brazil

Allan Ingelson Canadian Institute of Resources Law, University of Calgary, Faculty of Law at the University of Calgary, Calgary, Canada

Isabela Morbach Machado e Silva Energy Institute, University of São Paulo, Brazil

Yane Marcelle Pereira Silva Institute of Energy and Environment, University of São Paulo, São Paulo, Brazil

Teresa Castillo Quevedo COFECE, Offentlicher, Dienst, Benito Juarez, Ciudad de Mexico

Gustavo de Melo Ribeiro Institute of Energy and Environment, University of São Paulo, São Paulo, Brazil

Matthew Rimmer Faculty of Business and Law, Queensland University of Technology (QUT), Brisbane, Australia

Haline Rocha Institute of Energy and Environment, University of São Paulo, São Paulo, Brazil

Vitor Emanoel Siqueira Santos Institute of Energy and Environment, University of São Paulo, São Paulo, Brazil

Joshua D. Sarnoff Depaul University College of Law, Chicago, IL, United States

Paulo Negrais Seabra Institute of Energy and Environment, University of São Paulo, São Paulo, Brazil

Oscar W. Serrate Research Centre for Gas Innovation, Polytechnic School of University of São Paulo, São Paulo, Brazil

André Felipe Simões Institute of Energy and Environment, University of São Paulo, São Paulo, Brazil; School of Arts, Sciences and Humanities of the University of São Paulo, São Paulo, Brazil

Nathália Weber Research Centre for Gas Innovation, Polytechnic School of University of São Paulo, São Paulo, Brazil

Foreword

In several reports published over the years by the Intergovernmental Panel on Climate Change (IPCC), such as the 2018 *Summary for Policymakers of the IPCC Special Report on Global Warming of 1.5°C*, almost all the decarbonization pathways considered rely, to some extent, on carbon removal approaches to achieve net negative emissions after 2050. Likewise, recent extensive reports such as the International Energy Agency's Carbon Capture, Utilization, and Storage (CCUS) in *Clean Energy Transitions* (2020) (in a global context) and the National Academies of Sciences, Engineering, and Medicine on "Accelerating Decarbonization of the United States Energy System" (2021) (in the US context) also underscore the potential role of technology-based solutions, such as CCUS and Direct Air Capture and Storage (DACS), especially if emerging climate mitigation options would not be able to provide cost-effective alternatives in some of the difficult-to-decarbonize sectors of the economy.

The process of developing a complete end-to-end CCS system that captures carbon dioxide (CO_2) from large stationary point sources, and transports and injects the compressed CO_2 into a suitable deep underground geologic formation, is capital-intensive and requires a thorough understanding of the national and international legal, technical, project economics, and financial risks, as well as medium- to long-term environmental risk assessment issues. As excellently presented and discussed in this book on *"Carbon Capture and Storage (CCS) in International Energy Policy and Law: Perspectives on Sustainable Development, Climate Change, and Energy Transition"* the issues are often contextual and interdisciplinary. The deployment of

CCUS projects requires collaboration between private and public stakeholders, including governments in jurisdictions with rights to underground storage options, and an integrated approach that incorporates potential CO_2 and carbon utilization schemes. In the United States for instance, the current uses of captured CO_2 include Enhanced Oil Recovery processes, food and beverage processing, cleaning and solvent use, fumigants and herbicides, fire-fighting equipment, industrial and municipal water/wastewater treatment, pulp and paper, metal fabrication, synthetic fuels, etc. Other potential scalable applications are well-reported in the *Global Status of CCS Report 2020* published by the Global CCS Institute.

As research and development, as well as investment in the required technologies, gain traction, there is a key role to be played by international energy law and policy. Policymakers in relevant countries are also in a position to facilitate the required carbon reduction and net-zero objectives. The contributors and leading coeditors of this book have provided a comprehensive examination of the relevant international energy policy and legal issues. The authors rightly conclude that international energy law as an autonomous field can foster the incorporation of contemporary climate change mitigation policies in current regulatory policy and sustainable development scenarios.

Dr. Tade Oyewunmi[1,2]

[1]*Assistant Professor of Law and Senior Energy Research Fellow, Institute for Energy and the Environment, Vermont Law School, South Royalton, VT, United States* [2]*Visiting Assistant Professor of Law, Oregon University Law School, Eugene, OR, United States*

Acknowledgments

Hirdan's acknowledgments:

Hirdan Katarina de Medeiros Costa is grateful to the "Research Center for Gas Innovation—RCGI" (Fapesp Proc. 2014 / 50279–4), supported by FAPESP and Shell, organized by the University of São Paulo, and the strategic importance of the support granted by the ANP (National Agency of Petroleum, Natural Gas and Biofuels of Brazil) through the R&D clause. She also thanks to the support from the National Agency for Petroleum, Natural Gas and Biofuels Human Resources Program (PRH-ANP), funded by resources from the investment of oil companies qualified in the R, D&I clauses from ANP Resolution number no 50/2015 (PRH 33.1—Related to Call No 1/2018/PRH-ANP; Grant FINEP/FUSP/ USP Ref. 0443/19).

Carolina's acknowledgments:

Carolina Arlota is grateful to Professors Nuno Garoupa and Charlotte Ku, for their important support to her academic projects. Professors Brian McCall and Murray Tabb provided great encouragement. This work significantly benefited from conversations with Professors Peter Yu and Kim Talus.

Carolina Arlota acknowledges the research support granted by the Interim Dean of the University of Oklahoma College of Law, Kathleen Guzman, and the work of Professor Joel Wegemer in implementing such support. Natosha Greene provided excellent research assistance.

Editors' acknowledgments:

The Editors jointly acknowledge the comments of three anonymous referees in the initial phase of this project. The Editors appreciate the outstanding work of Nisbet Graham, Aleksandra Packowska, and Kamesh Ramajogi.

Carbon capture and storage in international energy policy and law: perspectives on sustainable development, climate change, and energy transition

The project of this book dates to late 2018, when we were initially planning to write an article on the prominent role of carbon capture and storage (CCS) in international energy law and policy. CCS has been the subject of Costa's research. Through her passion, CCS became a shared interest. Challenges appeared at the outset of our plan, however. First, there were some conceptual controversies about the denomination itself: CCS, carbon capture, use, and storage (CCUS), or what else? Their precise meaning was unclear. Second, existing outstanding books on international energy law either neglect CCS or barely address it.[1] Third, despite excellent books focusing on the technical aspects of CCS,[2] the interplay of implementation in Australia, and the use of CCS in specific ecosystems, such as permafrost, and in forests; *Carbon Capture and Storage: Technologies, Policies, Economics, and Implementation Strategies*, edited by Bash O. Dabboussi et al. (2011), which is divided in three parts addressing the conceptualization of CCS, the technology itself, and deployment drivers at the time; *Geological Storage of Carbon Dioxide (CO2): Geoscience, Technologies, Environmental Aspects and Legal Framework*, edited by Jon Gluyas and Simon Mathias (2013), which has a preponderantly technical emphasis but presents CCS case studies (including pilot projects) located in Germany, Australia, the North Sea, and the Netherlands; *Carbon Capture, Storage, and Utilization*, by Malti Goel (2014), which targets the technological aspects involved in light of decarbonization needs; *Introduction to Carbon Capture and Sequestration*, edited by Smit Berent et al. (2014), which discusses the technological aspects (mainly engineering and chemistry) and some energy policy implications of CCS; *Carbon Capture, Storage, and Usage*, edited by Wilhelm Kuckshinrichs and Jürgen-Friedric Hake (2015), which addresses the technological issues and industry sectors (energy, transportation) in German and EU contexts; *Climate Intervention: Carbon Dioxide Removal and Reliable Sequestration*, edited by the United States National Academy of Sciences (2015), which emphasizes the technological aspects of carbon dioxide removal (CDR), which is different from CCS, because the former removes carbon dioxide form the ambient air; *Recent Advances in Carbon Capture and Storage*, edited by Yongseung Yun (2017), is a technical engineering-oriented volume.

---

[1] See, e.g., *Research Handbook on International Energy Law* (ed. by Kim Talus, 2014); *International Energy Law*, by Rex Zedalis (2016); and *International Energy Law*, by Mohammad Naseem (2017). *International Law for the Energy and Environment*, by Patricia Park (2013), however, provides an overview of CCS.

[2] Among other leading research at the time, there were *Carbon Capture: Sequestration and Storage*, edited by R. E. Hester and R.M Harrison (2009), which was published by the Royal Society of Chemistry, with contributors discussing several topics, such as renewables, CCS

CCS (defined as the removal of gas from an emissions stream, such as a power plant, for instance, before such gas reaches the ambient air)[3] and the objectives of the Paris Agreement on Climate Change, which was signed on December 12, 2015, were still unaddressed. Accordingly, it became clear that we needed to expand our initial project to something more ambitious: a book investigating the role of CCS as a crucial technological innovation for implementing the principal goals of the Paris Agreement, namely, "holding the increase in the global average temperature to well below 2°C above preindustrial levels and pursuing efforts to limit the temperature increase to 1.5°C above preindustrial levels, recognizing that this would significantly reduce the risks and impacts of climate change" (The Paris Agreement on Climate Change, Art. 2, (a)).

Research about the effective implementation of the main objectives of the Paris Agreement on Climate Change is paramount because of current uncertainties in the international energy scenario, which were aggravated in the aftermath of the COVID-19 pandemic and increasing GHG (greenhouse gas) emissions in particular. International energy law and policies aiming to implement the goals established by the Paris Agreement should be based on mitigation because the World Bank has long defined mitigation action as the best insurance against an uncertain future.[4] Moreover, a recent report by the United

Nations Environment Programme found, "resulting atmospheric concentrations of major GHGs (carbon dioxide (CO_2), methane (CH_4), and nitrous oxide (N_2O)) continued to increase in both 2019 and 2020,"[5] concluding that to stabilize global warming, continuous reductions in emissions are required to achieve net-zero carbon dioxide (UNEP, 2020). Because CCS targets this type of mitigation, research on CCS is of immediate interest.

Increasing interest in CCS is also based on the extraordinary difficulties associated with reducing GHG emissions. GHGs spread globally and some—carbon dioxide, specifically—remain in the atmosphere for centuries; hence, the current GHG levels are the product of infinite sources throughout the past 200 years.[6] Global GHG emissions are the result of several factors, including technology, population size, and affluence, despite energy policies mainly focusing on technology.[7]

In this context, *Carbon Capture and Storage in International Energy Policy and Law: Perspectives on Sustainable Development, Climate Change, and Energy Transition* fills a void in the current literature.[8] Updated attempts to approach CCS from an international and comparative perspective are rare, if any. Emphasis on policies and

---

[3] Michael B. Gerrard, Introduction and Overview, in *Climate Engineering and the Law: Regulation and Liability for Solar Radiation Management and Carbon Dioxide Removal* (Michael B. Gerrard and Tracy Hester, (eds.), 2018), at 3.

[4] World Bank, Turn Down the Heat: Why a 4°C Warmer World Must Be Avoided 2 (2012), http://documents.worldbank.org/curated/en/865571468149107611/pdf/NonAsciiFileName0.pdf.

[5] U.N. Environmental Programme, Emissions Gap Report 2020: Executive Summary, at V (2020), https://wedocs.unep.org/bitstream/handle/20.500.11822/34438/EGR20ESE.pdf?sequence = 8.

[6] Gerrard (2018), at 6.

[7] Id.

[8] A notable exception and reference in the field is the second edition of *Carbon Capture and Storage: Emerging Legal and Regulatory Issues*, edited by Ian Havercroft et al. (2018), which covers technical and regulatory aspects of CCS targeting common law jurisdictions (Australia, India, United States, and UK), the European Union, in general, and China.

institutional approaches in the context of the Paris Agreement is scant, at best. Our review of the literature yields results based on particular sectors or country-specific studies.[9] Our project is different because its emphasis is on the policies involved, including the transboundary consequences of soaring GHG emissions.

This book advances the research on CCS on several fronts, and it is not restricted to a single technical audience or sector. On the contrary, the book benefits from an interdisciplinary approach, having authors who are engineers, energy professors, law professors, lawyers, and policymakers. The book also innovates in its transnational approach because its contributors are experts based in several countries around the world, such as Australia, Brazil, Canada, Mexico, and the United States. This is relevant because policy considerations are important for all actors involved in CCS research, regulation, and implementation, including industrial sectors. Regulators, scientists, lawyers, politicians, international actors, most notably, international institutions, and stakeholders around the globe will benefit from the comparative approach developed in this book. The policy analyses of CCS projects have a direct impact on those in the area where such projects are implemented and beyond. Hence, public participation is also discussed in the book.

The book presents its unique contributions in four main parts. Part I conceptualizes International Energy Law and CCS in light of climate change. In *International Energy Law: Still a Brave New World*, Carolina Arlota and Hirdan K.M. Costa discuss how the construction of an autonomous branch of legal science, called international energy law, deserves further study, specifically after the Paris Agreement (2015), while also looking at the current trends in modernizing the European Charter Treaty (ECT). Topics related to the energy transition, climate change, low-carbon economies, and international arbitration are also examined. The authors conclude that international energy law as an autonomous field may foster the incorporation of modern mitigation policies in current regulatory policy. In *The Energy of Cooperation*, Oscar W. Serrate and Nathália Weber present a compelling case for how cooperation on energy is key for a global energy policy that aims to reconcile competing national and transnational interests. *Climate Change Mitigation and*

---

[9] See, e.g., *The Social Dynamics of Carbon Capture and Storage*, by Nils Markusson et al. (2012), which discusses the perception and representation regarding CCS projects, focusing on public participation; *Carbon Capture and Storage*, by Michael Faure and Roy A. Partain (2017), which targets CCS's liability regimes and allocation of incentives for such regimes; *Carbon Capture*, by Howard J. Herzog (2018), which summarizes technological aspects of CCS in the larger context of decarbonization; *Biomass Energy with Carbon Capture and Storage (BECCS)*, edited by Clair Gough et al. (2018), which concentrates on biomass energy and CCS (and its technical and engineering capacity); *Carbon Capture, Storage, and Utilization: A Possible Climate Solution for Energy Industry*, edited by Malti Goel, M. Sudhakar, R.V. Shahi (2018), which addresses the technological aspects of CCS and its policy challenges specifically in coal-based economies, here with India as a paradigm; *Carbon Capture and Sequestration:* edited by Mai Bui and Niall McDowell (2019), which targets engineering and chemistry and briefly addresses energy policy; *Bioenergy with Carbon Capture and Storage*, edited by Jose Carlos Magalhaes Pires and Ana Luisa da Cunha Gonçalves (2019), which focuses on bioenergy, engineering and technology, with limited discussion on global policies; *Carbon Capture and Storage*, edited by Jose Carlos Magalhaes Pires (2019), is which technology and engineering oriented, and provides guidance on the environmental feasibility of CCS projects.

the Technological Specificities of CCS, coauthored by Romario Nunes, Vitor Emanoel Santos, and Hirdan K.M. Costa, provides a technical introduction to CCS, illustrating several uses of related technology; they further consider the role of CCS as an important alternative toward achieving decarbonization.

Part II presents case studies on CCS and related climate policies around the globe, including the specific challenges facing developing nations. Part II starts targeting the latter. In *The Institutional Approach of Climate Change in the Multinational Level: Lessons From the Brazilian Legislative Experience*, contributors Gustavo Ribeiro, Hirdan K.M. Costa, André Felipe Simões, and Carolina Arlota discuss the unique Brazilian federalism arrangement with three spheres of power and its consequences for climate policies. *CCS Technologies and Efforts on Climate Change in Latin American and Caribbean Countries*, jointly written by Romario Nunes and Hirdan K.M. Costa, investigates current CCS projects in the following developing countries: Colombia, Mexico, Brazil, and Trinidad and Tobago, discussing the main common challenges for effective CCS policies leading to their actual implementation in the Central and South American region.

In addition, Part II addresses the key common law jurisdictions and the supranational experience of the European Union. In *Geologic CO$_2$ Sequestration in the USA: The Allocation of Property Rights and Policy Implications Involved*, Owen Anderson updates the discussion on the legal models for securing the rights to use geologic pore space to sequester billions of metric tons of CO$_2$ deep underground as a way to impede climate change. The renewed interest in developing commercial-scale geologic carbon dioxide (CO$_2$) sequestration (GCS) has increased the urgency of resolving the

debate about the ownership of pore space rights in the United States for various commercial uses and the need for an appropriate regulatory framework in all countries. In *The United Kingdom's Experience in CCS Projects: The Current Regulatory Framework and Related Challenges*, Isabela M. Machado e Silva and Hirdan K.M. Costa analyze the challenges the United Kingdom faced when developing their earlier CCS projects, including their CCS legislation. The authors provide a critical view of the United Kingdom's CCS projects, drawing valuable lessons based on the UK regulatory framework and institutional design. Israel L. de Araújo, Vitor Emanoel Santos, and Hirdan K.M. Costa's *Carbon Capture, Utilization and Storage in Europe: A Regulatory Review and Specific Cases* overviews the main regulatory actions regarding CCUS in the European Union and Norway.

Part II also discusses the Australian experience in CCS because the country was among the pioneers in CCS. In *Australian Legislation on New Mitigation Technologies: The Case of CCS*, Israel L. de Araújo and Hirdan K.M. Costa survey the current Australian legislative framework on technologies aiming at the mitigation of GHG emissions. The chapter addresses the so-called Australian energy mix of energy resources, considering the incentives for key actors to engage in energy transition that favors diversified sources of energy. In *Carbon Capture and Storage: Intellectual Property, Innovation Policy, and Climate Change*, Matthew Rimmer focuses on technology development in respect to CCS—exploring the intellectual property, government funding, and innovation prizes. There has been a debate on whether CCS technologies should be classified as clean technologies for patent law and mechanisms such as green fast tracks. The technology has raised issues with respect to patent validity, patent infringement, patent licensing, and bankruptcy. There has also been discussion about the role of patent exceptions, such as technology transfer, public sector licensing, patent pools, compulsory licensing, and competition oversight. Apart from intellectual property, there has also been government funding of research into CCS technologies. In addition to intellectual property, there has also been extensive use of innovation prizes to encourage research and development regarding CCS. Overall, this chapter questions whether such incentives when it comes to CCS have been productive, especially given the rather lackluster outcomes in respect of research, development, and deployment thus far. In *Negative Emissions Technologies After COVID-19*, Joshua D. Sarnoff compares the climate emergency with the COVID-19 pandemic, focusing on the nature of negative emissions technologies (NETs) and of the IP rights that will be obtained to tackle our current climate emergency.

Part III presents comparative insights into CCS law and policy and their impact on the energy transition to a greener economy. Carolina Arlota and Hirdan K.M. Costa's *Who Is Taking Climate Change Seriously? Evidence From Evidence Based on a Comparative Analysis of the Carbon Capture and Storage (CCS) National Legal Framework in Brazil, Canada, the European Union and the United States* investigates the existing CCS legal framework in Brazil, Canada, the European Union, and the United States by using a comparative methodology. Gathering evidence of the different federal (and the EU's supranational) legislative experiences on CCS, the authors discuss the main incentives for pursuing carbon emissions reductions under the Paris Agreement. In *Reducing CO_2 Emissions Through Carbon Capture Use and Storage in Mexico and Alberta, Canada: Addressing the*

Legal and Regulatory Barriers, Allan Ingelson and Teresa Castillo Quevedo compare Mexican and Canadian experiences of CCS. Because of Canada's more extensive experience in developing and implementing CCUS projects in depleted oil and natural gas reservoirs, Mexico is now working with Canadians to develop a similar legal and regulatory framework to support enhanced oil recovery (EOR) for long-term CO_2 storage (CCS) projects in Mexico. The Canadian experience indicates clear property rights and certain regulatory systems are important in promoting CCUS and CCS projects, and a meaningful consultation program with project area residents and other stakeholders is essential to secure community support and the timely approval of CCS projects.

Part III proceeds to compare the Brazilian experience with the European Union's CCS Directive and its effectiveness toward large-scale CCUS deployment. Haline Rocha and Hirdan K.M. Costa's *Carbon Capture, Utilization and Storage (CCUS) Legal and Regulatory Barriers in Brazil: Lessons From the European Union* presents the main legal and regulatory barriers to large-scale CCUS deployment in Brazil, identifying the main gaps and regulatory barriers to CCUS in Brazil. Based on the EU experience, most of the regulatory challenges to CCUS are attributed to the geological CO_2 storage component, especially on how to manage CO_2 leakage risks and associated long-term liability. The lessons learned from the EU experience over the past 10 years since the CCS Directive was implemented are addressed.

Part III also includes an Asian perspective. In *An Overview of the Existing CCS Projects in Asia: Comparing Policy Choices and Their Consequences for Sustainable Development*, Romario Nunes, Haline Rocha, and Hirdan K.M. Costa provide an overview of the current CCUS projects and prospects in Asia. It starts by analyzing Asian countries' energy mix and CO_2 emissions, discussing the role of facing the continued reliance on fossil fuels and increasing energy demand. Further, the opportunities and barriers to CCUS commercial-scale deployment in China, Japan, Indonesia, the Philippines, Thailand, Vietnam, and India are identified and discussed. Finally, the chapter concludes by addressing the challenges of an absent CCUS regulatory regime, raising the lessons learned from comparative experiences.

Part III concludes with specific cross-country comparisons on liability for CCS. Silvia Cupertino, Romario Nunes, and Hirdan K.M. Costa's *Relevant Aspects of Carbon Storage Activities' Liability in Paradigmatic Countries: Australia, Brazil, Canada, Japan, the European Union, Norway, the United Kingdom, and the United States* studies such civil liability encompassing general principles, such as how the polluter pays all. It also addresses environmental civil liability in selected countries (Australia, Brazil, Canada, the United States, countries belonging to the European Union, the United Kingdom, Norway, and Japan) because in all the countries chosen, there is specific legislation on liability and projects in operation. Their research analyzes the following topics: the long-term civil liability for carbon dioxide storage activities in the selected countries, the prevention and control of environmental accidents, and the social risk management and safety of carbon storage activities.

Part IV targets the current picture and future developments in international energy law and policy in light of CCS and climate change in the context of energy transition. More specifically, it focuses on the current unsettled issues. Part IV starts with Alexandre Arlota's contribution

addressing the changing landscape of key energy actors in *A Transitioning Model: From Oil Companies to Energy Players*. The chapter discusses the phenomenon, currently in progress, by which major oil companies are being transformed into energy companies, thus diversifying their investments into gas, renewables, and CCS. The chapter also addresses how access to international funding via loans or equity influences (or even steers) the behavior of oil companies.

Part IV also considers fairness and equity as vectors in international energy law and policy. In *Sustainable Development and Its Linkage to CCS Technology: Toward an Equitable Energy Transition*, Hirdan K.M. Costa, Paulo Negrais Seabra, Carolina Arlota, and Edmilson Moutinho dos Santos research CCS technology as linked to the concept of sustainable development. Their contribution analyzes the current underpinnings of sustainable development, including the Sustainable Development Goals (SDGs) of Agenda 2030. It also discusses how sustainable development is currently understood as encompassing a triangular configuration that articulates economic development, environmental protection, and social progress; this complex configuration may foster a fair energy transition encompassing CCS technology as a measure to mitigate climate change and achieve net-zero emissions. Part IV also presents another topic based on equity and fairness. In *Why Is Social Acceptance Important for CCS?*, Yane Marcelle Pereira Silva and Hirdan K.M. Costa contextualize social acceptance and the related inclusionary goals for public participation in light of the literature, particularly in public policies, mainly on capture, storage, and transport carbon dioxide (CCS). The chapter develops a critical analysis of concepts such as social license, public acceptance, public perception, and public participation, critically assessing how social acceptance can be best implemented in CCS projects.

Part IV concludes with Carolina Arlota and Hirdan K.M. Costa's *Climate Change, Energy Transition, and Justice: Where We Are Now, and Where Are We (Should Be) Headed?* The chapter discusses energy transition and how humankind uses energy for their needs, reconciling such uses with social, environmental, and economic interests. As awareness about the consequences of climate change increases worldwide, so does the pressure on different society segments and critical stakeholders. In this context, the role assigned to each target group and, ultimately, to every one of us, including businesses, consumers, scholars, academics, lawyers, public opinion, and policymakers, has changed significantly. Nowadays, immediate calls for action on climate and environmental justice, access to energy, and a fair energy transition are recurrent topics in media outlets. These points of view are part of the same movement in joining the forces combating climate change. Defining who may be heard when determining energy transition policies is controversial domestically and internationally. Because this chapter concludes our book, it advances a modern approach discussing the interplay of climate change, energy transition, and its new perspectives on justice. Ultimately, this chapter shows how policymakers and scholars may be better equipped with the analysis and policies discussed throughout this work. To the extent that some of the challenges coming in their direction may be anticipated in the book, readers may behave (or change their behavior) accordingly. As for the unforeseeable challenges, the regulatory framework presented should also be determinative of new policies in the complex, unique, and exciting times to come. One thing is certain: justice and fairness are no longer abstract propositions; they can now be evaluated and incorporated into the domestic and international spheres. Accordingly, there are reasons to be cautiously optimistic about the future.

Carolina (Norman, OK) and
Hirdan K.M. Costa (São Paulo, S.P.)

PART I

Conceptualizing international energy law and carbon capture and storage (CCS) in light of climate change

International Energy Law: still a brave new world?

Carolina Arlota[1,2] and Hirdan Katarina de Medeiros Costa[3]

[1]Visiting Assistant Professor of Law at the University of Oklahoma, College of Law, Norman, OK, United States [2]Law and Economics Research Fellow, Capitalism and Rule of Law Project (Cap Law) at Antonin Scalia Law School, George Mason University, Fairfax, VA, United States [3]Institute of Energy and Environment, University of São Paulo, São Paulo, Brazil

1.1 Introduction

Access to energy has been a fundamental question through human history. As Heffron and Talus observe: "Energy has always been a central area of the global economy. Energy was central to the developments prior to industrialization, but with industrialization, access to energy sources became a central question for industry and governments around the world" (Heffron and Talus, 2016a, p. 2). Energy underpins several areas of knowledge within social sciences in general, and specifically within law and political science. Economics, engineering, and technology are evidence of the interdisciplinary nature of the field. Energy, thus, is essential to society, and laws to develop and explore its final use are intimately connected to different areas, requiring an autonomous field, namely, International Energy Law.

What is International Energy Law? Is it an autonomous field? What is its interaction with other fields of law? Why is it important to conceive its autonomy? What is the impact of International Energy Law to contemporary topics of study, such as climate change, carbon capture, use, and storage? Therefore this paper describes and analyzes the role of International Energy Law considering modern issues.

1.2 Theoretical framework

This overview starts with the discussion of Energy Law and why it should be a valued addition to law schools across the globe. This part also introduces the major concepts of International Energy Law.

The development of scientific knowledge has always been a goal of the University, as an institutional environment aimed at critical learning. Contemporaneously, this is verified in the spike of graduate and postgraduate courses, interdisciplinary research, and grant projects to investigate complex questions which do not fall under one single field of scientific knowledge. In this context, the human being is not merely understood as a biologic actor, or a political individual, but as a complex being who lives and interacts with the community and, in doing so, experiences itself in ways which reflect such exchanges and social interactions.

On one hand, our society has experienced amazing technological and intellectual development because of centuries of learning and exchanges engaged by previous generations. On the other hand, our society faces challenges which remain unaddressed as well as new questions posed by unprecedented intellectual and technological transformations. Hence, no single field of knowledge is completely capable of answering such unique questions presented nowadays. Should a potential isolated and single field approach be considered, it would be based on the pride and selfishness of human nature, and as such would be unreliable and untrustworthy. Therefore this article argues that knowledge of Energy Law must be aligned with an interdisciplinary approach, because it is the only one consistent with the nature of the field it investigates.

Energy Law has a distinct interdisciplinary nature, and according to Heffron and Talus: "Energy Law concerns the management of energy resources. This is a simple definition, and disguises that it is arguably one of the more complex areas of law. It demands that a scholar in the area engage with other disciplines to some degree, such as politics, economics, geography, environmental sciences and engineering" (Heffron and Talus, 2016b).

A review of the international literature shows that the first study systematically conceiving Energy Law as an autonomous branch of law was the article titled "Energy Law as an Academic Discipline" (Bradbrook, 1996). This article contends the inclusion of Energy Law in the law school curriculum as well as a research field in universities, due to its specificities and thematic pertinence with economic, social, and environmental questions. Following this line of reasoning, Zillman (2015) analyzes the juridicial and political experiences in the energy sector during the last 40 years, focusing on climate change and its impact for energy transition (Heffron et al., 2018).

In the last three decades, due to the privatization and liberalization of energy markets around the world, as Heffron et al. (2018) highlight, the necessity for studies and specific interaction with practitioners in the field of Energy Law are evidence of the scientific evolution experienced in Energy Law. Based on this premise, the authors stress that the economic and environmental impacts of this field have justified the creation of autonomous bureaucracies in governments and independent sections in law firms (Heffron et al., 2018). Evidently, there are several industries within national economies, but very few demand studies and analysis so specialized and independent as Energy Law does (Heffron et al., 2018; Talus, 2013).

Considering the above, it is relevant to understand Energy Law, its developments, and main challenges, specifically regarding the low-emission economies transformation, which is taking place in most countries. In addition, Heffron (2018) and Talus (2013) emphasize that Energy Law is a field of preeminent importance in several contexts, namely, in academia, in practice in individual litigation, class actions, and public advocacy.

What are the scientific premises of Energy Law? What are its methods and principles? Well, Heffron and Talus contend the following (2016b, p. 186):

> Energy Law and policy plays a vital role in the energy sector in the 21st century. It aims to ensure that societies meet their energy targets whether that is about the provision of increased energy security and/or economic benefits, and/or environmental goals. For many years, Energy Law has been developed to meet these societal aims. Yet, there has been little reflection by the legal community into aiming to understand Energy Law itself, and what it should aspire to.

Considering the above, Heffron et al. stress that due to its novelty, when energy transition is considered, very few scholars and practitioners alike are aware of Energy Law's existence (2016b). They eventually resort to traditional fields of law, such as property, contract, and tort law, for instance. Nevertheless, such areas are not equipped to solve the challenges and specificities required by the international law arena. Hence, improved legal framework and methodologies as well as clarification of its scientific object are due in order to provide systematization and to adequately answer the specificities of energy transition.

Heffron et al. (2016b) recommend the adoption of seven principles according to Table 1.1. Those principles provide guidance and coherence fostering the systematic knowledge of Energy Law (Heffron et al., 2018).

In developing those principles, Heffron et al. (2018) contend that their studies are premised on energy justice. In this context, their main goal is to contribute to the development of a field influenced, since its inception, by geopolitical circumstances as well as by raising environmental and economic considerations in a globalized world.

Moreover, renowned international law principles deserve mention. The principle of sovereignty over onshore and offshore energy resources, under the United Nations Resolution No. 3281/1974; and the principle of access to modern energy services, according to the Millennium Goals (2000) and reiterated in the Sustainable Development Goals (SDGs, 2030 Agenda); the principle of energy justice, which focuses on morality in decision-making, looking beyond traditional government and industry considerations on energy security, economic development, and technology (Heffron et al., 2018). These principles are premised on the sharing of costs and benefits of energy services aiming at a more inclusive and impartial decision process. These principles relate to environmental justice, atmospheric justice, and climate justice, all theoretically advanced recently.

TABLE 1.1 Principles of Energy Law.

Principles of Energy Law
1. The principle of national resource sovereignty
2. The principle of access to modern energy services
3. The principle of energy justice
4. The principle of prudent, rational, and sustainable use of natural resources
5. Principle of the protection of the environment, human health, and combatting climate change
6. Energy security and reliability principle
7. Principle of resilience

Data from Heffron, R.J., Ronne, A., Tomain, J.P., Bradbrook, A., Talus, K., 2018. A treatise for energy law. J. World Energy Law Bus. 11, 40.

The principle of energy justice is particularly interesting to scholars because it has implications for distributive political justice on energy taxation (Heffron, 2018). Energy justice aims at contributing to an equitable distribution of the natural resources of a given country (Heffron et al., 2018). As Heffron notes, distributive justice is currently a central issue and its scope and application are encompassing several analyses in the literature on Energy Law (Heffron et al., 2018).

Energy Law is subject to ongoing evolution (Heffron et al., 2018; Heffron and McCauley, 2017) (addressing energy justice and related principles such as the intergenerational equity) (see also Costa, 2012; Costa and Santos, 2012; Costa et al., 2017; Costa and Miranda, 2018; Costa et al., 2018; Weiss, 1993). Hence, energy taxation is framed as capable of providing effectiveness in distributive justice and energy justice (Heffron et al., note 6, 2018). Energy justice also fosters institutional action, which may minimize the impact of the so-called "cursing of natural resources" (Heffron et al., note 6, 2018).

Heffron et al. (2018) addresses procedural justice as among the corollaries of energy justice. Procedural justice concerns the equal ability of all social groups to participate in the decision-making processes of energy developments. Recognition justice, another corollary of the principle of energy justice, considers disparaging community opinions and perspectives based on race, gender, cultural background, and providing assurances that such groups (and places) are respected and not devalued (Heffron et al., 2018). Costa analyzes recognition justice as manifestations of world views included in integral sustainability (Costa, 2018).

Another energy principle commonly mentioned in the literature is prudent rational and sustainable use of natural resources (Freitas, 2011; Sachs, 2002, 2012; Santos, 1992; Sen, 2000; Veiga, 2006). This principle appears in several conventions, most notable in the Stockholm Declaration (1972) and Rio Declaration on Environment and Development (1992). This principle conditions that nonrenewable resources use must occur in ways to protect future exhaustion and to ensure the benefits of such use is shared by all mankind (Heffron et al., note 6, 2018; see also Costa, 2012, 2018; Costa et al., 2017, 2018; Costa and Musarra, 2018). This principle is also present in the goal of the United Nations Framework Convention on Climate Change (UNFCCC), promoting energy efficiency, energy conservation, and the use of renewable energy (Costa, 2018; Costa et al., 2018). The Kyoto Protocol (1997) and the Paris Agreement (2015) reaffirm such goals (Heffron et al., note 6, 2018).

Energy security and reliability also appear in the review by Heffron et al. (2018) This principle is the foundation of any modern political energy system. Additionally, the resilience principle is among the main concerns for the electric sector, due to potential interruptions which may impact distribution to consumers (Heffron et al., 2018; Dzedzej and Costa, 2018).

In light of the previous discussion about the principles of Energy Law, we join the call for more articles addressing the theoretical and practical applications of such principles, their corollaries and their methods in order to perfect them, and assure their applicability (Heffron et al., 2018). Ultimately, such principles would provide guidance for the policymakers, academics, lawyers, judges, and all involved in energy policy evaluation and application, including the society, in tune with the distributive justice principle (Heffron et al., 2018). Moreover, analysis of the existence of other principles and how those would interact in a society which is in constant evolution—as is the case with the present fourth revolution (Heffron et al., 2018).

The majority of the abovementioned principles already exist in Europe and the United States, guiding Energy Law applications in research and practice (Heffron et al., 2018). Moreover, these principles will foster the direct participation of practitioners and academics—even in related fields—to be more active in Energy Law, fostering interdisciplinary approaches (Heffron et al., 2018). As the world advances toward low-carbon emission economies, new principles and new applications of those Energy Law principles will ensure effectiveness and foster similar protections around the world. As Heffron (2018) urges, Energy Law principles provide guidance while orienting decision-making with *long-term* and *integrated* considerations.

1.3 Why International Energy Law is needed now?

Considering the interdisciplinary character of Energy Law and its developments, the field requires further conceptualization in its international dimension. Traditionally, regulation of energy resources and related activities were in the domestic sphere, exclusively (Bruce, 2014). As globalization and climate change became factors in Energy Law, more efforts focusing on the international dimension of Energy Law were needed. This work, thus, turns now to the analysis of International Energy Law as an autonomous branch, contributing to its systematic learning and accessibility of its specificities and applicability vis-a-vis the principles and interdisciplinary of Energy Law.

From a theoretical perspective, this article is premised on the fact that there is no single *International Energy Law* (Wawryk, 2014). This is a relevant point because it specifically addresses the interdisciplinary of Energy Law, as discussed in our previous section, and the incorporation of Energy Law principles into the international sphere. *Law* also refers to traditional sources of international law, namely, treaties and customary international law (Wawryk, 2014). *Law* encompasses the so-called mechanisms of soft law, that despite not being mandatory, provide guidance for private and public parties' behavior in the field (Wawryk, 2014). Previous efforts have focused on particular sectors, such as oil and gas (Wawryk, 2014).

There is a gap in US literature, despite many journals covering International Energy Law-related issues. In Europe and Brazil, there have been previous efforts at systematization (Talus, 2013). The European experience is quite developed due to the European Union (and related harmonization of laws) and the Energy Charter Treaty. Importantly, the project of European integration considers Energy Law and its security concerns and impacts on peace and domestic and regional markets (Craig and Burca, 2015).

This section addresses the need of International Energy Law as an autonomous branch of law. It is not intended to be an exhaustive survey of the field. Instead it shows previous efforts which have not been updated to deal with current challenges, namely, the US withdrawal from the Paris Agreement, the update about the Energy Charter Treaty, Carbon Capture Use and Stock (CCUS) developments, and ongoing financial and technological demands for sharing know-how and investments around the world. This section starts with climate change as a collective action problem and the specificities of international governance on this issue. Geopolitical considerations follow in light of the current world order. This section further reviews and proposes International Energy Law principles addressing the specificities of the field. This section concludes by reviewing current efforts in the literature and proposing common topics, in which International Energy Law features as a new field and branch of law.

1.3.1 Climate change as a collective action problem and international governance

Scientific consensus relates climate change to global warming, which has among its human-induced causes the accumulation of greenhouse gases (GHGs) in the atmosphere (Tol, 2014, 2019). The UNFCCC has, as its main goal, the stabilization of greenhouse gas emissions. In 2015 the 21st Conference of Parties of the UNFCCC enacted the Paris Agreement (Ari and Sari, 2017; Houghton, 2001), which aims to contain the rising global average temperature to well below 2°C above preindustrial levels, while advancing efforts to cap the temperature increase to 1.5°C above preindustrial levels. Because the agreement targets the reduction of GHGs, it has been a subject to controversies in the United States and abroad (Koh, 2019). The US Congress, however, has recognized climate change as a direct threat to US National Security (Defense Authorization Act of 2018 (2017)). Nevertheless, President Trump said he intended to withdraw from the Paris Agreement in June 1, 2017 (Trump, 2017). This withdrawal complicates the global challenge.

Climate change is the "quintessential global-scale collective action problem" (Etsy and Moffa, 2012). It is commonly defined as a collective action problem because the benefits of carbon abatement cannot be restricted to those who contributed to it; nor will climate change affect only those who contribute to create it (Hardin, 1968). Following this line of reasoning, the involved parties have incentives to free ride (Trebilcock, 2014). Two additional factors aggravate this complication. First, human behavior discounts the value of long-term challenges (Fishburn and Rubinstein, 1982; Giddens, 2011). Second, the majority of countries in the developed world are democracies based on electoral cycles which tend to reward short-term considerations (Giddens, 2011). Therefore climate change governance is notoriously difficult (Arlota, 2020).

The challenges are exacerbated by the division of powers in the international system, with top-down bargaining not being a realistic course of action (Sabel and Victor, 2017). Cognitive uncertainty about the feasibility of achieving policy outcomes, such as lowering carbon emissions at acceptable costs, also contribute to increased difficulties when bargaining (Sabel and Victor, 2017). Because the Paris Agreement reconciles elements of bottom-up measures, such as nationally determined contributions (NDCs), with the joint efforts of member states to actually commit to reduce carbon emissions (top-down mechanism), the Agreement is aligned with the modern framework for environmental protection (Morgan et al., 2015).

As mentioned previously, climate change has consequences that require policy coordination and multilevel governance (Etsy and Moffa, 2012). For instance, increasing state and local action has been argued as potentially capable of making the United States achieve its NDCs (Bloomberg, 2017). Meaningful climate change action has been occurring at the state and local level for some time (Avi-Yonah and Uhlmann, 2009). This should continue after the US withdraws, as states, cities, and business join forces to meet the goals of the Paris Agreement (Bloomberg, 2017). Nevertheless, those actions cannot substitute federal action, as the report of the America's Pledge initiative highlights (America's Pledge, 2017). Local and state actions have their own limitations and require significant legal expertise to be effective (Coglianese and Starobin, 2019). Moreover, subnational entities will consider transferring carbon-intensive industry to less stringent jurisdictions—so-called "leakage" (Coglianese and Starobin, 2019). Hence, it may even lead to greater harm if these jurisdictions have a less developed safety net for the population who may suffer direct damage (Wiener, 2007; Arlota, 2020).

Aiming to avoid leakage, economic notions are helpful. First, the notion that countries rely on coal, which is highly polluting, because the infrastructure is already in place. However, there is no need to build wind farms, extra grids, to invest in extra training or education (Eisen et al., 2015). Second, unregulated competitive markets can generate excessive amounts of air and water pollution, wastes, hazardous materials, and other forms of environmental degradation (Stewart, 2001). These market failures tend to be significantly more serious than those addressed by economic regulation, justifying strong regulatory measures (Stewart, 2001). Hence, the necessity to enact climate change regulation under international treaties is so strong.

Climate change is also unique because it mainly refers to the future and is global (Giddens, 2011). The Intergovernmental Panel on Climate Change estimates confidently that extreme events may be more important economically than climate change. Estimations include market damage (infrastructure, tourism, and increased energy demand) and nonmarket damage (ecological impact and culture values, often measured in terms of "willingness to pay") (Giddens, 2011). Predictions are affected perennially by uncertainty, speculation, and lack of information regarding future emission of GHGs, the effects of past and future emission on the climate system, the impact of changes in climate on the physical and biological environment, and the translation of such environmental impacts into economic damages (Interagency Working Group, 2016).

1.3.2 Geopolitical considerations

Geopolitical considerations refer to each country's domestic attitude toward regulation of the energy sector domestically and environmental protection. Challenges faced domestically are magnified in the international level, due to strategic considerations of other countries' behaviors and potential for free riding. A concept that could foster some international momentum is the social cost of carbon, which is the present value today of the damage of an additional ton of carbon dioxide in the atmosphere (Interagency Working Group, 2016; Nordhaus, 2017). This is an important concept for policy considerations and one that is germane to International Energy Law as an independent field. It addresses the concerns of each country regarding energy security and how they relate to each other and to global action (Heffron et al., 2018). The regressive nature of carbon-pricing led costs to be borne by consumers and, thus, those with lower income spend proportionally more on nondiscretionary goods and services and will be more affected (Trebilcock, 2014). If climate change does not affect people in isolation, it impacts the less well-off disproportionally, in a vicious cycle (Arlota, 2020, 2018; Nuttall, 2018).

Economic growth and preserving the environment are not necessarily exclusionary goals, as the UNFCCC treaty exemplifies (Articles 4 and 6). It is noteworthy that the United Nations Security Council recently recognized climate change as a "threat multiplier." Climate-related risks and conflicts are already a reality for millions of people around the globe as it impacts security and peace (Climate Change Recognized, 2019).

On a related point, international treaties aimed at curbing the effects of climate change need to include developing nations. The challenge is that governments in developing countries face increasing pressure to achieve economic prosperity, frequently at the expense of the environment (Samaan, 2011). More recently, the unlikely progress on renewable energy has displaced the once dominant assumption that economic growth and increasing greenhouse gas emissions must be tied (Deese, 2017). India is a great example of a country that transitioned from reluctant member of the Paris Agreement to vocal advocate that checks the compliance, particularly of developed countries, with their Paris' targets (Arlota, 2018; Westscott, 2017).

With the United States leaving the Paris Agreement, the window of opportunity is narrowing when it comes to climate change mitigation, particularly with research that shows the first decade after the agreement is crucial to achieve the objectives (Morgan et al., 2015; Hai-Bin et al., 2017). Relevant literature argues that mitigation is not only necessary but indispensable, because it is the sole alternative to effectively reduce carbon emissions (World Bank, 2012; Sovacool et al, 2016).

1.3.3 International Energy Law proposed principles: beyond mere overlap with International Environmental Law and Energy Law

Another reason for International Energy Law being an autonomous branch is the international legal order itself, because treaties are celebrated for having the protection of human dignity as the ultimate goal (Park, 2013). International human rights bodies have consistently contended that environmental harm can adversely affect human rights (Knox, 2016). The right to life can be threatened by natural events attributed to climate change, namely, floods, storms, droughts, crop failures, hunger, malnutrition, scarcity of water, proliferation of tropical deceases like malaria, the right to housing due to forced misplacement, and others (Knox, 2016). Because climate change is a type of environmental harm, human rights obligations that are applicable "in the context of environmental harm generally should apply to climate change as well" (Knox, 2016).

From the standpoint of the international order, the United Nations, has been very active in pursuing environmental protection, reducing the impact of climate change, and providing incentives to countries for global action. Recently, the United Nations enacted the SDGs, as part of the urgent call for action to all countries in the 2030 Agenda for the Sustainable Development for the planet. Among its 17 principles, the affordable and clean energy (goal 7), sustainable cities and communities (goal 11), and climate action (goal 13) are of particular interest (Sustainable Development Goals, 2015). Such goals show the necessity of all countries—developed and developing nations alike—to commit to effective and responsible actions to protect the environment and to curb global warming, specifically (Sustainable Development Goals, 2015).

Among the proposed principles for International Energy Law is the principle of common but differentiated responsibilities and respective capabilities' speaks specifically to different nations' responsibilities. According to this principle, responsibility for current and historical contributions need to be factored in (Arlota, 2020). Moreover, developing states have contributed less to the current concentration of GHGs and overall threshold on carbon saturation than developed states (Gillis and Popovich, 2017). Under the principle of common but differentiated responsibilities, developed countries provide climate financing to less developed countries (Etsy and Moffa, 2012; Hai-Bin et al., 2017). This is more than a moral commitment because, since 1800, the developed world has contributed more than 84% of the GHGs emissions up to 2002 (Brown, 2002). Importantly, Article 4 of the Paris Agreement determines that developed counties should remain leading the way through absolute emission reduction targets, and developing countries should reduce their emissions in accordance with different national circumstances.

There are principles derived from International Environmental Law that are also applicable to International Energy Law. Under the *Trail Smelter* case, the prohibition of transboundary harm is mandatory under International Law (Bratspies and Miller, 2006). This is particularly relevant, because the effects of global warming will be particularly dire after 2050 (Davenport and Landler, 2019). Therefore the prohibition of transboundary harm and the obligation of compensation ("polluter pays" principle) lead to state responsibility in a double-edged sword regime that assumes harm will occur while fostering prevention (Bratspies and Miller, 2006).

In this context, two international principles, based on two specificities of environmental protection, are also applicable to International Energy Law: the common good and the fact that damages are not limited to physical borders. This chapter, thus, turns to the precautionary principle followed by the principle of intra- and intergenerational equity.

The precautionary principle is found in Article 3 of the UNFCCC and states that parties should take precautionary measures to anticipate, prevent, or minimize the causes of climate change, and mitigate its adverse effects, while emphasizing that lack of scientific certainty should not be used to postpone such measures where threats of serious or irreversible damage exist. The provision also states that climate change policies and measures should be cost-effective to ensure global benefits at the lowest possible costs (UNFCCC).

Precautionary principle is determinative regarding which risks should be regulated, but not the stringency of regulation itself (Driesen, 2013; Farber, 2015). The precautionary principle has application in climate change policy because no definitive analysis exists to adequately capture the full costs of not taking regulatory measures to mitigate the risks (Kysar, 2004; Dana, 2009). Precaution is largely applicable to limit old technologies, such as fossil fuels (Weiner, 2016). Moreover, effective policymaking is based on the assessment of complete policy impacts, including ancillary costs, as well as ancillary benefits (Weiner, 2016; Revesz and Livermore, 2008). More significantly, perhaps, is that we may be beyond precaution now, as "we probably blew past our precautionary opportunity sometime in the 1980s. We are now, and have been for some time, in a postcautionary world" (Heinzerling, 2008).

In this proposed framework, the principle of intra- and intergenerational equity is highly relevant. This international principle is defined in the first part of Article 3 (1) of the UNFCCC. This principle defines rights and obligations regarding the use and enjoyment of natural and cultural resources as inherited by the present generation and to be passed to future generations in no worse condition than received (Redgwell, 2016). It is based on the moral duties owed by present generations to future ones (Intergovernmental Panel on Climate Change). The legal force of this principle is disputed, but it should be considered among the factors that will inform policy decisions regarding climate change (Redgwell, 2016; Kysar, 2010).

1.3.4 Contents of International Energy Law: Climate Governance, International Investment Treaties and its Arbitration, Nuclear Energy, and deep decarbonization

In light of the interdisciplinary feature of International Energy Law and the need for accommodating its specificities (International Law, National Security Concerns, Investment Arbitration, among others), this part addresses the specific contents of International Energy Law. It aims at establishing a core of legal issues which should be addressed in a course entitled International Energy Law. The issues proposed are the following: Climate Governance; Investment Treaties and International Investment Arbitration; Nuclear Energy; Renewable Resources and overview about pathways to deep decarbonization. In this last case, a Decarbonization Frameworking (Decarbonization Law) may emerge as an International Energy Law branch.

1.3.4.1 Climate Governance

The first suggested topic is Climate Governance. The first recommendation on this topic refers to international treaties aimed at environmental protection and energy resources. Essential treaties are the following: the UNFCCC, the Kyoto Protocol, the Montreal Protocol on Greenhouse Gases, the UNFCCC Treatment of Ozone Depleting Substances, the United Nations Convention on the Law of the Seas, and the Paris Agreement on Climate Change.

In addition, basic economic notions such as efficiency, the tragedy of the commons, market failure, and incentives for global action should be discussed. Basic concepts of strategic behavior of countries and jurisdictions would also benefit students. Incentives for free riding, the notion of leakage, and how game theory fosters cooperation on repetitive players in the international arena should also be discussed (Schenck, 2008; Chinen, 2001; Arlota, 2020).

Significant discussions should cover the Paris Agreement and its unique model of governance reconciling national units as well as subnational units to engage in efforts of mitigation and adaptation (Cooper, 2018). The Paris Agreement, signed by more than 175 countries, establishes limits on carbon emissions aiming to curb global warming, assuring that the raising global temperatures do not go above the 2°C until the year 2100.

After the thresholds were established in the Paris Agreement, funding to investments is changing (Barbiroglio, 2019). Moreover, mitigation technologies, such as CCUS, are a feasible alternative for the reduction of CO_2 emitted, particularly in the energy sector. Regulatory schemes of CCUS and related liability are developing in the world. Therefore it is crucial that additional principles of International Energy Law are developed to guide mitigation measures fostering the reduction of GHGs. This topic is present in international commitments, and is most prominently stated in the SDGs, having direct relevance in the prudent, rational, and sustainable use of natural resources and modern access to clean energy (SDG 7), sustainable cities and communities (SDG 11), responsible consume and share (SDG 12) and climate action (SDG 13), as discussed earlier.

Regarding policy evaluations, framework and premises determinations should be explicit (Sovaccol and Brown, 2015). Furthermore, policy objectives and basic assumptions, such as efficiency and cost-effectiveness, should be stated clearly (Felder, 2016).

1.3.4.2 Investment Treaties and Arbitration

The second suggested subject matter is international arbitration as an effective means to foster foreign investments. The core provisions on international arbitration in the Energy Charter Treaty, which encompass European nations mainly, and the International Center for the Settlement of International Disputes—the so-called ICSID Convention—are relevant for students and practitioners alike to visualize the necessity of alternative dispute resolution mechanisms in the field of International Energy Law (Energy Charter Treaty, 2019).

1.3.4.3 Activities in areas beyond national jurisdiction

The third suggested topic focuses on areas beyond the national jurisdiction of member states. For instance, Professor Zedalis makes a compelling case for the study of the 158 High Sea Convention, and Parts VII and XI of the 1982 Law of the Sea Convention (Zedalis, 2000). He also includes research on the Arctic and Antarctic in his book (Zedalis, 2000). Additional consideration for International Energy Law may also include the Moon Treaty of 1979 and developments, as extraterritoriality is considered from an energy security standpoint.

1.3.4.4 Nuclear energy

The fourth suggested issue is nuclear energy. Although, technically, it is not renewable (nuclear elements eventually stop producing energy) it is a zero emission resource, with minimal waste produced (Office of Nuclear Energy, 2018). Nonetheless, many countries have adopted nuclear energy as a major energy source, while others have consistently refrained from it (World Nuclear Power Association). The role of the International Atomic Energy Agency, its standard principles for safety, and the 1994 Convention on Nuclear Safety are of paramount interest for the International Energy Law field. Liability for nuclear damage is controversial. The Chernobyl case is paradigmatic, as no international state liability actually was pursued by individual states (Pelzer, 1987).

In many developed countries nuclear energy is now being phased out, while facing mixed prospects in developing jurisdictions due to its high initial costs, long lead times, and often delayed constructions (Khatib and Difiglio, 2016). Interestingly, nuclear energy's shares of global consumption is reducing while renewables are rising (Khatib and Difiglio, 2016).

1.3.4.5 Renewable resources

The fifth suggested topic is renewable resources. This topic is often an overlooked area of International Energy Law because countries tend to pursue their own domestic policies. More contemporaneously, however, renewable sources such as solar, wind, hydroelectric, and to some extent, biomass, have been considered beyond their domestic scenarios. Under the Paris Agreement and its efforts for transfer of technology, this article contends that renewable resources are expected to be crucial for mitigation efforts on carbon emissions and adaptation mechanisms for the impact of global warming.

International Energy Law findings conclude that countries cannot solely rely on nuclear energy or on renewable resources. A combination of portfolios is required to curb GHGs emissions (Khatib and Difiglio, 2016).

1.3.4.6 Pathways to deep decarbonization

The sixth recommended topic refers to the implementation techniques that could remove CO_2 from the atmosphere. From the creation of sinks—such as forests and plankton—to CCUS, decarbonization is necessary from an International Energy Law perspective. Topics relating to international cooperation (treaties, agreements and accords) to investments are all materialized in this discussion of international cooperation, investment, and liability. A United States initiative to cover such efforts comprehensively has begun. This legal framework could be adopted in other jurisdictions (Gerrard and DernBach, 2018). Issues relating to climate engineering and the law could also be addressed, specifically regarding regulation and liability for solar radiation management and carbon dioxide removal (Gerrard and Hester, 2018).

1.4 Conclusion

In light of the arguments, this article establishes the specificities of International Energy Law, its legal framework, and why International Energy Law is needed now as an autonomous legal branch of law. In addition, this article discusses the core principles of International Energy Law and ultimately suggests the foundation contents to be covered in an academic course about this new field. As with any new field, further academic discussions will advance the momentum for new courses on International Energy Law not only in law schools, but also in business schools, engineering, social sciences, and biology, among others. Those discussions will also foster a framework of the interdisciplinary which marks the topic and will contribute significantly to this promising field.

Acknowledgments

Hirdan Katarina de Medeiros Costa is grateful to the "Research Center for Gas Innovation—RCGI" (Fapesp Proc. 2014/50279-4), supported by FAPESP and Shell, organized by the University of São Paulo, and the strategic importance of the support granted by the ANP (National Agency of Petroleum, Natural Gas and Biofuels of Brazil) through the R&D clause. She also thanks to the support from the National Agency for Petroleum, Natural Gas and Biofuels Human Resources Program (PRH-ANP), funded by resources from the investment of oil companies qualified in the R,D&I clauses from ANP Resolution No. 50/2015 (PRH 33.1—related to Call No. 1/2018/PRH-ANP; Grant FINEP/FUSP/USP Ref. 0443/19). Carolina Arlota acknowledges the excellent research assistance of Natosha D. Greene.

References

America's Pledge, 2017. America's Pledge: Phase 1 Report: States, Cities and Business in the United States are Stepping up on Climate Action. Retrieved from: https://www.bbhub.io/dotorg/sites/28/2017/11/AmericasPledgePhaseOneReportWeb.pdf.

Ari, I., Sari, R., 2017. Differentiation of developed and developing countries for the Paris agreement. Energy Strategy Reviews 18, 175.

Arlota, C., 2020. Does the United States withdrawal from the Paris Agreement on climate change pass the cost−benefit analysis test. University of Pennsylvania Journal of International Law 41, 881−938.

Arlota, C., 2018. 'Cost & benefit analysis of the United States' withdrawal from the Paris agreement on climate change. Gujarat National Law University: Journal of Law & Economics. Received from: http://gjle.in/wp-content/uploads/2018/09/Arlota.pdf.

Avi-Yonah, R.S., David, M., Uhlmann, D.M., 2009. Combating global climate change: why a carbon tax is a better response to global warming than cap-and-trade. Standford Environmental Law Journal 28, 3.

Barbiroglio, E., 2019. European Investment Bank Will Stop Financing New Fossil Fuels Projects. Forbes.

Bloomberg, M., 2017. Declaration to United Nations Officials: We Are Still In (Jun. 5, 2017). Retrieved from: https://www.americaspledgeonclimate.com/about/.

Bradbrook, A.J., 1996. Energy Law as an academic discipline. Journal of Energy & Natural Resources Law 14.

Bratspies, R.M., Miller, R., 2006. Transboundary Harm in International Law. In: Bratspies, R.M., Miller, R. (Eds.), Transboundary Harm in International Law: Lessons from the Trail Smelter Arbitration. Cambridge University Press.

Brown, D.A., 2002. American Heat: Ethical Problems with the United States' Response to Global Warming. Rowman & Littlefield Publishers.

Bruce, S., 2014. Max Planck Encyclopedia on Public International Law. Retrieved from: https://opil.ouplaw.com/view/10.1093/law:epil/9780199231690/law-9780199231690-e2143#law-9780199231690-e2143-div1-2.

Chinen, M.A., 2001. Game theory and customary international law: a response to Professors Goldsmith and Posner. Michigan Journal of International Law 23, 143.

Climate Change Recognized as Threat Multiplier, 2019. United Nations Security Council Debates Its Impact on Peace. United Nations News.

Coglianese, C., Starobin, S., 2019. The legal risks of regulating climate change at the subnational level. The Regulatory Review 2.

Cooper, M., 2018. Governing the Global Commons: The Political Economy of State and Local Action, After the U.S. Flip-Flop on the Paris Agreement. Energy Policy 118, 440.

Costa, H.K.M., 2012. Princípio Da Justiça Intra e Intergeracional Como Elemento na Destinação Das Rendas de Hidrocarbonetos: Temática Energética Crítica na Análise Institucional Brasileira. Tese (Doutorado em Ciências) Programa de Pós-Graduação em Energia. Instituto de Eletrotécnica e Energia (IEE).

Costa, H.K.M., 2018. Royalties de petróleo: Justiça e Sustentabilidade. *Synergia, Rio de Janeiro*.

Costa, H.K.M., Miranda, M.F., 2018. first ed. Temas de Direito Ambiental: 30 Anos da Constituição, 1. Lumen Juris, Rio de Janeiro, pp. 151−172.

Costa, H.K.M., Musarra, R.M.L.M., 2018. Sustainable development and governance: natural gas consumption in the Amazon. IOSR Journal in Humanities and Social Science (IOSR-JHSS) 23, 72−81.

Costa, H.K.M., Musarra, R.M.L.M., Miranda, M.F., Moutinho Dos Santos, E., 2018. Environmental license for carbon capture and storage (CCS) projects in Brazil. Journal of Public Administration and Governance 8, 163−185.

Costa, H.K.M., Santos, E.M., 2012. Justiça e Sustentabilidade: Uma Contribuição na Reflexão Acerca da Destinação Dos Royalties de Petróleo. Revista de Estudos Avançados da Universidade de São Paulo.

Costa, H.K.M., Simoes, A., Santos, E.M., 2017. Integral sustainability as driving force for paradigmatic change in human lifestyle. Sustentabilidade em debate 8, 100−110.

Craig, P., Burca, G., 2015. EU Law. Oxford.

Dana, D., 2009. The contextual rationality of the precautionary principle. Queen's Law Journal 35, 67.

Farber, D.A., 2015. Coping with uncertainty: cost−benefit analysis, the precautionary principle, and climate change. Washington Law Review 90 (1659), 1721−1724.

Davenport, C., Landler, M., 2019. Trump Administration Hardens Its Attack on Climate Science. N.Y. Times.

Deese, B., 2017. Paris Isn't Burning: Why the Climate Agreement Will Survive Trump. Foreign Aff.

Driesen, D.M., 2013. Cost−benefit analysis and the precautionary principle: can they be reconciled? Michigan State Law Review 771, 791−812.

Dzedzej, M., Costa, H.K.M., 2018. Concepts and characteristics of complex systems and final energy usage. International Journal of Environment, Agriculture and Biotechnology 3, 1552−1561.

Eisen, J.B., Hammond, E., Rossi, J., Spence, D.B., Weaver, J.L., Wiseman, H.J., 2015. Energy. Economics and the Environment 729−730.

Defense Authorization Act of 2018 (2017). (H. R. 2810): Pub. L. No. 115-91, §355, 131 Stat. 1283, 1358.

Energy Charter Treaty Modernisation: Commission Welcomes Council's Mandate, 2019. EU News. Received from: http://trade.ec.europa.eu/doclib/press/index.cfm?id=2049.

Etsy, D.C., Moffa, A.L.I., 2012. Why climate change collective action has failed and What needs to be done within and without the Trade Regime. Journal of International Economic Law 15, 777.

Felder, F.A., 2016. "Why can't we all get along?" A conceptual analysis and case study of contentious energy problems. Energy Policy 96, 711.

Fishburn, P.C., Rubinstein, A., 1982. Time preference. International Economic Review 23, 677.

Freitas, J., 2011. Sustentabilidade: direito ao futuro. Editora Forum, Belo Horizonte.

Gerrard, M., DernBach, J., 2018. Legal Pathways to Deep Decarbonization in the United States: Summary and Key Recommendations. Environmental Law Institute.

Gerrard, M., Hester, T., 2018. Climate Engineering and the Law. *Cambridge University Press.*

Giddens, A., 2011. The Politics of Climate Change. Polity Press.

Gillis, J., Popovich, N., 2017. The U.S. Is the Biggest Carbon Polluter in History. It Just Walked Away From the Paris Climate Deal. N.Y. Times.

Hai-Bin, Z., Hancheng, D., Huaxia, L., Wentao, W., 2017. U.S. withdrawal from the Paris agreement: reasons, impacts, and China's response. Advances in Climate Change Research 8, 220.

Hardin, G., 1968. The tragedy of the commons. Science 162, 1243.

Heffron, R.J., 2018. The application of distributive justice to energy taxation utilizing sovereign wealth funds. Energy Policy 122, 649–654.

Heffron, R.J., McCauley, D., 2017. The concept of energy justice across the disciplines. Energy Policy 105.

Heffron, R.J., Talus, K., 2016a. The evolution of energy law and energy jurisprudence: insights for energy analysts and researchers. Energy Research and Social Science 19, 1.

Heffron, R.J., Talus, K., 2016b. The development of Energy Law in the 21st century: a paradigm shift? Journal of World Energy Law and Business 9 (3), 189–202.

Heffron, R.J., Ronne, A., Tomain, J.P., Bradbrook, A., Talus, K., 2018. A treatise for energy law. Journal of World Energy Law and Business 11, 34.

Heinzerling, L., 2008. Climate change, human health, and the post-cautionary principle. Georgetown Law Journal 96, 1565.

Houghton, J., 2001. *Science and International Environmental Policy: The Intergovernmental Panel on Climate Change*, In: Richard Revesz et al. eds., Environmental Law. the Economy and Sustainable Development 353, Cambridge University Press. 2008.

Interagency Working Grp. on Soc. Cost of Carbon, 2016. United States Government, Technical Support Document: Social Cost of Carbon for Regulatory Impact Analysis Under Executive Order 12866, 3.

Khatib, H., Difiglio, D., 2016. Economics of nuclear and renewables. Energy Policy 96, 740.

Knox, J.H., 2016. Human rights principles and climate change. Oxford Handbook of International Climate Change Law. Oxford University Press.

Koh, H.H., 2019. The Trump Administration and International Law. Oxford University Press, New York, NY.

Kysar, D.A., 2004. Climate change, cultural transformation, and comprehensive rationality. Boston College Environmental Affairs Law Review 31, 555.

Kysar, D., 2010. Regulating from Nowhere: Environmental Law and the Search for Objectivity. Yale University Press.

Nuttall, M., 2018. Environmental Institutions and Governance. Wiley Online Library.

Morgan, J., Dagnet, Y., Tirpak, D., 2015. Elements and Ideas for the 2015 Paris Agreement. World Resources Institute.

Nordhaus, W., 2017. Revisiting the social cost of carbon. Proceedings of the National Academy of Sciences of the United States of America 10, 1518.

Paris Agreement, 2015. Retrieved from https://unfccc.int/sites/default/files/english_paris_agreement.pdf. Accessed on: January 20, 2021.

Park, P., 2013. International Law for Energy and the Environment. Taylor & Francis Group, Boca Raton, FL.

Pelzer, N., 1987. The impact of the chernobyl accident on international nuclear energy law. International Nuclear Energy Law 294–311.

Redgwell, C., 2016. Principles and Emerging Norms in International Law: Intra-and-Inter-generational Equity. Oxford Handbook of International Climate Change Law. Oxford University Press.

Revesz, R.R., Livermore, M., 2008. Retaking Rationality: How Cost Benefit Analysis Can Better Protect the Environment and Our Health. Oxford University Press, pp. 151–170.

Rio Declaration on Environment and Development, 1992. Retrieved from https://www.un.org/en/development/desa/population/migration/generalassembly/docs/globalcompact/A_CONF.151_26_Vol.I_Declaration.pdf. Accessed on: January 20, 2021.

Sabel, C.F., Victor, D.G., 2017. Governing global problems under uncertainty: making bottom-up climate policy work. Climate Change 15, 144.

Sachs, I., 2002. Caminhos para o desenvolvimento sustentável. *Garamond, Rio de Janeiro.*

Sachs, I., 2012. De volta à mão visível: os desafios da Segunda Cúpula da Terra no Rio de Janeiro. Revista de Estudos Avançados, São Paulo 26, 74.

Samaan, A.A., 2011. Enforcement of international environmental treaties: at analysis. Fordham Envtl. L. Rev. 5, 261.

Santos, M., 1992. 1992: A redescoberta da Natureza. Revista de Estudos Avançados, São Paulo 6 (14), 96—106.

Schenck, L.M., 2008. Climate Change "crisis"— struggling for worldwide collective action. Colorado Journal of International Environmental Law and Policy 19, 319.

Sen, A., 2000. Desenvolvimento como liberdade. Cia. das. Let., São Paulo.

Sovaccol, B.K., Brown, M.A., 2015. Deconstructing facts and frames in energy research: maxims for evaluating contentious problems. Energy Policy 86, 36.

Sovacool, B.K., Brown, M.A., Valentine, S.V., 2016. Fact and Fiction in Global Energy Policy. Johns Hopkins University Press.

Stewart, R., 2001. *Economic Incentives for Environmental Protection: Opportunities and Obstacles*, In: Richard Revesz et al. eds.,Environmental Law, the Economy and Sustainable Development 353, Cambridge University Press, 2001.

Stockholm Declaration, 1972. Retrieved from https://www.ipcc.ch/apps/njlite/srex/njlite_download.php?id=6471. Accessed on: January 20, 2021.

Sustainable Development Goals, 2015. The United Nations. Retrieved from: https://www.un.org/sustainablede-velopment/sustainable-development-goals/.

Talus, K., 2013. EU Energy Law and Policy: A Critical Account. Oxford.

The Kyoto Protocol, 1997. Retrieved from https://unfccc.int/resource/docs/convkp/kpeng.pdf. Accessed on: January 20, 2021.

Tol, R.S.J., 2014. Quantifying the consensus on anthropogenic global warming in the literature: a re-analysis. Energy Policy 73, 701.

Tol, R.S.J., 2019. The Elusive Consensus on Climate Change, Working Paper Series 319. Department of Economics. University of Sussex Business School.

Trebilcock, M.J., 2014. Dealing with losers. The Political Economy of Policy Transition 120.

Trump, D., 2017. Statement by President Trump on the Paris Climate Accord. Retrieved from: https://www.whitehouse.gov/briefings-statements/statement-president-trump-paris-climate-accord/.

United Nations Framework Convention on Climate Change: Conference of the Parties. Paris, December 12, 2015.

United States Office of Nuclear Energy, 2018. 3 Reasons Why Nuclear Energy is Clean and Sustainable. Received from: https://www.energy.gov/ne/articles/3-reasons-why-nuclear-clean-and-sustainable.

Veiga, J.E., 2006. Desenvolvimento sustentável: o desafio do século XXI, second ed. *Garamond, Rio de Janeiro.*

Wawryk, A., 2014. International Energy Law: An Emerging Academic Discipline, in Law as Change: Engaging with the life and Scholarship of Adrian Bradbrook. University of Adelaide.

Weiss, E.B., 1993. Justice Pour Les Générations Futures. Editions Sang de la Terre, Paris.

Westscott, B., 2017. Reluctant Signatory India Takes Moral High-Ground on Paris Climate Deal. CNN.

Wiener, J.B., 2007. Think globally, act globally: limits local climate policies. University of Pennsylvania Law Review. 155, 1961.

Weiner, J.B., 2016. Precaution and climate change. Oxford Handbook of International Climate Change Law. Oxford University Press.

World Bank, 2012. Turn Down the Heat: Why a 4°C Warmer World Must Be Avoided. Retrieved from: http://documents.worldbank.org/curated/en/865571468149107611/pdf/NonAsciiFileName0.pdf.

World Nuclear Power Association. Nuclear Power in France. Received from: https://www.world-nuclear.org/information-library/country-profiles/countries-a-f/france.aspx.

Zedalis, R., 2000. International Energy Law. Routledge.

Zillman, D.N., 2015. Evolution of modern energy law: a personal retrospective. Journal of Energy & Natural Resources Law 30, 485.

I. Conceptualizing international energy law and carbon capture and storage (CCS) in light of climate change

The energy of cooperation

Oscar W. Serrate and Nathália Weber

Research Centre for Gas Innovation, Polytechnic School of University of São Paulo,
São Paulo, Brazil

2.1 Introduction

Energy is the result of work—human and social work—which requires and generates energy. It can be said that energy was the first good transacted in the primitive era. Families and groups united, to produce and benefit from the warmth of fires. Energy was born from the joint labor of these basic communities. The common good required working for energy. Energy provided for the common needs. Together with the wheels and the levers, energy was the component that provided calories and enhanced the initial physical barriers and limitations of mankind. People learned to use energy, to control it, and to multiply it. Energy created families, tribes, and communities. Before exchanges or money existed, life was sustained by energy, and cooperation became the way to produce and distribute it, but also it gave people the forces that united them in their quest for resilience and progress.

Societies developed based on the availability of energy. Energy was the fuel that moved the engines of history. Energy is the permanent presence in the continuous changes of life modes. It motivated the human actor to search for more, and better, forces to help execute simple and complex tasks. The other constant that accompanied this evolution was the intrinsic necessity of cooperation. The initial paradigm was based on the understanding that "uniting forces, energies can be multiplied." This chapter is about that kind of "cooperation": humankind's enormous opportunity to enhance building and improving communities based on the energy derived from its mental and physical work and from the natural, technological, economic, and social surroundings.

Energy and life have always coexisted as two sides of a same coin. By increasing the energy obtained from plants, animals, or rivers, people continually increased the quality and quantity of their lives, while simultaneously improving their physical and mental skills. Information and knowledge opened even more the channels for cooperation.

19

Mind-power became essential for new frontiers on the quest for improved livelihoods. Currently, human beings create and use multiple forms of "calories" in food, fuel, and other raw materials. The triangle energy—information—cooperation (EIC) leverages even more access to new forms of communities: larger cities providing services for more people. The vision of a larger home allowed humankind to consider longer-term targets and systemic and multifocal approaches. The vision awakened the necessity to anticipate processes for future needs, including poverty and environmental limits.

This chapter intends to contribute to the current global debate about the links between the three components of EIC. It highlights the positive actions of actors mainly using renewable technologies and the science of sustainability around the world trying to harmonize the EIC's three component's development. The working hypothesis adopted for this analysis is, precisely, that the EIC triangle is so intensely intertwined that any analysis of one aspect requires the systemic synthesis of the others.

2.2 Cooperation evolution

One hundred years ago, James Hayden Tufts was part of a great debate about "cooperation." He started his arguments citing Plato's famous myth of the gods that equipped humanity for living: "the one, arts and inventions to supply him with the means of livelihood; the other, reverence and justice to be the ordering principles of societies and the bonds of friendship and conciliation" (Tufts, 1918). He continued to say that "agencies for mastery over nature and agencies for cooperation among men remain the two great sources of human power" (Tufts, 1918). It was not yet the EIC triangle, it was the E&C duo. Both evolved gradually with mutual feedback, throughout history, until chemists and electricians became the protagonists of change. When steam opened the path for a new kind of energy, the evolution became a revolution. Before that innovation, progress was measured by the strength of the human association that sustained it. Since the "Industrial Revolution," however, it has been the mastery of new powers that has drastically modified the forms of human associations.

Political parties, corporations, labor unions, military groups, and others became dominant players of the new social landscape. And they contended for supremacy through "dominance [and] competition or cooperation all mean[ing] a meeting of human forces. They rest respectively on power, rivalry, and sympathetic interchange" (Tufts, 1918). Tufts thought the most important is cooperation and that it is largely the "touchtone for the others." Dominance implies inequality while cooperation implies some mutual relationship. In dominance, there is a top-down approach. In cooperation the processes are more horizontal; they include more sharing of common purposes and common needs, and benefit all. Yet, cooperation often requires competition, when common rules are applied, using the creative and not destructive potential of rivalry. Many times, competition increases cooperation if it avoids monopolies that erase both processes. Cooperation may also be harmful if it implies a negative sense of equality that prohibits new leaderships or new technologies, and instead promotes incompetence.

"A cooperating group has two working principles: first, common purpose and common good; second, that men can achieve by common effort what they cannot accomplish [alone]. Power is likewise a value in a cooperating group, but it must be power not merely used for the good of all, but to some extent controlled by all and thus actually shared" (Tufts, 1918). One of the main problems between cooperating groups is efficiency, given that public opinion or will is not necessarily followed by the corresponding action favoring the benefit for all. But it is better than the use of force or its extreme—war—which is always a lose—lose option.

A century later, the concerns from the dawn of the Industrial Revolution are similar to those at the infancy of the Information Revolution. This modern revolution is also characterized by simultaneous tremors in the world of energy and the world of human relations or perceptions. The new issue is that the dichotomy has been now amplified on a world scale.

History teaches that science and technology (S&T) revolutionized all human activities. However, history also teaches that if those revolutions are not accompanied by a parallel social and institutional progress they can be derailed and turned against themselves, or against important sectors of the communities that inspired them. In other terms, great changes cannot be sustained in the long run if their fruits are not shared or are not beneficial for all, or most, members of a given community. This was called by social scientists the "common good," a concept that was largely developed by many thinkers from multiple doctrines. They have agreed that there are objectives that can only be attained by collective will or action of communities, and that implies cooperation.

But a wider change happened by the end of the second millennium: what if there are many "common goods?" What if there are large tangible and nontangible goods that are essential for life but involve planetary vision? Global goods such as nature, air, climate, knowledge, and others. A lot of mindsets changed when mankind was able to see an image of Earth from space. It amplified the "zoom" and opened the doors to think about ourselves as a globe. Suddenly, our perspectives were transformed, and the world started to see itself as a planet, or what is now called a "Common Home." A common home requiring full cooperation from all of us. But the revolutionary wheels turned well beyond that spatial enlargement: time limits were also broken, and we all started to discuss our "Common Future." When asked the typical question of "how to," the answer was always: more of the same, meaning again, cooperation.

Some questions remain. For instance, is cooperation a tool to achieve a common purpose or a single good by itself. For example, knowledge is, without any doubt, the beginning and the end of cooperation—the cause and effect—and its "creation, acquisition, validation, and use, are common to all people as a collective social endeavor" (UN, 2016). Energy does not only require cooperation. Cooperation also generates new energies, new frontiers, new horizons, new societies, new human beings. That is how the original concept of "cooperation" has evolved since the days of cavemen; it is much more than simple "collaboration." The latest is mainly understood as a method for working together for a common objective, while "cooperation" is now accepted without the necessity of an exclusively material quid pro quo, with the basic idea that $1 + 1$ could be more than 2, if the community transcends the selfish walls of immediate returns. Is it possible? Yes, the connected world of the Third Millennium has shown that it is not only feasible but has learned that it is more than that: cooperation is the only way ahead of us, and the only way for future generations to come.

I. Conceptualizing international energy law and carbon capture and storage (CCS) in light of climate change

Gradually, this chapter will be demonstrating, with concrete evidence, that the EIC triangle (energy, information, cooperation = "SINERGY," the energy of many) includes the I as the modern bridge between E and C, and is the best framework to analyze and act in a cooperating world that is, more and more, ready to face the challenges of the 21st century, by delivering a better quality of life for most of the Earth's inhabitants.

2.3 War or cooperation?

Most of the 20th century consisted of global war. But it ended, inaugurating an era of global cooperation. How did we evolve from one extreme to the other? Historians highlight the dispute for the domination of vital spaces. But together there was also, in parallel, the dispute for the control of energy, to supply the growing demand of new industrial needs. With high differences on quantity and quality of works, infrastructure, transportation, resources, labor, cities, services, and countries were gradually creating a high wall separating the "developed" and "underdeveloped." The differences were measured and interpreted; thinkers widely disseminated ideas on why differences existed and how to modify them. "Center-periphery" was a typical metaphor that described graphically the abyss dividing those using more energy and resources and their distant suppliers.

The First World War ended with a weakened Europe, a wounded Russia, a colonized periphery, and dominance as the main paradigm for decision-makers. The wonderful perspective brought by mechanization, planes, automobiles, and railways were shaken by the earthquake forces of social, financial, and ideological struggles. It was the time of the largest divorce ever between E and C. No institutional or global arena was able to encompass the magnitude of rivalries and changes, even less be capable of recreating a credible climate for peace and justice. Technology became a synonym of secrecy and espionage; force was the symbol for controlling social diversity. Most of the calls for brotherhood or integration were discarded as utopias.

Then, it was a question of time before the horsemen of the apocalypse rode again. Two decades later, the main centers of power were ready to ignite an even wider world war. Millions killed, genocides, holocausts, destruction, and by the end the most extreme demonstration of badly used energy: atomic power vanishing cities from the face of Earth. Nevertheless, the image of the nuclear mushroom and the prospect of near-universal massive destruction, shook the conscience of the planet. Men learned that it was possible for the few to destroy the creation of all. All learned that people needed to create values and organizations that could establish guidelines and limits to the fruits of their own science.

The main lesson learned was that unity was essential, negotiations were required, participation was necessary, and mutual trust was a requirement for rebuilding some social fabric. The first measure adopted by the "winners" of a lose—lose war, was the Organization of the United Nations (UN), including all the independent countries of the world. It was not an easy task. The world had also created new divisions. An Iron Curtain was still separating ideological blocks; a Third World was being promoted and localized wars or guerrillas were also part of a "Cold" War that continued to corner cooperation within the limits of their security or political interests. Nevertheless, the UN achieved goals that were considered impossible before: an impressive process of decolonization, mainly in Africa, thousands of peacekeeping and peacemaking operations, humanitarian help for many critical spots, atomic weapons supervision, human rights agreements, trade rules, children and women support, among so many other achievements. Particularly, for the purpose of this analysis, it is necessary to highlight the promotion of a renewed framework for cooperation, with a lot of highs and lows.

2.4 Traditional cooperation

To the contrary of the initial perception that war had broken all the ties between nations on Earth, a counter example was shown with the reconstruction of Europe. The Marshall Plan was the biggest case in history that brought together partner countries, or Allies, and found a way of cooperating with so visible and impressive win—win results. The immediate after-war mechanisms mobilized people and tools to rebuild infrastructure, education, job structures, production, knowledge, sanitary conditions, culture, and a renewed sense of responsibility and belonging. The United States provided finance and obtained markets, shared technologies, and disseminated knowledge, cultural exchanges took place, and the Allies were strengthened until the wall of Cold War collapsed.

In the meantime, the Soviet "Union" gave another example of cooperation, but under the umbrella of domination and central planning. They strengthened their military forces but weakened their agriculture and industry; promoted socialism, but avoided nonpartisan social participation. The result was that in 1989 "the wall" collapsed. While this was happening, the most crucial confrontations were with regard to the fossil energies. A new Middle East became the center of conflicts and dictatorships. Oil countries were the focus of battles for leadership and control. Nevertheless, cooperation was opening some channels, even if still under the scope of the dominance paradigm.

Countries were shaking hands with a narrative of International Cooperation. That was cooperation between sovereign nations. A quick slide show of this period would show presidents closing deals with presidents, ministers with ministers, and bureaucrats with bureaucrats. They were all governmental agreements, most of them with small and controversial results. Most of them repeating the old donor—recipient paradigm, that aggravated, in many cases, the ancient dependence syndrome, trying to bridge the "haves" with the "have-nots," but provoking in many sectors a lack of trust, a feeling of beggars receiving alms, instead of a sense of partnership. Of course, many exceptions existed, but that kind of "cooperation" had its limits and left an environment of confrontation between those requiring more "aid" and the other side demanding a better use of "taxpayer's money" in the country of origin of the funds. There is no doubt also that some good examples of South—South cooperation achieved their objectives, but it was again, the case of cooperation between "poor" partners, the funds of which did not necessarily reach the pocket or hearts of most of the intended recipients. Cooperation is not only about "money."

Financial aid was the next or parallel phase of cooperation. It involved loans, credits, guarantees, and all sort of mechanisms used by banks. Development banks were the protagonists of these processes. But other private banks, and funds came also into the scene, offered "not-so-expensive" loans or grants with "not-so-complicated" procedures, with clear security priorities. A lot of these types of contracts multiplied around the "third" and "fourth" world countries, but they also lost credibility, and caused many debt crises, as well as inflationary pressures. In most cases, no capacities were developed to sustain the initial push that was supposed to ignite development.

Foreign Direct Investment was the other concept that did not completely achieve the expectations of all participants. It has limited impacts, confronts many cultural issues, leads to environmental complaints, has a lack of bilateral decision-making, profit transfers, a long list of judicial processes, and corruption accusations.

2.5 Global cooperation

The creation of the UN as a universal institution for peace and prosperity was the most important outcome demonstrating that mankind, despite all its deficiencies and dramas, has the means to improve itself. By cooperating with each other, people organized themselves to increase their energies and their well-being. After the Second World War a lot of progress was achieved, but the main challenge of unanimity was still a pending task. Pessimism about the future continued to be the trending mood because it looked like humanity could not control technology. The planet, for many decades, was still divided by the Iron Curtain and poverty. There was not a "Globe" as seen from the satellites: it was split by ideologies, preconceptions, and a dominance paradigm. The breakdown of the Berlin Wall changed everything. Some authors even thought it was the "End of History" (Fukuyama, 1989).

A new wave of hope ignited everywhere the birth of a new era of possibilities, a new era of cooperation. This time, a global cooperation. The planet, through the UN, called all sovereign countries to a roundtable that would change everything. The revolution had started. Its first publication was a symbol of the renewed quest: it was called "Our Common Future" (Brundtland, 1987), also known as "The Brundtland Report," because this work was produced by a UN Commission headed by Gro Brundtland, a former Norwegian Prime Minister. This document defined and popularized the term "sustainable development." It contained the basis for the gigantic step of looking to ourselves and our surroundings as a single species in charge of our own destiny.

The UN chose Brazil to be the site of a conference of all countries, a conference that generated the most impressive paradigm change in the history of science. The planet had looked to itself mainly from an economic perspective. It was necessary to open minds, to observe, and to act as an integrated and multidimensional system, with the human being at the center of all concerns. After a couple of years of rigorous preparation, Rio-92 happened, and the world would never again be the same. Global cooperation was on its way.

In 1992, at the Earth Summit in Rio de Janeiro, Brazil, the world reached a consensus, expressed in the "Rio Declaration," that summarized the new paradigm for cooperation and its action plan, called Agenda 21, which detailed the common tasks for all toward the 21st century. At that time, sustainable development included aspects such as[1]:

1. Changing the quality of growth
2. Reducing population growth

---

[1] Selected from the Brundtland Report.

3. Securing food supply
4. Maintaining biodiversity
5. Establishing safe energy
6. Ecological modernization of industry
7. Guiding land use and urbanization

The key highlights of the Rio Declaration were (UN—United Nations Conference on Environment and Development, 1992):

1. Human beings are at the center of the concern of sustainable development.
2. States have the right to use their own resources as they see fit.
3. Must integrate the environment into development plans.
4. States should enact environmental legislation and should cooperate where needed.
5. Should actively discourage or prevent relocation of activities or substances harmful to the environment or human health.
6. Apply the precautionary approach.
7. Internalize environmental costs and use economic instruments.
8. Environment Impact Assessment should be implemented everywhere.
9. Peace, development, and environmental protection are *interdependent and indivisible.*

"Sustainable Development" represented the beginning of a new era, because it adopted a holistic vision, and a multidimensional approach to the world issues. It combined initially the necessity of a viable natural environment, with the economic requirements and the social aspects of communities. For the first time, a global governmental gathering included "Nongovernmental-Organizations" (NGOs), a brand that was popularized and later evolved to the concept of "Civil Society Organizations" (CSOs), in order to define them for what they were, and not for what they were not. The idea was to introduce a third component in between the old dichotomy of public and private sectors, which had divided people and limited the access of communities in the spheres of decision-making and actions.

Agenda 21 was approved as a comprehensive plan that addressed the needed actions in 39 areas, spanning clusters of issues including economic and social, conservation and resource management, participation, and implementation. It was an *inclusive and universal* program, that placed a major emphasis on broad participation from governments, the UN, other intergovernmental organizations, and nine major groups in civil society. It included four sections (UN—United Nations Conference on Environment and Development, 1992):

1. Social and Economic Dimension (8 chapters)
2. Conservation and Management of Resources for Development (14 chapters)
3. Strengthening the Role of Major Groups (10 chapters)
4. Means of Implementation (8 chapters)

Two essential components promoted by this approach were first, that for cooperation to exist it requires "two-to-tango," meaning a win—win and equal rights for both parties involved. Second, that for development to be sustainable it demands local capacities to be developed, in order to transmit development from generation to generation. This was the reason to create the program Capacity 21, which gave a new motivation for the growth of capacity development. In the 1960s and 1970s most actors were pushing for training and skills, meaning individual capacities. Later in the 1980s, organizational capacities were the trend. During the 1990s, the concept had evolved to institutional reforms, policy changes, ethics, and others. After Rio-92 and during the 2000s, capacity development adopted a more integrated approach through partnerships.

Then, the scenario was set for a new style of cooperation.

2.6 The millennium development goals

The year 2000 was another turning point in the history of cooperation. The world was full of hope and the new millennium was opening doors of opportunities, made possible by peace, technology, and integration. All Heads of State called for a new summit at the UN, in New York, to analyze the important achievements of Agenda 21 that had produced more ups than downs and had created the way for even more ambitions and more focused targets. The renewed consensus was expressed in a document called "The United Nations Millennium Declaration" (United Nations General Assembly, 2000).

This Declaration included 60 goals regarding peace, development, the environment, human rights, the vulnerable, hunger and poverty, and Africa. It was a long document, not easy to digest, and even more difficult to be used as a day-to-day guidance for action. But Rio + 10 (2002) was approached with the hospitality of a new South Africa, and Agenda 21 was required to focus its already wide scope and establish the ways to measure its impacts. Then, in 2001, a group of experts[2] close to the UN Secretariat selected 18 targets from the Millennium Declaration and grouped them into 8 goals, with the objective of simplifying the Agenda and refocusing it around poverty reduction and other social goals. It became a very useful tool for monitoring and managing actions. It also contributed to target cooperation and incentivizing concrete outcomes. It diverted the debate from the issues being exclusively monetary, to an arena of building partnerships with mutual and global benefits. They were approved by all heads of state in the Johannesburg Summit of 2002, just a few months after the September 11 disaster.

The goals, later known as the Millennium Development Goals (MDGs), included a set of indicators of progress and for self-evaluation. The MDGs were:

1. Eradicate Extreme Poverty and Hunger
2. Achieve Universal Primary Education
3. Promote Gender Equality and Empower Women
4. Reduce Child Mortality
5. Improve Maternal Health
6. Combat HIV/AIDS, Malaria and other diseases
7. Ensure Environmental Sustainability
8. Global Partnership for Development

The goals were presented, accompanied by attractive icons that became famous, and, as soon as approved it was in the Conference, it went to the desks and trenches

---

[2] Oscar Serrate, coauthor of this chapter was part of that group.

of the main decision-makers around the planet, thus creating one of the most relevant platforms for development cooperation. This time, businesses became another pillar of the Global Partnership with the active and committed participation of their main leaders. The results of the MDGs were impressive. The following list presents some achievements (UN—United Nations Department of Economic and Social Affairs, 2016):

1. The number of people now living in extreme poverty has declined by more than half, falling from 1.9 billion in 1990 to 836 million in 2015.
2. The number of people in the working middle class—living on more than $4 a day— nearly tripled between 1991 and 2015.
3. The proportion of undernourished people in the developing regions dropped by almost half since 1990.
4. The number of out-of-school children of primary school age worldwide fell by almost half, to an estimated 57 million in 2015, down from 100 million in 2000.
5. Gender parity in primary school has been achieved in the majority of countries.
6. The mortality rate of children under-five was cut by more than half since 1990.
7. Since 1990, maternal mortality fell by 45% worldwide.
8. Over 6.2 million malaria deaths have been averted between 2000 and 2015.
9. New HIV infections fell by approximately 40% between 2000 and 2013.
10. By June 2014, 13.6 million people living with HIV were receiving antiretroviral therapy globally, an immense increase from just 800,000 in 2003.
11. Between 2000 and 2013, tuberculosis prevention, diagnosis, and treatment interventions saved an estimated 37 million lives.
12. Worldwide 2.1 billion people have gained access to improved sanitation.
13. Globally, 147 countries have met the MDG drinking water target, 95 countries have met the MDG sanitation target, and 77 countries have met both.
14. Official development assistance from developed countries increased 66% in real terms from 2000 and 2014, reaching $135.2 billion.

As Steve Jobs once said: "Everyone here has the sense that right now is one of those moments when we are influencing the future." He was probably thinking about the impacts of the wonderful technologies he helped create. The wonderful thing was that, at the same time, mankind was creating the institutions and organizations for human action. Among the thorns of terrorism and hunger, the roses of cooperation were blooming again. As never before, the planet started to directly feel the initial effects of global climate change and the impacts of a complex financial crisis, which erupted in 2008. Then, the Rio + 20 Conference of 2012 was called for, meeting again in Rio de Janeiro, and new energies were brought to face the renewed challenges. It was a conference that prepared the background for building an even more targeted action plan, the Agenda 2030 that was finally approved at the 2015 Summit. It included a larger audience and contributors, thanks to the global online conversations. Information was again transforming the way to build partnerships and, obviously, cooperation.

They delivered the Sustainable Development Goals (SDGs), the 17 SDGs, with their 169 indicators for follow-up and coordination. And this happened in the same year that witnessed the approval of the Addis Ababa Financing for Development Conference and the

Paris Agreement on Climate Change. Nelson Mandela had once said that "It always seems impossible until it's done."

2.7 The Sustainable Development Goals

The SDGs adopted were:

1. No poverty
2. Zero hunger
3. Good health and well-being
4. Quality education
5. Gender equality
6. Clear water and sanitation
7. Affordable and clean energy
8. Decent work and economic growth
9. Industry, innovation, and infrastructure
10. Reduced inequalities
11. Sustainable cities and communities
12. Responsible production and consumption
13. Climate action
14. Life below water
15. Life on land
16. Peace, justice, and strong institutions
17. Partnerships for the goals

Their 169 indicators detailed the principles of a systemic approach, a polytechnic methodology, human-centered priorities, and a tool capable of solving the issues of complexity or singularity. The SDGs became the cornerstone of the renewed science of sustainability, and the best narrative for communicating it.

This "Goals" script has already proven to be the most integrated, multidimensional, and effective common storyboard ever told for reconciling energy and the public. Many of these goals are directly or indirectly related with all kinds of energy. For Betz, a strategic narrative is a "concise statement of what it is doing, why and how that links to positive vision of the future with the individual actions of members of its own societies and member of other societies whom it wishes to influence" (Betz, 2008). Previously, other kinds of narratives had been used, but could not mobilize the required momentum for igniting cooperation. Maybe they were not completely strategic, as for example Bushell et al. (2017):

1. The "GORE" Narrative—it was about the evidence for climate change.
2. "The End of the World" and the alarmism—a catastrophic future.
3. "Every little helps" and the breakdown of complexity.
4. "Polar bear and distancing"—destructive effects of human action.
5. "Green Living"—drastic changes in lifestyles. Common sacrifices.
6. "Debate and Scam"—opposed to action on climate change.
7. "Carbon-fueled expansion"—mitigation will hinder economic growth.
8. The "Opportunities"—emphasizing good investments and jobs options.

2.8 The renewed cooperation agencies

Knowledge is power. Knowledge is energy. It has become in many cases the cornerstone of cooperation. "Knowledge shared is power multiplied," as Robert Noyce said. Information technologies are clearly contributing in that direction, despite abuses, multiple cases of alienation, and distortions. Examples abound, for example on how most of the multilateral organizations related to the development cooperation have been gradually joining this wave generated by the SDGs. Here follows some of their communications:

- *Word Bank Group*: "It is critical to do this work in close partnership with a broad and diverse group of global stakeholders. One indispensable partner is the UN, with which we recently signed a Strategic Partnership Framework to consolidate our joint commitment to help countries implement the 2030 Agenda" (World Bank Group, 2018).
- *Global Environment Facility*: "The *SDGs* hold the promise of a fresh start for our planet (...) The GEF is committed to help all actors deliver on that potential," Naoko Ishii, CEO of GEF (GEF—Global Environment Facility, 2015).
- *Green Climate Fund*: "Realizing the *SDGs* and lifting millions out of poverty will simply not be possible on a planet that is ravaged by climate change," Héla Cheikhrouhou, Executive Director of GCF (GCF—Green Climate Fund, 2016).
- *Economic Commission for Latin America and the Caribbean*: "The *2030 Agenda* represents the international community's response to the economic, distributive and environmental imbalances built up under the prevailing development pattern."
- *Interamerican Development Bank*: "The adoption of the *SDGs* at the UN General Assembly in late 2015 reinforced a global commitment "to shift the world onto a sustainable and resilient path" and set an ambitious agenda for the next 15 years," Luis Alberto Moreno, IDB President (IDB—Interamerican Development Bank, 2016).
- *Development Bank of Latin America*: "Common Assessment Framework (CAF) and United Nations Development Programme (UNDP) deepened their alliance with the signing of a cooperation agreement to work jointly in areas of mutual interest and to support the region's governments in achieving the SDGs for 2030" (CAF—Development Bank Of Latin America, 2016).

And so many other expressions of multilateral commitments and visions:

- The five UN priorities, according to the Secretary General: "From now on, my five priorities will be: ambition, ambition, ambition, ambition, ambition," Guterres said in a statement read to delegates at the conclusion of COP24, the Conference of the Parties. "Ambition in mitigation. Ambition in adaptation. Ambition in finance. Ambition in technical cooperation and capacity building. Ambition in technological innovation" (UN—United Nations, 2018).
- "The expansion of South—South and triangular cooperation in the science, technology and innovation realm creates new opportunities. South—South cooperation should build on its strengths in adapting appropriate, affordable technologies and solutions to local contexts, helping countries generate and sustain their own knowledge" (UNESCO—United Nations Economic and Social Council, 2018).

- "This Forum (DCF July 2018) has helped shape the role of development cooperation in the 2030 Agenda for Sustainable Development. Since 2015, it has advanced a practical, holistic concept of development cooperation that encompasses financial resources, capacity-building, technology development and transfer, policy change and multistakeholder partnerships" (UNESCO—United Nations Economic and Social Council, 2018).

2.9 The new players

A cooperation initiative always starts by finding a strategic partner and money. In order to build those strategic partnerships, it is necessary to analyze the current landscape of players that have drastically changed after the launch of the SDGs. Using the language of the SDGs became recurrent, but every partnership demands its own specificity to fit both parties and the mutual convenience.

The business partners are normally looking for new opportunities for income and job creation. According to the Business and Sustainable Development Commission, achieving the SDGs could unlock US$12 trillion a year in business value across four economic systems alone, by 2030. That would create more than 380 million jobs in food and agriculture, cities and urban mobility, energy and materials, and health and well-being. Oil companies are committing themselves to provide more and cleaner energy solutions, and a strong social license to operate.[3]

Networks are multiplying and acting directly in environmental issues, energy, economy, well-being, innovation models, and other areas. Civil Societies Organizations are multiplying and checking their performances based on the SDGs. Universities are providing innovations and solutions to the SDGs, helping create SDG implementers, and SDGs are contributing to focus on education, to evaluate the challenges of future jobs, and to support cooperation with new partners.

Necessarily, as always, the public, or the community, or just the citizen is always the most special partner, and probably the most difficult to engage in a trustful and sustainable association. The public perception is so fragmented given the inequalities, cultural differences, perspectives, experiences, etc. The point is that, again, we are so different and yet we share the same home. We can sit together and listen to common stories or narratives, to highlight the goals that are common to all of the participants, and to detail the common future.

The public perception, and not only the public interest, is crucial on really engaging in a mutually beneficial partnership. Each of the participants will always have a partial perception. Our human partner will need to be informed about the big and small picture of any initiative. Trust must be built bilaterally, and ethics is crucial for a sustainable engagement, and leadership needs to be developed to obtain the social license to operate.

The main problem with trust is that technology changes exponentially, but human organizations change logarithmically. For that reason, technology and society can no longer be

---

[3] SHELL Strategic Purpose. Available from: https://www.shell.com/investors/shell-and-our-strategy/ourstrategy.html.

divided as two separated parts of a plan. The new "public" is not only NGOs or CSOs. Everyone has his or her own cellular and is a protagonist, who can mobilize their peers with the click of a button. If cooperation is required, players need to listen to each other permanently, step by step, and at the very beginning of the relationship. Since the 1990s, humanity has taken what is now called a "Narrative Turn" (Mitchell, 1981). More and more, storytelling is becoming an effective way to talk and listen to one another.

Stories are even more useful if they are kept extremely simple, pleasant, and lovable. Leonardo da Vinci was quoted as saying that "Simplicity is the ultimate sophistication," so cooperation with the public requires the narration of the stories, the legends, the parables, or the metaphors of complexity, with extreme simplicity. A famous saying underlines that people fear what they do not understand and hate what they fear.

The public has become the protagonist of the new world of cooperation. Their engagement was solved by defining public policies and issuing the corresponding legal licenses, and then obtaining public engagement. Currently, in most of the cases, the process starts by engaging the public, then defining the policy, and lastly obtaining the legal licenses. Now, the best processes, normally, are initiated with the social licenses before going after the legal licenses. The traditional lines separating private and public sectors are becoming more and more blurred, because everything is becoming a "public issue," and the public is watching everyone. Again, this is a result of the permanently renewed Information and Communication Technologies.

2.10 Financing for sustainable development

Money isn't always the principal issue in cooperation. But resources are essential to mobilize energies. Financing for sustainable development has increased substantially since the frameworks for cooperation were improved, mobilized, and the sources multiplied their capacities. It was not a coincidence that while the planet was adopting Agenda 2030, at the same time, the Paris Agreement was finally approved by all countries, and the same year—2015—saw the global financing world adopt the Addis Ababa Action Agenda on financing for sustainable development.

Since then, the SDGs are being incorporated into public budgets, private investments are increasing, and a new conscience surged to make a fundamental shift at the international level. The UN is repeatedly calling for a "Strategy for Financing the 2030 Agenda for Sustainable Development, which sets out priority actions to align global economic policies and financial systems with the 2030 Agenda and seize the potential of financial innovation, new technologies and digitalization" (UN—United Nations, 2019).

A brief picture of how these processes are evolving can be seen by the figures in some specific areas. For example, "Climate finance flows increased 17% from 2013—14 to 2015—16 but are still below the commitment by developed countries to jointly mobilize $100 billion a year by 2020 from a wide variety of sources to address developing countries' climate financing needs" (UN—United Nations, 2019) Or the trend related to the Bilateral ODA (Official Development Assistance) for Gender equality and women's empowerment that more than doubled since 2012. And the growing pace in lending from multilateral development banks, which increased from around US$30 billion in 2005 up to more than US$60 billion in 2017 (UN—United Nations, 2019).

The National Development Cooperation Policies have been created by countries willing to focus on the objectives and commitments to help mobilize and align development cooperation with their SDGs. They normally cover more than the traditional ODAs, such as South—South cooperation and engagement of the private sector. And the procedures always involve a wide range of stakeholders, including community-based organizations, in order to ensure their adequate participation. In summary, there are resources available in the global and national arenas. The question is how to overcome the barriers that allow implementers of the SDGs to deliver by 2030.

Some of the barriers are very project-specific, but there are other big walls being raised that could limit access to funding or investments. Particularly, the UN Secretary General, expresses concern about anticipated low-growth figures, falling private investments in infrastructure, lowering spending on education, impacts of climate change, the renewed challenges to the international trading and financial systems, and many other corresponding risks (UN—United Nations, 2019). But many projects and initiatives are bridging those gaps or barriers.

2.11 The Brazilian case

One very important case study is the initiative of Ethanol in Brazil, that started 40 years ago as the ProAlcool Program. Its initial objectives were geopolitical and related to the issue of energy security and trade deficit concerns. Many sectors internally and externally looked at Brazil, at that time, as a potential superpower. The Program achieved success, in the initial years, multiplying its production and broadening its land coverage, and it brought environmental benefits. It is worth remembering that Brazil has pledged for 2030 to provide 45% of renewable energy and to increase the share of sustainable biofuel up to 18%, based on the capacities developed during this process.

This is a good example of a social innovation that prospered through internal cooperation that also been exported to other places and producers. Sugarcane ethanol reduces greenhouse gas (GHG) emissions by 90% on average compared to gasoline.

The mechanization involved ended former burning practices and people participated in its evolution by building trust and promoting emotional connections. Ethanol was seen as a green, clean, and renewable technology, and it was supported as a "silver bullet" to help the country. The sugarcane ethanol industry was widely known as a Brazilian success that also involved a wide partnership mobilizing resources and cooperation.

Brazil's current challenge now is how to fund and finance BECCS (BioEnergy with Carbon Capture and Storage), which is the new technological and social frontier to continue to improve the expansion of the climate and social benefits of this industry. For this purpose, studies are being made to evaluate the potential financial mechanisms that could support this kind of initiatives. Two specialized operating entities entrusted by the UNFCCC (the United Nations Framework Convention on Climate Change) are being analyzed: the GEF and GCF. The GEF (Global Environmental Facility) is an entity that has mobilized $18 billion in grants plus $93 billion in financing for 4500 projects. Previously, in 2009, with UNDP, the GEF created a $10.3 million project for BECCS for Brazil, through the SCCF (Special Climate Change Fund). The idea was to fund the capture of CO_2 from bioethanol production but has not yet prospered. The GCF (Green Climate Fund) was created in 2011 for loans and its current commitments are around $10 billion. They have multimillion-dollar capacities for projects and they also aim to leverage the private sector by using PPPs (Private–Public Partnerships) for infrastructure projects.

The cooperation architecture for this program is currently being built in Brazil, through the leadership of a research team located at the University of São Paulo (USP) at the Research Center for Gas Innovation (RCGI). RCGI is currently developing 46 projects in the areas of engineering, physical chemistry, energy policies and economics, and CO_2 abatement and geophysics; they all aim to contribute to the SDGs framework, incorporating gradually the use of biogas, solar, wind, and hydrogen. RCGI's initial founders were members of the FAPESP (São Paulo Research Foundation) and Shell, together with the USP. RCGI recognized the need to extend its multidisciplinary groups and go beyond technical solutions, reaching out to social sciences and the public.

2.12 Cooperation for climate change: the emblematic solution of CCS

The concentration of GHGs in the atmosphere poses a major challenge to mankind's future: the need to reduce carbon dioxide (CO_2) emissions. Taking into account the Paris Agreement's target to limit the rise of the global temperature to 2°C, compared to preindustrialization levels, the report World Energy Outlook 2017 establishes the Sustainable Development Scenario (SDC), which integrates a range of energy-related goals crucial for sustainable economic development. Those goals include climate stabilization, cleaner air, universal access to modern energy, and reduction of energy security risks. In order to achieve 1.7°C or 1.8°C reduction, the SDC requires a net zero CO_2 emission before the end of the century or even "negative" emissions, with CO_2 atmospheric sequestration. There is no single technology capable of meeting the targets, so a combination of technologies will be required, including renewable energies, nuclear power plants, and adjusting to different geographical, economic, and social constraints across countries (IEA—International Energy Agency, 2017b).

Carbon capture and storage (CCS) plays an important role in this scenario, especially for negative emissions, which can be achieved when these technologies are coupled with bioenergy (BECCS) or direct air capture. In the industry sector, CCS (or CCUS, from Carbon Capture Utilization and Storage) is key to cutting emissions in steel, cement, and chemical production, requiring a large increase in the level of CO_2 captured by 2060, as shown by an analysis of the International Energy Agency (IEA) in Energy Technology Perspectives 2017.

The World Bank has also exposed the importance of deploying CCS technologies to tackle CO_2 emissions:

> Carbon capture and storage (CCS) technologies are potentially capable of making a significant contribution to meeting global GHG mitigation objectives. CCS is also viewed as one of the key technologies to mitigate emissions from fossil fuel power stations by up to 90%. Other industrial sectors (cement, iron & steel, chemicals) and the fuel production and transformation sector, with substantial GHG emissions, could also benefit from CCS development. *World Bank Group (2017)*

Thus, to address hard-to-decarbonize sectors, a substantial increase in CCS will be needed in order to get the world on track to limit climate change's anthropogenic effects. Are we prepared to achieve this substantial increase?

This chapter now passes to the analysis of the challenges ahead. Challenges that have been studied can vary significantly, depending on assumptions of the projects' life cycle assessment or for the application of climate policies. Several authors have investigated the potential hurdles of CCS deployment to evaluate future projects' risks and constraints or to identify possible reasons for the mismatch between the so-called importance of technology and the low number of ongoing projects.

The Global CCS Institute released a report in 2018 claiming that capture and injection technologies are not technical barriers anymore and the underground storage availability is more than enough (GCCSI—Global CCS Institute, 2018). However, there are some technological pathways that can create constraints on the project's implementation. Pihkola et al. (2017) point out the technical aspects that can challenge the sustainability of CCS adoption in Finland, as an example. Postcombustion systems for retrofitting existing stationary emission sources are typically seen as not favorable from a stakeholder point of view, due to the considered low efficiency, the increase in the need for fossil fuels, and the consequent life cycle environmental impacts (Pihkola et al., 2017). Durmaz (2018) mentions the lack of technological maturity and observes the necessity of better understanding the geological storage risks (Durmaz, 2018). Even with the significant progress in the comprehension of the storage capacities and distribution of accessible storage sites, the author indicates the importance of incentives for further exploration and discovery of new storage sites.

Although technical solutions can be tangible, investment costs have been identified as one of the biggest concerns for CCS deployment and are strongly related to the technologies employed. Engineering innovations play an important role in reducing costs and making projects more attractive. Many authors have highlighted the problems of expressive costs and investments for demonstration plants and for technology diffusion, which can vary with capture routes, transportation system extensions, and the complexity of storage sites. Budinis et al. (2018) identify this topic as the major challenge preventing the widespread adoption of CCS in the short to medium term and show that the most expensive step is the capture process (Budinis et al., 2018). He also observes that in the long term CCS can be very cost-effective when compared with other mitigation options (Budinis et al., 2018).

The influence of economic circumstances on CCS implementation was explored for two demonstration projects, in the United Kingdom and Canada, by Kern et al. (2016). This evaluation is related to the general macroeconomic climate, established resource endowments, fossil fuel reserves, scale of fossil extraction in the economy, and public finance and export earnings (Kern et al., 2016). Substantial remaining fossil fuel reserves, fossil fuel extraction (assuming a prominent place in economic activity), government revenues, and exports are factors that can favor CCS demonstration projects. On the other hand, difficult economic circumstances, such as recession and low resource prices, can make CCS demonstration projects less likely to succeed. The absence of R&D funding, of all natures, also hinder the creation of new demonstration projects. Pihkola et al. (2017) have indicated that the lack of dedicated R&D funding in Finland for new technologies that could reduce costs compared to traditional capture techniques might prevent necessary demonstration and development activities (Pihkola et al., 2017). Funding origins were also analyzed in order to establish the relationship with demonstration projects' success. Thronicker et al. (2016) aimed to empirically identify the characteristics that render CCS projects more or less likely to become successfully operational, and one of the main findings was that public funding presented a negative impact, implying that such efforts have not achieved their objectives (Thronicker et al., 2016).

As Budinis et al. (2018) concluded in their review, no CCS barrier is exclusively technical. The abovementioned technical and economic barriers are intimately linked with—and in most cases dependent on—social sciences aspects (Budinis et al., 2018). Many more studies target political and legal dimensions and consider psychological and cultural factors related to society's values.

Probably the most highlighted point, that addresses the main barriers for CCS deployment, is the lack of effective policies addressing climate change and CCS activities. Difficulties to constitute comprehensive carbon policies and adequate policy instruments are identified by Durmaz (2018) as one of the major obstacles in CCS breakthroughs and reduce investor confidence to undertake projects (Durmaz, 2018). In addition, the author mentions the absence of well-defined regulations, related to short- and long-term responsibilities for CO_2 storage and transportation, as important challenges that must be solved in order to allow large-scale future deployment of the technology. Kern et al. (2016) also analyzed the role of government policy objectives, frameworks, and measures to CCS demonstration projects' success, through energy, climate, and CCS-specific policy perspectives (Kern et al., 2016). The study argued that strong climate policy objectives, a carbon pricing mechanism with a substantial incentive for emissions abatement, and regulatory clarity and simplicity favor CCS demonstration projects (Kern et al., 2016).

Issues regarding uncertainty and discontinuity in political decision-making are also relevant (Pihkola et al., 2017). Karimi (2017), from interviews with experts in Finland, Germany, and Norway, showed that a lack of legislation is one main obstacle. Most of the interviewees declared that CCS policies and regulations are not accompanying the evolution of technical development, and that this gap is delaying the progress of CCS. Despite some skepticism, most of those interviewed saw CCS as a key technology to tackle climate change, but more political concerns were raised by experts and by the author, such as lack of political will and incentives, institutional inertia, and the divergence of views between different government levels. By defining temporal features for policy making and deployment of large-scale CCS projects, Karimi (2017) concluded that these concerns combined with a poor temporal fit—due to the urgent need to cut emissions—can make CCS lose space to other solutions.

The critical mismatching, between political action (or total inertia) and the targets to deploy CCS technologies in order to tackle climate change, has its roots in the difficulty to engage stakeholders in a broader approach. Stakeholders' attitudes have also been investigated at different levels, and they involve governments, industry sectors, financial institutions, international organisms, the media, and the civil society [both organized (associations, NGOs, etc.) and nonorganized]. Besides the relevance of governments delineation of favorable political–institutional structures, two perspectives are raised as key factors for CCS diffusion.

The first perspective is related to the necessity to find solutions for industries and companies that adopt the technology without losing their market share. Onarheim et al. (2015) studied barriers and opportunities for the application of CCS in Nordic industries from a sectorial perspective, and they found significant potential for a decrease in the competitiveness of these industries in connection with the implementation of CCS, given that most of them compete in global markets. Furthermore, Braunreiter and Bennett (2017) focused on the importance of perceptions and positions of fossil fuel companies on CCS and how it informs decision-making on investment and advocacy. One of the main findings was that implementing CCS would require a significant change within the business strategy of fossil fuel companies, so companies regard governmental commitment and financial support as a prerequisite for their own investments.

The second perspective refers to the public acceptance of CCS, which is also crucial to CCS deployment, especially regarding CO_2 injection activities. Some studies have focused on understanding the factors that influence perception on CCS and technologies with similar characteristics (Huijts et al., 2012; Chalmers et al., 2013; Markusson et al., 2012; Tokushige et al., 2007). The major studies are related to the knowledge of the technologies in question, the risks and benefits perceived by the public, the trust with the organizations involved in the process of implementing the technologies, and the understanding of the population with regard to the fundamental problem addressed by CCS, for example, global warming.

This part now addresses the topic of building the future together, focusing on opportunities for cooperation on CCS. The readiness level creates a positive perspective for the technology diffusion. However, facing the climate problem requires time, scale, and the understanding that this is our common problem. It does not affect one single industry sector, or one country, so separated strategies or approaches can be not effective. Business as usual proved insufficient to reduce atmospheric CO_2 emission and the costs for the required adaptation are too large to be carried by government, industries, or one community alone. History lacks a similar precedent or an obvious way to find the solution, so new arrangements should be evaluated in order to find common ground.

Coninck et al. (2009) exposed the relevance of building strong international cooperation for CCS demonstration projects, pointing out global coordination, transparency, cost-sharing, and communication as guiding principles (Coninck et al., 2009). The authors claim international cooperation was based on the then necessity to scale up demonstration plants to assure and improve technology's maturity (Coninck et al., 2009). Even though the readiness level for some processes was enhanced, the arguments are still relevant for justifying the need for cooperation targeting innovations that can decrease the costs: (1) accelerating learning; (2) globalizing learning; (3) expanding social awareness of and discourse about the acceptability of CCS; and (4) ensuring consistency in safety and integrity of CCS projects.

However, as previously discussed, it is imperative to find solutions for social science-related identified barriers, if the objective is the large-scale deployment of CCS technologies within the required time window. Political—institutional structures give the background for the interactions between political and societal players, and they take into account characteristics such as policy legacies and paradigms, political culture, and constitutional allocations of power across branches of government and geographic spaces (federalism), as exposed by Kern et al. (2016). These structures are key influencing factors for decisions regarding the implementation of some demonstration projects. The institutional arrangement should comprise reasonable incentives to balance stakeholder needs to create the capacity of finding common ground with the development of climate change solutions. This balance needs economic tools with different levels of complexity; it also requires trust, commitment, and a sense of responsibility to promote a cooperation framework capable of building the necessary capacities to deploy CCS technologies with the right timing.

The IEA highlighted the importance of cooperation as one of the five keys to unlocking CCS investment (IEA, 2018). Reports point to strengthening partnerships as critical to supporting rapid and widespread uptake of CCS technologies across the globe: favoring technology transfers and sharing experiences aimed at reducing the costs for future projects. The main opportunities for cooperation listed by IEA include:

- identify and cultivate lower-cost CCS investment opportunities;
- map and assess CO_2 storage potential in key regions, using a consistent methodology and coordinating resource needs as required;
- undertake national and regional transport infrastructure audits;
- develop business models for CO_2 transport and storage infrastructure investment; and
- build institutional capacity, including among regulators and the financial community.

In addition to this list, the conclusions of identified challenges, previously discussed, points to the relevance of amplifying the voices of stakeholders considered in the CCS sphere. Government, industry, and civil society play essential roles in technology adoption, with regard to plant construction for demonstration projects or CO_2 injection sites, attitudes toward policies, and financial allocation priorities at the national level.

Another fundamental perspective is the importance of the international dimension of cooperation. International agreements related to climate change—or specifically focused on CCS deployment—between countries, companies, or other organizations are key to creating incentives and preventing market share losses for those who decide to adopt carbon reduction policies. Reliability and enforcement of agreements are equally relevant for the industrial and geographical context. One practical example of international cooperation opportunity was mentioned by the expert Ian Havercroft, who recommended the promotion of a "greater international cooperation around the transboundary shipping of CO_2 as well as pushing for greater regulatory coherence and allocation of risk for CCUS projects" (GCCSI—Global CCS Institute, 2018).

A related concern is securing new players, interests, and possibilities with regard to CCS. Cooperation agreements and partnerships imply multiple approaches and involve several stakeholders around a common objective. As discussed, deploying CCS at the right time and scale, while overcoming economical, technological, and political constraints, requires new arrangements between governments, business, industry, international organizations, financial institutions, research centers, universities, and hubs and networks to create strong collaborations. Efforts should concentrate on: (1) priority setting and coordination; (2) research and innovation; (3) technology transfer; (4) demonstration and pilot projects; (5) capacity building; (6) policy strategy; (7) provision of finance; and (8) supporting hubs and networks.

Collaboration programs, partnerships, and initiatives recently undertook great effort toward CCS deployment around the world. The Global CCS Institute is an international organization with the mission to accelerate CCS deployment and the commercial viability of CCS globally to tackle climate change. Members include governmental representatives, corporations, private industry, and academia. Since 2015, the organization has published The Global Status of CCS annually.

IEA heads one Technology Collaboration Program named Greenhouse Gas R&D, with the objective of examining technologies for reducing GHG emissions derived from fossil-fuel use. The focus is on CCS processes.[4] Created in 1991, the program now involves 15 member countries, the European Commission, the Organization of the Petroleum Exporting Countries, and 16 multinational sponsors, with worldwide activities that include conferences, expert networks, information papers, modeling and databases, summer school and student mentoring, technical evaluations and reports, and technical workshops.

Other important examples of initiatives aimed at stakeholder engagement are the Carbon Sequestration Leadership Forum, which includes 25 countries plus the European Commission as member governments. Also, the Mission Innovation, with 23 countries and the European Commission, focuses on accelerating global clean energy innovation.

There are also partnerships specifically in the industry sector. The Oil and Gas Climate Initiative has 13 big petroleum companies working on two major topics. The first is the Strategy and Policy and Climate Investments. The second topic is an initiative that promotes the CCUS investment day, with the objective of supporting commercial projects for CO_2 utilization or storage as well as technologies that can significantly lower the cost of CO_2 capture or can create products that utilize CO_2.

Another relevant partnership was The World Bank Carbon Capture and Storage Trust Fund (WB CCS TF), established in 2009, relying on financial support from the Norwegian government and the Global CCS Institute, which promotes capacity building in developing countries for CCS technologies. The CCS TF acts as a facilitator and a catalyst for CCS development in many countries. The main objectives are (1) to facilitate inclusion of CCS options into low-carbon growth strategies and policies developed by national institutions and supported by the World Bank Group interventions; (2) to support strengthening capacity- and knowledge-building to involve member countries in the international CCS forum; and (3) to create opportunities for member countries to explore CCS potential, access carbon markets, and realize the benefits of domestic CCS technology development (World Bank Group, 2017).

Considering the presented analysis, the following discussion brings some insights to identify potential opportunities for cooperation for CCS deployment in Brazil.

At first, it is important to evaluate the current state of CCS activities in the country. According to the Global CCS Map from Scottish Carbon Capture and Storage, there are two CCS projects operating in Brazil, in Lula and Buracica oil fields, both for enhanced oil recovery (EOR), which means injecting CO_2 during the declining phase of oil production to improve the process and have a better recovery rate. CO_2 injection is a well-known technology in Brazil, utilized specially for EOR, but other techniques must be covered with

[4] https://www.iea.org/tcp/ghg/.

regard to geological characteristics and reservoir suitability for carbon sequestration. There are only three demonstration projects and one is also for EOR purposes, but for considering postcombustion capture process, while the others are related to storage in coal seam and oxyfuel capture.

From the players' perspective, Brazil has some partnerships between oil companies, public foundations, and universities for CCS research. Research centers, such as the Centre of Excellence in Research on Carbon Storage[196] sponsored by Petrobras and Pontifícia Universidade Católica do Rio Grande do Sul, and RCGI,[5] sponsored by Shell, FAPESP, and USP are conducting studies concerning the technological development for CCS whole chain, policies, regulation, and environmental aspects.

From the institutional approach, there is a lack of regulations for CO_2 emissions, CCS activities, and carbon pricing mechanisms. This consolidates an important cooperation target of reducing uncertainties and stakeholders' risk perception, creating a better economic and political environment for demonstration, pilots, and commercial plants. Furthermore, another relevant condition for identifying opportunities for cooperation in Brazil is evaluating the country's particularities that can affect CCS chain features, considering technological, logistical, geological, and cultural aspects. For capture processes, CO_2 emissions from stationary sources are strongly concentrated in the South and Southeast of Brazil, including the industries of iron and steel, pulp and paper, and chemical and power plants. It is also important to consider oil and gas activities as emission sources.

Presalt petroleum activities now make up more than half of Brazilian production and more bidding rounds are planned (IBP, 2019); nevertheless, this production and projected growth must deal with a high proportion of associated CO_2 (ROCHEDO et al., 2016). Sinks for carbon, on the other hand, have a good potential for geological storage, because depleted oil and gas reservoirs are usually well-known structures with regard to their safety, and new geological possibilities for CO_2 injection are being massively studied, such as saline aquifers and artificial salt caverns. However, between the capture and storage processes, it is necessary to analyze the CO_2 transport infrastructure routes and logistics (considering technical, economic, environmental, and regulatory issues that can represent considerable challenges for long-distance transportation). In conclusion, Brazil's particularities offer important insights into the future cooperation for CCS deployment; the ethanol culture in Brazil can represent an important opportunity for the development of negative emissions' technologies, by investing in BECCS processes.

2.13 Conclusion

To cooperate, or not to cooperate, that is the question. This chapter was about the evidence that there is no other way for more and better energy than cooperation. The original meaning of "Development" implies the necessity for releasing internal energies that are "enveloped" in our mindset, in our culture, traditions, or inside the boxes of our old institutions. "Development," as such, is a firm belief in the potential of human beings and societies to internally generate the work required to counterbalance the unavoidable

---

[5] https://www.rcgi.poli.usp.br/.

deterioration of processes as they occur with the increasing entropy of closed systems. Cooperation is opening our arms and joining forces to multiply the quantity and quality of life for all. It is connecting ourselves with other people, through information and knowledge, to create and sustain new methods and new technologies.

Once a "social contract" for development is accepted by the participants, the first question is "how to sustain it," because cooperation implies considering the intergenerational ethics, meaning the necessity of including the rights of our sons, daughters, and grandkids on the contract. So, the "energy of cooperation" is required to ensure the sustainability of development. We have analyzed that mankind has been building common goals, has widely shared knowledge, and is solidifying bridges of communication aimed at all sectors and regions of our planet. We have discussed with some detail the partnerships that are growing with common purposes—eliminating mystical frontiers of private, public, social, economic, environmental, racial, gender, and language divisions.

We have verified Plato's thesis that "agencies for mastery over nature and agencies for cooperation among men remain the two great sources of human power," and most of our analysis can be summarized as a vision of those two parallel rails of the human quest for development and progress; about how they evolved, their current status, and the perspectives and challenges for their future evolution. We went into deeper considerations about the Brazilian case and a particular kind of technology related to carbon capture and storage, on both land and sea.

We chose probably the most challenging dilemma of our time, knowing that it is not possible to cover all the angles of such a controversial issue, but we have learned that it is possible to find emblematic solutions that can pave more and more kilometers (or miles, to be fully inclusive) of the road to development. Global knowledge has opened for us the guidelines of Agenda 2030 and our partners are synchronized to cooperate on most of its goals. Particularly on the issue of climate change, never has a scientific consensus on something's origins and consequences faced such a polarized public perception and engagement. Our analysis showed that there are many channels and examples of solutions, but all of them are under the frameworks of cooperation for the benefit of all.

In conclusion, we can affirm that cooperation brings hope, which is also a powerful energy, probably the most important of them. Optimism can be passive, such as waiting for things to happen, or it can be proactive, by working together toward our common future. Let's be active. Let's cooperate.

References

Betz, D., 2008. The virtual dimension of contemporary insurgency and counterinsurgency. Small Wars & Insurgencies 19 (4), 510–540. Available from: https://doi.org/10.1080/09592310802462273.

Braunreiter, L., Bennett, S.J., 2017. The neglected importance of corporate perceptions and positions for the long-term development of CCS. Energy Procedia 114, 7197–7204ISSN 1876-6102. Available from: https://doi.org/10.1016/j.egypro.2017.03.1825.

Brundtland, G., 1987. Report of the World Commission on Environment and Development: Our Common Future. United Nations General Assembly Document A/42/427.

Budinis, S., Krevor, S., Mac Dowell, N., Brandon, N., Hawkes, A., 2018. An assessment of CCS costs, barriers and potential. Energy Strategy Reviews 22, 61–81ISSN 2211-467X. Available from: https://doi.org/10.1016/j.esr.2018.08.003.

Bushell, S., Satre-Buisson, G., Workman, M.H.W., Colley, T., 2017. Strategic narratives in climate change: towards a unifying narrative to address the action gap on climate change. Energy Research and Social Science; Business and Sustainable Development Commission, January 2017. Better Business Better World. Available from: http://report.businesscommission.org/report.

CAF—Development Bank of Latin America, 2016. CAF and UNDP deepen their alliance to promote the Global Goals in Latin America and the Caribbean. September 22, 2016. Available from: https://www.caf.com/en/currently/news/2016/09/caf-and-undp-deepen-their-alliance-to-promote-the-global-goals-in-latin-america-and-the-caribbean/.

Chalmers, H., Gibbins, J., Gross, R., Haszeldine, S., Heptonstall, P., Kern, F., et al., 2013. Analysing uncertainties for CCS: from historical analogues to future deployment pathways in the UK. Energy Procedia 37, 7668–7679ISSN 1876-6102. Available from: https://doi.org/10.1016/j.egypro.2013.06.712.

Coninck, H., Stephens, J.C., Metz, B., 2009. Global learning on carbon capture and storage: a call for strong international cooperation on CCS demonstration. Energy Policy 37 (6), 2161–2165ISSN 0301-4215. Available from: https://doi.org/10.1016/j.enpol.2009.01.020.

Durmaz, T., 2018. The economics of CCS: why have CCS technologies not had an international breakthrough? Renewable and Sustainable Energy Reviews 95, 328-–340ISSN 1364-0321. Available from: https://doi.org/10.1016/j.rser.2018.07.007.

Fukuyama, F., 1989. The end of history? The National Interest (16), 3–18. Retrieved from: http://www.jstor.org/stable/24027184.

GCCSI—Global CCS Institute, 2018. Global Status of CCS 2018. Available at: https://www.globalccsinstitute.com/ (accessed 18.03.19).

GCF—Green Climate Fund, 2016. From the SDGs to the Paris Agreement—GCF is a Facilitator of Change. Remarks by Héla Cheikhrouhou. Statement 2016. March 31st, 2016. Available from: https://www.greenclimate.fund/.

GEF—Global Environment Facility, 2015. The GEF and the Sustainable Development Goals. September 2015. Available at: https://www.thegef.org/publications/.

Huijts, N.M.A., Molin, E.J.E., Steg, L., 2012. Psychological factors influencing sustainable energy technology acceptance: a review-based comprehensive framework. Renewable and Sustainable Energy Reviews 16 (1), 525–531ISSN 1364-0321. Available from: https://doi.org/10.1016/j.rser.2011.08.018.

IBP—Instituto Brasileiro de Petróleo, Gás e Biocombustíveis, 2019. Observatório do Setor, Número 7, Ano 1. Monitor IBP. August, 2019.

IDB—Interamerican Development Bank, 2016. Sustainability Report 2016. Available at: https://publications.iadb.org/.

IEA—International Energy Agency, 2017b. World Energy Outlook 2017. November 14, 2017.

IEA—International Energy Agency, 2018. Five keys to unlock CCS investment. IEA, Paris. Available at: https://www.iea.org/reports/five-keys-to-unlock-ccs-investment.

Karimi, F., 2017. Timescapes of CCS projects: is deferring projects and policies just kicking the can down the road? Energy Procedia 114, 7317–7325ISSN 1876-6102. Available from: https://doi.org/10.1016/j.egypro.2017.03.1862.

Kern, F., Gaede, J., Meadowcroft, J., Watson, J., 2016. The political economy of carbon capture and storage: an analysis of two demonstration projects. Technological Forecasting and Social Change 102, 250–260ISSN 0040-1625. Available from: https://doi.org/10.1016/j.techfore.2015.09.010.

Markusson, N., Kern, F., Watson, J., Arapostathis, S., Chalmers, H., Ghaleigh, N., et al., 2012. A socio-technical framework for assessing the viability of carbon capture and storage technology. Technological Forecasting and Social Change 79 (5), 903–918ISSN 0040-1625. Available from: https://doi.org/10.1016/j.techfore.2011.12.001.

Mitchell, W.J.T. (Ed.), 1981. On Narrative. University of Chicago Press, Chicago, IL.

Onarheim, K., Mathisen, A., Arasto, A., 2015. Barriers and opportunities for application of CCS in Nordic industry—a sectorial approach. International Journal of Greenhouse Gas Control 36, 93–105ISSN 1750-5836. Available from: https://doi.org/10.1016/j.ijggc.2015.02.009.

Pihkola, H., Tsupari, E., Kojo, M., Kujanpää, L., Nissilä, M., Sokka, L., et al., 2017. Integrated sustainability assessment of CCS—identifying non-technical barriers and drivers for CCS implementation in Finland. Energy Procedia 114, 7625–7637ISSN 1876-6102. Available from: https://doi.org/10.1016/j.egypro.2017.03.1895.

Rochedo, P., Costa, I., Império, M., Hoffmann, B., Merschmann, P., Oliveira, C., et al., 2016. Carbon capture potential and costs in Brazil. Journal of Cleaner Production 131, 280–295. Available from: https://doi.org/10.1016/j.jclepro.2016.05.033.

Thronicker, D., Lange, I., Pless, J., 2016. Determining the success of carbon capture and storage projects. The Electricity Journal 29 (7), 1–4ISSN 1040-6190. Available from: https://doi.org/10.1016/j.tej.2016.08.001.

Tokushige, K., Akimoto, K., Tomoda, T., 2007. Public perceptions on the acceptance of geological storage of carbon dioxide and information influencing the acceptance. International Journal of Greenhouse Gas Control 1 (1), 101–112ISSN 1750-5836. Available from: https://doi.org/10.1016/S1750-5836(07)00020-5.

Tufts, J.H., 1918. The Ethics of Cooperation. Houghton Mifflin company, Boston and New York.

UN—United Nations Conference on Environment and Development, 1992. Agenda 21, Rio Declaration, Forest Principles. United Nations, New York.

UN—United Nations General Assembly, 2000. United Nations Millennium Declaration, Resolution Adopted by the General Assembly, 18 September 2000, A/RES/55/2.

UN—United Nations, 2018. Greater Ambition in Mitigation, Finance, Technology Key to Defeating Climate Change, Secretary-General Says as Conference of Parties Concludes. December 15, 2018. Available at: https://www.un.org/press/en/2018/sgsm19409.doc.htm.

UN—United Nations, 2019. Financing for Sustainable Development Report 2019. Inter-agency task force on financing for development. Available at: https://www.un.org/development/desa/publications/financing-for-sustainable-development-report-2019.html.

UN—United Nations Department of Economic and Social Affairs, 2016, The Millennium Development Goals Report 2015, UN, New York, Available at: https://doi.org/10.18356/6cd11401-en.

UNESCO—United Nations Economic and Social Council, 2018. Trends and progress in international development cooperation. 2018 session July 21, 2015–July 27, 2016. Available at: https://www.un.org/ecosoc/en/.

World Bank Group, 2017. Directory of Programs Supported by Trust Funds 2017. Trust Funds & Partner Relations Development Finance. June 30, 2017. Available at: http://documents.worldbank.org.

World Bank Group, 2018. Implementing the 2030 Agenda. 2018 Update. Available at: http://pubdocs.worldbank.org/.

3

Climate change mitigation and the technological specificities of carbon capture and storage

Romario de Carvalho Nunes, Vitor Emanoel Siqueira Santos and Hirdan Katarina de Medeiros Costa

Institute of Energy and Environment, University of São Paulo, São Paulo, Brazil

3.1 Introduction

The problem of climate change is being effectively tackled through different strategies over decades. As climate models evolve and the opportunity window for action narrows, governments have progressively adopted multiple strategies to limit carbon dioxide emissions. The inevitable outcome is a profound change in the global energy mix. Global climate change is a long-term change in weather conditions identified by changes in temperatures, rainfall, winds, and other indicators. Medium conditions or changes in variability may include a higher incidence of extreme events (Solomon et al., 2007).

It is a well-known fact that temperature variation is related to the concentration of CO_2 in the atmosphere. Higher CO_2 (and other greenhouse gases—GHGs) concentrations contribute to the intensity of the global greenhouse effect (IPCC, 2013; European Comission, 2007; European Commission, 2011). The oceans are historically the major emitters of CO_2, followed by the breathing of living beings, the decomposition of organic matter, volcanic eruptions, and forest fires. There are also several contributing activities related to anthropogenic emissions, such as burning fossil fuels, changes in land use, industrial activities, and deforestation (IPCC, 2013; Le Quéré et al., 2013). The burning of fossil fuels is responsible for most anthropogenic emissions (87%), while deforestation and changes in land use account for 9% and industrial processes account for 4% (Le Quéré et al., 2013). Therefore, in order to cease the average temperature growth and mitigate global environmental change, it is necessary to halt anthropogenic CO_2 emissions.

Carbon Capture and Storage in International Energy Policy and Law
DOI: https://doi.org/10.1016/B978-0-323-85250-0.00004-9

Fossil fuels usually feed the industrial processes that generate high quantities of carbon dioxide. That is the case with thermoelectric power generation by coal, natural gas, or oil. The carbon in the CO_2 molecule expelled in these processes was initially attached to the fossil fuel source, which was stored underground and therefore was not a part of the natural carbon cycle (EASAC, 2013). When the CO_2 is captured and injected into a geological reservoir, it exits the carbon cycle, relieving the pressure on the natural processes to integrate it.

Carbon capture and storage (CCS) technologies contribute to the reduction of carbon dioxide in the atmosphere by separating and capturing the CO_2 formed in point sources and channeling it to a reservoir, which is usually a geological formation. The gas can also be transported and utilized in several ways, such as in the beverage industry, greenhouses, or enhanced petroleum and gas recovery. Some sources consider biological storage, such as through greenhouses or large-scale reforestation, a different category from geological storage. Nevertheless, geological storage presents higher capacities (EASAC, 2013).

According to studies published by the IPCC (2006) and the IEA (2010), there is a necessity for substantial CCS technology development associated with electricity generation by fossil fuels and industrial processes to minimize the costs expected to achieve the objectives of reducing gas emissions the greenhouse effect. The Energy Roadmap 2050 (European Commission, 2011) estimated the need for 3400 large-scale operational CCS projects by 2050 to achieve a 50% reduction in GHG emissions in the most optimistic estimates, in which case CCS would account for 20% of the necessary cuts.

Therefore, Section 3.2 explores CCS technologies and presents their several uses. Section 3.3 describes and analyzes CCS technological challenges, with positive and negative aspects of CCS. Section 3.4 brings the conclusion.

3.2 CCS technologies

Both separated and as an integrated process, the CCS technologies are already applied for different purposes. Since carbon dioxide is used in industry, such as in food and beverage processing, there are already commercially available processes to capture, channel, storage, and utilize the gas.

Capture processes will focus on separating and concentrating the CO_2 from the source, which is usually a flue gas from an industrial process, or a power plant fueled by coal, oil, or natural gas. Transport infrastructure is responsible for channeling the pressured carbon dioxide to the storage or utilization site. Underground storage technologies involve several processes widely applied in the oil and gas industry, such as reservoir engineering, drilling, and completing injection wells and reservoir monitoring (Bachu, 2008). This section will examine the CCS chain.

3.2.1 Capture

Most carbon capture processes currently in development are aimed at applications in fossil-fired power plants (Kuckshinrichs and Hake, 2015). The capture and separation of carbon dioxide are carried out in large flows, from 180 up to 400 tons of CO_2 per hour for a 500 MW power plant. However, these effluents typically have a low gas concentration, with about 4% in the exhaust gases of plants that use natural gas up to about 33% in the exhaust gases in cement plants (EASAC, 2013).

Among the leading CO_2 capture technologies are postcombustion, oxyfuel, and precombustion. These three are already in use in some capacity. Others in the experimental laboratory phase, although with high potential, are membrane separation, chemical looping, and carbonate looping.

The postcombustion capture technology consists of the installation of a separation system to treat the exhaust gases. In a fossil-fired power plant, air, and coal fuel a steam generator, which results in a CO_2-rich gas. Postcombustion capture is characterized by the carbon dioxide separation and purification after the combustion process. The flue gas treatment includes dedusting, desulfurization, and denitrification. Since the flue gas usually has low to average carbon dioxide concentration, a suitable solvent is used to absorb the CO_2 chemically. Appropriate solvents for this application include organic substances, like amines, and inorganic substances, such as alkaline earth solutions and ammonia (Kuckshinrichs and Hake, 2015). Following this, the CO_2-rich solvent is exposed to a temperature or pressure change, resulting in the desorbed CO_2. The following process is the compression of this carbon dioxide, in preparation for transport. After that, the near CO_2-free solvent is recycled and reused for other separation processes (Kuckshinrichs and Hake, 2015). This setup reduces the plant's necessary modifications, resulting in a lower cost compared to other technologies (EASAC, 2013). Postcombustion can be used in existing thermoelectric plants with low risk since it has been applied for more than 60 years (Duke et al., 2010). Furthermore, postcombustion processes achieve the highest degree of purity for carbon capture—above 99.99%. Postcombustion technology can have carbon capture efficiency losses from 9% to 14%, including compression, liquefaction, and conditioning.

Advantages to postcombustion carbon capture are the aforementioned low adaptation costs in the plant (since no profound changes to the power plant are required), the extensive experience with amine scrubbing from chemical industrial processes, the potential for efficiency improvements, and the commercial availability for all necessary components (Kuckshinrichs and Hake, 2015). The significant investment costs associated with the adaptation in power plants and the uncertainty about these plants' operational flexibility are notable drawbacks. Other drawbacks include high demand for space and water, especially in water-cooled plants (Kuckshinrichs and Hake, 2015).

The oxyfuel processes function by reducing the volume of exhaust gases and consequently increasing the CO_2 concentration by replacing air with oxygen gas in the combustion process (EASAC, 2013). Power plants with oxyfuel technology separate the oxygen from the air before burning it with the fuel, instead of burning air and fuel together as seen in the postcombustion process. The burning of the fuel with pure oxygen generates a CO_2-rich flue gas, with a concentration of around 89% (Kuckshinrichs and Hake, 2015). After that, carbon dioxide is separated from the steam in the flue gas. As the exhaust gases are recycled using CO_2 as an expansion medium (instead of air), desulfurization is no longer necessary. However, there is a disadvantage to the cost of producing oxygen gas, which is high, and also the need to recirculate large volumes of exhaust gases must also be considered (Kuckshinrichs and Hake, 2015).

I. Conceptualizing international energy law and carbon capture and storage (CCS) in light of climate change

The cryogenic air separation units are currently state-of-the-art for large-scale industrial processes. But this process is very energy-intensive, causing considerable efficiency losses (up to 11%), which is one of the current challenges for overall viability for this technology. All necessary components for the oxyfuel carbon capture process are commercially available.

In precombustion processes, there is a reduction in the production of CO_2 in combustion by replacing the carbon-rich fuel with a flow of gases containing hydrogen gas (which produces water when burned) (Kuckshinrichs and Hake, 2015). Coal is used in integrated gasification combined cycle power plants to generate other gases. It begins with its partial oxidation at high pressure, approximately 30 bar, and high temperatures. This process converts the coal into a raw gas composed of CO, H_2, and CO_2. Steam is then used as an oxidizing agent to convert CO into CO_2 and H (Kuckshinrichs and Hake, 2015). This downstream catalytic conversion results in high-pressure syngas, allowing carbon capture through physical absorption based on methanol or dimethyl ether/polyethylene glycol. This process leaves a substantial volume of hydrogen which is then used in a gas and steam process to provide power (Kuckshinrichs and Hake, 2015). Since CO_2 is separated before the burning process, after the gasification stage, where it has the highest concentration, the resulting gas has a purity of over 99% (EASAC, 2013). An advantage of this technology is the guarantee of higher concentrations and pressures at the outlet, favoring carbon dioxide separation. Another advantage is the use of equipment and technologies already known for producing syngas (either from partial reform and oxidation or from gasification of solid fuel) and for the separation of H_2/CO_2 resulting from water vapor reform. However, the need to convert fuel to hydrogen usually involves more radical changes in plant designs (Kuckshinrichs and Hake, 2015).

3.2.2 Transport

The main options for carbon dioxide transportation are pipelines, ships, or trucks (Metz et al., 2005). Gas pipelines are recommended for transport over long distances and large quantities, while maritime transport is typically used on smaller scales (Metz et al., 2005). Ground transportation has few advantages, as its relative costs are higher, being considered only in the initial phases of small-scale research and demonstration projects.

CO_2 can be commercially transported as a liquid or in gaseous form. When gas is transported in conditions close to atmospheric, it occupies a substantial volume, leading to the need for large installations. When transported by pipeline, CO_2 remains compressed. This volume can be further reduced if the CO_2 is liquefied (Metz et al., 2005). This process is very well-known, and its use is widespread.

The choice of CO_2 transportation technology is mostly economic, but other factors, such as local geology and infrastructure, can influence the method. According to Kuckshinrichs and Hake (2015), an attractive alternative for long-term storage projects is relocating emission sources closer to the storage sites, thus reducing the need for CO_2 transportation. Generally, a significant proportion of the pipeline costs are associated with planning measures and rights of way. Existing natural gas pipelines cannot transport CO_2, but the pipeline construction costs can be cut by sharing these existing infrastructure corridors for new pipelines (Kuckshinrichs and Hake, 2015).

Pipeline transportation efficiency depends on its pressure and the ambient temperature. Above the critical temperature of 304.2K (31.2°C), CO_2 is in the supercritical state, behaving mostly as a liquid during the flow. However, the carbon dioxide present in the flue gas, which is captured, is not at the supercritical state, and there can be phase changes during the CCS process chain. Min (2018) identified two transitions while investigating a CCS process from capture to injection: the first is between liquid and gaseous CO_2, which is a phase transition with potential flow consequences; the second is a transition from subcritical to supercritical states which, like the previous, can also change flow rates. Flow instability can easily be caused by changes in fluid properties, such as temperature and pressure. These properties are affected by oscillations in external environmental conditions and operating conditions (Min, 2018). Risks of CO_2 during transportation are mainly related to leakage (EASAC, 2013). Since carbon dioxide is heavier than air, it can accumulate in high concentration plumes near the ground, which can, in extreme cases, cause death in humans (in concentrations superior to 8% and exposure time higher than 30 min). There can also be adverse effects on flora and fauna and acidification of drinking water reserves (EASAC, 2013). With regard to CO_2 purity requirements, it is recommended to minimize transportation operational risks, as well as the implementation of leak monitoring and maintenance measures.

An essential characteristic of CO_2 transportation is the definition of the gas's purity level limits and overall quality criteria. When captured at the source, carbon dioxide can contain different impurities, depending on the origin and the carbon capture process used. The pipelines, the storage tanks on the ships and trucks, the geological storage facility, and the final utilization technique all have their limits. Typical impurities present in flue gases from fossil-fire power plants are SO_2, NO_x, H_2S, H_2, CO, CH_4, N_2, Ar, and O_2 (Kuckshinrichs and Hake, 2015). The European Union's CCS Directive, for example, does not establish concrete limits for the impurities but states that the gas should not adversely affect the integrity of the transportation structure or the storage facility.

3.2.3 Storage and utilization

Carbon dioxide can be stored in geological formations or assimilated through biological fixation. Plants process the gas through photosynthesis, biosynthesizing it in the form of carbohydrates and storing it in their cell wall (Herzog and Golomb, 2004). Geological storage can be achieved in different types of geological formations: deep saline aquifers, oil and gas reservoirs, and coalbeds. The world capacity for CO_2 storage in different geological reservoirs was estimated by Herzog and Golomb, 2004 as 100–10,000 gigatons of carbon in deep saline aquifers, 100–1000 gigatons of carbon in depleted oil, and gas reservoirs, and 10–1000 gigatons of carbon in coal seams. Geological storage is the most promising alternative for a carbon sink due to its capacity, well-known technology, and availability.

Inside these reservoirs, CO_2 molecules are trapped by physical or chemical processes. The porous rock layers, such as sandstones and carbonates, in depleted oil and gas reservoirs are surrounded by sealing rock layers (very low porosity and permeability formations), therefore constituting another trap mechanism for the CO_2 molecules (EASAC, 2013). When injected into saline aquifers and depleted oil reservoirs, carbon dioxide spreads laterally and rises, as it is lighter than the formation water (EASAC, 2013). This

I. Conceptualizing international energy law and carbon capture and storage (CCS) in light of climate change

pattern is interrupted by any sealing layer adjacent to the formation, characterizing structural or stratigraphic entrapment. Additionally, part of the supercritical CO_2 (s-CO_2) dissolves in the formation water, turning it denser and causing it to decant and acidifying the solution, which makes it able to react chemically with the adjacent rocks, leading to the dissolution of unstable minerals and, potentially, the precipitation of carbonated materials in other parts of the structure. In this way, carbon dioxide can be subject to mineral entrapment (Song, 2006). During the gas movement in the formation, capillary forces trap a fraction of it in the pores of the rocks, resulting in residual trapping (EASAC, 2013).

The storage reservoirs must be carefully selected according to their porosity, thickness, permeability, stability, and type of seal formation. Carbon dioxide injection processes are usually carried out under pressure and temperature conditions in which the gas is in a supercritical or, in some cases, liquid state (EASAC, 2013). However, the CO_2 injection in low-pressure oil and gas reservoirs is a particular case in which there is both storage and utilization at the same time. In these cases, it is also common to produce CO_2 after a while (Bachu, 2008). The enhanced oil and natural gas recovery (EOR or EGR) by CO_2 injection is a well-known oil industry technique that was first tested in 1972 in the Texas oil fields (EASAC, 2013). Inside the reservoir, carbon dioxide dissolves in the oil, reducing its viscosity and, consequently, increasing its flow through the reservoir.

The technologies involved in CO_2 storage are mostly well-known, and some have been used for decades. Since the processes are similar to those employed in oil and natural gas production, from surveying the area with 2D and 3D seismic analysis to drilling and completing the wells, everything is commercially available (EASAC, 2013). The newest addition to these is the creation of deep saline reservoirs, which can be formed by injecting water and pumping out the brine present in the salt rock deep underground. Carbon dioxide and methane are injected in the supercritical state, expelling the injected water's stabilizing brine. After a while, the methane returns to the surface through the well, and the CO_2 remains in place.

It is worth noting that reservoir monitoring technologies have become increasingly important for CCS projects. As leakage damage meets little tolerance from regulations, such as the European Directive, reservoir monitoring technologies are evolving fast (Jenkins et al., 2015). For this purpose, monitoring technologies vary with site logistics. Offshore wellbores are usually widely spaced and commonly not accessible. Therefore noninvasive and wide-area monitoring tools are appropriate, while onshore wellbore monitoring can employ more direct methods (CLSF, 2017). Jenkins et al. (2015) point out that public acceptance issues are much more acute onshore than offshore (save for a few exceptions like in Japan) and occasionally require modified or enhanced shallow monitoring, focused on soil gas and groundwater monitoring.

3.3 Technological challenges for wide CCS deployment

3.3.1 Advantages and challenges in gas pressure

Different carbon dioxide pressures can be set up for ship transportation, and each one poses related challenges and provides small advantages. At lower CO_2 pressures (6—8 bar at $-50°C$), the high density of the gas becomes an advantage, as well as the possibility to employ well-known technologies based on liquefied petroleum gas (LPG) ships (Norway, 2016). Moreover, the lower pressure allows the transporter to work with different volumes since both tank and ship sizes are scalable. Although these advantages improve viability, some challenges arise from the lower pressure: (1) small operational margin against freezing to dry ice for CO_2; (2) energy-intensive process; and (3) high insulation requirements (Norway, 2016).

Transport of CO_2 at medium pressure (15 bar at $-25°C$) is very well-known due to experience with food-grade quality CO_2 (Norway, 2016). This mature concept makes for an advantage in the utilization of this pressure. However, this requires a relatively high steel volume in the tank system, and the tank structure is technically challenging overall (Norway, 2016).

At high pressure (45 bar), and temperatures above $+10°C$, the transportation process in ships is less energy-intensive (Norway, 2016). From all alternatives, this has the lowest energy demand with direct injection and also has scalable tank capacity. But higher pressures demand much more space for the tank system. This is a challenging approach for CO_2 transport because it is a less mature concept and has some problems with the high steel weight for the tanks and the piping (Norway, 2016).

3.3.2 Storage

3.3.2.1 Nyos lake case

The case refers to the sudden leak of approximately 0.24 Mton of CO_2 that occurred in the early hours of August 21, 1986, on Lake Nyos, located in the northwest of the Republic of Cameroon (Damen, 2006).

The most accepted hypothesis states that the leakage of CO_2 occurred due to the lake's disturbance due to an undefined cause that moved water saturated with CO_2 of volcanic origin that had until then remained dissolved in the depths of the lake (Baxter, 1989). This movement caused an explosion in the lake by the rapid release of CO_2. This volume of CO_2 spread rapidly across the region, in a cloud that reached villages within a radius of approximately 20 km resulting in the instant loss of more than 1700 residents and thousands of domestic and wild animals. Records show that around 4000 survivors developed respiratory problems, injuries, and paralysis due to exposure to gases present in the cloud (Baxter, 1989).

Although the leak at Lake Nyos is not analogous to a leak from a geological reservoir, a similar situation can occur when an anthropogenic leak from a geological reservoir ends up in a deep lake with the accumulation of CO_2 at the bottom (Damen, 2006).

3.3.2.2 SCC-CO$_2$

Stress corrosion cracking (SCC) has been recognized as a severe concern in the oil and gas industry, as several SCC cases in pipelines associated with the presence of CO_2 in soils have been discovered (Majchrowicz, 2019). From the electrochemical corrosion analyses, it is known that CO_2 accelerates the general corrosion of carbon steels and the corrosion rate increases with the increase in the partial pressure of CO_2. Also, CO_2 becomes more aggressive in moisture, which leads to a decrease in the pH value and formation of carbonic acid (Majchrowicz, 2019).

Previous experience of long-term field operation in transporting high-density gas or liquid CO_2 for EOR has shown that corrosion of pipeline steels is not a safety concern as long as the transported gas or CO_2 liquid is relatively pure, and the impurity water content is controlled below 50 ppm (Majchrowicz, 2019). However, the EOR pipeline experience cannot simply be applied to the construction and operation of s-CO_2 pipelines due to the distinct difference in the physical and chemical properties of the medium transported in the two systems. In CCS, the s-CO_2 flows always contain specific amounts of aggressive impurities, such as H_2O, H_2, O_2, H_2S, SO_2, NO_x, CO, and acids, depending on the sources of CO_2 and the capture/separation technologies applied (Zeng and Li, 2020). The presence of these corrosive impurities can promote general/localized corrosion and increase the susceptibility to SCC of s-CO_2 tubes. Typical general corrosion and localized corrosion damage have been found on the internal surfaces of some s-CO_2 transport pipelines. To this day, corrosion of s-CO_2 steels for pipelines remains a significant challenge in implementing CCS technology (Zeng and Li, 2020).

3.3.2.3 Technology and tools for monitoring subsea storage facilities

Monitoring of the surface or close to the surface also needs to be performed before injection to provide reference data and during/after injection to detect any changes or impacts that may arise in the unlikely event of a leak (Ketzer et al., 2016). Several methods can be used for surface and subsurface environmental monitoring, such as chemical and biological analysis, markers, and remote sensing, among others (Ketzer et al., 2016).

Therefore the monitoring, measurement, and verification of CO_2 in CCS projects goes beyond the limits of the geological reservoir targeted by the injection, or the confinement seal rock, since all areas in which CO_2 may migrate must be considered, including soil, water bodies, and atmosphere (Ketzer et al., 2016).

Also, as provided by the International Energy Agency, for the CO_2 storage to be framed appropriately according to international standards, it is necessary to definitively trap the gas in an amount greater than 95% of the injected CO_2 (IEA, 2010).

The establishment and execution of monitoring criteria are essential to establish storage operations' security (IEA, 2010). Consequently, the establishment of monitoring requirements must be a vital component of the frameworks for CCS. Site-specific factors, such as depth, surface characteristics, and geology, will determine precise technologies, techniques, and application frequencies to be used in monitoring (IEA, 2010).

These components are essential to establish the security of storage operations (IEA, 2010). Consequently, the establishment of monitoring requirements must be a vital component of the frameworks for CCS. Site-specific factors, such as depth, surface characteristics, and geology, will determine precise technologies, techniques, and application frequencies to be used in monitoring (IEA, 2010).

The site-specific monitoring requirements (under the IPCC 2006 Guidelines for National Greenhouse Gas Inventories) have monitoring technologies that have been developed and refined over the past 30 years in the oil and gas industry, groundwater industries, and environmental monitoring. The suitability and effectiveness of these technologies can be strongly influenced by the geology paths and potential emissions at the storage sites; therefore, the choice of monitoring technologies will need to be made site by site (IPCC, 2006).

There are various CCS monitoring technologies designed to monitor the reservoir, overhead, seabed, or water column (IPCC, 2006). The common objective is to detect, characterize, and quantify any leakage of CO_2 from the intended storage location. Still, the choice of the right technical solution for a given project is not trivial. For example, seismic studies offer precious information on the migration and development of the CO_2 plume and changes in geophysical properties inside and above the reservoir (Grimstad et al., 2009). Still, they are expensive and rarely carried out. Electromagnetic and gravimetric surveys were also used to monitor the stored CO_2 plume, offering potentially useful but less detailed information (IPCC, 2006).

In this sense, the autonomous submarine vehicle (AUV) is an example of advances in monitoring technology. Due to the need to cover the storage reservoir area, in addition to taking into account the possible lateral migration of CO_2 into the storage complex and the additional lateral movement as the CO_2 goes through the overload (which is equivalent to potentially several hundred square kilometers in area), an unmanned system that can be deployed for long periods is required (Blackford et al., 2015). The AUV can be programmed to follow a predetermined research pattern in high resolution and house a range of relevant sensors for monitoring CCS leaks (e.g., chemical, acoustic, image products), having passive detection functionality (e.g., chemical sensors and passive hydrophones) that could last for months or active detection (e.g., images of acoustic sonar on the seabed or subsurface), lasting on the order of days (Blackford et al., 2015).

Technological limitations, costs, frequency (continuous, annual, etc.), need for mapping, and description of the storage location influence technology's choice to be used and the monitoring plan's development (IPCC, 2006).

Another relevant aspect for study and monitoring is the natural variation in the marine environment's condition since the biological activity, currents, turbidity, temperature, and water stratification cause the concentration of most substances in the water column to have natural fluctuations (Waarun, 2016). These fluctuations will result from several overlapping fluctuations linked to diurnal, lunar, or seasonal changes. This leads to an intricate pattern of variation in each parameter that makes it difficult to distinguish natural fluctuations from the initial conditions of a leak from CO_2 storage. To interpret CO_2, pH measurements, or to indicate leaks, it is necessary to have a baseline with natural fluctuations established over time, including daily and seasonal fluctuations (Waarun, 2016). Monitoring multiple parameters simultaneously can activate the identification of covariant patterns that characterize natural or leak-related changes and can be used to discriminate between them (Waarun, 2016).

3.3.2.4 Utilization

CO_2 is not only a GHG but also an essential source of carbon for the manufacture of organic chemicals, materials, and carbohydrates (e.g., food) (Chunshan, 2006).

In the carbon capture utilization and storage (CCUS) structure, CO_2 is used in several ways, mainly mineral carbonation and physical and chemical methods, such as fuels and chemicals, electrocatalytic conversion, a base for polymers in plastics production, urea production, mineralization processes (accelerated carbonation), beverage production, food processing, biological utilization and EOR, coal bed methane, and CO_2 fracking (Zhang, 2020).

According to Zhang (2020), although most R&D projects are in their initial stages, there are several projects under development worldwide and successful, such as the Port of Rotterdam CCUS project and Norway full-chain CCS. The author also highlights the growth of the CO_2 market worldwide, with an annual growth rate of more than 13% until 2022. Zhang (2020) recommends that the focus is on four categories of CO_2 use markets: building materials (such as carbonate and concrete aggregates); chemical intermediates (such as formic acid, methanol, and synthesis gas); fuels (such as methane and liquid fuels); and polymers (Zhang, 2020).

3.3.2.5 Development and research for conversion and optimization in the use of CO_2

According to Chunshan (2006), people and governments worldwide must realize that not only universities and research institutions but also industries are necessary to carry out real actions in the conversion and use of CO_2. The author also reinforces that industrial investments in the development of new CO_2-based technologies must be encouraged and facilitated by governments, including subsidies (Chunshan, 2006).

Chunshan (2006) lists some specific research directions and development issues, which involve new process concepts and the expansion of popular applications. Therefore the main trends cited by the author were selected and listed below in Table 3.1, as well as a possible motivator and practical example to be explored and developed.

3.4 Final remarks and conclusions

Carbon capture, storage, and utilization technologies are vital to the success of climate change mitigation. To achieve the goal of limiting the anthropogenic increase in the average global temperature by 2100 to 2°C, global CO_2 emissions must be cut by 50% by 2050 compared to levels in 1990. Capturing and storing CO_2 from power plants and extensive industrial processes such as cement production, steel production, refineries, and others is one of the necessary emissions mitigating efforts of the 21st century (Kuckshinrichs and Hake, 2015).

CCS technologies are in different stages of maturity, and the continued effort for improving some of them is significant for the broad viability of the CCS process chain. The most advanced capture technologies in commercial or pilot plants are precombustion, oxyfuel, and postcombustion, which is the most economically promising since it has reduced retrofitting costs for the plant. Other capture technologies in the experimental/ laboratory phase are membrane separation, chemical looping, and carbonate looping, all of which have high potential to improve the process (Kuckshinrichs and Hake, 2015).

Technologies for both onshore and offshore CO_2 transport are very well-known and commercially available due to similarities and long-term experience with carbon dioxide and other gases in average to long-distance transport. Pipeline transportation is the preferential method for its economic viability, large volume capacity, and long distances, but ships are also used in offshore operations. Geological storage technologies are also well developed, especially for EOR. In this case, the process is already commercially available and used in some oil fields with abundant CO_2 production. Storage in depleted oil and

TABLE 3.1   List of technologies for research and development.

Technology suggestion to be researched and developed	Motivating	Example
Use of CO_2 without synthesis gas separation: tri-reform against CO_2 reform	Exploring alternative processes for using CO_2 from industrial waste gases as a coreagent in the manufacture of chemicals is a great approach. Also, it has a relatively large market.	The production of synthesis gas $(CO + H_2)$ is a potential area for large-scale conversion and use of CO_2 in addition to the CH_4CO_2 reform being widely studied.
Use of the "Molecular Basket" adsorbent for CO_2 separation	The separation of CO_2 is often seen as the first step towards using chemicals.	It has been experimentally demonstrated that the adsorbent "molecular basket" can selectively capture CO_2 for its separation from natural gas and coal boilers' combustion gas.
Use of CO_2 to replace phosgene: synthesis of dimethyl carbonate	The traditional dimethyl carbonate (DMC) synthesis route uses phosgene. The use of CO_2 in the synthesis of DMC ecologically better since CO_2 can replace phosgene and chlorine (toxic chemicals).	It has numerous environmental benefits because it is better than existing industrial processes—based on phosgene or CO, toxic chemicals, and DMC itself can be used for a series of new ecological reactions to replace some of the environmentally harmful processes.
Use of CO_2 to extract supercritical fluid	CO_2 is much more benign for the environment than many of the existing solvents used in industries. Processes less harmful to the environment and with greater energy efficiency can be designed using CO_2 for separation and chemical reaction, and synthesis of materials.	The use of supercritical CO_2 (s-CO_2) allows for supercritical extraction free of contaminants of various substances, from beverages (such as caffeine from the coffee bean), foods (such as excess oil from fries), and organic and inorganic materials functional, herbal and pharmaceutical products.
Use of s-CO_2 as a solvent, medium, or reagent	s-CO_2 can be used as a solvent or reaction medium or coreagent for a series of organic reactions, leading to more ecological synthetic procedures and improved results (elimination of complicated product processing, greater selectivity, etc.).	Several studies have reported significant gains such as efficient production of formic acid in a supercritical mixture of CO_2 and hydrogen containing a catalytic ruthenium(II) phosphine complex, efficiency gains in processes, and creation of a method for fluorous biphasic catalysis.
Use of CO_2 as a mild oxidizer	The dissociation of CO_2 on the catalyst surface can produce active oxygen species. Some heterogeneous chemical reactions may benefit from using CO_2 as a mild oxidizer or as a selective source of "oxygen" atoms.	The use of CO_2 is beneficial for the selective dehydrogenation of ethylbenzene to form styrene and for the dehydrogenation of lower alkanes, such as ethane, propane, and butane to form ethylene, propylene, and butene, respectively. Also, adding CO_2 to the oxidation reaction of alkylaromatics with O_2 increases the reaction rate and improves selectivity in relation to aldehyde and oxygenated products.

(Continued)

I. Conceptualizing international energy law and carbon capture and storage (CCS) in light of climate change

TABLE 3.1   (Continued)

Technology suggestion to be researched and developed	Motivating	Example
Use of CO_2 to generate added value	Value-added CO_2 sequestration includes the storage of CO_2 in geological formations that also increase the recovery and production of oil or natural gas or CH_4 from the coal bed.	CO_2 in the flue gas of power plants can be used to enhance the recovery of oil and natural gas and enhance the recovery of methane from the coal bed.
Use of biochemical and geochemical environments for CO_2 conversion	The use of biochemical or geochemical environments to convert CO_2 is an alternative to the photochemical conversion of CO_2 (carbohydrate synthesis) under sunlight. Existing fossil fuels are believed to have originated from biochemical and geochemical transformations of organic substances that were initially present on the Earth's surface over millions of years, which can be seen as a deviation in the carbon cycle.	Studies have shown that coal was formed from biochemical degradation and geochemical maturation of higher plant materials that were initially formed on the earth's surface through CO_2 and H_2O photosynthesis.
Use of CO_2 in markets using gas	CO_2 can be converted into chemicals (urea, methanol, etc.) and synthetic polymeric materials (plastics, fibers, rubbers).	CO_2 can be used differently to make chemicals and materials. The global commercial demand for CO_2 is continuously increasing and can generate billions of dollars in revenue.

Adapted from Chunshan, S., June 2006. Global challenges and strategies for control, conversion and utilization of CO_2 for sustainable development involving energy, catalysis, adsorption and chemical processing. Catal. Today 115 (1–4), 2–32. https://doi.org/10.1016/j.cattod.2006.02.029.

gas reservoirs, as well as in saline aquifers, is possible but is more economically restricted. The significant challenges for storage are in coalbeds and natural gas reservoirs, which focus on enhanced recovery (Kuckshinrichs and Hake, 2015).

There are several applications for CO_2 utilization in different stages of innovation. While commercial use is widespread in the food industry, for example, some uses like photocatalytic/electrocatalytic activation of CO_2, oxalic acid production, and artificial photosynthesis are in the conceptual and laboratory stages. Carbon dioxide hydrogenation and polycarbonate production are uses that already have field studies in pilot projects. Finally, urea and methanol production through CO_2 is very well-known and commercially available (Kuckshinrichs and Hake, 2015).

Lastly, as discussed in Section 3.2, capture, transport, utilization, and storage technologies are, in the most part, commercially available due to their use in other processes and therefore do not need extensive development. Specific technological barriers were discussed on Section 3.3 and are mostly related to new capture technologies (or efficiency gains for existing ones) and the monitoring of geological storage sites, guaranteeing operational safety, and reducing liability risks for the operators. Therefore large-scale CCS operations are a realistic pathway, if not essential, to achieve the emission mitigation goals.

Acknowledgments

We are grateful to the "Research Center for Gas Innovation—RCGI" (Fapesp Proc. 2014/50279-4), supported by FAPESP and Shell, organized by the University of São Paulo, and the strategic importance of the support granted by the ANP (National Agency of Petroleum, Natural Gas and Biofuels of Brazil) through the R&D clause. We also thank the support from the National Agency for Petroleum, Natural Gas and Biofuels Human Resources Program (PRH-ANP), funded by resources from the investment of oil companies qualified in the R,D&I clauses from ANP Resolution No. 50/2015 (PRH 33.1—Related to Call No. 1/2018/PRH-ANP; Grant FINEP/FUSP/USP Ref. 0443/19).

References

Baxter, P.J., Kapila, M., Mfonfu, D., 1989. Lake Nyos disaster, Cameroon (1989) 1986: the medical effects of large-scale emission of carbon dioxide. British Medical Journal 298, 1437. Available from: https://doi.org/10.1136/bmj.298.6685.1437.

Bachu, S., 2008. CO_2 storage in geological media: role, means, status and barriers to deployment. Progress in Energy and Combustion Science 34, 254–273.

Blackford, J., et al., 2015. Marine baseline and monitoring strategies for carbon dioxide capture and storage (CCS). International Journal of Greenhouse Gas Control 38, 221–229. Available from: https://doi.org/10.1016/j.ijggc.2014.10.004.

Chunshan, S., 2006. Global challenges and strategies for control, conversion and utilization of CO_2 for sustainable development involving energy, catalysis, adsorption and chemical processing. Catalysis Today 115 (1–4), 2–32. Available from: https://doi.org/10.1016/j.cattod.2006.02.029.

CLSF, 2017. Enabling Large-Scale CCS Using Offshore CO_2, Utilization and Storage Infrastructure Developments. < https://www.cslforum.org/cslf/sites/default/files/documents/7thMinUAE2017/Offshore_CO2-EOR_Final_02_Dec_2017.pdf >. (accessed 01.04.20).

Damen, K., Faaij, A., Turkenburg, W., 2006. Health, safety and environmental risks of underground CO_2 storage—overview of mechanisms and current knowledge. Climatic Change 74, 289–318. Available from: https://doi.org/10.1007/s10584-005-0425-9.

Duke, M., Ladewig, B., Smart, S., Rudolph, V., Costa, J., 2010. Assessment of postcombustion carbon capture technologies for power generation. Frontiers of Chemical Engineering in China 4, 184–195. Available from: https://doi.org/10.1007/s11705-009-0234-1.

EASAC, 2013. European Academies Science Advisory Council. Carbon Capture and Storage in Europe. German National Academy of Sciences. ISBN 978-3-8047-3180-6.

European Comission, 2007. Communication from the commission. In: Europe 2020—A European Strategy for Smart, Sustainable and Inclusive Growth. COM (2010) 2020, Final, Brussels.

European Commission, 2011. Communication from the commission. In: Energy Roadmap 2050. COM (2011), Luxembourg.

Grimstad, A.A., Georgescu, S., Lindeberg, E., Vuillaume, J.F., 2009. Modelling and simulation of mechanisms for leakage of CO_2 from geological storage. Energy Procedia 1 (1), 2511–2518.

Herzog, Howard, Golomb, Dan, 2004. Carbon Capture and Storage from Fossil Fuel Use. Encyclopedia of Energy 1, 19. Available from: https://doi.org/10.1016/B0-12-176480-X/00422-8.

IEA, 2010. Carbon Capture and Storage—Model Regulatory Framework. OECD/IEA, Paris, p. 130.

IPCC, 2006. IPCC Guidelines for National Greenhouse Gas Inventories, Prepared by the National Greenhouse Gas Inventories Programme, Eggleston H.S., Buendia L., Miwa K., Ngara T. and Tanabe K. (eds). Published: IGES, Japan.

IPCC, 2013. Climate Change 2013: The Physical Science Basis. Contribution of Working Group I to the Fifth Assessment Report of the Intergovernmental Panel on Climate Change, Stocker, T.F., Qin, D., Plattner, G.-K., Tignor, M., Allen, S.K., Boschung, J., Nauels, A., Xia, Y., Bex, V. and Midgley, P.M. (eds.). Cambridge University Press, Cambridge, United Kingdom and New York, NY, USA, 1535 pp.

Jenkins, C., Chadwick, A., Hovorka, S.D., 2015. The state of the art in monitoring and verification—ten years on. International Journal of Greenhouse Gas Control 40, 312–349. Available from: https://doi.org/10.1016/j.ijggc.2015.05.00.

Ketzer, J.M.M., et al., 2016. Atlas brasileiro de captura e armazenamento geológico de CO_2. EDIPUCRS, 2016.

Kuckshinrichs, W., Hake, J.F., 2015. Carbon Capture, Storage and Use: Technical, Economic, Environmental and Societal Perspectives. Springer International Publishing.

Le Quéré, C., Andres, R. J., Boden, T., Conway, T., Houghton, R. A., House, J. I., Marland, G., Peters, G. P., van der Werf, G. R., Ahlström, A., Andrew, R. M., Bopp, L., Canadell, J. G., Ciais, P., Doney, S. C., Enright, C., Friedlingstein, P., Huntingford, C., Jain, A. K., Jourdain, C., Kato, E., Keeling, R. F., Klein Goldewijk, K., Levis, S., Levy, P., Lomas, M., Poulter, B., Raupach, M. R., Schwinger, J., Sitch, S., Stocker, B. D., Viovy, N., Zaehle, S., and Zeng, N., 2013. The global carbon budget 1959–2011, Earth System Science Data 5, 165–185, https://doi.org/10.5194/essd-5-165-2013.

Majchrowicz, K., et al., 2019. Exploring the susceptibility of P110 pipeline steel to stress corrosion cracking in CO_2-rich environments. Engineering Failure Analysis 104, 471–479. Available from: https://doi.org/10.1016/j.engfailanal.2019.06.016.

Metz, B., Davidson, O., Coninck, H., Loos, M., Meyer, L., 2005. Special Report on Carbon Dioxide Capture and Storage, United States of America. Cambridge University Press, Cambridge, NY, p. 443.

Min, I.H., Kang, S.G., Huh, C., 2018. Instability analysis of supercritical CO_2 during transportation and injection in carbon capture and storage systems. Energies 11 (8), 2040. Available from: https://doi.org/10.3390/en11082040.

Norway, 2016. Norwegian Ministry of Petroleum and Energy. Feasibility Study for Full-Scale CCS in Norway. < https://ccsnorway.com/wp-content/uploads/sites/6/2019/09/feasibilitystudy_fullscale_ccs_norway_2016.pdf >. (accessed 09/04/2020).

Solomon, S., Qin, D., Manning, M., Alley, R.B., Berntsen, N.L., Bindoff, Z., et al., 2007. Technical summary. In: Climate Change 2007: The Physical Science Basis. Contribution of Working Group I to the Fourth Assessment Report of the Intergovernmental Panel on Climate Change.

Song, C., 2006. Global challenges and strategies for control, conversion and utilization of CO_2 for sustainable development involving energy, catalysis, adsorption and chemical processing. Catalysis Today 115, 2–32. Available from: https://doi.org/10.1016/j.cattod.2006.02.029.

Waarun, I., et al., 2016. CCS leakage detection technology—industry needs, government regulations, and sensor performance. In: 13th International Conference on Greenhouse Gas Control Technologies, GHGT-13. November 2016, Lausanne, Switzerland.

Zeng, Y., Li, K., 2020. Influence of SO_2 on the corrosion and stress corrosion cracking susceptibility of supercritical CO_2 transportation pipelines. Corrosion Science 165. Available from: https://doi.org/10.1016/j.corsci.2019.108404.

Zhang, Z., et al., 2020. Recent advances in carbon dioxide utilization. Renewable and Sustainable Energy 125. Available from: https://doi.org/10.1016/j.rser.2020.109799.

Further reading

Abanades, J.C., Arias, B., Lyngfelt, A., Mattisson, T., Wiley, D.E., Li, H., et al., 2015. Emerging CO_2 capture systems. International Journal of Greenhouse Gas Control 40, 126–166. Available from: https://doi.org/10.1016/j.ijggc.2015.04.018.

Barros, N., Oliveira, G.M., Lemos de Sousa, M.J. 2012. Environmental impact assessment of carbon capture and sequestration: general overview. In: IAIA12 Conference Proceedings: Energy Future, The Role of Impact Assessment. 32nd Annual Meeting of the International Association for Impact Assessment, May 27–June 1, 2012, Porto—Portugal.

Bromhal, G.S., Birkholzer, J., Mohaghegh, S.D., Sahinidis, N., Wainwright, H., Zhang, Y., et al., 2014. Evaluation of rapid performance reservoir models for quantitative risk assessment. Energy Procedia 63, 3425–3431.

Gerstenberger, M.C., Christophersen, A., Buxton, R., Nicol, A., 2015. Bi-directional risk assessment in carbon capture and storage with Bayesian networks. International Journal of Greenhouse Gas Control 35, 150–159.

GCCSI, 2019. Global CCS Institute. Global Status of CCS 2019: Targeting Climate Change. Global Status Report, pp. 1–46.

Govindan, R., Babaei, M., Korre, A., Shi, J.Q., Durucan, S., Norden, B., et al., 2014. CO_2 storage uncertainty and risk assessment for the post-closure period at the Ketzin pilot site in Germany. Energy Procedia 63, 4758–4765.

Herzog, H., 2016. Lessons Learned from CCS Demonstration and Large Pilot Projects. Massachusetts Institute of Technology (MIT), Massachusetts.

Korre, A., Shi, J.Q., Imrie, C., Grattoni, C., Durucan, S., 2007. Coalbed methane reservoir data and simulator parameter uncertainty modeling for CO_2 storage performance assessment. International Journal of Greenhouse Gas Control 1, 492–501.

Metcalfe, R., Paulley, A., Suckling, P.M., Watson, C.E., 2013. A tool for integrating and communicating performance-relevant information in CO_2 storage projects: description and application to In Salah. Energy Procedia 37, 4741–4748.

Pawar, R.J., Bromhal, G.S., Carey, J.W., Foxall, W., Korre, A., Ringrose, P.S., et al., 2015. Recent advances in risk assessment and risk management of geologic CO_2 storage. International Journal of Greenhouse Gas Control 40, 292–311. Available from: https://doi.org/10.1016/j.ijggc.2015.06.014.

Rackley, S.A., 2017. Carbon Capture and Storage. Butterworth-Heinemann, Boston, MA.

Ramírez Ramírez, A., 2020. Chapter 13: Carbon Capture and Utilization. In: Carbon Capture and Storage. RSC Energy and Environment Series. RSC, pp. 426–446. Available from: https://doi.org/10.1039/9781788012744-00426.

Shi, J.-Q., Korre, A., Chen, D., Govindan, R., Durucan, S., 2014. A methodology for CO_2 storage system risk and uncertainty assessment. Energy Procedia 63, 4750–4757.

I. Conceptualizing international energy law and carbon capture and storage (CCS) in light of climate change

Case studies on CCS and related policies, and their consequences for climate change

The institutional approach of climate change at the multinational level: the new paradigm of the Brazilian legislative experience

Gustavo de Melo Ribeiro[1],
Hirdan Katarina de Medeiros Costa[1],
André Felipe Simões[1,4] and Carolina Arlota[2,3]

[1]Institute of Energy and Environment, University of São Paulo, São Paulo, Brazil [2]Visiting Assistant Professor of Law at the University of Oklahoma, College of Law, Norman, OK, United States [3]Law and Economics Research Fellow, Capitalism and Rule of Law Project (Cap Law) at Antonin Scalia Law School, George Mason University, Fairfax, VA, United States [4]School of Arts, Sciences and Humanities of the University of São Paulo, São Paulo, Brazil

4.1 Introduction

Initiatives of local governments related to climate policies have been observed since the late 1980s, when transnational networks of cities started to be formed for sharing good practices of urban public management, local environmental policies, and, later in the 1990s, climate change actions. In addition, the Brundtland Report of 1987 (WCED, 1987) and the reports released by the United Nations Conference on Environment and Development (1992, in Rio de Janeiro, Brazil) already highlighted the role of cities in tackling climate change (Harald Fuhr, 2018).

Transnational networks such as the International Council for Local Environmental Initiatives have been working in this direction since the early 1990s. Authors, such as Michele Betsill and Harriet Bulkeley, have discussed the implementation of local climate policies considering the challenges of multilevel governance since the early 20th century (Betsill, 2004, 2006, 2013; Betsill and Bulkeley, 2003).

The United Nations, through the Sustainable Development Goal (SDG) 11, point out the need to make cities and human settlements inclusive, safe, resilient, and sustainable. For this, public policies focused on urban life began to take place in public government policies, local society, and economic agents. The linkage among SDG 7, SDG 13, and SDG 11 deals with different possibilities to create conditions to improve cities' energy generation, supply, and infrastructure, as well as to tackle climate change.

The Paris Agreement ratifies the importance of subnational governments in bringing states and municipalities closer to global efforts to address climate change.[1] In addition, the national determined contributions (NDCs) are designed to involve policies of a wide variety of actors aiming at maximizing climate action (Paris Agreement, Article 3). Nonetheless, policies exclusively based on state and local actions will not be sufficient for any country to achieve significant reduction of greenhouse gases (GHGs) emissions.[2]

The Paris Agreement has bottom-up and top-down approaches. The latter include mandatory reporting of NDCs and record of emissions, for instance, according to Articles 3 and 4. In a bottom-up approach, subnational governments are taking part in decisions, commitments, and access to international cooperation mechanisms, such as technical assistance, financing, and technology transfer programs. Subnational entities have thus gained autonomy in positioning themselves in climate actions and commitments, often by facilitating transnational networks, bilateral agreements, and voluntary commitments (Anderton and Setzer, 2018; Macedo, 2017).

The Brazilian constitutional provisions and their unique federative arrangements coupled with the country's domestic policies and Brazil's long-standing international reputation present a new paradigm for climate action. Enabling multilevel strategies requires understanding the powers and competences of the different national entities, which in the Brazilian case are distributed in Union, States, Federal District, and Municipalities.[3] Therefore the purpose of the present chapter is to confront federal entities powers and competences with the main activities foreseen for climate change policies, aiming to prepare subnational climate change strategies for mitigation in the energy sector, aligned with national regulations and policies.

Considering the experience gained from the implementation of the Clean Development Mechanism projects in Brazil, the linkage between sustainable development and local stakeholders' engagement in the decision-making process is critical and was a determinant

---

[1] The preamble of the Paris Agreement determines the following: "Taking into account the imperatives of a just transition of the workforce and the creation of decent work and quality jobs in accordance with nationally defined development priorities; (...) Affirming the importance of education, training, public awareness, public participation, public access to information and cooperation at all levels on the matters addressed in this Agreement, recognizing the importance of the engagements of all levels of government and various actors, in accordance with respective national legislations of Parties, in addressing climate change."

[2] For instance, in the United States, states and local governments have traditionally been involved in climate action. The action of subnational units, however, is insufficient to achieve meaningful reduction of GHGs (Arlota, 2020).

[3] Brazilian Federal Constitution, 1988 (hereinafter FC).

input for the Paris Agreement design, encompassing both top-down and bottom-up strategies (Setzer, 2017; Benites-Lazaro et al., 2018).

In Brazil, our main Environmental law, from 1981, is the well-known Brazilian National Environmental Policy Act (Law 6938). It accounts for several advances regarding environmental protection and conservation (Costa et al., 2017).

In the last two decades, Brazil has concentrated climate mitigation efforts to curb deforestation, the country's main source of GHG emissions, but this strategy does not suffice to meet GHG reduction goals due to the rapid growth in fossil energy use in the last decade (Pepplow et al., 2019; Castro and Alves, 2018).

Thus considering the need to progressively expand the climate change mitigation actions in the scope of the Brazilian energy sector (considering supply and demand), we examine the powers and competences associated with the characteristic activities of this sector that permeate the different entities of the Federation, demonstrating the need for an effective articulated strategy.

The Brazilian State is a federation composed of legal entities endowed with autonomy, namely the Union, the States, the Municipalities, and the Federal District, with this autonomy being manifested by their respective own government bodies and by the possession of specific powers and competences (Mariotti, 2009). The latter can be defined as the faculty legally attributed to an entity, to an organ or agent of the public authority to issue decision, powers, and competences; they are the various forms of power used by state bodies or entities to perform their functions (Moraes, 2010).

The concept of federation is substantiated by the Federal Constitution (FC) of 1988 as a perpetuity clause in the process of democratization and decentralization in Brazil, manifested by the popular acts at the end of the Military Dictatorship (1964–85). The inclusion of municipalities as federated entities is the great differential of the 1988 FC, for although the federative model has been present in Brazil since 1889, it is usually adopted as a dual model that had only the central and regional sphere or only nominal, as is the in the case of the 1967 Constitution. Thus cooperative federalism, besides proposing a mechanism for distributing powers and competences, based on the German model, with the purpose of federative equilibrium, seeks to protect assets of national interest (Vargas and Rodrigues, 2016).

The States, the Federal District, and the Union seek to achieve the fundamental objectives of the Brazilian Republic[4] by three different branches of power: executive, legislative, and judiciary, while the municipalities act through two branches of power: executive and legislative. It is important to emphasize that the autonomy of the federal entities presupposes the distribution of legislative, administrative, and tax powers and competences (Moraes, 2010). A decisive aspect in the organization of the Brazilian State was established by the FC, which structured a system with different types of powers and competences: exclusive, privative, common, and concurrent (Silva, 2017).

The division of powers and competences of the Brazilian Federal Constitution is ruled by the principle of the predominance of interest, the Union being responsible for matters and issues of general interest, the States for regional interests and Municipalities for local

---

[4] The construction of a free, fair, and solidary society; to guarantee national development; to eradicate poverty and marginalization, to reduce social and regional differences and to promote the nation's welfare, with no prejudice in relation to origin, race, sex, color, age, or other forms of discrimination.

TABLE 4.1   Schematic panel for division of powers and competences.

Power and competence Type	Form	Application
Exclusive	Enumerated, explicitly established by the Federal Constitution (FC) or by the principle of predominance of interest. Not delegable.	Union, States, Federal District and Municipalities
Reserved or Remaining	Comprises all matters not assigned by FC.	States
Common, cumulative or parallel	Enumerated, explicitly established by the FC.	Union, States, Federal District and Municipalities
Private	Enumerated, explicitly established by the FC. Delegable.	Union
Remaining/ Reserved	Comprises all matters not assigned by FC.	States and Federal District
Concurrent	It may be established by more than one federative entity with primacy of the Union in the definition of general rules.	Union (general rules), States and Federal District
Supplementary	Establishes unfolding of general rules.	States, Federal District and Municipalities
Exclusive	Enumerated, explicitly established by the FC or by the principle of predominance of local interest. Not delegable.	Municipalities

interests. Based on this principle, the method adopted by the constituents in the division of powers was to list the powers of the Union and to indicate those of the Municipalities by means of the principle of local interest and by describing the common powers and delegating the remaining powers to the Member States. There is a possibility for the Union to delegate private legislative powers and competences by complementary laws. Administrative and tax powers, also named material powers, may be either exclusive or common.[5]

Therefore it is the competence of the States to carry out all the tasks that have not been enumerated and attributed by the FC to the Union and to the municipalities, in addition to the common powers. Table 4.1 presents a simplified schematic division proposed by Silva (2017).

4.2 Scope of the analysis

Considering our proposal, the analysis presents only the elements considered relevant for climate policies correlated with the energy sector, activities such as land use and

[5] Common powers and competences will be classified as those that concern all four federative entities and as those that are exclusive for the attributions enumerated and attributed specifically to each entity according to the FC.

change of land use and deforestation (Agriculture, Forestry and Other Land Use) were left out of this analysis, albeit very relevant for Brazilian climate policies.

4.3 Survey of competences focusing on the municipal level

4.3.1 Exclusive and private powers and competences of the Union

As observed in the previous section herein, the Union has the main administrative and legislative competences that are not delegable. Besides, considering the Brazilian Federal Constitution and respective Constitutional Amendments, as well as Law No. 9478, of August 6, 1997 (Petroleum Law) and its respective revisions, Law No. 11909 of March 4, 2009 (Law of Gas) and Law No. 12351 of December 22, 2010 provide for the sharing regime of the Pre-salt, thereby adjusting the exploration, production of oil and natural gas.

The energy sector is composed of the electric sector and the oil, natural gas, and bio-fuels sector in their different phases: exploration, production, refining, transportation, generation, transmission, and distribution. They are governed by the Union, except piped gas services, which is the attribution of the States.[6]

The Ministry of Mines and Energy, together with agencies, government bodies, and public, mixed, or private companies, has the power for energy planning jointly with the Energy Research Office (EPE); researching energy resources jointly with the National Agency of Electric Energy (ANEEL), the National Agency of Petroleum, Natural Gas and Biofuels (ANP), and public companies, such as Petrobras, Eletrobrás, and Eletronuclear; the exploration and production by public companies and private companies under a concession regime; the management (activities such as auctions, operations) by the ANP, ANEEL, Electric Energy Trading Chamber (CCEE), and the National Electric System Operator (ONS); the energy policies jointly with the National Energy Policy Council (CNPE).[7,8]

The Union powers also encompass the regulation, planning, and public policies of all the transport sectors and the administration (direct or indirect) of the following: the air and sea segments and their respective port infrastructures, also railroad, watercourses, and road modals that transpose the boundaries of the state or national territory. The Union is not involved in the administration of railroads, watercourses, and road transport which are restricted to the borders of the states or municipalities.

The planning of the industrial and agricultural and commercial sectors is also the attribution of the Union since it has power for developing the "national and regional plans for

[6] Pursuant to the second paragraph of article 25 of the Federal Constitution.

[7] In addition, see: First paragraph of article 20 of the Federal Constitution. It indicates that states, federal district, municipalities, and agencies under the direct administration of the Union have the right to participate in the results or financial compensation for the exploration of oil or natural gas, hydric resources for the purpose of generating electricity, and other mineral resources according to the territorial location of mineral resources, including mineral coal and the entire chain associated with nuclear mineral coal.

[8] Also, the fourth paragraph of article 176 of the FC stands out; it exempts from concession or authorization for the use of the renewable energy potential of reduced capacity.

economic and social development" and "national and regional plans for land use planning." The regulation and management of hydric resources are the power and competence of the Union, as well as the establishment of urban development policies for territorial occupation, housing, sanitation, traffic and urban transport, and finally, it is in the power of the Union to regulate expropriation and to carry out the planning of the civil defense.

In addition to the sectoral aspects, it is important to highlight the powers of the Union "for a financial operation, especially credit operations" involving all funded infrastructure works, be they associated with services of general (Union), regional (State), or local (Municipality) interest.

4.3.2 Reserved and concurrent powers of the States and Federal District

All the administrative and legislative matters that are not treated by the FC and matters with the predominance of regional interest are reserved to the States and to the Federal District. Article 25 describes States' powers in three paragraphs.[9] Article 26 presents the property of the States.[10]

In addition to the specified principles and matters, the States and Federal District have the legislative power and competence concurrent to the Union, in which the latter presents general precepts that will be specified by State legislation.

4.3.3 Administrative and legislative powers of municipalities

The FC of 1988 amplified the attributions of municipalities and increased their autonomy, giving powers of self-organization (through the organic law), self-government (election of mayors and municipal councils), extension of exclusive and supplementary powers (municipal legislation), and self-administration (own administration and to provide services of local interest) (Silva, 2017). The FC does not enumerate the exclusive powers of the municipalities in a discretionary manner, as carried out for the Union. The regulation is carried out based on the principle of predominance of interest, in the case of

[9] (1) All powers and competences that this Constitution does not forbid the states from exercising shall be conferred upon them; (2) the states shall have the power to operate, directly or by means of concession, the local services of piped gas, as provided for by the law, it being forbidden to issue any provisional measure for its regulation; and (3) the states may, by means of a supplementary law, establish metropolitan regions, urban agglomerations and microregions, formed by the grouping of adjacent municipalities, in order to integrate the organization, the planning and the operation of public functions of common interest.

[10] They are: (1) surface or subterranean waters, flowing, emerging or in deposit, with the exception, in this case, of those resulting from the work carried out by the Union, as provided by the law; (2) the areas, on the ocean and coastal islands, which are within their domain, excluding those under the domain of the Union, the municipalities or third parties; (3) the river and lake islands which do not belong to the Union; and (4) the unoccupied lands not included among those belonging to the Union.

municipalities, in addition to the common powers and competences described by Article 23 of the FC, we find in Article 30 the enumeration of the exclusive powers of the municipality.[11]

Municipalities constitute the sphere of governance, which is the closest to citizens, having as a legal instrument complementary to the FC, the City Statute (Law No. 10,257 of July 10, 2001), which regulates urban policies, which is the power and competence of municipalities (Article 182 of the Constitution); besides the instruments of territorial planning, namely the master plan, land use regulation and soil occupation, environmental zoning, participatory management, and economic and social development plans. It also includes progressive Urban Land and Building Tax (IPTU), grants contributions for improvement. and incentives for construction and alteration operations of land use.[12]

The City Statute itself recalls the Union's powers in relation to urban policies that must be observed by the municipal administration and city council, namely to legislate upon general norms of urban law and upon cooperation between different bodies of the federation in urban development; to establish guidelines for urban development, including housing, basic sanitation, transportation and urban mobility, to promote at the initiative of the Union or jointly with States, Federal District, and municipalities housing and sanitation programs, urban furniture and other spaces for public use; and to develop and to implement national and regional plans for land use and economic and social development.[13] It is also the competence of municipalities to legislate on direct urban development (Silva, 2017).

4.3.4 Powers and competences common to the Municipalities, States, Federal District, and Union

All the entities of the federation have common administrative powers and competences, as presented in Table 4.2. Note that legislative competences do not overlap, but rather are complementary, in the following forms: exclusive (local interest of Municipalities), privative (of the Union, which may be delegated to the States and Federal District), delegated (to the States and Federal District), remaining or reserved (all that is not defined by the FC being attributed to the States and Federal District), or supplementary (general guidelines given by the Union with regulations made by the States and Federal District).

Based on Article 23 of the FC, we realize that the preservation of life and protection of the environment is the attribution of all the federative entities; activities such as environmental licensing, damage and impact control, and the promotion of programs are the competence of the Federal Government, States and Federal District, and municipalities. Such

---

[11] They are: (1) legislate upon matters of local interest; (2) supplement federal and state legislations where pertinent; (3) institute and collect taxes within their jurisdiction, as well as apply their revenues (4) organize and render, directly or by concession or permission, the public services of local interest, including mass-transportation, which is of essential nature; and (5) promote, wherever pertinent, adequate territorial ordaining, by means of planning and control of use, apportionment, and occupation of the urban soil.

[12] Particularly noteworthy is constitutional amendment no. 39 of December 19th, 2002, which establishes street lighting as the municipality's powers and competence, allowing establishing the contribution to fund the Public Lighting Service Tax (COSIP).

[13] Brazil, City Estatute, Law 10.257, 2001.

TABLE 4.2 Table of common powers and competences of the Union, States, Federal District and Municipalities.

Competence	Relevant matter
Administrative Common (FC Art. (23) General Union Guidelines Specified by the States by complementary law.	1. To provide for health and public assistance; 2. To protect natural landscapes and to preserve forests; 3. To protect the environment and to fight pollution in any of its forms; 4. To provide the means of access to culture, education, science, technology, research and innovation; 5. To promote agriculture and organize the supply of foodstuff; 6. To promote housing construction programs and the improvement of housing and basic sanitation conditions; 7. To fight the causes of poverty and the factors leading to substandard living conditions, promoting the social integration of the unprivileged sectors of the population; 8. To register, monitor and control the concessions of rights to research and exploit hydric and mineral resources within their territories; 9. To establish and to implement an educational policy for traffic safety.

an attribution is regulated by Federal Laws 6803/80, 6938/81, decree 99274/90 and Resolutions CONAMA 001/86 and 237/97 and state complementary state laws (municipalities do not have the legislative power and competence on environmental licensing).

4.4 Brazilian institutional powers and competences to climate policies

To approximate powers and competences, analysis of the different entities of the federation to the debate on climate change, we will bind the distribution of competences according to the classification closest to the one adopted by the IPCC (2014), namely: (1) Energy, (2) Agriculture and Livestock, (3) Land Use and Land Use Change (including deforestation), (4) Disposal and waste management, and (5) Industrial Processes and Product Use. We highlight that the Union has exclusive administrative power and competence in the economic sectors and segments presented in Table 4.3.

It is important to emphasize that states and municipalities have power for implementing supplementary legislation to the federal legislation. Moreover, local governments, which are named municipalities in the FC, have been particularly powerful actors and may advance their own interests, even in cases they may conflict with national goals (Arlota, 2017). In this scenario, the fact that the FC established environmental protection as a goal embedded within the powers of municipalities, States, and the Union provides incentives for all these three spheres of power to meaningfully engage in such

TABLE 4.3 Tables of powers and competences and activities relevant for analysis.

Sector	Segment	Legislative	Administrative			
			Exploration	Guidelines	Planning	Management
Energy	Electricity sector[a]	U	U	U	U	U
	Nuclear industry	U	U	U	U	U
	Oil, NG, and liquid fuel industries[b]	U	U	U	U	U
	Piped gas	U	S	U	S	S
	Harbors and airports	U	U	U		U
	Air and sea transportation	U	U	U	U	U
	Interstate and international land and water transportation[c]	U	U	U	U	U
	State land transportation[d]	U	S	U	S	S
	Intermunicipal urban transport	U	S	U	S	S
	Municipal transport and urban mobility	U	M	U	M	M
	Consumption—agricultural sector	U	–	U	U; S	U; S
	Consumption—industrial sector	U	–	U	U; S	U; S
	Consumption—residential sector	U	–	U	U; S	U; S
	Consumption—services sector[e]	U	–	U	U; S	U; S
	Street lighting		M		M	M
Waste and sanitationWater	Sanitation and treatment of effluents	U; S; M	U; S; M	U; S; M	U; S; M	U; S; M
	Disposal of solid waste	U; S; M	U; S; M	U; S; M	U; S; M	U; S; M
	Hydric resources	U	U; S; M	U; S; M	U; S; M	U; S; M
Industrial processes	Mineral resources and metallurgy	U	U; S; M	U; S; M	U	U; S; M
	Production of goods and commodities	U; S	U; S; M	U; S; M	U	U; SM
	Stationary emissions	U; S; M	U; S; M	U; S; M	U; S; M	U; S; M
Agricultural	Production of goods and commodities	U; S	U; S; M	U; S; M	U	U; S; M
	Effluent treatment	U; S	U; S; M	U; S; M	U; S; M	U; S; M
	Waste treatment	U; S	U; S; M	U; S; M	U; S; M	U; S; M

(Continued)

TABLE 4.3 (Continued)

Sector	Segment	Legislative	Administrative			
			Exploration	Guidelines	Planning	Management
Environment	Forest conservation	U; S; M	U; S;M	U; S; M	U; S; M	U; S; M
	Environmental protection and pollution control	U; S; M	U; S; M	U; S; M	U; S; M	U; S; M
Urban development	Housing	U; M	U; S; M	U	U; S; M	U; S; M
	Civil defense	U; S	U	U	U; S; M	U; S; M
	Health	U; S; M	U; S; M	U; S; M	U; S; M	U; S; M
	Territorial ordaining	U; S; M	U; S; M	U; S; M	U; S; M	U; S; M
Financial operations	Credit policy, insurance and other operations	U	U	U	U	U

*a*Centralized power generation, transmission, distribution, operation and commercialization.
*b*Exploration, production, transportation, refining, distribution (except piped gas) and commercialization.
*c*Railway, road, waterway and pipeline models.
*d*The construction of a free, fair, and solidary society; to guarantee national development; to eradicate poverty and marginalization, to reduce social and regional differences and to promote the nation's welfare, with no prejudice in relation to origin, race, sex, color, age, or other forms of discrimination.
*e*Public Sector and Commerce.
U, Union; S, States and Federal District; M, Municipalities.

environmental actions as well as in advancing climate policies. This legislative framework will be detailed in the following section.

4.5 Institutional analysis and multinational level

The formulation of climate policies and their respective sectoral plans fall within the power of the Union, since the following acts are part of the matter, namely: "to elaborate and to execute national and regional plans of territorial ordaining and of economic and social development" of exclusive character, without the possibility of delegation to another federative entity. Thus it is the attribution of the Federation to prepare the national plans as it has been observed through the institution of the National Climate Change Policy (PNMC) and its respective Sector Plans, the National Adaptation Plan, the Brazilian NDC, the implementation strategy of the Brazilian NDC.

The Union has power for planning, managing, supervising, and legislating about the energy sector and associated GHG emissions, except for the distribution of natural gas. Also, to be highlighted are the public lighting services and potential energy efficiency actions, which according to constitutional amendment no. 39 of 2002, are the attribution of municipalities.

The states may legislate in a complementary manner and all the federation bodies have the power to register, monitor, and control the concessions of rights to research and exploit hydric and mineral resources within their territories, with the relevance of the theme being, according to local or regional conditions, decisive for the development of regulatory agencies.

When analyzing Article 24 of the FC, the role of municipalities for environmental legislation is not clear, since in this article the FC directly assigns the exercise of legislation to the States, the Federal District, and the Union. However, if we analyze subsections I and II of Article 30 of the FC, we will see that it is assigned to municipalities the supplementation of federal and state legislation in respect of matters of local interest, filling gaps regarding local peculiarities. In this context, the Federal Supreme Court in Special Appeal 58.6224/SP recognizes the municipalities as actors in environmental legislation by adopting as an official interpretation the combination of the two subsections and establishes that the municipal law is subject to suspension if contrary to the legal precepts of Article 24 of the FC and if it disrespects the cooperative federalism established in Brazil (Santos and Costa, 2018).

The Brazilian transportation sector, comprising both freight and passenger transportation, accounts for 46% of all the GHG emissions related to the energy sector. Note that it is the attribution of the municipalities to manage the urban transport service (contracting and inspecting the services provided), although it is legislated by the Union, that has also power for drawing up plans and guidelines, and it is the task of the state to manage the intermunicipal urban transport. Several alternative technologies to diesel engines have been tested in different cities in the country: full electric buses, hybrid buses, and natural gas engines as a transition fuel (Galbieri, 2018; Brito, 2017).

Basic sanitation presupposes a broad spectrum of integrated measures that involves the different phases of the water cycle and are matters of common interest, being the powers of all the federation bodies (Barroso, 2002).

From the point of view of legislative power, note that all matters within the powers of the Union are susceptible to supplementary regulation, except as provided in Articles 49, 51, and 52 (Silva, 2017). Thus we can infer that the guidelines established by the PNMC and the Brazilian NDC should be regulated regionally and locally by the States, Federal District, and Municipalities.

4.6 Conclusion and policy implications

Brazil has a unique federal system, because under its FC environmental protection policies (and related climate policies) are shared among the three spheres of power, namely, the federal union, the states, and local governments (municipalities). As discussed, the country has a long-established international reputation as environmentally conscious, being a signatory of all the major treaties on environmental protection and climate change, including the UNFCCC (1992) and its umbrella treaties, namely, Kyoto and the Paris Agreement on Climate Change. Brazil's unprecedented federative arrangements coupled with its specific constitutional provisions and international actions provide a new paradigm for action fostering climate goals.

Brazil plays an important role in global climate change mitigation efforts; after all, almost 50% of its energy matrix is based on renewable energies, which are usually seen as a "natural" GHG mitigation strategy. However, it conversely has one of the highest deforestation rates in the world, which considerably increases the contribution of the country in terms of global GHG emissions.

In this context, it is also opportune to mention the growing challenges for Brazil to continue relying largely on the generation of electricity from large hydropower plants due to the increase in water evaporation in the reservoirs and changes in the hydrological regime reducing the water storage. According to the Statistical Yearbook of Electric Energy of 2017, there was a 101.9% growth in coal-fired thermoelectric generation between 2012 and 2016; and a 20.8% growth in natural gas-fired thermoelectric plants. In this context, in recent years, to meet the national demand for electricity and even for maintaining energy security, it has been increasingly necessary to activate the coal-fired and natural gas thermoelectric plants installed in the country (EPE, 2017).

As subnational governments gain relevance in the global efforts to curb GHG emissions, understanding their attributions becomes a key factor for an effective deployment of international resources and design of new instruments that will leverage local action towards climate change. Most states and municipalities in Brazil have not defined climate mitigation commitments and could benefit from the present analysis for developing climate mitigation plans, establishment of GHG reduction targets, development of sectoral plans resulting in effective climate policies aligned with regional and local reality (Aylett, 2014). Moreover, large metropolises located in the developed world should also consider adding financial contributions to the Green Climate Fund established under the Paris Agreement. Paris, for instance, has already done so (Green Climate Fund, 2020).

States and municipal leaderships can also play a key political role, strengthening the climate policies agenda that can be disrupted by political changes with conservative parties that have a skeptical perspective towards the climate change agenda. An emblematic example was the reaction of the US states to the departure of the United States from the Paris Agreement led by the US Government of Donald Trump: 14 US states (The US Climate Alliance), together with 383 cities (US Climate Mayors), committed to maintain the commitments of the Paris Agreement, regardless of the position of the US federal government (Bloomberg, 2017).

Brazil, in its position as a developing country and with a strong industrial base, needs to realign its style of economic development in order not to increase the emitted amount of GHG in the atmosphere. It is therefore necessary to plan and to implement practices and policies aimed at a new economy, namely an economy based on lower use of fossil fuels, that is, a "low-carbon economy." The present work rests on this principle, insofar as it uncovers the controversial definition of powers and competences among the different public administrative spheres, regarding the climatic policies implemented or to be implemented in Brazil, as broadly mentioned.

In addition to assisting developers and promoters of public policies and programs to mitigate and to adapt to climate change, the article's results will also serve to guide actions for legal adaptation of economic agents in the carbon-intensive sectors especially for the energy sector—which is mostly regulated by the direct and indirect public administration of the Union.

In this context, the present study serves as a map of powers and competences per the economic sector, according to the IPCC classification (IPCC, 2014). The evaluation undertaken will clarify the roles and attributions of the different government spheres, assisting the policy makers of subnational climate policies, who have gained prominence in the fight against climate change, after the Paris Agreement. This global climate agreement, directly and/or indirectly, calls on state, municipal governments, and the private sector to jointly perform the transnational efforts agreed upon at the Conferences of the Parties to the United Nations Framework Convention on Climate Change.

Acknowledgments

We are grateful to the "Research Center for Gas Innovation" (Fapesp Proc. 2014/50279-4), supported by FAPESP and Shell, organized by the University of São Paulo, and the strategic importance of the support granted by the ANP (National Agency of Petroleum, Natural Gas and Biofuels of Brazil) through the R&D clause. We also thank the support from the National Agency for Petroleum, Natural Gas and Biofuels Human Resources Program (PRH-ANP), funded by resources from the investment of oil companies qualified in the R, D&I clauses from ANP Resolution No. 50/2015 (PRH 33.1—Related to Call No. 1/2018/PRH-ANP; Grant FINEP/FUSP/USP Ref. 0443/19). Co-author André Felipe Simões thanks CAPES (Coordination for the Improvement of Higher Education Personnel) for supporting him, in 2020, in acting as Visitant Professor at the Climate and Energy College of the University of Melbourne, Australia (PRINT CAPES 09/2019).

References

Anderton, K., Setzer, J. 2018. Correction to: Subnational climate entrepreneurship: innovative climate action in California and São Paulo. Reg Environ Change 18, 1285. Available from: https://doi.org/10.1007/s10113-017-1238-x.

Arlota, C., 2020. Does the United States withdrawal from the Paris Agreement pass the cost–benefit analysis test? The University of Pennsylvania Journal of International Law. Available from: https://scholarship.law.upenn.edu/jil/vol41/iss4/1/.

Arlota, C., 2017. Should local governments be included in the constitution? A comparative analysis between the U.S. and Brazilian supreme courts' reasoning regarding annexation laws. The University of Bologna Law Review. Available from: https://bolognalawreview.unibo.it/article/view/7493/7214.

Aylett, A., 2014. Progress and challenges in the urban governance of climate change: results of a global survey. MIT, Cambridge, MA.

Barroso, L.R., 2002. Saneamento Básico: Competências Constitucionais da União, Estados e Municípios. Revista de Informação Legislativa Fonte. Available from: http://www2.senado.leg.br/bdsf/bitstream/handle/id/762/R153-19.pdf?sequence=4.

Benites-Lazaro, L.L.-T., et al., 2018. Governança e desenvolvimento sustentável: a participação dos stakeholders locais nos projetos de Mecanismos de Desenvolvimento Limpo no Brasil. Cuadernos de Geografía-Revista Colombiana de Geografía 27 (2). Available from: https://doi.org/10.15446/rcdg.v27n2.66336.

Betsill, M.M., Bulkeley, H., 2003. Cities and climate change: urban sustainability and global environmental governance. Routledge, London.

Betsill, M.A., 2004. Transnational networks and global environmental governance: the cities for climate protection program. International Studies Quarterly 48 (2), 471–493. Available from: https://doi.org/10.1111/j.0020-8833.2004.00310.x.

Betsill, M.M., 2006. Cities and the multilevel governance of global climate change. Fonte 12, 141–159. Available from: https://heinonline.org/HOL/P?h=hein.journals/glogo12&i=152.

Betsill, M.M., 2013. Revisiting the urban politics of climate change. Environmental Politics 22, 136–154. Available from: https://doi.org/10.1080/09644016.2013.755797.

Bloomberg, 2017. Recommendations of the Task Force on Climate-related Financial Disclosures, Final Report. Available at https://assets.bbhub.io/company/sites/60/2020/10/FINAL-2017-TCFD-Report-11052018.pdf (accessed on May 5, 2021).

Brasil, 2001. Estatuto da Cidade, Lei n° 10.257 de 10 de julho de 2001. Brasília. Fonte: http://www.planalto.gov.br/ccivil_03/LEIS/LEIS_2001/L10257.htm.

Brito, T.L., 2017. Qualitative comparative analysis of cities that introduced compressed natural gas to their urban bus fleet. Renewable and Sustainable Energy 71. Available from: https://doi.org/10.1016/j.rser.2016.12.077.

Castro, A.S., Alves, J.D., 2018. Condicionantes das emissões de dióxido de carbono (CO_2) No Brasil: evidências empíricas de uma curva no formato de "N". Rev. de. Economia Contemporânea, 22. Available from: https://doi.org/10.1590/198055272236.

Costa, H.K.M., Simoes, A.F., Moutinho dos santos, E., Silva, I.M.M.E., 2017. Modificações legislativas e impactos nos royalties e na participação especial destinados aos órgãos da administração direta. Revista de Politica Públicas da UFMA, 21, 959–981.

EPE, 2017. Anuário Estatístico de Energia Elétrica 2017—Ano Base 2016 v3 Dados preliminares. Empresa de Pesquisa Energética, Rio de Janeiro.

Galbieri, R.B., 2018. Bus fleet emissions: new strategies for mitigation. Mitigation and Adaptation Strategies for Global Change 23 (7). Available from: https://doi.org/10.1007/s11027-017-9771-y.

Green Climate Fund, 2020. Status of Pledges and Contributions made to the Green Climate Fund. https://www.greenclimate.fund/sites/default/files/document/status-pledges-irm_1.pdf.

Harald Fuhr, T.H., 2018. The role of cities in multi-level climate governance: local climate policies and the 1.5°C target. In: Srge-Vorsatz, K. (Ed.), Current Opinion in Environmental Sustainability. 30, pp. 1–6. Available from: https://doi.org/10.1016/j.cosust.2017.10.006.

IPCC, 2014. Summary for policymakers. In: Climate Change 2014: Impacts, Adaptation, and Vulnerability. Part A: Global and Sectoral Aspects. Contribution of Working Group II to the Fifth Assessment Report of the Cambridge. Cambridge University Press, United Kingdom.

Macedo, L.S., 2017. Participação de cidades brasileiras na governança multinível das mudanças climáticas. Instituto de Energia e Ambiente, São Paulo. Available from: https://doi.org/10.11606/T.106.2017.tde-18102017-203603.

Mariotti, A., 2009. Leituras do direito constitucional. EDIPUCRS, Porto Alegre.

Moraes, A., 2010. Direito Constitucional, twenty sixth ed. Atlas, São Paulo.

Pepplow, L.A., Silva, V.L., Betini, R.C., 2019. Evaluation of global heating reduction potential with the replacement of electricity supplied by the local concessionaire via solar renewable source. Brazilian Archives of Biology and Technology 62. Available from: https://doi.org/10.1590/1678-4324-smart-2019190003.

Santos, S.F., Costa, B.S., 2018. Competência legislativa do município em matéria ambiental: uma análise crítica do re 58.6224/sp conforme a doutrina e o supremo tribunal federal. Revista Gestão & Sustentabilidade Ambiental 7 (1). Available from: https://doi.org/10.19177/rgsa.v.

Setzer, J., 2017. How subnational governments are rescaling environmental governance: the case of the Brazilian state of São Paulo. Journal of Environmental Policy & Planning 19. Available from: https://doi.org/10.1080/1523908X.2014.984669.

Silva, J.A., 2017. Curso de Direito Constitucional Positivo, fortyth ed. Malheiros, São Paulo.

United Nations Conference on Environment and Development, 1992. Agenda, 21. Rio Declaration, Forest Principles, United Nations, New York.

Vargas, B.R., Rodrigues, J.W., 2016. XXV Encontro Nacional do CONPEDI—Constituição e Democracia I. DIREITO E DESIGUALDADES: Diagnósticos e Perspectivas para um Brasil Justo. v. 1. Florianópolis: CONPEDI. Fonte: https://www.conpedi.org.br/publicacoes/y0ii48h0/509my5cz/Ho722tL4dGkF2tWJ.pdf.

WCED - World Commission on Environment and Development, 1987. Report of the World Commission on Environment and Development: Our Common Future. Available at https://sustainabledevelopment.un.org/content/documents/5987our-common-future.pdf (accessed on May 5, 2021).

5

Carbon capture and storage technologies and efforts on climate change in Latin American and Caribbean countries

Romario de Carvalho Nunes and
Hirdan Katarina de Medeiros Costa

Institute of Energy and Environment, University of São Paulo, São Paulo, Brazil

5.1 Introduction

In the contemporary context, the theme of climate change mitigation is the subject of international consideration. In the meantime, companies and governments seek alternatives to address climate change, mainly caused by global warming, which has occurred by the intensification of greenhouse gas (GHG) emissions. Among the projects capable of softening the emission of these gases is the technology called Carbon Capture and Storage (CCS) (Nunes and Costa, 2019).

The Intergovernmental Panel on Climate Change (IPCC) defines the capture and storage of CO_2 in geological reservoirs as a process consisting of CO_2 separation, emitted by stationary sources related to energy production and also industrial plants, in the transport of this CO_2 and its long-term storage in geological reservoirs, isolating it from the atmosphere. It is possible to separate the CO_2 emitted in the burning of fossil fuels, process it to its liquid form, and transport it by pipelines, highways, or by sea to geological reservoirs, such as deactivated mines, oil fields, or other places where CO_2 can be stored (Câmara, 2011).

Studies by the IPCC (2005), the Forum of Major Economies for Energy and Climate—MEF (2009), and the International Energy Agency (IEA) pointed to the need for

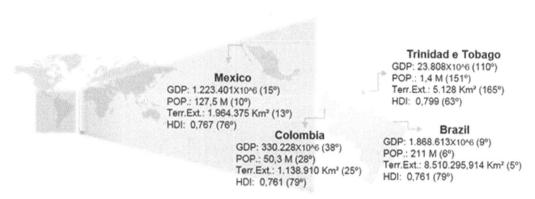

Mexico
GDP: 1.223.401X10^6 (15°)
POP.: 127,5 M (10°)
Terr.Ext.: 1.964.375 Km² (13°)
HDI: 0,767 (76°)

Trinidad e Tobago
GDP: 23.808X10^6 (110°)
POP.: 1,4 M (151°)
Terr.Ext.: 5.128 Km² (165°)
HDI: 0,799 (63°)

Colombia
GDP: 330.228X10^6 (38°)
POP.: 50,3 M (28°)
Terr.Ext.: 1.138.910 Km² (25°)
HDI: 0,761 (79°)

Brazil
GDP: 1.868.613X10^6 (9°)
POP.: 211 M (6°)
Terr.Ext.: 8.510.295,914 Km² (5°)
HDI: 0,761 (79°)

FIGURE 5.1 Location of selected countries and main comparative data. GDP, Gross domestic product; HDI, human development index; POP, Population. Source: *Human Development Reports; IBGE. Comparar Paı́ses.*

substantially implementing CCS for fossil-powered energy generation and industrial processes to minimize the costs foreseen for meeting the GHG reduction targets (EASAC, 2013).

Within this context, this chapter investigates climate change policies related to CCS technologies in Colombia, Mexico, Brazil, and Trinidad and Tobago https://www.globalccsinstitute.com/archive/hub/publications/190128/fact-sheet-ccs-ccus-the-americas.pdf. Considering the case studies analyzed from a descriptive methodology, the chapter discusses the main common challenges for effective CCS policies leading to its actual implementation in the Central and South American region.

Preliminarily, it is necessary to geographically locate and collect macroeconomic and social data from the countries to be addressed in this chapter, as can be seen in Fig. 5.1, which contains data related to the size of the economy, population, territorial extension and human development index.

5.2 Macroscenario for climate policy implementation

Although some initiatives seek to unlink the increase in the concentration of carbon dioxide in the atmosphere from the systemic warming of the planet, a relatively much larger portion of the scientific community accepts the correlation between the two factors as a causal link (Nunes and Costa, 2020).

In 1988 this analysis of climate change became the subject of a global research initiative entitled the Intergovernmental Panel on Climate Change, better known by the acronym IPCC, the result of an initiative by the United Nations Environment Programme and the World Meteorological Organization (Nunes and Costa, 2020).

The finding of global atmospheric warming can be evidenced by analyzing the historical variation of CO_2 concentration (ppm) over the last 800,000 years, observing the discrepant increase of this concentration of 1950, as can be seen in Fig. 5.2. The peaks

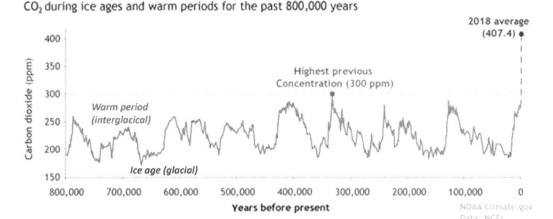

FIGURE 5.2 Global concentrations of carbon dioxide in the atmosphere (CO_2) in parts per million (ppm) over the past 800,000 years. *NOAA*, National Oceanic and Atmospheric Administration.

and valleys accompany the ice age (low CO_2) and warmer interglacials (higher CO_2). During these cycles, CO_2 was never more than 300 ppm. In 2018 it reached 407.4 ppm. On the geological timescale, the increase (blue dashed line) seems virtually instantaneous (NOAA, 2020).

According to Huaman (2014), the total emission reductions compared to the baseline scenario in 2050 are 43 $GtCO_2$, and the share of emissions reductions of electricity generation technologies is 19% by CCS, 17% by renewables, 6% by nuclear, and 15% by improvements in efficiency and fuel exchange between fossil fuels, respectively (Huaman, 2014). Large emission reductions are also needed in the energy end-use sectors (industrial sector, transportation, and residential and commercial sector), these sectors contribute about 38% of the total emission reductions (Huaman, 2014).

In Latin America, although all the countries selected are signatories to the Paris Agreement, only Brazil has carbon storage facilities, as shown in Table 5.1.

As regards macrodata, Brazil, Mexico, and Colombia are countries with relatively similar characteristics, with a considerable economy and population, vast territory and similar human development index. The exception is Trinidad and Tobago; despite having a very small population, territory, and economy, it has high gas production, according to Table 5.2, being one of the reasons for which it was chosen for analysis in this chapter.

The energy characteristics aforementioned provide an initial overview for the analysis of conditions that would enable the implementation of carbon dioxide capture and storage technology, such as carbon dioxide emissions, oil and gas production, oil refining and electricity generation facilities, mainly powered by coal and natural gas. Table 5.2 also shows the use of renewable energy for energy generation, which may be indicative of a transition to the implementation of more robust energy policies committed to meeting the nationally determined contributions (NDCs) of the Paris Agreement, including decarbonization targets.

TABLE 5.1 Carbon Capture and Storage (CCS) facilities and Paris Agreement signatory.

Country	CCS facilities	Signatory Paris Agreement
Brazil	2	Yes
Colombia	0	Yes
Mexico	0	Yes
Trinidad and Tobago	0	Yes

TABLE 5.2 Energy characteristics of countries and their current use of renewable energy.

Country	Carbon dioxide emissions	Gas production	Oil production	CO_2 emissions— power plants	Oil: refinery throughput	Renewables power
Brazil	Very high	High	Very high	Very high	Very high	Very high
Colombia	High	Medium	High	Medium	High	Medium
Mexico	Very high	Very high	Very high	Very high	High	Medium
Trinidad and Tobago	Medium	Very high	Low	Very low	Very low	Very low

5.2.1 Paris Agreement and decarbonization targets

The Paris Agreement uses NDCs as a short-term basis to achieve its long-term goal of limiting the rise in global average temperature below 2°C above preindustrial levels and pursuing efforts to limit temperature increases to 1.5°C above preindustrial levels (Binsted, 2020). NDC targets are initially set for the period 2025 or 2030 and vary widely from country to country, reflecting the unique challenges and mitigation opportunities of countries and social, economic, and political circumstances. The Paris temperature targets require zero global carbon dioxide (CO_2) emissions before the end of the century (Binsted, 2020).

Data show that selected countries and Latin America and the Caribbean (LAC) have been decreasing their carbon dioxide emissions since 2016, except for Colombia, as shown in Graph 5.1, drawn from the most recent data collected in the British Petroleum's Statistical Review of World Energy.

The reduction in carbon dioxide emissions may be related to external factors, such as the economic crisis in 2014, which affected several countries—especially hydrocarbon producers or the beginning of compliance with the NDCs of the Paris Agreement. According to Graph 5.2, a very sharp growth is observed in the use of renewable energies for energy generation, especially from the 2000s.

LAC have the lowest carbon electricity sector in all the regions of the world, thanks to its largest share of hydroelectricity (González et al., 2019). Yet that is changing. Hydroelectric generation reduced its percentage in the energy mix from 58% in 2009 to

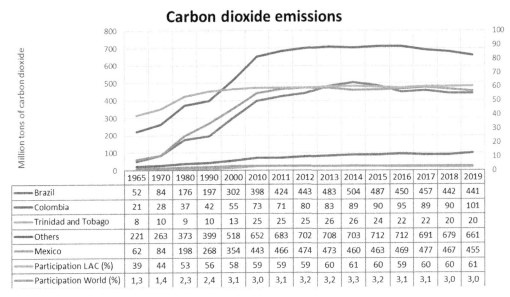

Carbon dioxide emissions

	1965	1970	1980	1990	2000	2010	2011	2012	2013	2014	2015	2016	2017	2018	2019
Brazil	52	84	176	197	302	398	424	443	483	504	487	450	457	442	441
Colombia	21	28	37	42	55	73	71	80	83	89	90	95	89	90	101
Trinidad and Tobago	8	10	9	10	13	25	25	25	26	26	24	22	22	20	20
Others	221	263	373	399	518	652	683	702	708	703	712	712	691	679	661
Mexico	62	84	198	268	354	443	466	474	473	460	463	469	477	467	455
Participation LAC (%)	39	44	53	56	58	59	59	59	60	61	60	59	60	60	61
Participation World (%)	1,3	1,4	2,3	2,4	3,1	3,0	3,1	3,2	3,2	3,3	3,2	3,1	3,1	3,0	3,0

GRAPH 5.1 Historical survey of carbon dioxide emissions. Source: *Elaborated by authors from British Petroleum, 2020. Statistical Review of World Energy (accessed 15.07.20).*

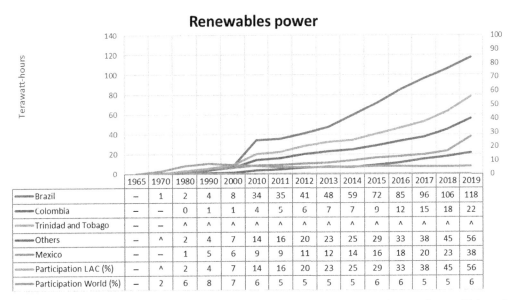

Renewables power

	1965	1970	1980	1990	2000	2010	2011	2012	2013	2014	2015	2016	2017	2018	2019
Brazil	—	1	2	4	8	34	35	41	48	59	72	85	96	106	118
Colombia	—	—	0	1	1	4	5	6	7	7	9	12	15	18	22
Trinidad and Tobago	—	—	^	^	^	^	^	^	^	^	^	^	^	^	^
Others	—	^	2	4	7	14	16	20	23	25	29	33	38	45	56
Mexico	—	—	1	5	6	9	9	11	12	14	16	18	20	23	38
Participation LAC (%)	—	^	2	4	7	14	16	20	23	25	29	33	38	45	56
Participation World (%)	—	2	6	8	7	6	5	5	5	5	6	6	5	5	6

GRAPH 5.2 Historical survey on the use of renewable energies for energy production. Source: *Elaborated by authors from British Petroleum, 2020. Statistical Review of World Energy (accessed 15.07.20).*

II. Case studies on CCS and related policies, and their consequences for climate change

50% in 2016 (González et al., 2019). Utilization rates have been reduced by droughts and capacity additions have decreased due to social and environmental concerns and increased capital costs (González et al., 2019).

However, to achieve the objectives proposed in the NCDs of the Paris Agreement, it is necessary to implement policies aimed at changing the current methods of energy production from carbon dioxide-intensive sources to low or no carbon dioxide sources. However, this change, in turn, may result in the devaluation or deactivation of carbon-intensive assets before the end of their useful life, called "stranding" of assets, a concept briefly explained as follows.

The concept of idle assets has been explored by experts in various disciplines in contexts ranging from fossil fuel reserves to electricity, natural gas liquefaction, and agriculture (Binsted, 2020). They can be considered assets that suffer early or unforeseen phasing out. In the context of climate change mitigation, idle assets can manifest in various ways, such as fossil fuel resources that cannot be burned in order to maintain a long-term temperature target or the anticipated decommissioning of capital assets due to climate policies (Binsted, 2020).

The issue of idle assets is important because they can result in financial market and political instability due to a rapid loss of wealth for the owners of the affected capital assets (Binsted, 2020). Also, financial institutions in LAC are not as robust as in other regions, which can undermine the countries' ability to cope with the instability created by idle assets (Binsted, 2020).

Long-term decarbonization goals are important for energy infrastructure planning because a power plant life can range from 30 to 50 years (González et al., 2019). These findings suggest that to meet the average allowed carbon budget of 2°C (6.2 GtCO$_2$) or 1.5°C (5.8 GtCO$_2$), utilities in the region would need to prematurely close 10%−16% of the existing fossil fuels capacity, respectively, or reduce the utilization rate of existing plants for the same purpose (González et al., 2019). Doing so can be politically difficult, because policies that result in the closure or reduced use of the plant would decrease the financial value of these assets, that is, they would create idle assets (González et al., 2019). Closing power plants would also result in sudden job losses for workers and communities that depend on these assets (González et al., 2019).

Idle assets are a key issue for LAC countries, although the region accounts for less than 10% of the global carbon dioxide (CO$_2$) emissions and already generates more than half of its electricity from renewable sources. For example, in a recent analysis, Binsted (2020) found that the region ranks second (only behind the Middle East) in terms of total volume of unburned oil and gas reserves (Binsted, 2020).

It is necessary to verify what the implications are in terms of idle assets of the energy sector and the opportunity of implementing CCS technology for meeting NDCs for LAC to meet the objectives of the Paris Agreement.

5.3 Carbon storage facilities and use of CCS technologies

Recent data on carbon storage facilities provided by the Global CCS Institute (2017) show that 18 CCS projects are currently in operation worldwide, being associated with a capture of 37 million tons of CO_2 per year (Mtpa), equivalent to the activity of 8 million cars. Also, according to the Global status of CCS Institute (2017), more than 220 million tons of CO_2 were stored cumulatively by 2017.

The storage potential is not yet entirely clear; doubts exist about determining how large the storage capacity is globally (Ketzer, 2016). In deactivated oil and gas reservoirs there is an estimated storage capacity between 675 and 900 Gton of CO_2. In saline formations, estimates point to a capacity of at least 1000 Gton, and these locations may represent a capacity 10 times higher (Ketzer, 2016).

According to data collected from the Global CCS Institute CO2RE database in October 2018, LAC is one of the regions with the fewest carbon dioxide storage facilities. Fig. 5.3 shows that only two facilities—equivalent to 1% of the world's facilities—are in the countries of this region.

According to the IEA, globally, more than 30 million tons of CO_2 are captured in large-scale CCS facilities for use or storage, more than 70% in North America. Fig. 5.4 shows the dispersion of this capture, with highlights for the United States [24.85] Mtpa—10 projects (56% = about all projets 71%= about all capture (Mtpa))] and Australia (4 Mtpa—1 project). Brazil has one project in operation, which captures 1 Mtpa annually.

According to the data issued by The Global Status of CCS (2018), Petrobras stored 2.5 Mtpa in the Santos Basin, accumulating 7 Mt of CO_2 injected through the EOR method since 2013.

Analyzing the data obtained, it is possible to verify that in the implementation and operation of the facilities in countries, there is a trend to produce a considerable amount of hydrocarbons (Canada, United States, Brazil, China, Saudi Arabia, United Arab Emirates, Algeria, United Kingdom, and Norway), as well as those that refine it in considerable quantities (Canada, United States, Brazil, China, Saudi Arabia, and Japan) (Nunes and Costa, 2020). This would mainly be due to the infrastructure

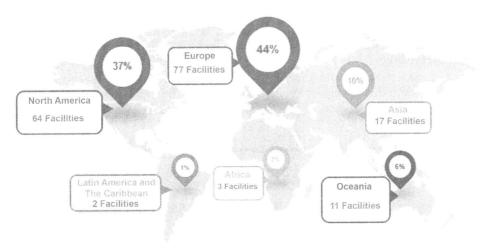

FIGURE 5.3 CO_2 storage facilities by region. Source: *Elaborated by author from Global CCS Institute (2020). CO2RE., https://co2re.co/ (accessed 15.07.20).*

FIGURE 5.4   Amount of CO_2 stored per year per country. Source: *Elaborated by authors from the Global CCS Institute (2020). CO2RE., https://co2re.co/ (accessed 15.07.20).*

already installed on-site in the case of producers, as well as considerable quantities of CO_2 emissions during refining. Japan would be an example of this case (Nunes and Costa, 2020).

The above finding can be ratified by Graphs 5.3 and 5.4, showing the number of CCS storage facilities in general as well as the number of installations in operation in the world, respectively:

Graph 5.5 shows in detail the operability of CO_2 installations worldwide:

Note the relatively low number of facilities in operation in the world, which corresponds to only 11%. However, there are numerous facilities under construction and development, which demonstrates a trend toward increased CO_2 capture and storage in the coming decades.

Brazil has a modest stake in this list, with only one installation in operation, although it is a large-scale facility.

Table 5.3, based on data from the Global CCS Institute, contains the most up-to-date data on CCS facilities distributed across the planet, including operational facilities, but not necessarily currently in operation, as well as projects under development.

5.4 Technical analysis of the potential use of CCS

Geological storage of CO_2 involves the injection of gas into geological formations at great depths using the CCS technology, with emphasis on gas injection in oil and gas fields, deep saline formations, and coal layers (IEA, 2010).

The known storage sites are both on land and at sea and are scattered around the globe, including Mexico, Colombia, Brazil, and Trinidad and Tobago, as shown in Fig. 5.5.

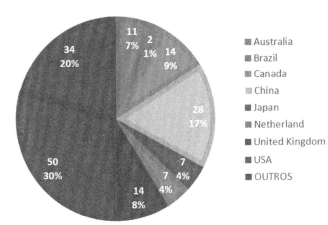

Australia
Brazil
Canada
China
Japan
Netherland
United Kingdom
USA
OUTROS

GRAPH 5.3 Countries with the largest number of carbon dioxide storage facilities. Source: *Elaborated by authors from the Global CCS Institute (2020). CO2RE., https://co2re. co/ (accessed 15.07.20).*

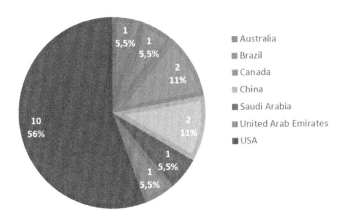

Australia
Brazil
Canada
China
Saudi Arabia
United Arab Emirates
USA

GRAPH 5.4 Countries with facilities currently in operation. Source: *Elaborated by authors from the Global CCS Institute (2020). CO2RE., https://co2re.co/ (accessed 15.07.20).*

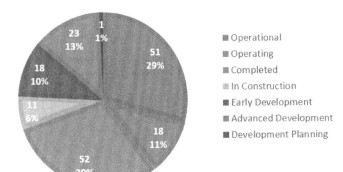

Operational
Operating
Completed
In Construction
Early Development
Advanced Development
Development Planning

GRAPH 5.5 Operationality of carbon dioxide storage facilities. Source: *Elaborated by authors from the Global CCS Institute (2020). CO2RE., https://co2re.co/ (accessed 15.07.20).*

TABLE 5.3 Complete list with all carbon dioxide storage facilities in the world.

Country	Facility	Operational	Operating	Completed	In construction	Early development	Advanced development	Development planning
						Facility status		
Algeria	1			1				
Australia	11	4	1	2		1	3	
Belgium	2				2			
Brazil	2		1	1				
Canada	14	4	2	3	3		2	
China	28	11	2	3	4	5	3	
Croatia	1	1						
Denmark	2			2				
France	4	1		2			1	
Germany	3			3				
Iceland	1	1						
India	2	1		1				
Indonesia	1						1	
Ireland	1					1		

Country								
Italy	1			1				
Japan	7	3		2	2			
Netherland	7	3		2		1	1	
Norway	7	3		3			1	
Saudi Arabia	2	1	1					
South Africa	2						1	1
South Korea	3			2		1		
Spain	4	1		3				
Sweden	2	1		1				
United Arab Emirates	2		1			1		
United Kingdom	14	2		5		6	1	
United States	50	14	10	15		3	8	
Total	174	51	18	52	11	18	23	1

Elaborated by authors from the Global Institute CCS.

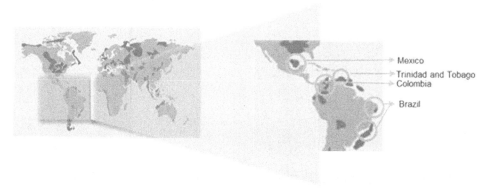

FIGURE 5.5 Propitious locations for geological storage. *Source: Adapted from Gaspar, H., Alexandre C., 2014. Captura e Armazenamento de CO$_2$. Faculdade de Ciências e Tecnologia Universidade Nova de Lisboa, Lisboa.*

In the oil and gas industry, CCS technology is developed in oil and gas reservoirs to stimulate the production of hydrocarbons in mature fields, or those requiring increased pressure in the reservoir, to increase the productivity of extraction wells (Goraieb, 2005). Sometimes, the injection may not only be carbon dioxide at high concentrations, but also mixtures composed mostly of atmospheric air and even natural gas, called underground storage of natural gas (Goraieb, 2005). In the oil energy system, CO$_2$ storage consists of injecting CO$_2$ into the reservoir rock, representing one of the special secondary oil recovery methods. This special recovery is also known as advanced or improved and is referred to as enhanced oil recovery (EOR). From the 55 largest CCS projects in the world, the Global CCS Institute states that the EOR-CO$_2$ mechanism would have a stake in about 58% globally.

This CO$_2$ injection technology has been eventually considered a safe alternative for storing this gas and can be proven by operating dozens of cases for hydrocarbon exploration in various geological contexts (Zoback, 2007). However, this technology developed for specific purposes and with little prominence eventually had a great global projection, due to recent (implementation) initiatives to stock CO$_2$ aiming exclusively at storing GHGs, due to climate change mitigation technologies.

We thus highlight projects, initiatives, and studies regarding the use and implementation of CCS in the countries analyzed in this chapter.

5.4.1 Brazil

Brazil has the National Plan for Climate Change based on the general guidelines of the National Climate Change Policy, approved in December 2009 under Federal Law No. 12,187 (Clarke, 2016). The plan main objectives are to: (1) stimulate efficiency increase in the constant search for best practices in economic sectors; (2) maintain the high share of renewable energy in the electric matrix, preserving the important position Brazil has always occupied; (3) encourage the sustainable increase in the participation of biofuels in the national transport matrix and work toward structuring an international market for sustainable biofuels; (4) seek sustained reductions in deforestation rates, throughout the

Brazilian biome, to achieve zero illegal deforestation; (5) eliminate the net loss of forest cover in Brazil by 2015; (6) strengthen intersectoral actions aimed at reducing the vulnerabilities of populations; and (7) identify the environmental impacts resulting from climate change and stimulate scientific research that can outline a strategy to minimize the socioeconomic costs of adaptation in the country (Clarke, 2016).

The implementation of CCS emphasizes the characteristics and resources of the basins, since it constitutes most of the South American Platform, a section of the continental crust that is the stablest part of the South American tectonic plate (Milani et al., 2007). Ancient igneous and metamorphic rocks formed during the Precambrian period (over 590 million years ago) form a basement complex that underlies the entire platform, including Brazil (Milani et al., 2007). Subsequent complex movements of the Earth's crust since Precambrian times have created a series of depressions or basins in different parts of Brazil. In total, 31 sedimentary basins occur within the Brazilian territory, covering an area of approximately 6.4 million km^2, more than 50% of the Brazilian territory, 75% of which is located in a terrestrial environment (Milani et al., 2007).

In Brazil, the Campos Basin has significant potential for CO_2 storage, with a theoretically estimated capacity of 950 $MtCO_2$, with about 75% of this capacity in the Roncador (28%), Marlim (18%), Albacora (17%), and Barracuda (12%), as shown in Fig. 5.6 (Ketzer et al., 2016). This capacity would be sufficient to store the equivalent of 3.5 years of the total emissions from Brazilian stationary sources (Ketzer et al., 2016).

The Brazilian Atlas of Capture and Geological Storage of CO_2 (Ketzer et al., 2016) analyzed all the sedimentary basins according to seven criteria related to the stages of a CCS project. The study took into account the following parameters: occurrence of coal deposits, active hydrocarbon production, existence of saline formation data, theoretical capacity for CO_2 storage, existence of mature oil and gas fields, associated issuing sources, and existence of transport infrastructure (pipelines and terminals) (Ketzer et al., 2016).

From this, a prospectivity map was generated through a basin-by-basin analysis, classifying them into three main groups: low, medium, or high prospect for storage, as shown in Fig. 5.7.

The Paraná, Campos, Santos, Potiguar, and Recôncavo sedimentary basins are classified as those with the greatest prospects for CO_2 storage in Brazil, mainly due to their outstanding production of hydrocarbons and the presence of mature fields and, in the case of the Paraná Basin, the occurrence of coal deposits (Ketzer et al., 2016). The Campos Basin is the largest oil-producing basin in the country (Ketzer et al., 2016). Also, these basins have a good association between sources and sinks, and a network of pipelines for transporting CO_2, which increases their prospects (Ketzer et al., 2016). Emissions from sources within these basins (up to the 300 km zone) reach around 368 Mt/year (Ketzer et al., 2016).

According to studies by Rockett et al. (2011), among the main stationary sources, the highest CO_2 emissions are concentrated in the southeast region. The authors also identified the main CO_2 emitting sectors, being cement, steel, refineries, thermopower, ethylene, ethanol, biomass, and ammonia production plants (Rockett et al., 2011).

In terms of transport infrastructure, the basins in this group are served by approximately 14,300 km of pipelines, which transport oil, gas, minerals, and CO_2. Preliminary assessments indicate that the Campos Basin has the highest theoretical capacity for storing

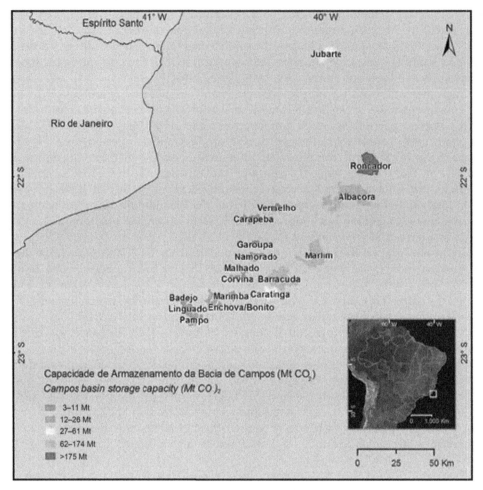

FIGURE 5.6 Campos basin storage capacity (MtCO$_2$), Brazil. Source: *From Ketzer, J.M.M., et al., 2016. Atlas brasileiro de captura e armazenamento geológico de CO$_2$. Edipucrs, Porto Alegre.*

CO$_2$ in oil fields (Ketzer et al., 2016). The other basins of the group have a large capacity associated with the production of hydrocarbons, such as the Santos Basin (including the presalt reservoirs). In the Paraná Basin, despite the occurrence of coal deposits, the high capacity of the basin is due to the existing deep salt formations (Ketzer et al., 2016).

According to Beck et al. (2011) there are some smaller CCS projects, both in the demonstration phase and in operation. Among them, the Petrobras Miranga Project, focused on EOR, storage in depleted gas reservoirs and saline aquifers, and the Centre of Excellence in Research on Carbon Storage Carbometano Porto Batista Project, developed in the scenario of methane gas associated with Charqueadas coal (RS–Brazil) from CO$_2$ injection and recovery of adsorbed gas. There is also a consortium in which Petrobras would seek to develop a pilot project for capturing CO$_2$ using oxyfuel. Still according to Beck et al.

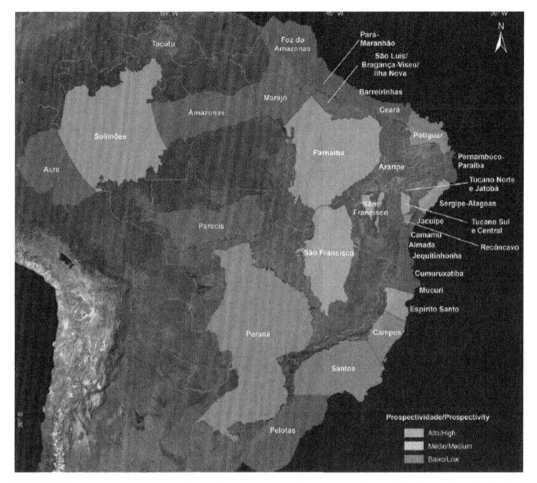

FIGURE 5.7 Prospectivity of Brazilian sedimentary basins. Source: *From Ketzer, J.M.M., et al., 2016. Atlas brasileiro de captura e armazenamento geológico de CO_2. Edipucrs, Porto Alegre.*

(2011), other options would be under technical and economic analysis, such as EOR in the presalt areas, CO_2 storage in saline aquifers, salt caves, and in depleted gas fields.

Moreover, as part of the Brazilian experimental CO_2 capture in Project-Phase 3 (CCP3), Petrobras updated its fluid catalytic cracking (FCC) unit on a pilot scale to demonstrate the technical feasibility of operating the unit in oxyfuel conditions (Mello, 2013). The FCC process can represent up to 30% of a refinery's total CO_2 emission, with oxyfuel technology being a promising option to reduce these emissions (Mello, 2013).

5.4.2 Colombia

Colombia has developed and implemented the Colombian Low Carbon Development Strategy (CLCDS) which includes sectoral and total emissions for the country climate policy and the implementation of 80 reduction measures in the sectors analyzed (Clarke, 2016). The CLCDS is a program focused on developing the country's short, medium, and long-term planning (Clarke, 2016). The sectors participating in the CLCDS are industry, energy, mining, transportation, housing, waste, and agriculture and have three main objectives: (1) to identify and to value actions aimed at preventing the accelerated growth of GHG emissions while the productive sector grows; (2) to develop reduction action plans for each productive sector in the country; and (3) to create and to promote tools for implementing reduction action plans, including monitoring and reporting. Such measures have the greatest potential for reducing GHG emissions (Clarke, 2016).

The reductions include the potential for GHG reductions, including CO_2, but do not include reductions in land use or land use change, which represent 20% of Colombia's emissions (Clarke, 2016). The baseline policy, built on the expected reductions in these policies and measures, is implemented as an emission limit from 2015 to 2050. In 2030, the policy limits CO_2 emissions to about 60% above the 2010 levels or 33% for all GHG (Clarke, 2016).

Colombia currently comprises a small portion of GHG emissions in Latin America, mainly due to its low levels of energy consumption and a high share of clean electricity production (Calderón et al., 2016). However, this low carbon economy may not be sustainable in the future. The country shows strong macroeconomic growth and stability, and a higher income is expected to increase the demand for fossil fuels (Calderón et al., 2016). The potential increase in carbon intensity in Colombia requires special attention to climate policy. The Colombian economy has experienced steady growth over the past decade, with an average annual growth in gross domestic product per capita of 3.5% from 2005 to 2012, above the Latin American average (Calderón et al., 2016). Higher income increased the demand for fossil fuels, especially in the transportation, manufacturing, and power generation sectors. The use of fossil fuels for electricity generation remains low, yet natural gas has gained prominence in the last decade (Calderón et al., 2016).

Also, according to Calderón et al. (2016), Colombia has a substantial coal resource base, with reserves forecast for 92 years at current production levels. The use of coal, therefore, may increase in the future with low energy prices and reductions in water resources due to climate change. Thus the potential for large increases in Colombia's carbon emissions in the future means that a renewed focus on efforts to reduce GHG emissions in Colombia may be necessary (Calderón et al., 2016).

Studies carried out by Calderón et al (2016), also suggest that Colombia could become CO_2 neutral by 2050, or even negative, with the adoption of biomass CCS, which captures CO_2 emissions in the generation of electricity through biomass. About two-thirds of the CO_2 captured in 2050 would be processed in conversion technologies based on biomass, mainly for producing synthetic fuels and hydrogen. These fuels would then be used in the transport sector, contributing to the decarbonization of the transport sector and further improvements in energy efficiency. Electricity would also play an important role in the decarbonization of energy demand sectors, mainly with the implementation of hydroelectric, wind, and solar electricity generation (Calderón et al., 2016).

Technically, Colombia has a high potential for mitigating CO_2 emissions through CCS-EOR, which has not yet been fully explored as a mitigation strategy for its oil industry (Yanez, 2020). An integrated CCS-EOR project considers capturing CO_2 at the emitting points and then transporting it through pipelines dedicated to the oil fields to the EOR (Yanez, 2020).

The location of industrial CO_2 sources is located mainly in the central and northern regions of the country along the valley of the Magdalena River and between the central and eastern regions of the Andes. Four potential clusters for the implementation of CO_2-EOR projects in Colombia have been identified (Yanez, 2020). The CO_2 capture potential is estimated at 5.9 $MtCO_2$, representing approximately 32% of the emission inventories (18 $MtCO_2$) (Yanez, 2020).

The CCS-EOR can be strategic for Colombia as it represents a considerable amount of incremental oil recovery. In total, the additional oil recovery potential was estimated at 487 MMbbl (Yanez, 2020).

5.4.3 Mexico

Although Mexico's emissions contribute only to approximately 2% of the global GHG emissions, the country has set ambitious targets, based on its reduction capacity (Clarke, 2016). For 2020 Mexico's emission reduction target is to reduce its baseline emissions by 30% and, by 2050, the country promises a potential reduction of 50% compared to 2010 levels, provided that developed countries finance adaptation actions, clean energy, and support technology transfers (Clarke, 2016). In 2012 Mexico enacted its General Law on Climate Change, which establishes the legal bases for the advancement of climate policy (Clarke, 2016).

Mexico's ambitious long-term strategy will require a major transformation of the energy sector. For example, the National Commission on Climate Change targets emissions from power generation to achieve a 35% share of clean (nonfossil) energy by 2024, 40% by 2030, and 50% by 2050 (Clarke, 2016). Annual GHG emissions in Mexico have increased sharply in the past two decades, growing 33% between 1990 and 2010, with demographic and economic growth playing an important role in increasing emissions. The increase in emissions from energy production and use accounts for most of the growth and today constitutes two-thirds of the national total, as shown in Fig. 5.8 (Veysey, 2016).

Thus models that opt for CCS instead of renewables estimate that it is a more economical way to achieve deep emission reductions (accounting for issues such as intermittency from renewable sources, transmission, and distribution costs and so on). The possibility of negative net GHG emissions with biomass CCS is also highlighted by Veysey (2016), as it has a significant option value, and provides flexibility in sectors where reduction is difficult (e.g., transport).

A key issue in this context is the long-term perspective for CCS in general and biomass in particular. Many of the models rely on CCS to reduce the carbon intensity of electricity in the 50% reduction scenario, and several use the biomass CCS to create new carbon sinks, offsetting emissions in other areas (Veysey, 2016).

In Mexico, there are also opportunities to use CCS from new projects for power plants with combined natural gas cycle (NGCC), which started operating between 2016 and 2020, as shown in Fig. 5.9 (González-Díaz, 2017). These projects are suitable to incorporate the process of carbon capture. For that, it is necessary to determine how far from the oil fields these plants would be located, and how much CO_2 would be generated (González-Díaz, 2017). The identification of suitable future plants to incorporate CO_2 capture is important for Mexico to achieve its emissions mitigation goal (González-Díaz, 2017).

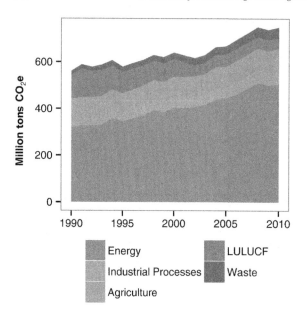

FIGURE 5.8 Mexican greenhouse gas emissions by source. *Source: From Veysey, J., et al. (2016). Pathways to Mexico's climate change mitigation targets: a multi-model analysis. Energy Economics. 56, 587–599.*

FIGURE 5.9 Location of the new natural gas combined cycle power plant projects, whose operation started in 2016–20. *Source: Adapted from González, D.A., et al. (2017). Priority projects for the implementation of CCS power generation with enhanced oil recovery in Mexico. International Journal of Greenhouse Gas Control 64, 119–125.*

The amount of CO_2 that would be produced only by the new NGCC projects located within the carbon capture use and storage (CCUS) inclusion zone would be 29.64 $MtCO_2$/year, with 90% of the CO_2 generated by these plants being abated at the point of emission if a process to capture CO_2 were incorporated. Other plants are also candidates for a CCUS retrofit or to be prepared for CCUS (González-Díaz, 2017).

In Mexico, there are still other plants located a short distance from oil fields in the Gulf of Mexico. Capturing 90% of the CO_2 emitted by these plants would result in approximately 13.35 $MtCO_2$/year, which can be supplied to EOR (González-Díaz, 2017). The remaining CO_2 emissions generated by the plants located farther from these oil fields could then be connected to existing EOR projects in a second phase, or geological storage in nonhydrocarbon reservoirs. In effect, such plants could be considered priority CCS-EOR projects, as they would provide economic benefits from additional oil production and would provide experience and infrastructure for future CO_2 storage (González-Díaz, 2017).

There is also an emission of 20.1 $MtCO_2$/year from existing plants, industries, and refineries as potential primary sources, which, if added to the 29.64 $MtCO_2$/year of new NGCC plants located in the inclusion zone, could be used for EOR projects (González-Díaz, 2017). Evidently, therefore, the demand for CO_2-EOR in oil fields in the Gulf of Mexico, estimated at approximately 50 $MtCO_2$/year could be met (González-Díaz, 2017).

5.4.4 Trinidad and Tobago

Trinidad and Tobago (T&T) are the southernmost islands in the Caribbean, located close to the coast of Venezuela. The country's economy is heavily based on oil and gas, being one of the largest producers of fossil fuels in the Caribbean. T&T also emits millions of tons of carbon dioxide a year. This value is expected to increase due to the growth of the manufacturing sector in the country (Alexander, 2011). The country's relatively small population places T&T among the world leaders with regard to CO_2 emitted per capita (Alexander, 2011). Also, small island states, such as T&T and the Caribbean in general, are the most vulnerable to coastal erosion, rising sea levels, and floods due to intense rains, expected to be more severe if the climate warms up (Alexander, 2011).

A survey conducted in 2010 indicated that, although a larger number of participants felt that CCS was safer than dangerous (28% vs 25%) and the associated risks were known to science, a significant majority (90%) was still very concerned about leakage and environmental issues (Alexander, 2011).

This technology may have practical relevance for T&T since the main sources of pure CO_2 emissions within the country are located at Point Lisas Industrial Estate with several ammonia production facilities (Boodlal, 2014). These sources are relatively close to potential geological storage sites, occurring in land in Fyzabad and Oropouche (less than 50 km south from Point Lisas) and off the southeast coast of the island (less than 150 km southeast of Point Lisas) (Boodlal, 2014).

The latest T&T inventory suggests that most T&T CO_2 emissions emanate from the petrochemical and power generation sector, with the petrochemical sector accounting for around 56% (Boodlal, 2014). The entire petrochemical sector is concentrated in a central location in the industrial estate of Point Lisas (Boodlal, 2014). This geographic area can be considered the main "area of origin" for CO_2 emissions in T&T and this "area of origin" is relatively close to possible geological storage sites. These potential locations include oil fields and gas fields (Boodlal, 2014). Also, transportation and compression costs would be relatively lower in T&T due to the relatively inexpensive cost of electricity and the proximity to storage locations (Boodlal, 2014). Also, the country has considerable experience in the production of hydrocarbons and more than 30 years of experience in CO_2 injection for improved oil recovery (Boodlal, 2014).

Boodlal and Taweel suggested that the total volume of annual CO_2 emissions available for use at Point Lisas industrial estate of the 11 ammonia plants is around 6 million tons (including the reused quantities under consideration) (Boodlal, 2014). We here assume one-third of that amount or 2 million tons of CO_2 (6.25%) of the total annually (Boodlal, 2014).

The fact that CO_2 generation is concentrated in close locations and is generated by ammonia production facilities is advantageous, since all related CO_2 emissions generated are already "captured" or separated, as this step is inherent in the process of ammonia synthesis (Seetahal, 2014). Therefore about 80% of the investment required for a CCS is eliminated. Mahogany Field is a deep saline aquifer located on the southeast coast of T&T and is less than 155 km from the Point Lisas estate, further reducing the costs of implementing this technology (Seetahal, 2014).

5.5 Efforts and opportunities to implement CCS projects

The Global CCS Institute periodically analyzes some processes and the degree of adherence of each country to the CCS technology can be verified, be it in the political, technical, legal, and environmental spheres. These analyses allow, for instance, correlating how close or far countries are to use and to develop CCS as a technology for reducing GHG emissions and meeting their commitments under the Paris Agreement. In this vein, Table 5.4 provides a brief definition of each of these parameters.

From the information collected on the website, a heat map was produced, as shown in Fig. 5.10, which shows the trend in the implementation and use of CCS in each country selected according to the parameters analyzed by the Global CCS Institute. The selected countries are those that have carbon storage facilities in operation in addition to the countries covered in this chapter. The warm colors indicate that the country has a high score on the parameter analyzed, while the cold colors indicate the opposite.

Analyzing the heat map, Mexico and Brazil are observed to have a reasonably good level of compliance with the parameters compared to other countries, demonstrating that the use and implementation of CCS projects are possible and viable given the parameters analyzed. Nevertheless, it is important to note that, although Trinidad and Tobago has a low score for most parameters, there are chances of successfully using and implementing CCS in the country, as shown in other points in this chapter. However, there is no doubt that investments and development are necessary, such as the implementation of stronger and more forceful policies.

From this analysis, we selected five challenges and opportunities for implementing CO_2 storage facilities in the selected countries:

5.5.1 Specific legislation for using CCS technology

In 2018 the Global CCS Institute revised the CCUS-Legal and Regulatory Indicator, an index that offers a detailed assessment of the legal and regulatory frameworks and classification of each country, focusing on a broad spectrum of administrative and permission agreements throughout the year (Havercroft, 2018). Project life cycle includes issues related to public consultations, environmental impacts, and long-term responsibility, in which out of the 55 countries analyzed, Mexico ranks 26th while Brazil ranks 36th (Havercroft, 2018).

As an evaluation methodology, the Global CCS Institute indicator uses a comparison between the various models and contrasts national circumstances to determine the scope of an individual structural legal, and regulatory jurisdiction for implementing CCUS Projects.

TABLE 5.4 Definition of each process monitored by the Global CCS Institute.

Monitored process	Description
Policy Indicator Database	Carbon Capture and Storage (CCS) Policy Indicator records an individual nation's CCS policy development. The Indicator tracks a broad spectrum of policies ranging from direct support for CCS to broader implicit climate change and emission reduction policies. The resulting Indicator score represents a comprehensive model for tracking progress and opportunities for the development of policies to support CCS deployment.
Storage Indicator Database (Index)	CCS Storage Indicator records an individual nation's development of its storage resources. The Indicator evaluates a country's geological storage potential, maturity of their storage assessments, and progress in the deployment of CO_2 injection sites.
Legal and Regulatory Database (Index)	The CSS Legal and Regulatory Indicator offers a detailed examination and assessment of a nation's legal and regulatory frameworks. The Indicator focuses upon a broad spectrum of administrative and permitting arrangements across the project life cycle, including issues related to environmental assessments, public consultation, and long-term liability.
CCS Requirement Indicator (Index)	CCS Requirement Indicator is a relative index based on the global share of fossil production and consumption. The Indicator provides one guide of a country's need to deploy CCS to reduce their emissions from fossil fuels.
CCS Readiness Requirement (Index)	The CCS Readiness Index actively monitors the progress of CCS deployment. The Index tracks a country's requirement for CCS, its policy, law and regulation, and storage resource development. Through these indicators, the Rl identifies those nations which are leaders in the creation of an enabling environment for the commercial deployment of CCS.
Storage Resource Database (GtCO$_2$)	Storage resources map defines the broad overview of national storage estimates around the world. Published estimates include deep saline formations, depleted oil and gas fields, and CO_2-EOR estimates. Resource estimates are not standardized.

Elaborated by authors from the Global CCS Institute.

Monitored Process Country	Policy Indicator Database (Index)		Storage Indicator Database (Index)		Legal & Regulatory Database (Index)		CCS Requirement Indicator (Index)		CCS Readiness Requirement (Index)		Storage Resource Database (GtCO2)
	2018	2019	2018	2019	2018	2019	2018	2019	2018	2019	2020
Australia	23	22	36	36	77	77	57	57	62	61	100+
Brazil	9	10	86	86	34	34	43	42	43	43	100+
Canada	40	40	36	35	75	75	48	48	71	71	100+
China	40	40	87	91	32	32	86	86	53	54	100+
Colombia	NA	NA	NA	NA	NA	NA	40	40	NA	NA	NA
Mexico	16	16	54	61	47	47	42	42	39	41	100+
Norway	56	56	96	96	45	45	26	26	65	65	10-100
Saudi Arabia	11	11	79	79	12	12	23	23	34	34	10-100
Trinidad and Tobago	1	1	26	26	37	37	14	14	21	21	NA
United States	41	41	96	96	73	73	82	82	70	70	100+
Scale	Min.0	Max.100	Min.0	Max.100	Min.0	Max.100	Min.0	Max.86	Min.0	Max.100	Min.<1 Max.>100

FIGURE 5.10 Heat map of Carbon Capture and Storage usage and deployment trend in selected countries. Source: *Elaborated by authors from the Global CCS Institute.*

Mexico is thus observed to be developing a policy dedicated to CCS. In March 2014 it launched its Roadmap for CCUS, as can be seen in Fig. 5.11. This policy includes, for example, the adoption of a national policy to make the new large CO_2-emitting industrial facilities ready for CCUS (González-Díaz, 2017).

II. Case studies on CCS and related policies, and their consequences for climate change

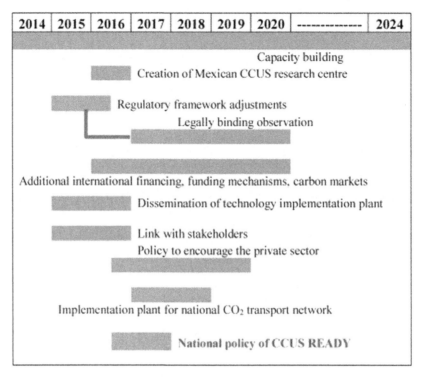

| 2014 | 2015 | 2016 | 2017 | 2018 | 2019 | 2020 | -------------- | 2024 |

Capacity building
Creation of Mexican CCUS research centre

Regulatory framework adjustments
Legally binding observation

Additional international financing, funding mechanisms, carbon markets

Dissemination of technology implementation plant

Link with stakeholders
Policy to encourage the private sector

Implementation plant for national CO$_2$ transport network

National policy of CCUS READY

FIGURE 5.11 Public policy action taken from 2014 to 2024. *CCUS*, carbon capture use and storage. *Source: From González, D.A., et al. (2017). Priority projects for the implementation of CCS power generation with enhanced oil recovery in Mexico. International Journal of Greenhouse Gas Control 64, 119–125.*

5.5.2 Idle assets

In this case, the focus would be on idle assets and investments in the energy sector, an important one in the context of climate change mitigation. Natural gas and oil plants without carbon capture and storage (CCS) represent the largest fraction of the decommissioned capacity prematurely in LAC (Binsted, 2020). Natural gas without CCS represents about 45% of the idle capacity in the NDCs scenario at 2°C and about 54% of the idle capacity in the NDCs scenario at 1.5°C, questioning the role of natural gas as a "transition fuel." To meet the growing demand for electricity, between 751 and 967 GW of new capacity are to be installed over the 30-year period, from 2021 to 2050 (Binsted, 2020). These capacity additions are approximately 1.9–2.5 times the total generation capacity of electricity in LAC in 2015 (Binsted, 2020). As expected, the scenarios with the cumulative 1.5°C emission budget require more capacity additions than the 2°C scenarios. This is because the stricter budget reach in the 1°C requires the electricity sector to decarbonize more quickly, replacing carbon-intensive plants with new low-carbon capacity and producing more electricity in general (Binsted, 2020). Therefore energy-intensive sectors can reduce emissions by changing their energy use to electricity (Binsted, 2020).

5.5.3 Committed emissions from operational and planned generators

A study completed in 2019 found that there are about 4146 electric power generating facilities in LAC that use coal, peat, and oil shale (coal for short), natural gas, or oil as the main fuel. This comprises 169 GW of the fossil-based capacity (González et al., 2019). Mexico and Argentina lead the natural gas capacity with 44 and 23 GW, respectively (González et al., 2019). Mexico and Chile have the largest coal capacity with 6 and 4.9 GW, respectively (González et al., 2019). Brazil and Mexico lead the oil capacity with 11 and 6.7 GW, respectively. Fig. 5.12 shows the technology planned and operational capacity in LAC (González et al., 2019).

The figure also shows large quantities of fossil fuel plants, mainly based on natural gas, which will start operating between 2019 and 2022. The peak in 2022 results from the plants in the bidding process starting operations, especially in Brazil (31 GW) (González et al., 2019).

The study reports 456 planned fossil-based generators, adding up to 102 GW or 61% of the current fossil fuel capacity in the region. Most of the planned fossil fuel plants are natural gas ones (87 GW), followed by coal shale, peat, and oil (13.5 GW), and oil (2.1 GW) (González et al., 2019). Brazil leads the fossil-based plant, with 38 GW of natural gas, 4.8 GW of coal, and 0.9 GW of oil. Mexico and Chile have 22 and 6.7 GW of natural gas capacity in their pipelines, respectively (González et al., 2019). Compromised gas pipeline emissions are dominated by natural gas (63%), followed by coal (26%) (González et al., 2019).

Fig. 5.13 provides details of the committed emissions from existing and planned plants by country (González et al., 2019). Mexico, Argentina, and Brazil lead the committed emissions from operating generators, with 1.8, 1, and 0.9 GtCO$_2$, respectively (González et al., 2019). If planned plants are built, Brazil would become the main contributor to committed emissions in the region, with 2.7 GtCO$_2$, almost tripling the committed emissions from its operational generators (González et al., 2019).

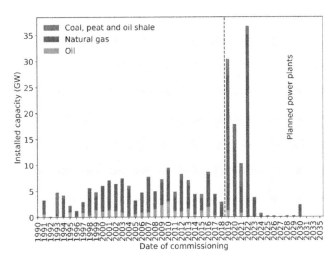

FIGURE 5.12 Capacity by date of commissioning. The bars in 2019 and later correspond to the planned plants. We trace data from 1990; however, the database includes units that started operating before that. Source: *From González, E., et al., 2019. Committed emissions and the risk of stranded assets from power plants in Latin America and the Caribbean. Environmental Research Letters 14 (12).*

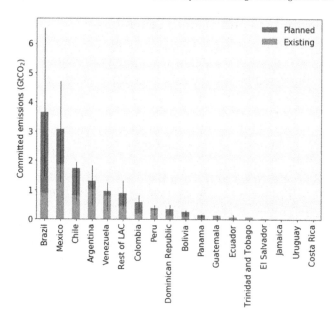

FIGURE 5.13 Committed emissions from existing and planned power plants by country (2019−64). Source: *From González, E., et al., 2019. Committed emissions and the risk of stranded assets from power plants in Latin America and the Caribbean. Environmental Research Letters 14 (12).*

Committed emissions from plants under construction and the bidding status add up to 4.3 $GtCO_2$, while authorized and announced would add 1.4 and 0.9 $GtCO_2$, respectively (González et al., 2019). More than half (62%) of the committed emissions from the planned plants come from natural gas generators, which would add 4.2 $GtCO_2$ (González et al., 2019). The largest pieces would be added by Brazil (1.9 $GtCO_2$) and Mexico (1.1 $GtCO_2$) (González et al., 2019).

Mexico consumes the largest number of fossil fuels of all Latin American countries, contributing about 1.4% of total global GHG emissions (González et al., 2019). The Mexican government has proposed actions to monitor and to reduce its net GHG emissions into the atmosphere (González et al., 2019). In 2012 Mexico established a comprehensive General Law on Climate Change, which determined the design and implementation of a national monitoring, reporting, and verification system (Olguin, 2018).

5.5.4 Presalt and Salt Cave

The CCS technology in Brazil will be relevant to the industry and can be crucial for the development of some fields in the Presalt, since some wells showed CO_2 concentrations well above those found in the Campos Basin (Beck et al., 2011). In general, the presence of CO_2 in the hydrocarbon found in the fields of the Presalt Cluster is between 8% and 12%, considered significant in comparison with the composition of hydrocarbons in other locations (Câmara, 2011). Fig. 5.14 shows the presalt areas in the Campos and Santos Basins.

The oil in the presalt reservoirs in Brazil has a very high gas−oil ratio, with a high carbon dioxide content. Some of these gases are treated and separated on the platform using the membrane technology, reducing the CO_2 content to 3%, and it is possible to transfer

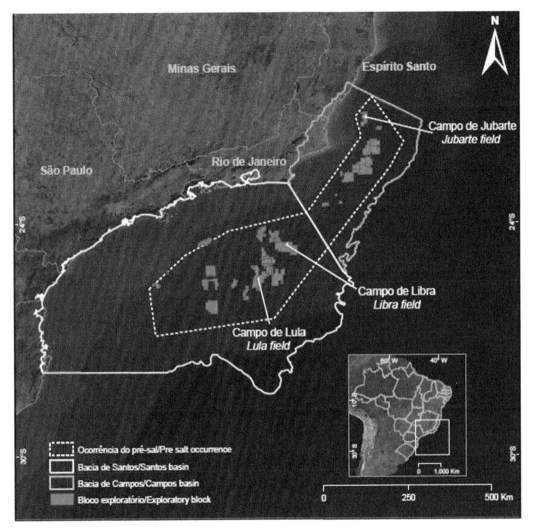

FIGURE 5.14 Presalt area in the Santos and Campos Basins. Source: *From Ketzer, J.M.M. et al., 2016. Atlas brasileiro de captura e armazenamento geológico de CO_2. Edipucrs, Porto Alegre.*

part of the natural gas to the coast through carbon steel pipes (González et al., 2019). The remainder, containing mainly CO_2, comes back and is reinjected into the reservoir. At the beginning of the useful life of the fields, the reinjection of this gas is used as an EOR; however, as the same CO_2 molecules are recycled several times in the same drainage radius of the wells, the CO_2 content begins to increase significantly, hindering the treatment in the platform, which may force the closure of the production wells (González et al., 2019). In this scenario, the salt rock serves as a strategic geomaterial for the process of confining the gas stream with a high CO_2 content, since this gas can be injected and contained in salt caves, instead of being reinjected into the reservoirs (Maia, 2019).

Due to the expected increase in the production of gaseous hydrocarbons in the coastal region, in the Santos, Campos, and Espírito Santo Basins, besides the characteristics of certain fields, the use of the gaseous portion of which is not economically viable only through Natural Gas Processing Unit processing, it is necessary to process it in a technically safe way at viable costs (Costa, 2019). Thus in fields with a high content of CO_2 in the hydrocarbon, it is possible to carry out the production, separate the liquid and gaseous portions, and inject the latter, under conditions that physically separate chemically and at low cost carbon dioxide and methane, for example Costa (2019). The storage of the gaseous portion of sedimentary formations close to the producing fields is among the eligible technical solutions (Costa, 2019).

An emerging proposal seeks the location of structurally sf places viable for installing water injectors for dissolving part of the salt structure, accounting for the artificial formation of the salt cave (Costa, 2019). Then, the aforementioned gas is injected so as to maintain the stress conditions before dissolution as a necessary condition for the newly constructed structure not to collapse (Costa, 2019). As designed, salt caves can promote better use of energy resources (Costa, 2019).

Considering that a cave holds 1.5 million cubic meters of standard gas per day (3.840 billion kilos of CO_2 in a supercritical state), it could support the disposal of CO_2 from an oil production field, for example, for up to 7 years until its consequent closure (Costa, 2019). Thus a central disposal station that contains the order of 15 caves—and is located in an unstructured saline dome—could accumulate the CO_2 emitted from up to 15 fields in the same period, or fewer fields in a longer period, according to the convenience demanded by the surplus of the productive processes. The set would then contain more than 108 $MtCO_2$ (Maia, 2019).

Such storage capacity for the set of 15 caves demands occupying an area of approximately 8 km^2 (2.85 km × 2.85 km), since the safe distance between the caves is 250 m (Maia, 2019). There would then be 900 m between axes and 750 m between separation pillars, and each cave would be approximately 450 m high by 150 m in diameter (Maia, 2019). Thus, an appropriate location for the disposal center would be the Lula Dome, near Tupi, 10 km from the Lula discoverer well, serving the Santos and Sapinhoá Basin, 300 km away from the coast and 3440 m deep (3890 m) if the base with 450 m of height of each cave considered) (Maia, 2019).

5.5.5 Increasing production and refining of hydrocarbons

The countries selected have a common challenge: to manage carbon dioxide emissions from hydrocarbon production and refining.

Refining is responsible for 6% of the total global CO_2 emissions and represents one of the largest energy consumers. Although the goals of reducing carbon emissions in transport fuels have affected the operational strategies of many facilities, in addition to stimulating new investments in biorefineries, the sector needs to undergo major changes to face the challenges of decarbonization (Sachs, 2018). CCS is also considered to play a key role in this transition. In this sense, the risk aversion of investors makes it difficult to spread new and expensive technologies (Sachs, 2018).

Despite not having high production or oil refining, Trinidad and Tobago stands out in the production of natural gas, presenting a relatively high production, even higher than Brazil, as shown in Graph 5.6. Except for Mexico, Brazil, Trinidad and Tobago, and Colombia have prospects for increasing natural gas production. As Trinidad and Tobago exports most of its natural gas production, the greatest challenge would be to Brazil and Colombia, as the trend is for this gas to be used in electricity generation facilities, contributing to a larger amount of carbon dioxide emissions if CCS is not implemented.

As to oil production, the Brazilian production stands out, having become the largest oil producer in LAC, as shown in Graph 5.7. The Brazilian production derives mostly from the presalt fields, which have a considerably higher CO_2 compared to the postsalt. The implementation of CCS technologies, including as an EOR method, will be necessary for Brazil to reach the NDCs stipulated in the Paris Agreement.

Globally, 4% of the total anthropogenic CO_2 emissions are released by the oil refining sector. CO_2 capture and storage (CCS) is a technology option with recognized potential for mitigating CO emissions (Yanez, 2020).

Regarding oil refining, as shown in Graph 5.8, Brazil also stands out for two reasons. The first is related to the increase in recent decades—with a decline in recent years due to the economic crisis experienced in the country—but which is tending to increase again, mainly due to the construction of new refineries, such as Abreu e Lima and Petrochemical Complex of Rio de Janeiro. The second point refers to the privatization of part of the Brazilian refineries, currently monopolized by Petrobras. From the privatizations, with the decentralization of state power—according to the incentives given by the government to the private sector—there is a trend to invest in improving the operational performance of refineries, through retrofits, and implementation of new projects such as

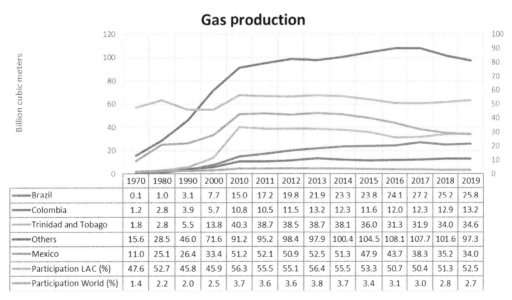

Gas production

	1970	1980	1990	2000	2010	2011	2012	2013	2014	2015	2016	2017	2018	2019
Brazil	0.1	1.0	3.1	7.7	15.0	17.2	19.8	21.9	23.3	23.8	24.1	27.2	25.2	25.8
Colombia	1.2	2.8	3.9	5.7	10.8	10.5	11.5	13.2	12.3	11.6	12.0	12.3	12.9	13.2
Trinidad and Tobago	1.8	2.8	5.5	13.8	40.3	38.7	38.5	38.7	38.1	36.0	31.3	31.9	34.0	34.6
Others	15.6	28.5	46.0	71.6	91.2	95.2	98.4	97.9	100.4	104.5	108.1	107.7	101.6	97.3
Mexico	11.0	25.1	26.4	33.4	51.2	52.1	50.9	52.5	51.3	47.9	43.7	38.3	35.2	34.0
Participation LAC (%)	47.6	52.7	45.8	45.9	56.3	55.5	55.1	56.4	55.5	53.3	50.7	50.4	51.3	52.5
Participation World (%)	1.4	2.2	2.0	2.5	3.7	3.6	3.6	3.8	3.7	3.4	3.1	3.0	2.8	2.7

GRAPH 5.6 Historical survey of gas production. Source: *Elaborated by authors from British Petroleum, 2020. Statistical Review of World Energy (accessed 15.07.20).*

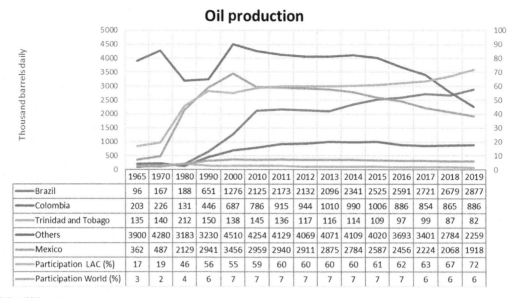

	1965	1970	1980	1990	2000	2010	2011	2012	2013	2014	2015	2016	2017	2018	2019
Brazil	96	167	188	651	1276	2125	2173	2132	2096	2341	2525	2591	2721	2679	2877
Colombia	203	226	131	446	687	786	915	944	1010	990	1006	886	854	865	886
Trinidad and Tobago	135	140	212	150	138	145	136	117	116	114	109	97	99	87	82
Others	3900	4280	3183	3230	4510	4254	4129	4069	4071	4109	4020	3693	3401	2784	2259
Mexico	362	487	2129	2941	3456	2959	2940	2911	2875	2784	2587	2456	2224	2068	1918
Participation LAC (%)	17	19	46	56	55	59	60	60	60	60	61	62	63	67	72
Participation World (%)	3	2	4	6	7	7	7	7	7	7	7	7	6	6	6

GRAPH 5.7 Historical survey of oil production. Source: *Elaborated by authors from British Petroleum, 2020. Statistical Review of World Energy (accessed 15.07.20).*

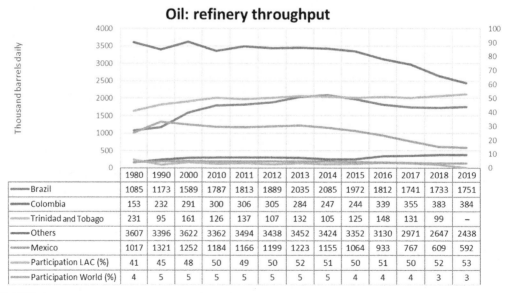

	1980	1990	2000	2010	2011	2012	2013	2014	2015	2016	2017	2018	2019
Brazil	1085	1173	1589	1787	1813	1889	2035	2085	1972	1812	1741	1733	1751
Colombia	153	232	291	300	306	305	284	247	244	339	355	383	384
Trinidad and Tobago	231	95	161	126	137	107	132	105	125	148	131	99	–
Others	3607	3396	3622	3362	3494	3438	3452	3424	3352	3130	2971	2647	2438
Mexico	1017	1321	1252	1184	1166	1199	1223	1155	1064	933	767	609	592
Participation LAC (%)	41	45	48	50	49	50	52	51	50	51	50	52	53
Participation World (%)	4	5	5	5	5	5	5	5	4	4	4	3	3

GRAPH 5.8 Historical survey of refinery throughput. Source: *Elaborated by authors from British Petroleum, 2020. Statistical Review of World Energy (accessed 15.07.20).*

CCS, since the focus of these companies currently tends to remain downstream, in the refinery itself, contrary to what is practiced by Petrobras. This understanding is ratified by Eakin et al. (2006), which associates the development and dissemination of new technologies in countries, such as those in LAC, to private international investment. However, the literature focused on environmental technologies (e.g., "specific adaptations") suggests that, for these private initiatives to be successful, a "generic" enabling environment must be created by a strong state.

The transfer of technology to address climate change concerns involves not only the movement of knowhow from one place to another but also the managerial, technical, and economic skills of all the actors involved, including public bodies (Eakin et al., 2006). Government responsibility is particularly important especially as regards vulnerable sectors that may not be able to independently control technologies suited to their specific circumstances and needs (Eakin et al., 2006).

5.6 Paths built and directions followed by countries in meeting the Paris Agreement goals and conclusions

As noted throughout the chapter, all four countries signed the Paris Agreement, implemented national policies, and built clear plans to reduce CO_2 emissions, but without taking into account clearly and emphatically the use of CCS as a relevant factor for the compliance with the stated goals, as verified in the data in Fig. 5.10. The fact is confirmed when analyzing Graph 5.1, to which the small reduction in CO_2 emissions seems to be much more related to the decrease in economic activity in recent years in the region than to the compliance with the Paris Agreement NDCs.

All four countries are apathetic about CCS—they know the benefits of using the technology and its importance but are not willing to develop and deploy the technology in a broad and definitive way.

The opportunities of these countries arising from the share of almost three-quarters of oil production (Graph 5.7) and more than half of gas production in LAC (Graph 5.6), as well as the geological possibility of CO_2 storage (Fig. 5.5) and preliminary studies described in this chapter—which present several options for the implementation of CCS—show that there is an exceptional path under construction in these countries that provide a possibility of success in the implementation of CCS.

However, to follow the correct direction of this path—in the effective and consistent reduction of CO_2 emissions—it is necessary for governments to regulate and encourage CCS activities; stakeholders to invest and develop projects; society to know the importance of this technology in the context of global warming and the security involved in CO_2 storage; and the academy to explore and break the frontier of knowledge as a way to promote CCS, which has been the objective of this chapter.

5.7 Final considerations

From the data collected and analyses carried out, it can be seen that Colombia, Brazil, Mexico, and Trinidad and Tobago have unique opportunities for developing projects and policies using CCS technology, as well as presenting several challenges to be analyzed and resolved. One of the biggest challenges will be to meet the targets ratified by each of these countries in the Paris Agreement regarding the reduction in CO_2 emissions, to which the CCS may be a protagonist in meeting these goals. As verified, the policies and programs implemented to fulfill the goals of the Paris Agreement are not sufficient for the success of the presented NDCs, therefore it is necessary to develop and implement CCS quickly.

Colombia is perhaps one of the countries that most need to develop and use CCS technology due to its constant oil production and refining, in addition to its increasing CO_2 emissions into the atmosphere. The country also has considerable geological CO_2 storage capacity, as well as a network of relatively interconnected and nearby CO_2 emitters and storage sites.

Brazil has unique opportunities in terms of CO_2 storage capacity in the presalt, either via EOR or in salt caves. Due to the territorial configuration, with the relative proximity between large CO_2 emitting centers and infrastructure already installed, it would facilitate the implementation of this technology, including the large oil refineries currently in the process of privatization. There is a clear need to implement specific policies and regulations so that the capture and carbon activity can be developed.

In this context, Mexico has already taken an important step toward regulating CO_2 capture and storage activity, taking advantage of three great opportunities available and covered in this chapter: large CO_2 storage capacity, adequacy, and construction of power plants already adapted to the use of CCS technology and taking advantage of the reduction of CO_2 emissions in the atmosphere to meet its goals.

Trinidad and Tobago has a high production of natural gas and a very reasonable geological capacity for CO_2 storage, even though it is an extremely small country, making this an unique opportunity to develop the application of CCS technology. Although the use is not globally representative, the implementation of this technology in the country would be dramatically impactful, considerably reducing its CO_2 emissions into the atmosphere, and could serve as a model for other smaller countries with large hydrocarbon production, such as Brunei and Qatar.

The construction of a policy and regulation geared to the activity of capturing and storing CO_2 is undoubtedly one of the great challenges presented by all four countries analyzed in this chapter. The need for public incentives, be it through subsidies or laws, can be an opportunity for increasing the number of CO_2 storage facilities, especially for using EOR.

There are several other opportunities to be explored, such as further study of CO_2 storage capacities in reservoirs, promoting knowledge among the general public and stakeholders about the technology, among others, which can be explored and developed throughout the region.

Acknowledgments

We are grateful to the "Research Center for Gas Innovation—RCGI" (Fapesp Proc. 2014/50279-4), supported by FAPESP and Shell, organized by the University of São Paulo, and the strategic importance of the support granted by the ANP (National Agency of Petroleum, Natural Gas and Biofuels of Brazil) through the R&D clause. We also thank the support from the National Agency for Petroleum, Natural Gas and Biofuels Human Resources Program (PRH-ANP), funded by resources from the investment of oil companies qualified in the R, D&I clauses from ANP Resolution No. 50/2015 (PRH 33.1—Related to Call No. 1/2018/PRH-ANP; Grant FINEP/FUSP/USP Ref. 0443/19).

References

Alexander, D., et al., 2011. Employing CCS technologies in the Caribbean: a case study for Trinidad and Tobago. Energy Procedia 4, 6273−6279.

Beck, B., et al., 2011. The current status of CCS development in Brazil, Energy Procedia, 4.

Binsted, M., et al., 2020. Stranded asset implications of the Paris Agreement in Latin America and the Caribbean. Environmental Research Letters 15 (4).

Boodlal, D., Alexander, D., 2014. The impact of the clean development mechanism and enhanced oil recovery on the economics of carbon capture and geological storage for Trinidad and Tobago. Energy Procedia 63, 6420−6427.

British Petroleum. 2020. Statistical Review of World Energy (accessed 15.07.20).

Calderón, S., et al., 2016. Achieving CO_2 reductions in Colombia: effects of carbon taxes and abatement targets. Energy Economics 56, 575−586.

Câmara, G., et al., 2011. Tecnologia de armazenamento geológico de dióxido de carbono: panorama mundial e situação brasileira. Revista Eletrônica Sistenas & Gestã 6 (3), 238−253.

Clarke, L., et al., 2016. Long-term abatement potential and current policy trajectories in Latin American countries. Energy Economics 56, 513−525.

Costa, H.K.M., et al., 2019. Legal aspects of offshore CCS: case study—Salt Cavern. Polytechnica 2 (1−2), 87−96.

Eakin, H., Lemos, M.C., 2006. Adaptation and the state: Latin America and the challenge of capacity-building under globalization. Global Environmental Change 16 (1), 7−18.

EASAC, 2013. European Academies Science Advisory Council—Carbon Capture and Storage in Europe. EASAC policy report 20. ISBN: 978-3-8047-3180-6. May 2013. Halle.

Gaspar, H., Alexandre, C., 2014. Captura e Armazenamento de CO_2. Faculdade de Ciências e Tecnologia Universidade Nova de Lisboa, Lisboa.

Goraieb, C.L., Iyomassa, W.S., Appi, C.J., 2005. Estocagem Subterrânea de Gás Natural: Tecnologia Para Suporte ao Crescimento do Setor de Gás Natural no Brasil. Editor: IPT, São Paulo, SP, p. 226.

Huaman, R.N.E., Jun, T.X., 2014. Energy related CO_2 emissions and the progress on CCS projects: a review. Renewable and Sustainable Energy Reviews 31, 368−385.

González, E., et al., 2019. Committed emissions and the risk of stranded assets from power plants in Latin America and the Caribbean. Environmental Research Letters 14 (12).

González, D.A., et al., 2017. Priority projects for the implementation of CCS power generation with enhanced oil recovery in Mexico. International Journal of Greenhouse Gas Control 64, 119−125.

Havercroft, I., 2018. CCS Legal and Regulatory Indicator (CCS-LRI). Global Institute.

IEA, 2010. Carbon Capture and Storage—Model Regulatory Framework. OECD/IEA, Paris, 130pp.

IPCC, 2005. In: Metz, B., et al., (Eds.), Carbon Dioxide Capture and Storage, Special Report. Cambridge University Press, United Kingdom.

Ketzer, J.M.M. et al., 2016. Atlas brasileiro de captura e armazenamento geológico de CO_2. Edipucrs, Porto Alegre.

Maia, A.C., et al., 2019. Experimental salt cavern in offshore ultra-deep water and well design evaluation for CO_2 abatement. International Journal of Mining Science and Technology 29 (5), 641−656.

Mello, L., et al., 2013. Oxycombustion technology development for fluid catalytic crackers (FCC)—large pilot scale demonstration. Energy Procedia 37, 7815−7824.

Milani, E.K., et al., 2007. Bacia do Paraná. Boletim de Geociências da Petrobras 15 (2), 265−287.

NOAA—National Oceanic and Atmospheric Administration, 2020. Climate Change: Atmospheric Carbon Dioxide. < https://www.climate.gov/news-features/understanding-climate/climate-change-atmospheric-carbon-dioxide > (accessed 15.07.20).

Nunes, R.C., Costa, H.K.M., 2020. An overview of international practices for authorization and monitoring CO_2 storage facilities. International Journal of Advanced Engineering Research and Science 7 (6).

Nunes, R., Costa, H.K.M., 2019. Operação e fechamento de instalações de armazenamento para atividades de CCS no Brasil. In: Costa, H.K.M. (Ed.), Aspectos Jurídicos da Captura e Armazenamento de Carbono. Lumen Juris, Rio de Janeiro.

Olguin, M., et al., 2018. Applying a systems approach to assess carbon emission reductions from climate change mitigation in Mexico's forest sector. Environmental Research Letters 13 (3).

Rockett, G.C., et al., 2011. Brazilian renewable carbon capture and geological storage map: possibilities for the Paraná Basin. Energy Procedia 37, 6105–6111.

Sachs, J., et al., 2018. The role of CCS and biomass-based processes in the refinery sector for different carbon scenarios. Computer Aided Chemical Engineering 43, 1365–1370.

Seetahal, S., Alexander, D., 2014. A real options analysis of carbon dioxide sequestration for Trinidad and Tobago: a case study of the Mahogany field. Energy Procedia 63, 7242–7246.

The Global Status of CCS, 2017. Disponível em: <https://www.globalccsinstitute.com/wp-content/uploads/2018/12/2017-GlobalStatus-Report.pdf> (Acessado em: 01.03.20).

The Global Status of CCS, 2018. <https://www.globalccsinstitute.com/wp-content/uploads/2018/12/Global-Status-of-CCS-Report-2018_FINAL.pdf> (accessed 01.07.20).

Veysey, J., et al., 2016. Pathways to Mexico's climate change mitigation targets: a multi-model analysis. Energy Economics 56, 587–599.

Yanez, E., 2020. Exploring the potential of carbon capture and storage-enhanced oil recovery as a mitigation strategy in the Colombian oil industry. International Journal of Greenhouse Gas Control 94.

Zoback, M., 2007. Reservoir Geomechanics. Cambridge University Press, Cambridge. Available from: http://doi.org/10.1017/CBO9780511586477.

Further reading

Global CCS Institute. CO2RE. <https://co2re.co/> (accessed 15.07.20).

Human Development Reports. <http://www.hdr.undp.org/en/countries> (accessed 15.07.20).

IBGE. Comparar Países. <https://paises.ibge.gov.br/#/mapa/comparar/brasil?lang=pt> (accessed 15.07.20).

Geologic CO_2 sequestration in the United States of America*

Owen L. Anderson

The University of Texas at Austin School of Law, Austin, TX, United States

Unlike more "green" outlooks, this article assumes that demand for oil and gas will continue to increase, although at a somewhat slower rate, even if personal vehicles continue to transition away from gasoline to electric energy.[1] Indeed, oil production is projected to increase by 12% to 2030 when it needs to decline by 40% to meet the 1.5°C global warming target without efforts to capture CO_2 (IEA, 2018). Thus, if this oil projection holds true, carbon capture, utilization, and storage (CCUS) and other efforts to reduce CO_2 will be needed on a massive scale if the limit on temperature increase is to be achieved. As with all new ventures, cost and risk are huge concerns. Although "fossil fuel companies cannot afford carbon capture in the short-term, ... they know they need it to

---

* This chapter is an adaptation and update of prior publications originally coauthored by Owen L. Anderson and R. Lee Gresham.
Owen L. Anderson is a Professor and Distinguished Oil and Gas Scholar at The University of Texas School of Law. He is the Eugene Kuntz Chair in Oil, Gas and Natural Resources Emeritus and George Lynn Cross Professor Emeritus at The University of Oklahoma College of Law. He is also a distinguished visiting professor at the University of Melbourne and the University of Dundee.

[1] A rapid transition away from oil and gas is far from assured. On January 22, 2019, Fatih Birol, Executive Director of the International Energy Agency (IEA), explained during a panel on "Strategic Outlook on Energy" at the World Economic Forum in Davos, that electric vehicles (EVs) today will not end the growth in world oil and gas demand and will not be they principle solution to reducing carbon emissions because much of the electricity used to charge them comes from fossil fuels. In its 2018 Global EV Outlook (IEA, 2018), the IEA projected that the number of electric vehicles could 125 million by 2030 under its New Policies Scenario. But this compares with over 1 billion vehicles on the road—a number that is also projected to increase. Moreover, real growth in demand for oil are trucks, petrochemical plants, and aircraft.

Carbon Capture and Storage in International Energy Policy and Law
DOI: https://doi.org/10.1016/B978-0-323-85250-0.00003-7

survive in the long term" (Tomlinson, 2020). In the United States, CCUS is being encouraged by what are called "45Q" tax credits, available to taxpayers who implement qualified CCUS projects.[2]

CO$_2$ is geologically sequestered by capturing[3] and separating it[4] from other substances, purifying and compressing it into to a supercritical fluid, which is then transported and injected approximately a kilometer or deeper into microscopic pore spaces in deep subsurface rock matrixes, such as those already retaining saltwater, or into petroleum reservoirs where it is utilized to facilitate enhanced recovery. The injected CO$_2$ flows through and fills the pore spaces in permeable layers of the rock matrix, while less permeable rock layers prevent its upward and downward migration. Depending on the formation's geology and the depth, porosity, and permeability of the injection zone, sequestered CO$_2$ from a single project potentially could spread over hundreds to thousands of square kilometers (see ExxonMobil, 2020; Pruess et al., 2001; Brennan and Burruss, 2006, p. 182; Gresham et al., 2010, p. 2900). Indeed, some CCUS projects could cross international or state/provincial boundaries. The former would require a treaty and the latter would, at a minimum, require cooperation, if not a formal agreement. The subsurface pressure effects could be felt over a much greater area, potentially raising earthquake concerns, like those in Oklahoma and Texas that result from saltwater injection operations associated with unconventional oil and gas exploitation. Deep coal veins and carbonaceous shales, such as the Marcellus in the Northeast United States, may also prove suitable for CCUS. Because carbon adsorbs carbon dioxide at a greater rate than methane, CO$_2$ injected into a natural gas reservoir for CCUS might simultaneously be used to recover additional natural gas in a process analogous to enhanced coalbed methane recovery; however, the practical value of this technique is not yet known (Wickstrom et al., 2008). Scientists believe that adsorption would allow sequestration at shallower depths than absorption in deep saline formations, which must be at least 800 m below the surface to maintain liquid CO$_2$ in a supercritical state.

Current resource estimates for sequestration in deep saline formations do not account for reduced capacity due to conflicting uses of pore space, such as fluid waste disposal. Thus, the commercial viability of CO$_2$ sequestration depends, in part, on how issues related to property rights in, and competing uses of, the subsurface are resolved.[5] Several commentators have weighed in on this issue.[6] In an effort to bring a theoretical basis for all subsurface uses,

---

[2] The credits increase to up to US\$50 per metric ton of CO$_2$ for qualified geologic storage or up to US\$35 per metric ton of CO$_2$ for utilizing it in a qualified matter, such as for enhanced oil recovery. These credits are discussed in more detail in Hutt et al. (2020).

[3] CO$_2$ can be captured by postcombustion, precombustion, or oxy-fuel combustion methods. The latter method allows for easier purification of the CO$_2$.

[4] In July 2020, ExxonMobil reported a significant breakthrough in the capture and separation process (ExxonMobil, 2020).

[5] Another key issue not addressed in this paper is placing a "value" on injected CO$_2$ to make it commercial.

[6] See generally Anderson and Gresham (2015), Anderson (2009, 2010). See also Gray (2015), Morgan (2013) (addressing English law); Havercroft et al. (2011), Fish and Martin (2010). For an article arguing that the mineral owner should own the pore space, see Kramer (2014).

Professor David Pierce has articulated a "community" view of the subsurface (Pierce, 2011a) that is consistent with the previously stated views of the author of this chapter. In the United States, prior to injecting CO_2 into the subsurface for permanent geologic sequestration, the injector must have a legal right to use the pore space that largely shields the injector from potential liability for trespass, nuisance, pollution, and other claims. In some countries, especially in the United States, takings law is implicated if a CCUS project developer receives regulatory authorization to use pore space that shields the developer from liability.

To facilitate CCUS as well as other useful subsurface enterprises on a worldwide basis, host governments (HGs) must codify a formal process for permitting access to, and use of, pore space for CCUS beneath both public and private lands. The law should limit the holder's liability for harm except for any actual and substantial damages caused directly and relatively promptly by the CCUS operations. To assure the long-term integrity of carbon sequestration reservoirs, HGs in some countries should allow CCUS developers to acquire a robust servitude property interest to secure the geologic integrity of their projects throughout the entire sequestration reservoir. Some HGs will need to support their permitting framework with eminent domain legislation akin to how some HGs have facilitated the underground storage of natural gas. Such a framework should facilitate the rapid development of commercial-scale CCUS projects by both standardizing procedures for acquiring the authorization to use pore space as well as by constraining acquisition costs.

6.1 Subsurface property rights and competing uses of the subsurface

For CCUS to enable the continued use of fossil fuels and simultaneous deep emission reductions, it must be widely deployed. To do this, the technology must be integrated into a larger commercial, legal, and regulatory scheme. Key considerations are (1) the amount of CO_2 to be injected is huge, for example, a one-gigawatt (GW) coal-fired power plant typically produces roughly six to eight million metric tons of CO_2[7] annually; (2) the areal footprint over which the injected CO_2 will migrate will be hundreds to thousands of square kilometers; and (3) the injected CO_2 must remain in the subsurface indefinitely—effectively occupying the subsurface pore space in perpetuity.

Throughout the world, subsurface activities vary extensively, as do the depths at which these industrial and commercial enterprises are carried out. CCUS would occur far deeper than ordinary subsurface uses, such as basements, buried power lines and pipelines, and tunnels. However, CCUS could interfere with oil and gas development, natural gas storage, fluid waste disposal, and deep groundwater recovery and storage (see U.S. Department of Interior, 2009).[8] Injected CO_2 is likely to spread throughout an

---

[7] Corresponds to 1 kg/kWh captured at 60% and 90% capture efficiency, respectively.

[8] See U.S. Department of Interior (2009) Framework for Geological Carbon Sequestration on Public Land. Report Submitted to the Committee on Natural Resources of the House of Representatives and the Committee on Energy and Natural Resources of the Senate in Compliance with Section 714 of the Energy Independence and Security Act of 2007 (P.L. 110−140, H.R.6). Washington, DC, 22 p. ("[C]arbon sequestration may potentially conflict with other land uses including existing and future mines, oil and gas fields, coal resources, geothermal fields, and drinking water sources," p. 1).

immense area (Gresham et al., 2010, p. 2900)—causing some CCUS projects to overlap with each other and with other subsurface uses. A key issue is how to accommodate competing and overlapping uses without compromising the long-term integrity of the CO$_2$ storage reservoir. Many initial CCUS projects are likely to occur as part of enhanced hydrocarbon recovery[9] projects.[10]

Subsurface formations with hydrocarbon-bearing strata are typically well-characterized and are often stacked between nonhydrocarbon-bearing saline aquifers (see, e.g., Doughty and Pruess, 2004). Thus CCUS will likely occur at depths like that of hydrocarbon reservoirs (U.S. Energy Information Administration, 2020a). The possibility of developing a CO$_2$ sequestration site above or below oil or natural gas reservoirs or within depleted reservoirs has the advantage of reducing characterization and capital costs compared to an uncharacterized site, but doing so could also create potential interference with hydrocarbon development.[11] State legislatures in several oil and gas producing US states[12] have attempted to create CCUS-specific law aimed, in part, in avoiding conflicts with other commercial uses of the subsurface, but some of these give CCUS a lower priority over oil and gas operations.[13]

The potential subsurface impacts of CO$_2$ injection are varied. In a reservoir with active hydrocarbon resource production, particularly natural gas, migrating CO$_2$ could comingle directly with the resource and require removal of the CO$_2$ from the production stream and reinjection (Zillman et al., 2014). Soluble CO$_2$ could cause the precipitation of carbonate minerals and plug flow paths, which would reduce the extraction efficiency for existing hydrocarbon production facilities

---

[9] In "enhanced" or "secondary" recovery operations, oil and gas producers inject fluids into the subsurface to repressurize the reservoir or to otherwise increase ultimate oil and gas production.

[10] When used to enhance oil recovery, an oilfield operator injects compressed CO$_2$ into an oil reservoir. The injected CO$_2$ acts like a solvent, which causes the oil to expand and flow more easily through the reservoir rock into production wells. For example, in the Weyburn oilfield in Saskatchewan, Canada, the operator injects CO$_2$ for several days and then injects water into about 70 injection wells in the field. The injected CO$_2$ allows the oil to expand and more readily flow through the reservoir rock into production wells. The injected water increases the reservoir pressure to help push the oil toward the production wells. Production in the Weyburn oilfield increased from 8000 barrels of oil per day in 1990 to nearly 30,000 barrels by 2005. Because the CO$_2$ is miscible with the oil, some CO$_2$ returns to the surface during oil production, but the operator separates the CO$_2$ from the oil and then compresses and reinjects it. Most of the CO$_2$ used in the injection operations is captured at a coal gasification facility in the State of North Dakota in the US. See Global CCS Institute (2014). For many years, naturally occurring CO$_2$ has been produced from wells in the San Juan Basin of New Mexico and Colorado and transported by pipeline to southwest Texas for use in enhanced hydrocarbon recovery.

[11] Benson et al. (2005, p. 210) (stating that the presence of CO$_2$ in the basin can lead to corrosion problems and can change the composition such that plugging, erosion, and processing problems arise).

[12] See, for example, Wyo. Stat. Ann. § 34-1-152 (2009), Wyo. Stat. Ann. §§ 34-1-152 and Wyo. Stat. Ann. §§ 34-1-313 − 34-1-317 (2009) 153. See generally Task Force on Carbon Capture and Geologic Storage (2007a). The Task Force has also drafted model regulations (Task Force on Carbon Capture and Geologic Storage, 2007a, pp. 36−47).

[13] For an international perspective, see generally Zillman et al. (2014).

(Benson and Cole, 2008, p. 328).[14] The pressure effects from the injection operation could adversely affect other injection operations by potentially altering the ability to inject, plume size and shape, and associated monitoring—particularly if multiple sites are used to inject CO_2 into a single basin (WRI, 2008).

Use of the subsurface for CCUS may require coordination with underground natural gas storage operations. Gas storage has helped to balance the supply and demand fluctuations of natural gas around the world for nearly 100 years. Underground gas storage is an analog to CCUS (WRI, 2008, p. 211) and might serve as a model for how to regulate and commercialize CCUS. Depleted hydrocarbon fields and saline aquifers are commonly used for natural gas storage (WRI, 2008, p. 211). Injected CO_2 will readily mix with natural gas. If natural gas storage and CO_2 sequestration are operated in close proximity within the same geologic formation, the two substances might comingle and degrade the quality of the natural gas (Bachu, 2000, p. 964). In the United States, on average, over 1500 billion cubic feet of natural gas is stored in over 400 active storage facilities in the lower 48 states (U.S. Energy Information Administration, 2020b). About 80% of these storage facilities consist of depleted oil or natural gas fields (Energy Infrastructure).

Another regulatory and commercial analog to CCUS is fluid waste disposal. Hazardous and nonhazardous fluid wastes and municipal wastewater are often disposed of below the lowest underground source of drinking water by injecting them deep into subsurface formations, where, like CO_2, they are intended to remain indefinitely (U.S. Environmental Protection Agency(a); Argonne National Laboratory et al., 2004). The US Environmental Protection Agency (EPA) and delegated state agencies have regulated the underground injection of fluid wastes under the Underground Injection Control (UIC) program by creating "classes" of injection wells and setting standards for injection to protect underground sources of drinking water. These waste reservoirs are formations where fresh water is protected from the injection zone by an impermeable cap rock or confining layer, much like what would be used for CO_2 sequestration. Injection zones typically range from slightly over 500 m to more than 3000 m in depth (see Office of Water, 2001). There are approximately 800 Class I wells in the United States, mostly located in the sedimentary basins of the Gulf Coast and Great Lakes regions (U.S. Environmental Protection Agency(a)). About 17% of Class I wells are used for hazardous waste disposal, about 53% of the Class I wells are for nonhazardous wastes, and the rest are for municipal wastewater disposal in the State of Florida (U.S. Environmental Protection Agency(a)).[15] Since 1983, annual injection volumes range from five to seven billion gallons in Texas. The cumulative total volumes in Texas since the 1950s is approaching 300 billion gallons (U.S. Environmental Protection Agency(b)). CCUS wells would be either Class II if the CO_2 is used for enhanced hydrocarbon recovery or Class VI if the CO_2 is simply sequestered. Class VI wells require a specific permit from the US EPA and require

---

[14] See also Gunter et al. (2004).

[15] See also Keith et al. (2005).

special financial assurances, more robust plugging and abandonment operations, continued monitoring, and integrity testing (U.S. Environmental Protection Agency(b)).

Hydrocarbon production produces large amounts of saltwater. The operator separates out the saltwater from the petroleum resources and then disposes of the saltwater by injecting it into underground rock formations that are receptive to large amounts of fluids. In the United States, the UIC program of the Safe Drinking Water Act (SDWA) regulates the disposal of wastes from oil and gas wells (see U.S. Environmental Protection Agency (c)).[16] "It is estimated that over two billion gallons of fluids are injected in the United States every day. Most oil and gas injection wells are in Texas, California, Oklahoma, and Kansas" (U.S. Environmental Protection Agency(d)). About one-quarter of this saltwater is injected to enhance oil production (Argonne National Laboratory et al., 2004, p. 49). The total volume of saltwater injected annually is roughly equivalent to the volume that 2 gigatons (Gt) of CO_2 would occupy at a depth of 1 km (Benson and Cole, 2008, p. 212). The practice of disposing of produced saltwater is like, but much smaller in scale than, CCUS (Benson and Cole, 2008, p. 234).[17]

Compressed air energy storage (CAES) is another potentially competing subsurface use.[18] CAES could help manage the intermittency of large-scale electricity produced by wind and solar. Surplus electricity produced by wind that would otherwise flow into the electric grid could instead be used to compress air that is pumped and stored in deep geologic reservoirs to be used later to make natural gas turbines operate more efficiently (Denholm and Sioshansi, 2009, pp. 3149–3150). Currently, only two such facilities are operational. A 290-megawatt (MW) CAES plant operating in Germany has been compressing roughly 300,000 cubic meters of air in a natural gas storage reservoir roughly 600–800 m below the surface (see Crotogino et al., 2001). A 110 MW CAES plant is also currently operating in the United States in McIntosh, Alabama (Crotogino et al., 2001, p. 1). The Battelle Memorial Institute suggested that future United States compressed air storage projects should be located in formations roughly 650–850 m below the surface and at least 100 m away from any dissimilar geologic formation (Allen et al., 1982).

Underground aquifer storage and recovery (ASR) of ground water is another growing subsurface use. ASR involves injecting water into underground reservoirs for later retrieval (State of Washington Department of Ecology). ASR is used in Colorado, Florida, Oregon, Texas, and Australia. Where the geology is suitable, ASR may be a promising solution for the future of freshwater management (Kiel and Thomas, 2003, p. 25). While some ASR operations would be too shallow to compete with CCUS, deep ASR may grow in the future.

The previous discussion illustrates that CCUS operations have the potential to interfere with various actual or foreseeable uses of subsurface pore space. Currently, there is little

---

[16] Such wells are categorized as Class II wells.

[17] In 2018, over 37 gigatons of CO_2 emissions occurred, about one-quarter from the burning of fossil fuels. For further discussion see generally Keller et al. (2003).

[18] See Energy Storage Association (concluding that "a the net present value of CO_2 sequestration yields an expression for the efficiency factor which seems preferable to previous carbon accounting methods," that a subsidy for initially noncompetitive CO_2 sequestration technology can be sound economic policy, and that CO_2 sequestration "presents a potential low-cost solution to the greenhouse gas problem").

to no federal, and limited state, statutory authority governing subsurface property rights issues in the context of CO_2 sequestration. Various US federal and state agencies regulate existing subsurface injection activities; however, these agencies largely ignore the issue of property rights by citing jurisdictional limits. The federal SDWA gives the EPA authority to manage the UIC program, which regulates underground fluid waste injection and enhanced oil recovery, but not natural gas storage (40 C.F.R §§ 144-146, 2011). Many states have asserted primacy to administer the UIC program, but some do so through more than a single state agency. For example, in petroleum producing states, oil and gas conservation agencies manage the UIC program regarding saltwater disposal and injections related to petroleum operations, but one or more other agencies manage other UIC matters.

The EPA has determined that the SDWA confers to the Agency the authority to regulate geologic sequestration of CO_2[19]; however, states may seek primacy to regulate CCUS. In 2010, the EPA published its final rules for managing the injection of CO_2 for geologic sequestration under the UIC program (U.S. Environmental Protection Agency(b); Federal Requirements Under the Underground Injection Control UIC Program for Carbon Dioxide CO_2 Geologic Sequestration GC Wells, 2010). The rules create a new class of wells, Class VI, and regulate on-site characterization, well construction and operation, postinjection monitoring, and postclosure stewardship (U.S. Environmental Protection Agency(b); Federal Requirements Under the Underground Injection Control UIC Program for Carbon Dioxide CO_2 Geologic Sequestration GC Wells, 2010). However, property rights are not addressed (U.S. Environmental Protection Agency(b); Federal Requirements Under the Underground Injection Control UIC Program for Carbon Dioxide CO2 Geologic Sequestration GC Wells, 2010). The EPA has issued several CCUS permits.

6.1.1 Who owns pore space in the United States?

The United States is among a few countries that follow the accession theory of mineral ownership—that is, the surface owner of land owns the mineral rights. In the United States, property rights historically have been defined by state law. Common law property rights are generally viewed as a bundle of rights that can be subdivided among various private interests. A portion of this bundle includes rights to subsurface pore space. In general, almost any distinct property right may be burdened by licenses, leases, servitudes, and security interests. In the United States, the property right to inject various fluids into deep subsurface pore space is governed by state, not federal, law.

Under the common-law maxim *cujus est solum, ejus est usque ad coelum et ad inferos* (commonly known as the *"ad coelum* doctrine") (1 Coke, 1832; 2 Blackstone, 1902; 3 Kent, 1896), a fee simple[20] owner of land holds title to the entire tract from the heavens to the depths of the earth. Under this maxim, a fee simple owner would own the subsurface pore space. The question of pore-space ownership most commonly arises when the fee-simple interest is severed into a surface estate and one or more separate mineral interests. As between the

---

[19] See Safe Drinking Water Act, 42 U.S.C. § 300hd (2006); see also U.S. Environmental Protection Agency(b).

[20] Fee simple (2004), Black's Law Dictionary (2004). Means the mineral and surface interests are held by a single owner (p. 648).

surface owner and mineral owner, few states have determined the ownership of pore space. To date, only a handful of cases across the country have addressed this issue, but the vast majority of, and better-reasoned, decisions hold that the surface owner owns the pore space.[21]

[21] See, for example, Int'l Salt Co. v. Geostow (1989) (cavity created by mining of salt in New York belonged to surface owner, subject to salt miner's right to use the cavity for salt mining purposes); Miss. River Transmission Corp. v. Tabor (1985) (surface owner owns storage rights in Louisiana, p. 672); Ellis v. Arkansas Louisiana Gas Company (1979) (surface owner owns storage rights in Oklahoma); Emeny v. United States (1969) (United States acquired certain mineral rights but surface owner owns storage rights, p. 1324); Fisher v. Continental Resources, Inc. (2015) (accepting that surface owner owns pore space subject to right of oil and gas lessee to use the subsurface for salt water disposal operations related to unit in which the land in question had been included); United States v. 43.42 Acres of Land (1981) (stated "the mineral owner cannot be considered to have ownership of the subsurface strata containing the spaces where the minerals are found," pp. 1043, 1046); Rook v. James E. Russell Petroleum, Inc. (1984) (although Kansas has not directly addressed pore-space ownership, such rights are considered severable from the right to produce oil and gas, pp. 166–167); Cent. Ky. Natural Gas Co. v. Smallwood (1952) (the owner of the oil and gas rights, not the surface owner, has the authority to grant a gas-storage lease); (Texas American Energy Corp. v. Citizens Fidelity Bank & Trust Co. (1987), Dep't of Transp. v. Goike (1996) (court held that "the storage space, once it has been evacuated of the minerals and gas, belongs to the surface owner, p. 365); Jones-Noland Drilling Co. v. Bixby (1929) (dicta: "The fee in the soil, except the oil and gas, remains in the lessor unencumbered with those rights of the lessee. The lessee is not the owner of the solids of the earth to which the pumping and other equipment is annexed. He, at most, is the owner of the oil and gas, in place, and merely has the right to use the solid portion so far as necessary to bore for, discover, and bring to the surface the oil and gas," p. 383); Sunray Oil v. Cortez Oil Co. (1941) (surface owner has the right to grant permission to inject wastewater into the subsurface as long as there is not interference with the mineral estate's recovery of oil and gas); Pomposini v. T.W. Phillips Gas & Oil Co. (1990) (surface owner owns storage rights in Pennsylvania); Lightning Oil Co. v. Anadarko E&P Offshore, LLC (2017) (surface owner owns and controls the subsurface while the mineral owner owns the severed minerals and the right to access them in Texas); Humble Oil & Ref. Co v. West (1974a) (surface owner owns storage rights in Texas); and Tate v. United Fuel Gas Co. (1952) (based on exception language in the deed, the oil and gas owner did not own "clay, sand, or stone," and thus surface owner owned the spaces in the storage formation that were devoid of recoverable gas). But see, Grynberg v. City of Northglenn (1987) (surface owner did not have a right to test for coal to determine suitability of land for a surface water reservoir suggests the possibility that mineral owner controls access to pore spaces); Gray-Mellon Oil Co. v. Fairchild (1927) (dicta: "While the oil is fugitive, the sand-bearing oil is as stationary as a bank of coal. The only practical use to which the oil-bearing sand can be put is to get the oil out of it. The exclusive, permanent right to get the oil from the sand is necessarily a right to a part of the land, for to use the sand in any other way would be to destroy the right to extract the oil from it, as the sand must be allowed to remain as it is for the oil to flow through it," p. 745); United States Steel Corp. v. Hoge (1983) (holding that coal owner owned rights to coalbed methane within the coal seam, arguably implying that pore space within coal seams is owned by mineral owner in Pennsylvania); Lillibridge v. Lackawanna Coal Co. (1891) (holding that coal owners, not surface owners, owned coal-mining shafts in Pennsylvania); and Faith United Methodist Church and Cemetery of Terra Alta v. Morgan (2013) [construing a "surface only" deed and stating in dicta that surface "generally means the exposed area of land, improvements on the land, and any part of the underground used by a surface owner as an adjunct to surface use (e.g., medium for the roots of growing plants, groundwater, water wells, roads, basements, or construction footings), pp. 480–481"]. In Staben v. Anterra Energy Services, Inc. (2012), a trial court refused to decide the issue of pore space ownership, which is treated as an issue of first impression in California, because the party asserting the issue failed to raise it in a timely fashion.

Assuming that the surface owner owns pore spaces, the mineral owner has an implicit right of reasonable use to facilitate enjoyment of mineral rights, including the right to use the subsurface, including pore space. This implicit right to use is akin to a servitude that is tied to, and limited by, the scope of the severed mineral right.[22] A study of pore-space ownership in the United Kingdom has rejected the notion of a so-called English Rule— that English cases somehow supported the notion that the mineral owner owned pore spaces (Barton, 2014). The author of this study concludes that in the United Kingdom, pore space is generally owned and possessed by the landowner, not the mineral owner (Barton, 2014, p. 34). After canvassing English case law on this matter, he concludes: "There is no English Rule to the Contrary (Barton, 2014, p. 36)." In Bocardo SA v. Star Energy UK Onshore Ltd. (2011), the Supreme Court held that the surface owner owns the subsurface, subject to the rights of lawful transferees.

Mineral rights may be severed by specific substances or subdivided by areas or by depth, they are frequently fractionalized among multiple parties, and they are often made subject to a multitude of transactions and property interests. Thus a single tract of land may be subject to a variety of mineral claims held by dozens or hundreds of claimants. Because the nature of a carbon-sequestration right is most closely analogous to an easement, if mineral owners were found to own the pore-space, a carbon sequestration right would likely have to be secured from every fractional mineral owner (Elliott v. Elliott, 1980). Thus, for purposes of utility and efficiency, ownership of pore-space is best left in the surface owner in the United States.[23]

For CCUS to be commercial, a regulatory and legal framework is needed.[24] The United Kingdom has a regime in place for offshore CCUS,[25] which has been adapted

---

[22] See, for example, Burlington Resources Oil and Gas Co., LP v. Lang & Sons Inc. (2011) (holding that mineral owner had right to use pore space for disposal of wastewater produced from oil well on the same tract).

[23] Recent white papers and law review articles have analyzed whether, in the first instance, the surface owner or the mineral owner on split-estate land has property rights in the pore space. While most of these papers and articles conclude that the surface owner would prevail over the mineral owner in most cases, the issue is far from resolved. See Cooney (2005), de Figueiredo (2007), Wilson and de Figueiredo (2006, pp. 10121–10122) (stating that most courts have held that after the removal of underground minerals, oil, or gas, the surface owner retains the right to use the remaining space for storage, but that mineral rights holders often retain some rights to access the pore space for continued exploration or extraction of minerals in other areas); Anderson (2009, pp. 99–109) (containing a now outdated discussion of Texas law due to the more recent decision in Lightning Oil Co. v. Anadarko E&P Offshore, LLC (2017) that the surface owner owns and controls the subsurface while the mineral owner owns the severed minerals and the right to access them in Texas); and Havercroft et al. (2011, p. 366) (stating "there are protectable property interests in pore space that are vested in the surface owner, the mineral owner, or both").

[24] This is needed in any country desirous of facilitating CCUS. In addition to our opinion, see generally McHarg and Poutie (2014). The authors discuss the risks associated with CCUS and how the United Kingdom has attempted to mitigate these risk (McHarg and Poutie, 2014, pp. 253–264) and how CCUS may be incentivized (McHarg and Poutie, 2014, pp. 264–265).

[25] Jill Morgan has observed that the "UK's onshore capacity for CCS is limited. With one exception (the Wytch Farm Field), its oil and gas fields are too small, its major aquifers are widely used for potable water extraction onshore, and its onshore coal seams have low permeability (Morgan, 2013, p. 813)."

II. Case studies on CCS and related policies, and their consequences for climate change

and adopted by the European Union.[26] Several US states have regulatory frameworks to manage geologic sequestration of CO$_2$. Some of these states have passed legislation explicitly defining pore-space ownership.[27] For example, a Wyoming statute addresses the issue of property rights by stating that "[t]he ownership of all pore space in all strata below the surface lands and waters of this state is declared to be vested in the several owners of the surface above the strata" (Wyo. Stat. Ann. § 34-1-152a, 2009). However, the statute was amended to clarify that the mineral estate is still dominant over the surface estate (Wyo. Stat. Ann. § 34-1-152a, 2009). A North Dakota statute similarly proclaims that "[t]itle to pore space in all strata underlying the surface lands and waters vested in the owner of the overlying surface estate" (N.D. Cent. Code § 47-31-04, 2009). North Dakota's law further prohibits the complete severance of pore space from surface estate (N.D. Cent. Code § 47-31-05, 2009). A Montana statute creates a presumption that the surface owner owns subsurface pore space if deeds or other severance documents do not demonstrate otherwise (Mont. Code Ann. § 82-11-180, 2009). Like Wyoming and North Dakota, Montana's statute explicitly does not interfere with common law or the dominance of the mineral estate.

6.1.2 Does the use of pore space for CCUS require landowner compensation?

For CCUS to be commercial and to operate efficiently and effectively, HGs will need to address the ownership of pore space. Most HGs will likely pass legislation asserting government ownership[28] or control of deep pore space and then adopt a regulatory process for implementing CCUS. In the United States, however, landowners will likely assert a right to compensation for the taking of pore space and mineral owners may seek compensation to the extent that a CCUS operation interferes with their right of reasonable use. Our prior paper (Anderson and Gresham, 2015) discusses the need to compensate for this taking. Whether a taking occurs depends upon whether the courts will analogize deep pore-space use with the use of airspace by aircraft, thus limiting the law of trespass.[29] The Second Restatement of Torts makes no express distinction between surface trespass and harmless subsurface intrusions,[30] stating that "a trespass may be committed on, beneath, or above the surface of the earth" (Restatement Second of Torts § 159, 1965a). The only

---

[26] Morgan (2013, p. 250) (citing Directive 2009/31/EC [2009] OJ L140/114, April 23, 2009).

[27] See Mont. Code Ann. § 82-11-180 (2009), N.D. Cent. Code § 47-31-02 (2009), Wyo. Stat. Ann. § 34-1-152 (2009).

[28] Alberta passed legislation in 2010 proclaiming that the Crown owns, and always has owned, the pore space (Bill 24, Carbon Capture and Storage Statutes Amendment Act, 2010). For a discussion of similar legislation in several Australian states, see Hepburn (2014).

[29] See Hinman v. Pac. Air Lines Transp. Corp. (1936) (holding that the use of airspace is not unlawful without proof of actual injury, pp. 758–759) and United States v. Causby (1946) (recognizing that airplanes may freely navigate airspace unless the flights are so low and constant as to make it impossible for the true owner to fully enjoy and use the surface estate). For further discussion, see Anderson and Gresham (2015).

[30] The focus of Wickstrom et al. (2008), Anderson (2010) is to criticize the Second Restatement, to illustrate that it no longer reflects the weight of case law, and to call upon the American Law Institute to rewrite trespass law to reflect modern case law that limits subsurface trespass claims to situations where a subsurface intrusion causes actual and substantial damages.

exception to this broad categorization relates to airspace intrusions (Restatement Second of Torts § 159, 1965a) by aircraft that are beyond the immediate reaches of the surface of the earth (Restatement Second of Torts § 159 cmt. G., 1965b).[31] The author of this chapter has argued that the Restatement should adopt an exception for deep subsurface intrusions like the exception it already recognized for aircraft.[32]

6.2 Liability models for the use of pore space for industrial and commercial underground fluid injection activities

Regarding traditional surface trespass, *Loretto v. Teleprompter Manhattan CATV Corp.* is perhaps the seminal case in which the US Supreme Court addressed the issue of permanent physical occupation of a surface owner's property (Loretto v. Teleprompter Manhattan CATV Corp., 1982). The specific issue before the Court was whether a New York law requiring a landlord to allow the installation of a cable company's cables in rental properties to facilitate cable television services to tenants rises to the level of a taking without just compensation (Loretto v. Teleprompter Manhattan CATV Corp., 1982, p. 421). The Supreme Court ruled that the state statute amounted to a taking of a portion of the plaintiff's property—around one and one-half cubic feet on the outside of the rental building, entitling her to just compensation (Loretto v. Teleprompter Manhattan CATV Corp., 1982, p. 441). On remand, however, the Court of Appeals of New York ruled that the amount of compensation awarded could be nominal (i.e., one dollar) and predetermined (Loretto v. Teleprompter Manhattan CATV Corp., 1982, pp. 423–424), provided landowners had a process available through which to seek more compensation by proving special circumstances (Loretto v. Teleprompter Manhattan CATV Corp., 1982, p. 421).[33]

Apart from a trespass case involving subsurface mining (Del Monte Mining & Milling Co. v. Last Chance Mining & Milling Co., 1898), the US Supreme Court has not addressed subsurface trespass; however, state and other federal courts have ruled on this issue in the context of various commercial and industrial subsurface injection activities that gave rise to both takings and trespass claims. US courts generally balance competing interests in the subsurface and have placed great weight on the public interest and regulatory approval associated with certain activities. Specifically, courts have given surface and mineral owners only limited protection against subsurface intrusions resulting from the migration of fluids injected underground.

In the cases addressing the use of, or intrusions into, subsurface pore space, US courts have generally modified the *ad coelum* doctrine by limiting the ability of surface

---

[31] Though this exception would technically apply to any airspace, it more generally applies to the airspace extending upward beyond the immediate reaches of the surface, as aircraft rarely come within the useable reaches of most land. In its seminal *Causby* decision, the court granted relief to a chicken farmer who suffered actual harm resulting from low-flying aircraft (United States v. Causby, 1946, pp. 266–267).

[32] For further discussion, see Anderson (2010).

[33] That is unless the property owner meets the burden of proof for establishing that the diminution in value of the property was materially different than the general assumption and is therefore entitled to receive greater compensation (Loretto v. Teleprompter Manhattan CATV Corp., 1983, pp. 432–433).

owners and mineral owners to recover money damages. Courts have considered subsurface trespass in the context of five subsurface injection activities that are somewhat analogous to CCUS: (1) licensed subsurface storage of natural gas[34]; (2) licensed subsurface injection and disposal of fluid waste; (3) state-authorized subsurface injections to enhance hydrocarbon recovery (unitization)[35]; (4) injections of fluids and prop pants to facilitate petroleum production (hydraulic fracturing); and (5) subsurface storage and recharge of fresh water. In each of these groups, the courts have balanced the need to protect a landowner's right to exclusive possession and enjoyment with the public interest to facilitate valuable enterprises that meet important societal needs. Although not unified, consistent, or large in number, the great weight of case law supports the proposition that takings and trespass claims for deep subsurface uses will not stand absent actual and substantial damages.

The most analogous body of US case law concerns the licensed subsurface injection and disposal of fluid wastes,[36] which, like CCUS, is done to sequester fluids indefinitely. Underground waste-injection cases show that most courts have rejected the notion that property owners have the absolute right to prevent the underground migration of fluid waste into their pore space or that they are entitled to compensation for such use, at least where the waste injection is conducted pursuant to regulatory approval.

In each fluid-waste case, the courts have placed great weight on the public interest and regulatory approval associated with the underground injection of fluid wastes, modifying common law relating to subsurface property rights accordingly. At the same time, though, each of the courts held open the possibility that a plaintiff could recover damages if it could show that the migration of injected waste caused actual harm to, or interference with the use of, her property. In much of the nation, it appears that most underground waste-injection operations conducted pursuant to federal or state authorization under the UIC program that do not cause actual harm to adjacent properties may be carried out without compensation being paid to the surrounding landowners because the activity is considered necessary and in the public interest.[37]

---

[34] Most decisions reject trespass claims, but some do so by referencing eminent domain authority. Lone Star Gas Co. v. Murchison (1962, pp. 879–880), White v. N.Y. St. Natural Gas Corp. (1960), Humble Oil & Ref. v. West, 1974b, p. 817), ANR Pipeline Co. v. 60 Acres of Land (2006, p. 940). One case dodge that trespass claim by holding that the underground injection of gas for storage constitutes abandonment of the gas, which allows others to capture it (Hammonds v. Central Kentucky Natural Gas Co., 1934, pp. 205–206). Kansas follows *Hammonds* where no storage permit has been issued and eminent domain is not used (Anderson v. Beech Aircraft Corp., 1985, p. 1032); see also Union Gas Sys., Inc. v. Carnahan (1989, p. 967). These cases were distinguished in Reese Exploration, Inc. v. Williams Natural Gas Co. (1993), p. 1523). An Oklahoma statute provides that an injector does not lose title to gas injected for storage even if it migrates to other lands [Okla. Stat. Ann. tit. 52, §§ 36.1-36.7 (1951), confirmed in Oklahoma Natural Gas Co., a Division of ONEOK Inc. v. Mahan & Rowsey, Inc. (1986)].

[35] This process can cause migration of the injected fluid, or the native oil and gas sought to be produced, into a neighboring production field and inhibit another producer's ability to recover oil or gas resources.

[36] For detailed discussion of this case law, see Anderson and Gresham (2015).

[37] Three billion tons/year is about the same as the mass of CO$_2$ produced/year by between 750 and 1000 medium-sized (500 Mw) coal-fired power plants [see Keith et al. (2005, p. 501A)].

Absent substantial harm, attempts by landowners to prevent subsurface intrusions of waste have not been successful where the disposal operator is conducting its activities consistent with a regulatory permit.[38] The leading case is Chance v. BP Chemicals, Inc. (1996). BP Chemicals secured a Class I underground injection well permit to dispose of hazardous chemical waste (Chance v. BP Chemicals, Inc., 1996). Subsequently, neighboring landowners initiated a class-action suit wherein they asserted that the injected waste trespassed into their subsurface pore space (Chance v. BP Chemicals, Inc., 1996, p. 986). The *Chance* court concluded that a landowner's subsurface right to exclude others extends only to invasions that "actually interfere with the [landowner's] reasonable and foreseeable use of the subsurface" (Chance v. BP Chemicals, Inc., 1996, p. 992). The court expressly found that the trial court did not err in refusing to allow the plaintiff to present evidence that "environmental stigma associated with the deep wells had a negative effect on appellants' property values due to the public perception there may have been injectate under appellants' property and that the injectate may be dangerous" (Chance v. BP Chemicals, Inc., 1996, p. 993). The Ohio Supreme Court placed significant weight on the fact that the plaintiffs had no specific evidence that defendant's wells were causing any problems, only opinion testimony that problems may arise in the future (Chance v. BP Chemicals, Inc., 1996, p. 993). In other words, a landowner may not recover damages for speculative harm or for the loss of speculative value. Since the injection of hazardous waste by BP Chemicals was not interfering with any reasonable and foreseeable use of plaintiff's property, the court held that migration of the waste into neighboring pore space was not compensable. Although BP Chemicals was operating pursuant to valid state and federal permits, the court did state that its permit did not shield the company from liability, indicating that one class member might have a valid claim because the migration of the subsurface waste may have forced that member to abandon plans to drill for natural gas (Chance v. BP Chemicals, Inc., 1996, p. 994).

Use of the same stratum at issue in *Chance* has been proposed for CCUS. Since sequestered CO_2 could migrate laterally over a very sizeable area (e.g., hundreds of thousands of square miles),[39] requiring project developers to obtain consent from all pore-space owners within the migratory path of the CO_2 plume could have the practical effect of prohibiting the development of many CCUS projects due to the potentially crippling cost of such an obligation. *Chance* seems to relieve potential CCUS developers from this burden.

The full body of case law shows that courts have generally held certain underground fluid injection activities—that is, enhanced hydrocarbon recovery, underground waste disposal, and freshwater storage and recharge—to be in the public interest and thus protected from claims of subsurface trespass: (1) when the activity is licensed under a state or federal regulatory program, and (2) where the property owner is unable to prove actual harm to, or interference with use and enjoyment of, the land resulting from the injection

---

[38] See, for example, Boudreaux v. Jefferson Island Storage & Hub, L.L.C. (2001, p. 274), Mongrue v. Monsanto Co. (2001, p. 432; Cassinos v. Union Oil Co. (1993) (finding actual harm), Raymond v. Union Tex. Petroleum Corp. (1988), West Edmond Hunton Lime Unit v. Lillard (1954, p. 731) (finding actual harm), West Edmond Salt Water Disposal Ass'n v. Rosecrans (1950, p. 970), Sunray Oil v. Cortez Oil Co. (1941). A key jurisdiction where the issue remains unresolved is Texas. For a review of Texas case law, see Anderson and Gresham (2015).

[39] See Pruess et al. (2001), Brennan and Burruss (2006), Gresham et al. (2010, p. 2900).

operations. Thus, while courts have rejected any absolute protection of rights in the subsurface on behalf of landowners, they have preserved limited landowner rights to use and exploit the subsurface and to recover money damages for actual and substantial harm caused by subsurface invasions. Similarly, courts will entertain landowner suits in trespass and nuisance when airborne particles and pollution invade the landowner's airspace and cause actual and substantial harm.[40] In airspace pollution cases, courts consider whether the invasion resulted in an actual interference with the plaintiff's use and enjoyment of property or caused actual harm (Henderson et al., 2007, pp. 400–401). In subsurface invasion cases, courts consider essentially these same factors and generally reach similar conclusions. In both lines of cases, courts apply a "liability rule"[41]—which permits an airspace or subsurface intrusion without permission of property owners so long as the violator pays damages for any harm caused. This contrasts with the "property rule"—which would allow such an intrusion only with prior permission of the property owner.[42]

Nevertheless, CCUS developers will likely have to compensate owners of subsurface rights, including owners of mineral rights, whether severed or not, for actual harm to, or for interfering with, their right of reasonable use and exploitation. In these scenarios, the amount of compensation due will depend on the market value of an existing use precluded by CCUS. In other circumstances, where the geologic formation is appropriate for CO$_2$ sequestration but not for other commercial uses, the costs associated with acquiring pore-space rights might be nominal, at most, where no other economic use is precluded or impaired. Any significant compensation might be limited to CCUS operations conducted on the surface.[43] This limited liability approach appears to be the intent of the North Dakota statute enacted in 2019, which provides: "Injection or migration of substances into pore space for disposal operations, for secondary or tertiary oil recovery operations, or otherwise to facilitate production of oil, gas, or other minerals is not unlawful and, by itself, does not constitute trespass, nuisance, or other tort" (N.D. Cent. Code § 47-31-09, 2009). While this statute does not address direct CCUS operations, it does address the subsurface storage or disposal of substances in connection with oil and gas operations.

The Interstate Oil and Gas Compact Commission (IOGCC) issued a model statute for CCUS based on existing state laws for natural gas storage (Task Force on Carbon Capture and Geologic Storage, 2007a). The IOGCC Model Statute recommends provides: "The Model General Rules and Regulations propose the required acquisition of these storage

---

[40] See, for example, Davis v. Georgia-Pacific Corp. (1968), holding that intrusion of fumes, gases, and microscopic particles on the property of another can constitute a trespass in addition to nuisance (p. 483); and Henderson et al. (2007), discussing how some courts have allowed claims for trespass, in addition to nuisance, for claims based on the intrusion of smoke, gases, or odors (pp. 402–403).

[41] See Calabresi and Melamed (1972) (reasoning that some entitlements are protected by a "liability rule" (i.e., damages) which permits violation of the entitlement without permission of the owner so long as the violator pays damages, p. 1092).

[42] Calabresi and Melamed (1972) (reasoning that some entitlements are protected by a "property rule" (i.e., an injunction) which permits violation of the entitlement only with permission of the property owner, p. 1092).

[43] See Brown v. Legal Found. of Wash. (2003), holding that the state's taking of private property did not violate the Fifth Amendment because the value of the property, measured by the owner's pecuniary loss, was zero (p. 240).

rights and contemplate use of state natural gas storage eminent domain powers or oil and gas unitization processes to gain control of the entire storage reservoir" (Task Force on Carbon Capture and Geologic Storage, 2007a, p. 11). The IOGCC Model Statute does not authorize CCUS in formations containing commercial quantities of oil, gas, or other valuable resources, and it requires the CCUS project developer to identify and negotiate in good faith with all property owners "having property interests affected by the storage facility" (Task Force on Carbon Capture and Geologic Storage, 2007c). All property owners within one-half mile of the proposed project boundary must be notified by first-class mail and given an opportunity to participate in hearings (Task Force on Carbon Capture and Geologic Storage, 2007c,d). For a CCUS project covering hundreds or thousands of square miles and much larger number of affected landowners, providing notice would be very difficult and expensive. Thus the required statutory amendments, individual landowner negotiations, and subsequent condemnation proceedings required under the IOGCC Model Statute are not likely to facilitate large-scale CCUS. More robust legislation, either state or federal, is needed to provide eminent domain for CCUS projects. The United Kingdom has published a guidebook for regulatory best practices (Global CCS Institute).

6.3 Potential legal and commercial frameworks for permitting access and use of pore space for geologic CO_2 sequestration

Article 4(2) of the United Nations Framework Convention on Climate Change provides that signatory developed countries and economies in transition:

> . . .commit themselves specifically as provided for in the following:
> (a) Each of these Parties shall adopt national policies and take corresponding measures on the mitigation of climate change, by limiting its anthropogenic emissions of greenhouse gases and protecting and enhancing its greenhouse gas sinks and reservoirs . . . .
> [CCUS] is a mitigation step and thus one means of meeting this commitment.

For an overview of HG regulatory frameworks for CCUS, readers are referred to the periodically updated publication of the International Energy Agency (IEA, 2016). The fifth edition, published in 2016, reviews the CCUS permitting processes in The Netherlands, United Kingdom, United States, and Alberta, Canada.[44] It also discusses the EU framework. Components of a comprehensive regulatory framework include permitting of the CO_2 capture facility, the CO_2 pipeline facility, and the CO_2 storage facility, related operational regulations, liability limits, and CO_2 valuation. Given the newness of CCUS, the regulations must be flexible to allow for innovation but certain enough to instill investor confidence. Moreover, the regulatory framework needs a means of discouraging CCUS projects that may substantially harm the environment.[45]

[44] For a discussion of CCUS frameworks in the EU and Australia, see Dixon et al. (2015). For a discussion of the recommended regulatory framework for Japan, see Komatsu et al. (2018). For South Korea, see Park (2018). For a discussion of Norway's offshore CCUS initiatives, see Bankes (2012).

[45] See generally Helman et al. (2014).

At present, there is limited US state-level authority and practically no federal-level authority in the United States for handling the subsurface property rights issues associated with the use of pore space for CO_2 sequestration as well as other commercial and industrial subsurface injection of fluids. Although state law determines underlying property rights to both state-owned and privately-owned land, federal law will need to address the acquisition of CCUS rights regarding federal land. While the Environmental Protection Agency has the authority under the Underground Injection Control program to permit and regulate geologic sequestration of CO_2, the Agency currently has no authority to consider the subsurface property issues attendant to permanent geologic CO_2 sequestration. A legal framework for CCUS is needed that balances the interests of private property owners with the public benefit of sequestration and reduces the possibility of interference with other commercial uses of the subsurface that are also in the public interest. To be workable, such a framework must allow regulators to consider the trade-offs between private interests and the public benefit of a proposed CCUS project and determine the most equitable and beneficial use of the pore space. A successful framework would increase the potential for either avoiding most subsurface property disputes outright or resolving them at the outset in a stable and predictable environment. Where possible the optimal approach for federations, such as the United States and Brazil, would be a unified federal approach. While Brazil could simply declare federal control of pore space, such a declaration by the US Congress would surely result in a constitutional challenge. Thus, pore-space property rights in the United States are likely to be determined by state law, except for federal public domain lands and perhaps for federally reserved rights under the Stock Raising Homestead Act, which potentially could be construed as reserving pore space in the federal government (43 U.S.C. § 299, 1993).

Where no vested private rights in pore space have been previously recognized, governments would have the option of declaring them to be government controlled. For ease of implementing CCUS projects, this would best be done at the federal level by a single, unified statutory framework. Unfortunately, however, this is likely to be accomplished in the United States for private lands on a case-by-case basis in state, not federal courts. Judicial reconciliation may result in a patchwork of different rights from state to state. On the other hand, legislative action to fix the existence and nature of property rights would have the benefit of establishing clear and uniform principles that would yield predictable outcomes, particularly if done at the federal level in federations where such authority is likely to be recognized. Federal legislation would foster coordination when geologic basins underlie several different states. And federal control would facilitate cross-border treaties for CCUS projects.

Beyond issues of ownership, both US federal and state governments have intervened to regulate land use and could do the same for the deep subsurface uses. Should lawmakers choose to regulate, the question becomes whether regulatory intervention meets the constitutional standard for a taking of private property within the meaning of the Fifth Amendment to the US Constitution (U.S. Const. amend. V.). Assuming no taking, this approach would have the simplifying effect of allowing the governments, preferably the federal government, to establish a workable framework for use of the deep subsurface. On the other hand, should the US Supreme Court decide (or let stand a lower court ruling to the same effect) that use or regulation of pore space for CCUS constitutes a Fifth Amendment taking, then the government must decide (and courts must ultimately pass on the legality of) whether and how to fix the constitutionally required just compensation for such a taking. If courts determined that the degree of taking associated with using pore space at depths of a kilometer or more imposes a negligible burden on the use rights of surface owners, then the courts could set compensation at a nominal level—perhaps even zero. A decision not to regulate would not, in itself, preclude sequestration projects, but developers of such projects would then be burdened with the tasks of identifying owners and negotiating contracts to secure long-term robust sequestration rights without regulatory guidance and certainty. In the rural parts of the western United States, where much of the land is federal and private holdings are often large, this might not be terribly difficult. In the eastern United States, where small tract, fractional ownership of real property predominates, acquiring rights for secure sequestration could entail transactions with thousands of individual landowners.

If the US federal or state governments do not intervene to manage or limit the protection of private property rights for the use of pore space for geologic CO$_2$ sequestration, then the cost of acquiring pore-space rights could adversely affect CCUS project economics and greatly increase the cost of generating electricity.[46] For instance, if the use of relatively thin sandstones with low mass-to-volume storage capacities for CCUS result in CO$_2$ plumes that migrate over hundreds to thousands of square-kilometers, then the cost of acquiring the right to use pore space, should compensation be required, could make the project uneconomical (Gresham, 2010, pp. 143−171). Such costs might compare with a sequestration project's operational capital cost that, conservatively, will also be very high (McCoy and Rubin, 2009).

Research conducted in 2010 suggests that the cost of acquiring the right to use pore space for CCUS in the United States. may not be trivial and could increase the overall cost of generating electricity with CO$_2$ capture by a fraction of a percent to up to 40% and account for up to 20% of the total levelized cost of electricity when all CCUS-related costs are factored into generation facilities' revenue requirements (see Table 6.1) (Gresham, 2010, pp. 159−163). This could render electric generation facilities' capture of CO$_2$ for permanent geologic sequestration in saline aquifers unprofitable (Gresham, 2010, pp. 163−165). Thin formations with low porosities will likely lead to high injection costs, large CO$_2$ plumes, and high costs for acquiring necessary property rights—all of which will increase the levelized cost of operating a power plant with CO$_2$ capture and, in turn, customer electricity prices. For example, the analysis conducted in 2010 suggests that the combined annualized cost of sequestration (i.e., injection field characterization and injection facility/field capital and operational costs) and property rights acquisition alone could increase the levelized cost of electricity for an integrated gasification combined cycle (IGCC) electric generation facility with CO$_2$ capture by as much as nearly $30 per MWh (see Table 6.1) (Gresham, 2010, pp. 159−163). While the underlying capital and operating costs assumed for the various components along the CCUS-chain (i.e., IGCC facility with CO$_2$ capture capability, CO$_2$ pipeline infrastructure, and geologic sequestration

---

[46] See Gresham et al. (2010, pp. 2900−2901), Gresham (2010, pp. 143−171).

TABLE 6.1 CO_2 plume size estimate and present value cost (2009$) of electricity generation and geologic carbon dioxide sequestration for an integrated gasification combined cycle facility under the imposition of a $50/t$CO_2$ price.[47]

	Frio Sandstone (TX)	Mt. Simon Sandstone (IL)	Oriskany Sandstone (PA)	Medina Sandstone (PA)	Source
Facility net electric output (MW)	530	530	530	530	CMU CEES (2010) and Gresham (2010, p. 159)
Operational time horizon (years)	30	30	30	30	
CO_2 sequestered (metric tons CO_2/year)	3.9×10^6	3.9×10^6	3.9×10^6	3.9×10^6	CMU CEES (2010) and Gresham (2010, p. 159)
CO_2 emitted (metric tons CO_2/year)	1.8×10^5	1.8×10^5	1.8×10^5	1.8×10^5	CMU CEES (2010) and Gresham (2010, p. 159)
Levelized cost of generation ($/year)[a]	306×10^6	306×10^6	306×10^6	306×10^6	Gresham (2010, p. 160)
Levelized cost of electricity ($/MWh)	88	88	88	88	Gresham (2010)
Wholesale electricity price (¢/kWh)[b]	7.1	6.3	5.7	5.7	U.S. Energy Information Administration (2020c), EIA (2008) and Gresham (2010, p. 160)
CO_2 emission cost ($/year)[c]	8.9×10^6	8.9×10^6	8.9×10^6	8.9×10^6	Gresham (2010, p. 160)
CO_2 pipeline levelized cost ($/year)[d]	8.2×10^6	10×10^6	10×10^6	10×10^6	Gresham (2010, p. 160)
CO_2 sequestration levelized cost ($/year)[e]	8.7×10^6	7.4×10^6	91×10^6	44×10^6	Gresham (2010, p. 160)
Pore-space lease term (years)	100	100	100	100	
Areal extent of CO_2 plume (km^2)	440	330	4900	2400	Gresham (2010, p. 160)
Annual pore-space lease cost ($/year)[f]	590,000–11×10^6	440,000–8×10^6	6.7×10^6 –120×10^6	3.2×10^6 –60×10^6	Gresham (2010, p. 160)
Generation facility revenue requirement ($/MWh)	96–99	95–98	107–123	121–152	Gresham (2010, p. 160)

[a]*Includes cost of capture.*
[b]*Equals the wholesale price of electricity reported by the Energy Information Agency for the North American Reliability Corporation region corresponding to the geographic location where each of the sandstone formations are situated.*
[c]*Based on $50 cost per metric ton of CO_2 emitted.*
[d]*Based on 100 km pipeline length.*
[e]*Includes annualized injection facility/field capital cost, annualized injection field characterization capital cost, and injection facility/field annual operating cost.*
[f]*Annual lease rate range $5–100/acre/year.*

facility) and the assumed cost of electricity used in the 2010 analysis may not precisely align with 2015 estimates and reported values, the findings still reasonably demonstrate that the potential cost of being required to acquire pore-space rights prior to commencing CCUS operations could constitute a significant portion of the overall cost to capture and geologically sequester CO_2.

If neither the US federal government or the states act to establish a framework for managing access and use of deep pore space for geologic CO_2 sequestration, then

[47] See generally Gresham (2010, pp. 143–171), see also Gresham et al. (2010, pp. 2900–2901).

resolution will be left to the courts to sort out under tort, property, and contract[48] law. Relying on the courts to adjudicate disagreements about subsurface property rights and contractual obligations between CCUS project developers/operators and private property owners could significantly delay, if not permanently halt, the development of many CCUS projects. This discussion of judicial barriers to CCUS development assumes, of course, that the appropriate property owners will be amenable to the use of the deep pore space in which they hold a vested property interest. Hold-out landowners could prevent development of CCUS projects, especially in the eastern United States where there are innumerable small private land holdings. Absent eminent domain or unitization authority, hold-outs could effectively stymie implementation of CCUS projects or make them very expensive, such as by bargaining for an ongoing time and volume-based storage fee. However, the cost of acquiring pore-space rights via condemnation or unitization could be economically prohibitive. Faced with highly variable and unpredictable acquisition costs, would-be CCUS developers might be discouraged from moving forward with a project before even attempting negotiations with the appropriate surface owners and mineral owners. Assuming the urgent need to address climate change, this is clearly an outcome that must be avoided.

If pore space is determined to be protected by a property rule, then the owner's lack of prior actual use, or the lack of any reasonable, future nonspeculative, investment-backed uses,[49] for the pore space would not dissuade U.S. courts from concluding that all affected surface owners (and in some cases mineral owners) have the right to exclude others from using pore space for geologic CO_2 sequestration and are entitled to compensation if the pore space is used.[50] Such would be the result if the U.S. natural gas storage model is followed. Under this approach, existing and nonspeculative, investment-backed future uses of the pore space would be relevant only in determining the amount of just compensation due.[51] This approach may be a natural extension of the Montana, North Dakota, and Wyoming legislation, which declare that surface owners hold title to subsurface pore space.[52] Thus, this approach would most likely require compensation to surface owners

---

[48] Certainly, contracts will play a huge role in commercializing CCUS. A typical CCUS project will involve several parties, including the CO_2 producer, capturer, transporter, and user/sequester. Although a single party may play two or more of these roles, there may often be multiple producer parties and capturer parties. All parties will need to make suitable contracts to allocate costs, risks, and benefits. These contracts are likely to contain numerous representations, warranties, and indemnities to help assure the commercial success of a project. The contracts will have to consistently identify the party who will receive available tax credits. For further discussion, see generally Hutt et al. (2020).

[49] For the purpose of this paper, "nonspeculative, investment-backed use" and "nonspeculative, investment-backed expectation" mean the ability to recover actual mineral resources or engage in current or imminent subsurface activities that have substantial economic value.

[50] See Loretto v. Teleprompter Manhattan CATV Corp. (1982, pp. 436–438), Tahoe-Sierra Pres. Council, Inc. v. Tahoe Reg'l Planning Agency (2002, p. 322).

[51] For a discussion of approaches to determine value and just compensation, see Wilson and de Figueiredo (2006, p. 10122), Havercroft et al. (2011, pp. 417–423).

[52] See Mont. Code Ann. § 82-11-180 (2009), N.D. Cent. Code §§ 47-31-01 − 47-41-08 (2009), Wyo. Stat. Ann. § 34-1-152 (2009).

for any CO_2 sequestration, either by bilateral agreement or through the exercise of eminent domain.

Alternatively, compulsory-unitization legislation like that used to facilitate the enhanced recovery of oil and gas in the United States could be adapted for CCUS.[53] Unitization for CCUS would essentially be a contractual instrument backed by statutory authority that would force nonconsenting landowners to allow the use of their pore space for CO_2 sequestration if a statutorily specified majority of the pore-space owners, on an acreage basis, have voluntarily agreed to the creation of a sequestration unit. Note that compulsory unitization, in the context of a CCUS project, is not fundamentally different from eminent domain. The difference lies in the fact that a statutory percentage of owners, on an acreage basis, must first agree voluntarily to the establishment of a CCUS project before nonconsenting owners could be forced to transfer sequestration rights for just compensation through a regulatory proceeding that is, practically speaking, like a court proceeding in eminent domain, although administered by a regulatory agency instead of by courts. In other countries, although voluntary agreements are encouraged, the HG may act to force unitization on all parties with or without an agreement.

As discussed in Section 6.2, above, the IOGCC issued a model statute for CCUS that contemplates the acquisition of property rights in the same manner as for achieving compulsory unitization for EOR or, alternatively, through eminent domain as is done for natural gas storage projects: "The Model General Rules and Regulations propose the required acquisition of these storage rights and contemplate the use of state natural gas storage eminent domain powers or oil and gas unitization processes to gain control of the entire storage reservoir" (Task Force on Carbon Capture and Geologic Storage, 2007b). The IOGCC Model Statute has the advantage of working through the well-established mechanisms of state oil and gas agencies that are familiar with drilling and reservoir regulations. The IOGCC Model Statute is not without its shortcomings, however. One limitation of the IOGCC's Model Statute is that it does not authorize CO_2 sequestration to be conducted in formations containing economically recoverable amounts of oil and gas or other valuable resources (Task Force on Carbon Capture and Geologic Storage, 2007c). Under the IOGCC's model, a CCUS developer could not acquire (either by unitization or eminent domain) the right to use a formation that had once produced minerals without first establishing, by agreement or otherwise, that the minerals are exhausted. Over the large areas that could be affected by a CO_2 sequestration operation, this may be difficult to prove. Moreover, the question of just how depleted mineral-bearing strata must be before the pore space can be used for CO_2 sequestration may be addressed differently from state to state, and some states view the issue of mineral exhaustion very favorably to the mineral estate owner, taking into account potential new production methods and technologies.[54] This means that if a CCUS project was developed and it was later discovered that CO_2 was being injected into, or was migrating into, strata containing recoverable resources, then the only option may be to discontinue sequestration operations. Thus, project developers will be motivated to identify formations, such as saline aquifers, that have never yielded valuable minerals and have little prospect of doing so.

[53] Several states have adopted this approach. See, e.g., Ky. Rev. Stat. Ann. §§ 353.800-353.812 (2011), Mont. Code Ann. § 82-11-204 (1977), Wyo. Stat. Ann. § 35-11-316 (2009).

[54] See Int'l Salt v. Geostow (1988, p. 1270), *accord* Dept. of Transp. v. Goike (1996).

Another important question to consider regarding the IOGCC Model Statute's unitization alternative is whether the proposition of having to acquire property rights in the subsurface from the required threshold of landowners—perhaps thousands of landowners—to develop a CCUS site will make the unitization model administratively unwieldy and economically unattractive. The IOGCC unitization model requires a CCUS project developer to identify and negotiate in good faith with all property owners "having property interests affected by the storage facility" (Task Force on Carbon Capture and Geologic Storage, 2007c). This statement is broad enough to include both surface and severed mineral owners. Due to generational transfers, severed mineral ownership becomes highly fractionalized. Consequently, the unitization model would likely result in higher costs associated with acquiring the pore-space rights necessary for geologic CO_2 sequestration than would be realized under the simpler eminent domain approach that limits required compensation to only those instances where the injection and migration of CO_2 materially impairs current or nonspeculative, investment-backed future uses of the subsurface. An expansive view of subsurface property rights modeled after unitization or, for that matter, under the natural-gas-storage eminent domain model, that appears to entitle all affected landowners to compensation, regardless of any actual harm, could greatly discourage the development of CCUS due to the potentially overwhelming cost of having to compensate all property owners overlying the CO_2 sequestration reservoir.[55] With CO_2 sequestration, this could be true even if the value of just compensation is clearly defined and constrained, as in *Loretto*,[56] because the CO_2 plume could migrate over hundreds to thousands of square kilometers.[57] Of course, the nature and extent of the money expended and the infrastructure needed will depend on how widely CCUS is deployed, where it is deployed, and how integral a technostrategy CCUS regulatory approach becomes in the United States approach to limiting GHG emissions.

On balance, the IOGCC's eminent-domain model, properly constrained to situations where a regulatory permit is insufficient to safeguard CCUS, is preferable to the unitization model. Unitization is necessary to facilitate the enhanced recovery of hydrocarbons that must be allocated to resource owners, and unitization has always been controversial.[58] The controversy arises because not all operators and royalty owners have faith that the proposed unit operations will result in greater hydrocarbon recovery and profits. Moreover, all working-interest owners are responsible, at least as a matter of accounting, for their respective share of the costs of unit operations. Finally, because an entire oil and gas reservoir, or substantial portion thereof, is unitized and managed by a single unit operator, the production and costs must be allocated on a fair-share basis. Determining the appropriate allocation of costs and production is highly contentious and often defeats unitization efforts.

On the other hand, CCUS is essentially a disposal operation, and the costs of such an operation are likely to be borne by the party or parties that secure the necessary permits. While there may be profits resulting from disposal fees, carbon credits, and other

---

[55] Gresham et al. (2010, pp. 2900–2901), Gresham (2010, pp. 143–171).

[56] See Loretto v. Teleprompter Manhattan CATV Corp. (1983) (fixing compensation for the taking at one dollar, p. 435).

[57] Gresham et al. (2010, p. 2900), Gresham (2010, p. 160).

[58] See generally Anderson (1985, 2009).

incentives, the business seems more analogous to waste-disposal operations than it is to unitization for enhanced hydrocarbon recovery. One important exception would be where CCUS is done in random with enhanced hydrocarbon recovery. But again, the need for unitization in this circumstance arises from the production of hydrocarbons, not from the carbon-sequestration objective.

Just as *Causby* (United States v. Causby, 1946) confirmed that federal legislation cut off property interests in the higher airspace, in theory, legislation authorizing the use of the deep subsurface for CO$_2$ sequestration could similarly truncate property interests in the deep subsurface, except in connection with those uses that are currently in existence or subject to nonspeculative, investment-backed expectations. This is the approach that will likely be taken outside the United States. Even in the United States, numerous courts have held that a surface owner's interest in the subsurface is "limited" at best, relying on *Causby* and other cases limiting the surface owner's right to control the airspace.[59] Arguably, even if states expressly provide by statute that a surface owner has a property right in the pore space, as Wyoming, North Dakota, and Montana have done, such a state-created property interest may be limited by the judicial application of *Causby* to subsurface rights that places "objective" limits on rights to the subsurface (Chance v. BP Chemicals, Inc., 1996, p. 992). In other words, the argument would be that, just as Wyoming could not vest in surface owners the right to the airspace far above their property as a result of the objective, background principles expressed in *Causby*, Wyoming cannot vest in surface owners the right to the deep subsurface as a result of courts' application of *Causby* to the subsurface (Chance v. BP Chemicals, Inc., 1996, p. 992).

This argument is consistent with a 2008 article by Professor John Sprankling in the UCLA Law Review entitled *Owning the Center of the Earth* (Sprankling, 2008). In this article, Professor Sprankling argues that private property rights to land should not extend more than 300 m below the surface of the earth and that the subsurface beneath that threshold, subject to an exception for valuable mineral rights, should belong to the federal government (Sprankling, 2008, p. 982). The article did not focus on geologic CO$_2$ sequestration specifically, but instead focused on the issue of subsurface ownership in connection with today's technological ability to develop various energy and climate change technologies, including CCUS, that must make use of the subsurface in ways not contemplated in the past (Sprankling, 2008, pp. 1029–1032). Professor Sprankling contends that, based on case law involving subsurface water, oil and gas development, and hazardous waste injection, among others, American law has never determined whether a landowner's rights extend more than two miles below the surface, and that the law regarding rights within two miles of the surface is largely inconsistent (Sprankling, 2008, p. 1020). He concludes that property owners should have some rights below the surface to accommodate foundations, trees, and other normal surface facilities, subject to the 300-m limit (Sprankling, 2008, pp. 1026–1028, 1031).

Following Professor Sprankling's argument, the US federal or state governments could enact legislation declaring a "public highway" of sorts in the subsurface at a specified

---

[59] See, for example, Chance v. BP Chemicals, Inc. (1996, p. 992); Coastal Oil & Gas Corp. v. Garza Energy Trust (2008, p. 11).

depth below the surface of the earth just as has been done with navigable airspace.[60] Such an easement would not need to transfer ownership of subsurface minerals to the federal government, and the easement could even be expressly subordinate to mineral rights, but require accommodation[61] by mineral owners to minimize conflicts. The legislation could also establish a system for compensating property owners with existing uses of the subsurface below that depth if they are harmed by CO_2 injection and truncate the establishment of future private property rights and expectations going forward.[62] This regime would need to formalize and standardize procedures for authorizing access and use of pore space for CO_2 sequestration. Such a framework would obviate many of the property rights conflicts that might arise when CO_2 sequestration projects involve the use of pore space in more than one state.

While this approach would almost certainly facilitate the development of CO_2 sequestration by simplifying the process of accessing the right to use pore space for CCUS and constraining the cost of acquiring subsurface property rights, it would almost certainly invite takings challenges. Oil and gas resources are being developed onshore at depths far below 300 m, and this development is occurring based upon an assumption of ownership rights in those resources to all depths and reasonable use of the subsurface to facilitate development. Moreover, given the weight of and factual variety of case law addressing subsurface uses, the need for such a public easement is questionable in terms of facilitating appropriate uses of pore spaces in the public interest. The U.S. Congress has not found it necessary to declare an airspace easement; yet courts have had no hesitation to safeguard the use of airspace by aircraft by limiting trespass and related claims to circumstances where the landowner suffers actual harm. Lastly, but most importantly, it is doubtful that it will ever be politically feasible for the Congress or state legislatures to implement such a legal framework, which would certainly depend on the political climate and attitudes of the courts over the coming decades.

6.3.1 Application of a liability rule—limiting the protection of pore-space rights for geologic CO_2 sequestration based on existing uses and nonspeculative, investment-backed expectations

While vesting deep subsurface property in the federal or state governments may seem appealing to facilitate new technologies like CCUS, such an approach fails to recognize the realities of how the subsurface has historically been used and is used today. In many regions of the United States, subsurface property rights below 300 m include coal production, oil and natural gas exploration, production, and storage, freshwater production and storage, fluid waste and wastewater injection, and compressed air energy storage. The US Congress has chosen to implicitly recognize subsurface property rights under some

---

[60] See 49 U.S.C. § 40103 (1994).

[61] See, for example, Getty Oil Co. v. Jones (1971, pp. 621–622); Sun Oil Co. v. Whitaker (1972, pp. 810–811).

[62] Sprankling does recognize the potential need to acknowledge and honor "all existing rights to extract specific valuable minerals, at least to the extent appropriate to ensure a reasonable return on prior investments" (Sprankling, 2008, pp. 1037–1038).

circumstances, such as through the eminent domain provisions of the Natural Gas Act,[63] and courts have recognized those rights by allowing for claims of trespass and nuisance in cases of actual interference or harm.[64] Courts also have created mechanisms to compute just compensation when subsurface areas are needed for a public use such as natural gas storage.[65] Thus, the country's history of the use of the subsurface is in fact different from its use of the airspace.

Even though common law limits property rights in airspace other than those used in connection with the surface, the same is far less true regarding the subsurface. To date, there has been no federal declaration of a "public highway" in the subsurface, and any future declaration along those lines would come into conflict with vested economic interests in the subsurface in many areas of the United States. Economic use of the subsurface may end at a certain depth, for instance any deeper than is necessary for existing and future natural gas storage, waste injection, and oil and gas development, but this depth is subject to changing technology and economics. And to the extent that CO₂ sequestration will be at depths that are currently subject to existing or nonspeculative, investment-backed uses (and it appears that it will be), there do not appear to be any background principles of common law that would prevent state legislatures from vesting those property rights in surface owners or perhaps in mineral owners if they choose to do so—at least where the courts have not previously determined ownership—or prevent courts from limiting those rights or from recognizing contrary rights in the exercise of their jurisdiction.

An approach based on existing and nonspeculative, investment-backed uses would likely result in the protection of subsurface property rights in some circumstances but not in others, based on whether the geology is suitable for CO₂ sequestration, as well as whether sequestration might compete with oil and gas development, natural gas storage, and the like. Protecting subsurface property rights based on existing uses and nonspeculative, investment-backed expectations would provide a middle-ground approach to property rights that makes geologic CO₂ sequestration somewhat more expensive to implement, but would recognize, value, and compensate for competing economic uses that would be impaired by CCUS and, consequently, lessen opposition to such projects. Moreover, this approach is firmly grounded in common law. In the context of enhanced recovery of oil and gas, fluid waste injection, and freshwater storage and recovery, the courts have refused any absolute protection of property rights in the deep subsurface but have retained limited protection of property rights that would allow property owners to recover monetary compensation for damage to property caused by actual and substantial harm or interference. Allowing recovery only for actual damage to property is different from finding that a landowner possesses the type of property right in the subsurface that empowers the landowner to prevent others from injecting fluids into the pore space underlying the landowner's property (i.e., a property rule); it is this type of absolute protection of subsurface property rights the courts seem to have clearly rejected in the context of enhanced hydrocarbon recovery, underground fluid waste injection, and freshwater storage.

---

[63] See Natural Gas Act, 15 U.S.C. § 717fh (1988).

[64] See Section 6.1 (Subsurface Property Right and Competing Uses of the Subsurface) of this paper.

[65] See Columbia Gas Transmission Corp. v. Exclusive Natural Gas Storage Easement (1992, p. 1199).

One application of this standard is in essence a "first-in-time, first-in-right" approach to the use of pore space for CO$_2$ sequestration, where neither the government nor its agencies would oversee and manage the right to access and use subsurface pore space for CO$_2$ sequestration. The "first in time, first in right" theory, also referred to as prior appropriation, has been used in the United States to encourage and give a legal framework for other commercial activities. Prior appropriation water rights, sometimes known as the Colorado Doctrine as referred to by the US Supreme Court case *Wyoming v. Colorado* (State of Wyoming v. State of Colorado, 1922), is a system of allocating water rights based on the general principle that water rights are unconnected to land ownership and can be sold or mortgaged like other property. The first person to use a quantity of water from a water source for a beneficial use has the right to continue to use that quantity of water for that purpose, subject to abandonment for nonuse. Subsequent users can use the remaining water for their own beneficial purposes provided they do not impinge on the rights of previous users. The early prospectors and miners in the California Gold Rush of 1849, and later gold and silver rushes in the western United States, also applied "first in time, first in right" theory to mineral deposits. The first one to discover and begin mining a deposit was acknowledged to have a legal right to mine. As with water rights, mining rights could be forfeited by nonuse. The miner's codes were later legalized by the federal government in Mining Act of 1866, and then in the Mining Law of 1872. Similarly, the Homestead Act of 1862 granted legal title to the first farmer to put public land into agricultural production.

Under this approach, a CCUS operator would possess the privilege to inject CO$_2$ into subsurface pore space with the knowledge that it will migrate through the targeted geologic strata provided the injection operation complies with regulations. Under this approach, uncompensated use of pore space would be permissible only if it does not interfere with a verified existing or nonspeculative, investment-backed use of the subsurface that has been asserted by a property owner—that is, a liability rule—the rule that generally governs the use of pore space for underground disposal of fluid wastes.

A second, and our recommended, application of this standard would be for federal or state governments to codify a formal process for managing the access and use of pore space for geologic CO$_2$ sequestration, wherein the project developer acquires a permit to use the pore space for CCUS from the appropriate UIC permitting agency. In certain cases, it might be necessary for the CCUS project developer to injunctive relief to assure that the integrity of the sequestration field is not compromised. In such cases, the grant of a permit to use pore space should also convey the right to invoke eminent domain, where necessary. For such an approach to pass constitutional muster, legislation or the courts must specifically authorize the injection of CO$_2$ for the purpose of permanent sequestration into designated underground geologic reservoirs and declare that CCUS for the purpose of mitigating climate change is a public use carried out in the public interest. Additionally, US courts must authorize, or legislation governing CCUS must include provisions authorizing, UIC permitting agencies to issue permits granting CCUS developers the right to access and use pore space for the injection and sequestration of CO$_2$.

Common or statutory law should also create a presumption that the regulatory grant of the right to access and use pore space for geologic CO$_2$ sequestration does not amount to a compensable taking because the issuance of a permit to use pore space

(1) is not a confiscation of property and (2) is not the first step in a regulatory taking since most pore-space owners will unlikely suffer either an actual loss or an interference with any investment-backed expectation. However, the framework for managing the access and use of pore space for CCUS should provide property owners with an opportunity to rebut this presumption by presenting evidence in an administrative permitting proceeding that demonstrates that CO_2 sequestration will result in a material impairment to a current or nonspeculative, investment-backed future use of the subsurface and that the property owner will suffer a consequent economic loss requiring just compensation. If it is demonstrated that a preexisting interest would be materially impaired by CO_2 injection, then the geologic CO_2 sequestration project should be permitted only upon (1) a modification of the project that avoids the impairment; (2) a contractual agreement between the owner of the preexisting interest and the project developer; or (3) a finding by the permitting agency that the condemnation of the preexisting interest through the exercise of eminent domain, with appropriate compensation, is necessary for the proper operation of the CCUS project. Lastly, legislatures or the courts should establish subsurface trespass and nuisance liability standards whereby the use of pore space for the permanent geologic sequestration of CO_2 by a valid permit holder is not compensable unless the owner of the pore space or other affected property owners suffer actual and substantial damages.

Professor David Pierce has suggested a property-law framework for all reservoir rock structures that this author believes would work well with our suggested regulatory approach to CCUS and the deep subsurface. He argues that, unlike the surface, a landowner cannot construct a fence around the boundaries of a subsurface reservoir rock structure because each owner's interest is structurally connected. Indeed, a landowner cannot easily monitor the deep subsurface for possible trespassers. Professor Pierce argues that property rights to affirmatively use strata, reservoirs, and pore spaces should be governed by the doctrine of correlative rights, a doctrine that has long been applied to subsurface oil and gas resources and has been long recognized at common law as well as in oil and gas conservation regulatory law. A correlative-rights approach would recognize that each landowner would have a legally protected opportunity to use the subsurface correlatively with other landowners as part of a community of subsurface owners (see Pierce, 2009, pp. 768–772; Pierce, 2011a, pp. 255–264; Pierce, 2011b, pp. 693–695; Pierce, 2012, pp. 255–263).[66]

In the CCUS context, this community approach would balance a landowner's right to use the subsurface with neighboring landowners' equal and correlative *opportunity* to make productive use of the rock structure held in common with others. For example, a subsurface owner's desire *not* to use a connected subsurface structure would not necessarily limit use by other members of Professor Pierce's "reservoir community." If the activity is beneficial to the "community," then it can be pursued, regardless of dissenting community members. Because in most cases the only valuable use of a subsurface rock structure will be its commercial value for an industrial activity, development will be viewed as a desirable community activity. Although Professor Pierce views the reservoir community

[66] Professor Pierce has written a series of articles on this: (Pierce, 2009, pp. 768–772; Pierce, 2011a, pp. 255–264; Pierce, 2011b, pp. 693–695; Pierce, 2012, pp. 255–263).

as a self-regulating system governed by property law, this community approach could be subjected to regulation designed to prevent wasteful and inefficient subsurface uses and prioritize uses while promoting the affirmative exercise of correlative rights to fully develop connected subsurface rock structures.

6.4 Conclusion

If lawmakers commit to the widespread deployment of CCUS, then project developers will need authorization to access and use subsurface pore space to avoid liability for subsurface trespass and nuisance. If implemented unwisely, the result could be protracted negotiations with hundreds, if not thousands, of individual property owners for each CCUS project sought to be developed. On the other hand, implementation could also be as straightforward as receiving the appropriate regulatory approval to inject CO_2 for the purpose of permanent geologic CO_2 sequestration. US case law arising from industrial and commercial underground fluid injection operations is instructive of how subsurface property rights might be dealt with in the context of CCUS. As is discussed in Section 6.1, this body of case law demonstrates that courts have typically held certain underground fluid injection activities—for example, enhanced hydrocarbon recovery, underground waste disposal, and groundwater storage and recharge—to be in the public interest and thus shielded from claims of trespass when (1) the operation is licensed under a state or federal regulatory program, and (2) the property owner could not demonstrate actual harm to, or interference with use and enjoyment of, the land as a result of injection and migration of fluids. In this line of cases, US courts have adhered to a liability rule, not a property rule, when deciding subsurface trespass and nuisance disputes. Whether some legislatures or some courts will choose to apply a liability rule to geologic CO_2 sequestration remains an open question. Revising the Restatement of Torts such that it includes an exception for deep subsurface intrusions like those recognized for aircraft would help resolve this debate and, most importantly, reflect the prevailing judicial trend. And adopting the property rule advocated by Professor Pierce would facilitate subsurface use and permitting and lessen liability concerns.

The legal complexity associated with acquiring pore-space rights for CCUS may be further exacerbated by the fact that subsurface CO_2 plumes could be very large in size—on the order of hundreds of thousands of square kilometers in areal extent. By targeting sequestration formations that are favorable to limiting CO_2 plume migration—for example, thick formations with high porosity—property rights acquisition costs, regardless of whether the use of pore space for CCUS is subject to a liability or property rule, may be considerably constrained. On the other hand, the use of thin formations for CCUS with low porosities will likely lead to large CO_2 plumes and high property rights acquisition costs, which in certain instances could render a project uneconomical, especially under a strict property rule. Moreover, to the extent CCUS projects are developed and operated in areas where the subsurface is already being used commercially for natural gas storage, enhanced oil and gas production, hydraulic fracturing, or other uses, the cost of obtaining the rights to use pore space could be substantial. In situations where a CCUS project comes into conflict with competing uses of the subsurface, the value of the right to use pore space for sequestration will be derived from the value of those rights as a function of the existing or future investment-backed uses of the subsurface that would be precluded by

CCUS. In those circumstances where the geologic formation is appropriate for CCUS but not for other commercial uses, the compensatory cost associated with acquiring the rights to use pore space might be nominal, perhaps even zero, because no plausible economic use would be precluded or impaired. Therefore, careful site selection is paramount from both a legal and economic perspective.

To be sure, the application of a strict property rule to geologic CO_2 sequestration may very well foster public acceptance and appease staunch advocates of private-property rights. However, no demonstrable legal or economic rationale exists for recognizing such a rule. Thus, requiring compensation to property owners who have no current or nonspeculative, investment-backed future use of the subsurface seems both unwise and unnecessary when using pore spaces for CCUS. This author believes that the application of a liability rule to CCUS would be a pragmatic approach to mitigating the potential legal hurdles and negative economic effects associated with acquiring pore-space rights for CO_2 sequestration. As discussed above, this would require a formal process for regulating the access and use of pore space for geologic CO_2 sequestration. Under such a framework, the project developer would apply for a permit to use the pore space for CCUS, as well as the right to invoke eminent domain in those specific instances where condemnation of the pore space is necessary to protect the integrity of the sequestration field or where compensation is proven to be necessary. Generally, however, the use of pore space for the permanent geologic sequestration of CO_2 by a valid permit holder would not be compensable unless the injection and migration of CO_2 materially impairs current or nonspeculative investment-backed existing or future uses of the subsurface. This particular approach could facilitate more rapid development of commercial-scale CCUS in response to climate change by both standardizing a procedure for acquiring pore space and constraining acquisition costs.

References

1 Coke (19th ed.) Institutes ch. 1, § 1(4a), 1832.
2 Blackstone (Lewis ed.) Commentaries 18, 1902.
3 Kent (Gould ed.) Commentaries 621, 1896.
40 C.F.R §§ 144–146, 2011.
43 U.S.C. § 299, 1993.
49 U.S.C. § 40103, 1994.
Allen, R.D., Doherty, T.J., Fossum, A.F., April 1982. Geotechnical Issues and Guidelines for Storage of Compressed Air in Excavated Hard Rock Caverns. Battelle Memorial Institute. doi: 10.212/5437632.
Anderson v. Beech Aircraft Corp. 699 P.2d 1023 (Kan. 1985).
Anderson, O.L., 1985. Mutiny: the revolt against unsuccessful unit operations. In: Proceedings of 30th Annual Rocky Mountain Mineral Law Institute. Rocky Mountain Mineral Law Institute.
Anderson, O.L., 2009. Geologic CO₂ sequestration: who owns the pore space? Wyoming Law Review 9 (1), 97–138.
Anderson, O.L., 2010. Lord Coke, The Restatement, and Modern Subsurface Trespass Law. Annual Institute on Mineral Law 57, Reprinted from: Texas Journal of Oil, Gas & Energy Law, 6, 203 (2011).
Anderson, O., Gresham, R.L., 2015. Legal and commercial models for pore-space access and use for geologic CO₂ sequestration. Enhanced Oil Recovery: Legal Framework for Sustainable Management of Mature Oil Fields. Rocky Mountain Mineral Law Foundation.
ANR Pipeline Co. v. 60 Acres of Land, 418 F. Supp. 2d 933 (W.D. Mich. 2006).
Argonne National Laboratory, Veil, J.A., Puder, M.G., Elcock, D., Redweik, R.J., Jr., January 2004. Injection of produced water. In: A White Paper Describing Produced Water From Production of Crude Oil, Natural Gas, and Coal Bed Methane, pp. 33–34. Retrieved from: https://publications.anl.gov/anlpubs/2004/02/49109.pdf.

Bachu, S., 2000. Sequestration of CO_2 in geological media: criteria and approach for site selection in response to climate change. Energy Conversion and Management 41 (9), 953–970. Available from: https://doi.org/10.1016/S0196-8904(99)00149-1.

Bankes, N., 2012. The legal and regulatory issues associated with carbon capture and storage in Arctic states. Carbon & Climate Law Review 6, 21–23. Available from: https://doi.org/10.21552/CCLR/2012/1/206.

Barton, B., 2014. The common law of subsurface activity: general principle and current problems. In: Zillman, D. N., McHarg, A., Bradbrook, A., Barrera-Hernandez, L. (Eds.), The Law of Energy Underground: Understanding New Developments in Subsurface Production, Transmission, and Storage. Oxford University Press, p. 21., Chapter 1.

Benson, S.M., Cole, D.R., 2008. CO_2 sequestration in deep sedimentary formations. Elements 4 (5), 325–331. Available from: https://doi.org/10.2113/gselements.4.5.325.

Benson, S., Cook, P., Anderson, J., Bachu, S., Nimir, H.B., Basu, B., et al., 2005. Underground geological storage. In: Metz, B., Davidson, O., de Coninck, H., Loos, M., Meyer, L. (Eds.), Carbon Dioxide Capture and Storage. IPCC, p. 195.

Bill 24, Carbon Capture and Storage Statutes Amendment Act, 2010. 3rd Sess., 27th Leg., Alberta, 2010, amending the Mines and Minerals Act, R.S.A. 2000, c. M-17.

Bocardo SA v. Star Energy UK Onshore Ltd, 2011, 1 AC 380.

Boudreaux v. Jefferson Island Storage & Hub, L.L.C., 255 F.3d 271 (5th Cir. 2001).

Brennan, S.T., Burruss, R.C., 2006. Specific storage volumes: a useful tool for CO_2 storage capacity assessment. Natural Resources Research 15, 165–182. Available from: https://doi.org/10.1007/s11053-006-9019-0.

Brown v. Legal Found. of Wash., 538 U.S. 216, 240, 2003.

Burlington Resources Oil and Gas Co., LP v. Lang & Sons Inc., 361 Mont. 407, 259 P.3d 766, 2011.

Calabresi, G., Melamed, A.D., 1972. Property rules, liability rules, and inalienability: one view of the cathedral. Harvard Law Review 85 (6), 1089–1128.

Carnegie Mellon University Center for Energy and Environmental Studies (CMU CEES), May 7, 2010. Integrated Control Model Carbon Sequestration Addition (Version 6.2.4) (Software). Carnegie Mellon University, Pittsburgh, PA.

Cassinos v. Union Oil Co., 18 Cal. Rptr. 2d 574 (Cal. Ct. App. 1993).

Cent. Ky. Natural Gas Co. v. Smallwood, 252 S.W.2d 866 (Ky. Ct. App. 1952), overruled on other grounds.

Chance v. BP Chemicals, Inc., 670 N.E.2d 985 (Ohio 1996).

Coastal Oil & Gas Corp. v. Garza Energy Trust, 268 S.W.3d 1 (Tex. 2008).

Columbia Gas Transmission Corp. v. Exclusive Natural Gas Storage Easement, 962 F.2d 1192 (6th Cir. 1992).

Cooney, D., 2005. Part 2: Analysis of Property Rights Issues Related to Underground Space Used for Geologic Storage of Carbon Dioxide. Interstate Oil and Gas Compact Commission Task Force on Carbon Capture and Geologic Storage, Subgroup of State Oil and Gas Attorneys.

Crotogino, F., Mhmeye, K.U., Scharf, R., 2001. Huntorf CAES: More Than 20 Years of Successful Operation. Saarland University. Retrieved from: http://www.fze.uni-saarland.de/AKE_Archiv/AKE2003H/AKE2003H_Vortraege/AKE2003H03c_Crotogino_ea_HuntorfCAES_CompressedAirEnergyStorage.pdf.

Davis v. Georgia-Pacific Corp., 445 P.2d 481 (Or. 1968).

de Figueiredo, M.A., 2007. The Liability of Carbon Dioxide Storage (Doctoral thesis, Massachusetts Institute of Technology, Cambridge, Massachusetts). Retrieved from: https://sequestration.mit.edu/pdf/Mark_de_Figueiredo_PhD_Dissertation.pdf.

Del Monte Mining & Milling Co. v. Last Chance Mining & Milling Co., 171 U.S. 55, 1898.

Denholm, P., Sioshansi, R., 2009. The value of compressed air energy storage with wind in transmission-contrained electric power systems. Energy Policy 37 (8), 3149–3158. Available from: https://doi.org/10.1016/j.enpol.2009.04.002.

Dep't of Transp. v. Goike, 560 N.W.2d 365 (Mich. App. 1996).

Dept. of Transp. v. Goike, 560 N.W.2d 365 (Mich. Ct. App. 1996).

Dixon, T., McCoy, S.T., Havercroft, I., 2015. Legal and regulatory developments on CCS. International Journal of Greenhouse Gas Control 40, 431–448. Available from: https://doi.org/10.1016/j.ijggc.2015.05.024.

Doughty, C., Pruess, K., 2004. Modeling supercritical carbon dioxide injection in heterogeneous porous media. Vadose Zone Journal 3 (3), 837–847. Available from: https://doi.org/10.2113/3.3.837.

Elliott v. Elliott, 597 S.W.2d 795, 802 (Tex. Ct. App.—Corpus Christi 1980).

Ellis v. Arkansas Louisiana Gas Company, 609 F.2d 436 (10th Cir. 1979).

Emeny v. United States, 412 F.2d 1319 (Ct. Cl. 1969).

Energy Infrastructure. Underground natural gas storage. Retrieved from: https://energyinfrastructure.org/energy-101/natural-gas-storage.

Energy Storage Association. Mechanical energy storage: compressed air energy storage (CAES). Retrieved from: https://energystorage.org/why-energy-storage/technologies/mechanical-energy-storage.

ExxonMobil, July 24, 2020. ExxonMobil collaborates on discovery of new material to enhance carbon capture technology. Retrieved from: https://corporate.exxonmobil.com/News/Newsroom/News-releases/2020/0724_ExxonMobil-collaborates-on-discovery-of-new-material-to-enhance-carbon-capture-technology.

Faith United Methodist Church and Cemetery of Terra Alta v. Morgan, 231 W.Va. 423, 745 S.E.2d 461, 2013.

Federal Requirements Under the Underground Injection Control (UIC) Program for Carbon Dioxide (CO2) Geologic Sequestration (GC) Wells. Environmental Protection Agency (EPA), 75 Fed. Reg. 77230-303 (December 10, 2010) (to be codified at 40 C.F.R. pts. 124, 144, 145, 146, and 147).

Fee simple, 2004. Black's Law Dictionary, eighth ed. West Group.

Fish, J.R., Martin, E.L., 2010. Technical Advisory Committee: Approaches to Pore Space Rights. California Carbon Capture and Storage Review Panel.

Fisher v. Continental Resources, Inc., 49 F. Supp. 3d 637 (D. N.D. 2015).

Getty Oil Co. v. Jones, 470 S.W.2d 618 (Tex. 1971).

Global CCS Institute. Carbon capture and storage regulatory test toolkit. Retrieved from: https://www2.gov.scot/Resource/Doc/917/0113634.pdf.

Global CCS Institute, April 16, 2014. What happens when CO₂ is stored underground? Q&A from the IEAGHG Weyborn-Midale CO₂ monitoring and storage project. Retrieved from: https://www.globalccsinstitute.com/resources/publications-reports-research/what-happens-when-co2-is-stored-underground-qa-from-the-ieaghg-weyburn-midale-co2-monitoring-and-storage-project.

Gray, T., 2015. A 2015 analysis and updated on U.S. pore space law—the necessity of proceeding cautiously with respect to the "stick" known as pore space. Oil and Gas, Natural Resourecs, and Energy Journal 1 (3), 277.

Gray-Mellon Oil Co. v. Fairchild, 219 Ky. 143, 292 S.W. 743, 1927.

Gresham, R.L., December 2010. Geologic CO₂ sequestration and subsurface property rights: a legal and economic analysis (Doctoral thesis, Carnegie Mellon University, Pittsburgh, Pennsylvania). Retrieved from: https://kilthub.cmu.edu/articles/thesis/Geologic_CO_sub_2_sub_Sequestration_and_Subsurface_Property_Rights_A_Legal_and_Economic_Analysis/6718082/1.

Gresham, R.L., McCoy, S., Apt, J., Morgan, M.G., 2010. Implications of compensating property owners for geologic sequestration of CO₂. Environmental Science & Technology 44 (8), 2897–2903. Available from: https://doi.org/10.1021/es902948u.

Grynberg v. City of Northglenn, 739 P.2d 230 (Colo. 1987).

Gunter, W.D., Bachu, S., Benson, S., 2004. The role of hydrogeological and geochemical trapping in sedimentary basins for secure geological storage of carbon dioxide. Geological Society, London, Special Publications 233 (1), 129–145. Available from: https://doi.org/10.1144/gsl/sp/2004.233.01.09.

Hammonds v. Central Kentucky Natural Gas Co., 75 S.W.2d 204 (Ky. 1934), limited by Texas American Energy Corp. v. Citizens Fidelity Bank & Trust Co., 736 S.W.2d 25, 28 (Ky. 1987).

Havercroft, I., Macrory, R., Stewart, R.B., 2011. Carbon Capture and Storage: Emerging Legal and Regulatory Issues. Hart Publishing, Wilson, E., Klass, A., 2010. Climate Change, Carbon Sequestration, and Property Rights. University of Illinois Law Review, pp. 363–428.

Helman, L., Parchomovsky, G., Stavang, E., 2014. Dynamic regulation and technological competition: a new legal approach to carbon capture and storage. In: Zillman, D.N., McHarg, A., Bradbrook, A., Barrera-Hernandez, L. (Eds.), The Law of Energy Underground: Understanding New Developments in Subsurface Production, Transmission, and Storage. Oxford University Press, p. 295. , Chapter 15.

Henderson, J.A., Pearson, R.N., Kysar, D.A., Siliciano, J.A., 2007. The Tort's Process, seventh ed. Aspen Publishers.

Hepburn, S., 2014. Ownership models for geological sequestration: a comparison of the emergent regulatory models in Australia and the United States. Environmental Law Reporter 44 (4), 10310–10325.

Hinman v. Pac. Air Lines Transp. Corp., 84 F.2d 755 (9th Cir. 1936).

Humble Oil & Ref. Co v. West, 508 S.W.2d 812 (Tex. 1974a).

Humble Oil & Ref. v. West, 508 S.W.2d 812 (Tex. 1974b).

Hutt, J., Lee, A., McAnelly, J., McGinley, E., Gamble, J., 2020. The way forward: a legal and commercial primer on carbon capture, utilization and sequestration. Texas Journal of Oil, Gas and Energy Law 16 (1).

IEA, 2016. Carbon Capture and Storage: Legal and Regulatory Review. IEA, Paris. Retrieved from: https://www.iea.org/reports/carbon-capture-and-storage-legal-and-regulatory-review.

IEA. Global EV Outlook 2018. IEA, Paris. Retrieved from: https://www.iea.org/reports/global-ev-outlook-2018.

Int'l Salt Co. v. Geostow, 878 F.2d 570 (2d Cir. 1989).

Int'l Salt v. Geostow, 697 F. Supp. 1258 (W.D.N.Y. 1988).

Jones-Noland Drilling Co. v. Bixby, 282 P. 382 (N.M. 1929).

Keith, D.W., Giardina, J.A., Morgan, M.G., Wilson, E.J., 2005. Regulating the underground injection of CO_2. Environmental Science & Technology 39 (24), 499A–505A. Available from: https://doi.org/10.1021/es0534203.

Keller, K., Yang, Z., Hall, M., Bradford, D., September 2003. Carbon Dioxide Sequestration: When and How Much? (Working Paper No. 94). Retrieved from Princeton University, Center for Economic Policy Studies (CEPS) website: https://gceps.princeton.edu/wp-content/uploads/2017/01/94bradford.pdf.

Kiel, P.J., Thomas, G.A., 2003. Banking groundwater in California: who owns the aquifer storage space? Natural Resources & Environment 18 (2), 25–30.

Komatsu, E., Yanagi, K., Nakamura, A., October 2018. Policy strategy and scenario for long-term CCS development in Japan. In: 14th Greenhouse Gas Control Technologies Conference Melbourne, October 21–26, 2018 (GHGT-14).

Kramer, B., 2014. Horizontal drilling and trespass: a challenge to the norms of property and tort law. Colorado Natural Resources, Energy & Environmental Law Review 25 (2), 291–338.

Ky. Rev. Stat. Ann. §§ 353.800-353.812, 2011.

Lightning Oil Co. v. Anadarko E&P Offshore, LLC, 520 S.W.3d 39 (Tex. 2017).

Lillibridge v. Lackawanna Coal Co., 22 A. 1035 (Pa. 1891).

Lone Star Gas Co. v. Murchison, 353 S.W.2d 870 (Tex. Civ. App. 1962).

Loretto v. Teleprompter Manhattan CATV Corp., 458 U.S. 419, 1982.

Loretto v. Teleprompter Manhattan CATV Corp., 446 N.E.2d 428, 432 – 33 (N.Y. 1983).

McCoy, S.T., Rubin, E.D., 2009. Variability and uncertainty in the costs of saline formation storage. Energy Procedia 1 (1), 4151–4158. Available from: https://doi.org/10.1016/j.egypro.2009.02.224.

McHarg, A., Poutie, M., 2014. Risk, regulation, and carbon capture and storage: the United Kingdom experience. In: Zillman, D.N., McHarg, A., Bradbrook, A., Barrera-Hernandez, L. (Eds.), The Law of Energy Underground: Understanding New Developments in Subsurface Production, Transmission, and Storage. Oxford University Press, p. 249. , Chapter 13.

Miss. River Transmission Corp. v. Tabor, 757 F.2d 662 (5th Cir. 1985).

Mongrue v. Monsanto Co., 249 F.3d 422 (5th Cir. 2001).

Mont. Code Ann. § 82-11-180, 2009.

Mont. Code Ann. § 82-11-204, 1977.

Morgan, J., 2013. Digging deep: property rights in subterranean space and the challenge of carbon capture and storage. International Comparative Law Quarterly 62 (4), 813–837. Available from: https://doi.org/10.1017/S0020589313000353.

N.D. Cent. Code § 47-31-02, 2009.

N.D. Cent. Code § 47-31-04, 2009.

N.D. Cent. Code § 47-31-05, 2009.

N.D. Cent. Code § 47-31-09, 2009.

N.D. Cent. Code §§ 47-31-01 – 47-41-08, 2009.

Natural Gas Act, 15 U.S.C. § 717f(h), 1988.

Office of Water, March 2001. Class I Underground Injection Control Program: STUDY of the Risks Associated With Class I Underground Injection Wells (Report No. EPA 816-R-01-007). U.S. Environmental Protection Agency. Retrieved from: https://www.epa.gov/sites/production/files/2015-07/documents/study_uic-class1_study_risks_class1.pdf.

Okla. Stat. Ann. tit. 52, §§ 36.1-36.7 (West 1951).

Oklahoma Natural Gas Co., a Division of ONEOK Inc. v. Mahan & Rowsey, Inc., 786 F.2d 1004 (10th Cir. 1986).

Park, M., 2018. South Korea's legal and regulatory system for carbon capture and sequestration: backgrounds, current circumstances, and recommendations. Journal of Korean Law 18, 235–268.

Pierce, D.E., 2009. Minimizing the environmental impact of oil and gas development by maximizing production conservation. North Dakota Law Review 85 (4), 759–779.

Pierce, D.E., 2011a. Carol Rose comes to the oil patch: modern property analysis applied to modern reservoir problems. Penn State Environmental Law Review 19 (2), 241–265.

Pierce, D.E., 2011b. Developing a common law of hydraulic fracturing. University of Pittsburg Law Review 72, 685−699.

Pierce, D.E., 2012. Oil and gas easements. Energy and Mineral Law Institute 34, 318−353.

Pomposini v. T.W. Phillips Gas & Oil Co., 580 A.2d. 776 (Pa. Super. 1990).

Pruess, K., Xu, T., Apps, J., Garcia, J., 2001. Numerical modeling of aquifer disposal of CO$_2$. SPE Journal 8 (1), 49. Available from: https://doi.org/10.2118/66537-MS.

Raymond v. Union Tex. Petroleum Corp., 697 F. Supp. 270 (E.D. La. 1988).

Reese Exploration, Inc. v. Williams Natural Gas Co., 983 F.2d 1514 (10th Cir. 1993).

Restatement (Second) of Torts § 159, 1965a.

Restatement (Second) of Torts § 159 cmt. G.,1965b.

Rook v. James E. Russell Petroleum, Inc., 679 P.2d 158 (Kan. 1984).

Safe Drinking Water Act, 42 U.S.C. §300h(d), 2006.

Section five: resolving underground resource conflicts around the world. In: Zillman, D.N., McHarg, A., Bradbrook, A., Barrera-Hernandez, L. (Eds.), The Law of Energy Underground: Understanding New Developments in Subsurface Production, Transmission, and Storage. Oxford University Press.

Sprankling, J.G., 2008. Owning the center of the Earth. UCLA Law Review 55, 979−1040.

Staben v. Anterra Energy Services, Inc., No. 56-2011-00394152-CU-FR-VTA, 2012 WL 10235812 (Cal. Super. November 13, 2012).

State of Washington Department of Ecology. Aquifer storage, recovery, & recharge. Retrieved from: https://ecology.wa.gov/Water-Shorelines/Water-supply/Water-recovery-solutions/Aquifer-storage-recovery-recharge.

State of Wyoming v. State of Colorado, 259 U.S. 419, 1922.

Sun Oil Co. v. Whitaker, 483 S.W.2d 808 (Tex. 1972).

Sunray Oil v. Cortez Oil Co., 112 P.2d 792 (Okla. 1941).

Tahoe-Sierra Pres. Council, Inc. v. Tahoe Reg'l Planning Agency, 535 U.S. 302, 2002.

Task Force on Carbon Capture and Geologic Storage, 2007a. Appendix I: model statute for geologic storage of carbon dioxide. Storage of Carbon Dioxide in Geologic Structures: A Legal and Regulatory Guide for States and Provinces. Interstate Oil and Gas Compact Commission, pp. 31−35.

Task Force on Carbon Capture and Geologic Storage, 2007b. Appendix II: model general rules and regulations. Storage of Carbon Dioxide in Geologic Structures: A Legal and Regulatory Guide for States and Provinces. Interstate Oil and Gas Compact Commission, pp. 36−47.

Task Force on Carbon Capture and Geologic Storage, 2007c. Appendix II: model general rules and regulations: § 3 (a). Storage of Carbon Dioxide in Geologic Structures: A Legal and Regulatory Guide for States and Provinces. Interstate Oil and Gas Compact Commission, p. 33.

Task Force on Carbon Capture and Geologic Storage, 2007d. Appendix II: model general rules and regulations: § 5(b)(3). Storage of Carbon Dioxide in Geologic Structures: A Legal and Regulatory Guide for States and Provinces. Interstate Oil and Gas Compact Commission, p. 42.

Tate v. United Fuel Gas Co., 71 S.E.2d 65 (W.Va. 1952).

Texas American Energy Corp. v. Citizens Fidelity Bank & Trust Co., 736 S.W.2d 25 (Ky. 1987).

Tomlinson, C., July 31, 2020. Oil industry could lose by gambling on carbon capture. Houston Chronicle. Retrieved from: https://www.houstonchronicle.com/business/columnists/tomlinson/article/Oil-industry-could-lose-by-gambling-on-carbon-15446419.php.

U.S. Const. amend. V.

U.S. Department of Interior, 2009. Framework for Geological Carbon Sequestration on Public Land. Report Submitted to the Committee on Natural Resources of the House of Representatives and the Committee on Energy and Natural Resources of the Senate in Compliance With Section 714 of the Energy Independence and Security Act of 2007 (P.L. 110-140, H.R.6). Washington, DC, 22p.

U.S. Energy Information Administration, January 10, 2020a. Average depth of crude oil and natural gas wells. Retrieved from: http://www.eia.gov/dnav/pet/pet_crd_welldep_s1_a.htm.

U.S. Energy Information Administration, August 20, 2020b. Weekly natural gas storage report. Retrieved from: http://ir.eia.gov/ngs/ngs.html.

U.S. Energy Information Administration, August 20, 2020c. Current issues & trends. Retrieved from: https://www.eia.gov/electricity/.

U.S. Environmental Protection Agency(a). Underground injection control (UIC): class I industrial and municipal waste disposal wells. Retrieved from: http://water.epa.gov/type/groundwater/uic/wells_class1.cfm.

U.S. Environmental Protection Agency(b). Underground injection control (UIC): class VI—wells used for geologic sequestration of CO_2. Retrieved from: https://www.epa.gov/uic/class-vi-wells-used-geologic-sequestration-co2.

U.S. Environmental Protection Agency(c). Protecting underground sources of drinking water from underground injection (UIC). Retrieved from: http://water.epa.gov/type/groundwater/uic.

U.S. Environmental Protection Agency(d). Underground injection control (UIC): class II oil and gas related injection wells. Retrieved from: https://www.epa.gov/uic/class-ii-oil-and-gas-related-injection-wells.

Union Gas Sys., Inc. v. Carnahan, 774 P.2d 962 (Kan. 1989).

United States Steel Corp. v. Hoge, 468 A.2d 1380 (Pa. 1983).

United States v. 43.42 Acres of Land, 520 F. Supp. 1042 (W.D. La. 1981).

United States v. Causby, 328 U.S. 256, 1946.

West Edmond Hunton Lime Unit v. Lillard, 265 P.2d 730 (Okla. 1954).

West Edmond Salt Water Disposal Ass'n v. Rosecrans, 226 P.2d 965 (Okla. 1950).

White v. N.Y. St. Natural Gas Corp., 190 F. Supp. 342 (E.D. Pa. 1960).

Wickstrom, L.H., Slucher, E.R., Baranoski, M.T., Mullett, D.J., 2008. Geologic Assessment of the Burger Power Plant and Surrounding Vicinity for Potential Injection of Carbon Dioxide. Ohio Department of Natural Resources.

Wilson, E., de Figueiredo, M., 2006. Geologic carbon dioxide sequestration: an analysis of subsurface property law. Environmental Law Reporter News & Analysis 36, 10114.

World Resources Institute (WRI), 2008. Integration among storage projects: basin-scale management. CCS Guidelines: Guidelines for Carbon Dioxide Capture, Transport, and Storage. WRI, Washington, DC, pp. 62–63.

Wyo. Stat. Ann. § 34-1-152(a), 2009.

Wyo. Stat. Ann. §§ 34-1-313 – 34-1-317, 2009.

Wyo. Stat. Ann. § 34-1-152, 2009.

Wyo. Stat. Ann. § 35-11-316, 2009.

Further reading

26 U.S.C. § 45Q(f)(4), 2018.

Council, L., Bergel, S., February 25, 2016. Everything you always wanted to know about class I injection wells in Texas [PowerPoint slides]. Retrieved from: http://www.gwpc.org/sites/default/files/event-sessions/Council_Lorrie_0.pdf.

Financial Times, April 23, 2019. Big oil's $5tn investment is incompatible with Paris deal. Retrieved from: https://www.ft.com/content/08453afc-61e8-11e9-b285-3acd5d43599e?segmentId = a7371401-027d-d8bf-8a7f-2a746e767d56.

Goldthorpe, S., 2017. Potential for very deep ocean storage of CO_2 without ocean acidification: a discussion paper. Energy Procedia 114, 5417–5429. Available from: https://doi.org/10.1016/j.egypro.2017.03.1686.

The United Kingdom's experience in Carbon Capture and Storage projects: the current regulatory framework and related challenges

Isabela Morbach Machado e Silva[1] and Hirdan Katarinade Medeiros Costa[2]

[1]Energy Institute, University of São Paulo, Brazil [2]Institute of Energy and Environment, University of São Paulo, São Paulo, Brazil

7.1 Introduction

In a political and social context in which there is a clear commitment to limit the global temperature increase to 1.5°C, developing technologies to mitigate greenhouse gas emissions is a central and urgent issue (IPCC, 2014). Carbon Capture, Utilization, and Storage (CCUS)[1] technologies are often highlighted as a critical component of the future low-carbon energy system (IEA, 2018).

Providing law and regulation has been one of critical elements for Carbon Capture and Storage (CCS) deployment. The reason is that CCS deployment raises several hidden issues, and the lack of responses creates legal uncertainty with ensuing economic risks (Global CCS Institute, 2018a). In the past two decades, national regulators and regional legislatures worldwide have established legal and regulatory frameworks to support CCS large-scale facilities and to pilot projects for demonstration, including public participation

---

[1] In that chapter, both technologies, CCUS and CCS, will be referred to as CCS. Note that CCUS is often presented as an alternative to CCS. In brief, the concept of CCUS includes utilization, when CO_2 can be to be recycled for further usage. Products—such as plastics and fuels—can be derived from CO_2, in a circular manner, as a replacement for fossil fuels. There is considerable debate about the difference between CCUS and CCS in terms of permanent geological CO_2 storage.

rules for decreasing concerns on lack of information and clarifying the risks of this new technology. However, defining a robust and complete regulatory rule remains a challenge.

The United Kingdom has made a recognized effort to establish a comprehensive legal framework for CCS (UK Department of Energy and Climate Change, 2012). According to the Global CCS Institute (2018a, p. 67), in 2018 the United Kingdom already had 65 CCS-specific laws or existing laws applicable across most of the CCS project cycle.

Like other countries, the United Kingdom has adopted a net-zero emissions target by 2050, following the Committee on Climate Change (IEA, 2020). Several decarbonization options are available to achieve this target, including resource and energy efficiency, extensive electrification, and a hydrogen economy development. The use of CCUS is strongly recommended by the Committee (Climate Change Committee, 2019).

After introducing the UK's energy sector and its mitigation plans to achieve net zero emissions, this chapter presents the core legislation on CCS applied to the United Kingdom, and, in the last section, it focuses on projects and efforts made by the United Kingdom government during the past two decades to implement CCS technologies. Note that the study is based on a literature review and cannot provide a jurisprudence analysis.

7.2 UK's energy sector

The industrial revolution in the United Kingdom had fertile land to grow, and to better understand a century of dependence on coal as a fuel, which considerably changed under Mrs. Thatcher's administration and the following prime ministers. This process of improving energy mix could be considered a result of decreasing internal coal production, as well as from energy import to meet the needs since the 1970s (Pearson and Watson, 2012).

Analyzing the UK's energy profile, BEIS (2019b) pointed out the following supply aspects: primary oil (crude oil and natural gas liquids) accounted for 43% of the total production, natural gas 30%, primary electricity (consisting of nuclear, wind, solar, and natural flow hydro) for 16%, followed by bioenergy and waste 10%, while coal accounted for the remaining 1%. As observed, hydrocarbons are still an important energy source in the United Kingdom.

Moreover, during the last 30 years, natural gas and primary electricity consumption have risen considerably, and bioenergy and waste have also grown. However, the consumption of oil and coal has fallen (BEIS, 2019b).

In the late 20th century, institutional reforms, legal arrangements improved by the Parliament, and the economic conditions on energy commodities allowed an increase in the combined cycle gas turbine plant (CCGT), nuclear energy generation, and a step-by-step reduction of conventional coal thermopower plants (Pearson and Watson, 2012). In addition, new renewable energy sources in the last decade, such as wind, support energy security even more reducing coal and nuclear facilities.

Therefore changing the UK's demand profile is still a significant challenge considering natural gas consumption, but the decrease in coal demand gives the United Kingdom credibility to meet its climate goals.

Regarding the UK's carbon dioxide emissions, BEIS (2019b) estimated that about 81% of the total UK anthropogenic greenhouse gas emissions in 2018 came from fossil fuels, that is, an estimated 448.5 million tons. This number is 2.5% lower than in 2017 and 44% lower than in 1990

(BEIS, 2019b). The decrease may be explained by a replacement in the fuel mix with electricity generation, with reduced use of coal and gas and increased use of renewables (BEIS, 2019b; IEA, 2019).

One may say that the United Kingdom is committed to an energy transition and to the adoption of low-carbon measures. According to the IEA, the United Kingdom has an integrated energy and climate policy framework. It keeps a strong international leadership in climate change, electricity (and gas) market reforms, and supply security (IEA, 2019). The country has been in the first line of recognizing climate change threats and adopting greenhouse gas (GHG) emission reductions by at least 80% by 2050 over the 1990 levels. Great efforts should be made at innovation, productivity, and competitiveness of the industry, combined with new institutional governance to meet this goal.

As the IEA report says, under the Climate Change Act, the UK government "has to set legally binding five-year caps on emissions—carbon budgets—12 years in advance and then publish a report with policies and measures to meet that budget and the previous ones. Five carbon budgets have been set to date; the fifth one was set in July 2016 and covers the period 2028–32" (IEA, 2019, p. 26).

In 2017 the Clean Growth Strategy was released to achieve its goals, forecasting a £2.5-billion investment in low-carbon innovation over 5 years. Among the key aspects highlighted were:

- Accelerating the shift to low-carbon transport
- Improving energy efficiency in business and industry
- Improving the energy efficiency of homes
- Fostering the rollout of low-carbon heating
- Delivering clean, smart, and flexible power

In 2019 the Climate Change Act 2008 was amended to set a new target of achieving net-zero GHG emissions by 2050 (Department for Business Energy and Industrial Strategy, UK, 2019a).

Recently, the UK government announced a 10-point plan for a green industrial revolution, including CCS investment as one of the keys for net zero, and proposed to provide public support via the CCUS Infrastructure Fund to establish two operational "SuperPlace" clusters of CCUS.

Therefore the challenge is to implement the real support for cleaning technologies, such as CCS, to achieve net-zero emissions linked to a creative economic scenario post-Covid, where recovery will be a priority.

7.3 UK's CCS regulatory framework

As aforementioned, the United Kingdom[2] is one of the most advanced countries in terms of CCS regulation. There are more than 60 laws regarding CCS. This study is unable

---

[2] It is crucial to notice that from January 31, 2020, the United Kingdom is no longer a EU Member State. Its exit from the EU has several impacts in terms of legislation and regulations. Following the *Withdrawal Agreement*, the United Kingdom has entered an implementation period, during which it remains subject to EU law. That is why European Instruments are considered for the proposal of this work.

to encompass all the established rules. In that context, the present analysis focuses on the core regulation, including national and international instruments.

First, the United Kingdom has been participated in several international instruments that prescribe how CO_2 storage must be regulated, especially offshore storage. Besides international instruments, there are also national Act and specific regulations which address CO_2 storage. The primary legislation identified is the Convention on the Prevention of Marine Pollution by Dumping of Wastes and Other Matter 1972 (also known as "London Convention"), amended by the London Protocol; the Convention for the Protection of the Marine Environment of the North-East Atlantic (the "OSPAR Convention"); EC Directive on Geological Storage of Carbon Dioxide (Directive 2009/31/EC).[3] National legislation is the Energy Act 2008; Planning Act 2008; and Petroleum Act 1998.

The London Convention is one of the first global conventions established to protect the marine environment from human activities. Currently, 87 States are Parties to this Convention. According to Article 2, the main objectives are to "protect and preserve the marine environment from all sources of pollution and take effective measures (...) to prevent, reduce, and where practicable eliminate pollution caused by dumping or incineration at sea of wastes or other matter." Resulting from some amendments to the London Protocol, the CO_2 streams matters have been included as waste that may be considered for dumping in Annex 1[4]:

> Annex 1
> 1.8 Carbon dioxide streams from carbon dioxide capture processes for sequestration. (...)
> 4 Carbon dioxide streams referred to in paragraph 1.8 may only be considered for dumping, if:
> 1 disposal is into a sub-seabed geological formation; and
> 2 they consist overwhelmingly of carbon dioxide. They may contain incidental associated substances derived from the source material and the capture and sequestration processes used; and
> 3 no wastes or other matter are added for the purpose of disposing of those wastes or other matter.

Another legislation is the Convention for the Protection of the Marine Environment of the North-East Atlantic signed in 1992 (also known as "OSPAR Convention") and entered into force on March 25, 1998. After the amendments adopted in 2007 by the OSPAR Commission, those changes were designed to allow offshore CO_2 storage activities. As reported by Milligan (2014, p. 165), the amendment established four conditions to permit the dumping of CO_2 streams:

> the first three of these conditions are identical in substance to those found in the 1996 London Protocol. The fourth condition is considerably more restrictive—CO_2 streams must be 'intended to be retained' permanently in subsoil geological formations and must not 'lead to significant adverse consequences for the marine environment, human health, and other legitimate uses of the maritime area.

---

[3] Directive 2009/31/EC has been applied to the United Kingdom because of its past membership of the EU.

[4] Report of the 28th Consultative Meeting of the Contracting Parties to the London Convention and the 1st Meeting of Contracting Parties to the London. Protocol (IMO Document LC 28/15; October 30–November 3, 2006). Annex 6: "Resolution LP.1(1) on the amendment to include CO_2 sequestration in the subseabed geological formations in Annex 1 to the London Protocol."

In 2009, the European Parliament and the European Union (EU) Council adopted Directive 2009/31/EC (CCS Directive) on carbon dioxide geological storage. As defined in Article 1, the Directive "establishes a legal framework for the environmentally safe geological storage of carbon dioxide (CO_2) to contribute to the fight against climate change". Member States from the EU were requested to comply with the Directive by June 25, 2011, and a large majority have completed transposition. As stated by Haan-Kamminga et al. (2017, p. 241),

> Measures to be adopted by the Community should comply with the principles of subsidiarity and proportionality. Consequently, the Directive does not go beyond what is necessary to achieve the objective of setting a legal framework for the storage of CO. The implication is that the Member States still have ample opportunity to go beyond the measures taken by the Community.

At the national level, the regulatory framework regarding licensing offshore CCS has been introduced in the United Kingdom through the Energy Act 2008. This legislation has transformed several energy infrastructure aspects and market regulation in the United Kingdom.

The Law declares the Crown's rights over the Exclusive Economic Zone (EEZ) to perform "gas storage" (with or without the purpose of enhanced oil recovery—EOR) in compliance with the provisions of the 1982 United Nations Convention on the Law of the Sea. It also determines that the Government can also designate "Gas Importation and Storage Zones" within the EEZ. A concession will be required from operators who intend to carry out CCS activities within the newly appointed EEZ.

Subsequently, the legal framework introduced a regime based on licensing activities, which requires from all activities related to CO_2 storage (aiming at elimination–storage–permanent) a license emitted from the competent authority. These activities also include the conversion, exploration, and maintenance of the controlled site that will be used for CO_2 storage. The exact place refers to a location that can be found "within, below, or over" the territorial sea or waters inside a Gas Importation and Storage Zone.

The Law also provides more detailed rules about granting licenses, content, terms, and conditions attached to them. It establishes that the licensing authority responsible for granting licenses will be at a political decision level, such as the Secretary of State or the Ministers, depending on the place the license is granted to.

When granting a license, the authority can define requirements for a specific case or for the license to be granted. A license may include provisions regarding financial security concerning future obligations and obligations between the closure of an installation and the license termination. Each licensing authority can make further regulations regarding the terms and conditions of the licenses it grants.

There is also a detailed section on the execution of licenses and practices considered criminal and introducing sanctions in cases where activities are carried out without a license or when the license holder does not comply with the prescribed conditions. The licensing authority can provide guidance, a kind of conduct adjustment requiring the licensee to take the necessary measures. To assist in the Law's functions, inspectors may be appointed by the Secretary of State or by Ministers.

The United Kingdom Planning Act (2008) defines general rules on infrastructure and may be applied to CCS pipelines and facilities. For instance, the National Policy Statements states that projects must contain an "explanation of how the policy set out in the statement takes account of Government policy relating to the mitigation of, and adaptation to, climate change"; and the Nationally Significant Infrastructure Projects (NSIP) include "underground gas storage facilities" as well as the "construction of a pipeline."

Regarding offshore installations for carbon storage installations, the Law applies the Petroleum Act 1998. A detailed program of plans and approvals is foreseen, determining who intends to terminate the use of an offshore installation, providing an "abandonment program" that establishes the measures to be taken in connection with the abandonment of an offshore installation or subsea pipeline. In addition, the Energy Act 2008 provides for the licensing authority to formulate regulations on license termination, which may include provisions on financial security.

In September 2010, the UK Parliament passed the Carbon Dioxide Storage Regulations. These Regulations were approved under the Energy Act 2008 and introduce a licensing regime for offshore CCS activity, describing the requirements to be met to grant a storage license or authorization by the Secretary of State and the scope of these permissions.

Such regulations partially comply with the UK's past obligation to implement the EU Directive on carbon dioxide geological storage (Directive 31/2009/EC) for UK domestic law. Among the issues covered by the EU Directive, the regulation covered: the conditions for granting licenses and operating licenses, the storage operator's commitment, the closure of the storage location; the postclosure period; and financial security. However, the regulations do not cover the minister's licenses for CCS activities carried out in adjacent territorial waters.

As can be seen, most of the concerns focus on trying to equip mechanisms to predict and to manage possible environmental damage, not only those already expected but also those not yet studied, ensuring a time guarantee for monitoring and accountability of the parties exploring the CCS activity. However, for this activity to be viable, conditions that do not prevent it have to be imposed.

Note that the use of CO_2 for EOR is not submitted to the regulatory framework established under the Energy Act 2008. The CO_2 storage from EOR operations remains unregulated under the Energy Act 2008.

In turn, the CO_2 storage because of EOR operation is regulated under the Petroleum Act 1998. Unfortunately, the Petroleum Act 1998 and associated regulations do not establish detailed provisions concerning CO_2 storage. It provides a detailed basis for regulating these activities to the extent that they are used to improve oil exploitation during EOR operations.

In sum, after analyzing the UK framework, we find a sense of creating a regulatory framework (for instance, a General CCS Law), followed by the current sparse regulation. However, it aims to build a safe and foreseeable legal and institutional environment. This contradictory aspect may be seen in cases explored ahead.

7.4 UK's CCS projects

According to Hawkes (2019), there are 574 possible storage sites offering 78 Gt or 200 years of the UK's total carbon dioxide emissions at 2016 levels. If one consider the projects and efforts made by the UK government during the past two decades to implement CCS technologies, one may observe an irregular or shifting behavior regarding stimulus policy for CCS projects.

The termination of several projects (e.g., Longannet,[5] the total cost of which was estimated at £1,9 billion in July 2010) shows some of the challenges, such as securing financing for large-scale CCS projects; policy uncertainties are not issues easy to solve. For instance, Hawkes (2019) pointed out that barriers to Kingsnorth and Longannet projects, such as the 1-year delay in inviting to negotiate (July 2008–09), as well as BP's withdrawal.

The National Audit Office (2012) report listed a number of challenges to these projects: cost of operational support not dealt with early enough, rapidly evolving policy/regulatory scenario, allocation of risk to Government required by the industry, too restrictive (only 300–400 MW postcombustion coal), poorly managed procurement, lack of fit with strategic vision, and the costly liability for leakage.

Hawkes (2019) classified those projects as the first wave of CCS in the United Kingdom. In 2007, the competition launched, followed by an award (2009) and by an expectation of operation (2014).

Pearson and Watson (2012, p. 1) examined a long-term perspective on UK energy policies from 1980[6] to 2010, aiming to "offer some reflections on what has changed (and what has not), and what lessons might be learned." They noted that in 2003, despite the UK government's announcement[7] that they would support "relevant research projects" for low carbon emission, technologies such as CCS were not seen as centrally necessary in the short- and medium-term. The White Paper of 2007 showed a change in perspective because it was announced that "the United Kingdom was going to take the next step with respect to the emerging field of CCS technologies. A competition to build the UK's first full-scale demonstration project for CCS was announced" (Pearson and Watson, 2012, p. 27).

The next step was the CCS Roadmap publishing in 2012 by the UK government. It established a program of measures and interventions to enable a CCS reduction in costs. In that context, the United Kingdom Government announced a "CCS Commercialization Programme with £1 billion in capital funding to support commercial-scale CCS, targeted specifically to learn by doing and to share resulting knowledge to reduce the cost of CCS such that it can be commercially deployed in the 2020s" (UK Department of Energy and Climate Change, 2012, p. 7). This moment was characterized as a second phase (according to Hawkes, 2019). It included a CCS commercialization program, cost reduction taskforce, operational support through contracts for differences, R&D program, supply chain support, and international engagement.

---

[5] NAO, 2012. Carbon capture and storage: lessons from the competition for the first UK demonstration. Report HC 1829, Session 2011–12. National Audit Office, London.

[6] The year in which the Parliamentary Group for Energy Studies was founded.

[7] Department of Trade and Industry, 2003. Energy White Paper: Our Energy Future—Creating a Low Carbon Economy. HMSO, London, p. 12.

According to Spence et al., the mentioned program was "a Competition for FEED funding, a capital grant, sharing CCS risks, and a negotiated contract for difference. The competition was launched in April 2012, and there were eight bids. Two bids were selected to go forward to the FEED stage" (2014, p. 6259). One was the White Rose CCS Project, and the other was the Peterhead project.

The White Rose CCS Project was planned to be a 450 MW coal-based power station fitted with oxy-combustion CCS technology capable of providing clean power to 630,000 homes. According to Hackett and Ficheme, "approximately 2 million tons per annum of CO_2 were to be transported and stored in an offshore geological saline formation (Endurance) located offshore in the Southern North Sea" (2016, p. 2).

The other project was called *Peterhead Project*. Located next to the village of Boddam in Aberdeenshire, the project involved the capture of CO_2 from industrial-scale gas-fired power. After being captured, it would be transported through a pipeline and stored at the Goldeneye Platform (Spence et al., 2014).

Despite those prominent projects, the Government stepped back one more time, and its initiative did not reach its goal. In 2015 the UK government canceled that £1 billion competition. It happened less than 1 year before it was due to be awarded. Two projects had been in the process of constructing a CCS demonstration plant at commercial scale.

The cancelation of the UK competition represented a withdrawal of government financial support. This inefficient decision-making caused the industry to mistrust of the Government and may have had a negative impact on private investment participation for future government programs to support CCS. In that sense, Gross explains:

> The cancellation of the planned CCS Commercialisation Programme was a significant setback to CCS development in the United Kingdom. There is substantial scope to improve the allocation of risks, ensuring they are allocated to the party best-placed to absorb or manage them, thus allowing industry participants to access lower capital costs. A new CCS strategy requires detailed consideration of ownership structures, policy instruments, and the allocation of risk between public and private sector participants. It will be particularly important for the Government to consider how to best manage the risks associated with stores. Therefore there is a need for action on three main fronts: separate funding for near-term capture and infrastructure projects, an approach to risk allocation for CO_2 storage sites, and strategy and regulation to allow the industry to invest in T&S infrastructure.

> There is now an opportunity to progress an alternative CCS development strategy that can build on lessons from the competition and provide a pathway to cost-effective CCS deployment during the mid-2020s. However, there is significant urgency for the Government to progress an alternative route to real CCS deployment by the mid-2020s, as failure to prove the deployment of CCS within these timescales will increase the costs and risks of meeting carbon targets. Gross (2016)

In late 2017, as part of the ups and downs regarding UK CCS projects support, the Government stated its new approach in the Clean Growth Strategy (Department for Business Energy and Industrial Strategy, UK, 2019a). This third phase was characterized as a broad aspect that included CO_2 uses. Thus a CCUS cost challenge taskforce around pipelines of financeable projects was structured.

It reveals one more push trying to enable the United Kingdom to become a global technology leader for CCUS and scale up the CCUS deployment during the 2030s. The Government presented a report to the UK Parliament regarding reducing UK emissions.

It was stated that "As set out in our ambitious CCUS Action Plan, we are shifting our focus to initial deployment, with an ambition to deliver the UK's first CCUS facility in the mid-2020s and to have the option to deploy at scale during the 2030s" (Department for Business Energy and Industrial Strategy, UK, 2019a, p. 58).

This strategy sets out the British proposals for decarbonizing all sectors of the UK economy through the 2020s. It explains how the whole country can benefit from low-carbon opportunities while meeting national and international commitments to tackle climate change (Department for Business Energy and Industrial Strategy, UK, 2019a).

According to the Global CCS Institute, "In its ambitious industrial policy, the UK government recognized CCUS as an important technology in its transition to a low-carbon economy" (Global CCS Institute, 2018b, p. 68).

Although the UK policy for CCS had not been prosperous and stable during the past decades, the current approach reaffirms its aspiration of enabling the United Kingdom to become a global CCS leader.

In a research project funded by the UK Energy Research Centre, Chalmers et al. (2013) analyzed "the main uncertainties facing potential investors in CCS and policymakers wishing." The paper presents historical case studies about technology implementation and try to "develop three potential pathways for CCS deployment in the United Kingdom over the period to 2030" (2013b, p. 7668). The pathways presented were:

Pathway 1: "On track" is a broadly successful pathway, with a plausibly high level of CCS deployment. By 2030, CCS has an established position as a technically proven and financially viable option, and is competitive with other low-carbon electricity generation technologies.

Pathway 2, Variant A: "Momentum lost." Commercial-scale demonstration of CCS goes ahead, and is quickly followed by further deployment up to the mid-2020s. By this time, CCS has established itself as technically viable, but from the mid-2020s onwards, it is not generally a preferred option as part of the low-carbon generation mix in the United Kingdom. Financial viability ends up being marginal.

Pathway 2, Variant B: "Slow and sporadic." Commercial-scale demonstration of CCS does go ahead, followed by limited further deployment up to 2030. CCS has established itself as technically viable, but it is not generally a preferred option as part of the low-carbon generation mix in the United Kingdom. Financial viability remains marginal with deployment in particular market niches only.

Pathway 3: 'Failure' No CCS deployment beyond a limited demonstration programme. (2013, pp. 7675–7676).

Suppose we apply the methodology described by Chalmers et al. (2013) to the current UK scenario; we will find something between Pathway 2 variant A and variant B. There are not several commercial-scale demonstrations of CCS yet. Still, the UK government keeps trying to support and deploy CCS as an indispensable technology with the potential to decarbonize the economy, including the most recent The Ten Point Plan for a Green Industrial Revolution, which still needs more detailed public policy steps by the Government to demonstrate whether 10 points are feasible regarding economic constraints.

7.5 Public participation

The extensive international literature has pointed out that public acceptance in its broad concept needs to be part of the design of CCS projects (OECD/IEA, 2005; Malone, 2005;

Curry et al., 2005; Reiner et al., 2006; Van Alphen et al., 2007; Shackley et al., 2009; Markusson et al., 2012; Karimi and Toikka, 2014). Mainly because there is mistrust regarding the possible environmental impacts of CCS technology (IPCC, 2005). Thus during the environmental license process, public participation rules may help increase public acceptance of CCS projects (Musarra and Costa, 2019).

Considering the information made available by authorities in the United Kingdom Energy Act (2008), there is a provision that the Secretary of State must maintain a register with information related to the sequestration licenses granted; however, a no disclosure rule exists when the information involves national security, or when it may damage commercial interests. Moreover, it is essential to highlight that the register is available for inspection by the public free of charge, and that members of the public facilities may obtain copies of entries upon the payment of a fee.

According to Lewis and Westaway (2020), under the analysis of the Planning Act 2008 and Energy Act 2008, there is tension between the implementation of CCS projects and the need for public participation. For instance, under the Planning Act, "permission for new carbon capture facilities and onshore CO_2 pipelines falls within the streamlined consent procedure for NSIP" (Lewis and Westaway, 2020, p. 140).

Thus, considering the simplified system within the Planning Act, public participation will be related to a public consultation on the draft National Policy Statement. Its content reflects details of a specific project that should be approved within 12 months; a consultation of the interest during the preparation of the project applied to the Secretary of State; and the making of representations and the submission of evidence to, or appearance at hearings to consider the NSIP, all before an examining authority is appointed by the Secretary of State (Lewis and Westaway, 2020, p. 140).

Although the rules on public participation can be improved, Lewis and Westaway (2020) concluded that in the CCS projects analyzed, such as the White Rose CCS Project and the Yorkshire–Humber Pipeline, those provisions from the Planning Act worked well. We can say that they are valuable examples of consenting procedures with little dissent. According to Lewis and Westaway (2020), the UK's CCS experience is mainly based on financial difficulties and policy barriers rather than on public concerns. However, public participation has still not been precisely tested in the United Kingdom because of the absence of executed projects.

7.6 Final considerations

This chapter put forth that CCS has become not just a prominent but a critical instrument for reducing carbon emissions in both international and UK policy debates. In that perspective, the UK government stated through a report that it led the world in developing a regulatory framework to enable CCS implementation (UK Department of Energy and Climate Change 2012, p. 36).

Besides the UK government's aims, deployment a technology as CCS shall be more challenging than initially planned. Many supportive initiatives to deploy large-scale CCS facilities were launched. However, none has met its final goal so far; more than that, there was a signal of instability in terms of government financial support over the past decade.

In terms of regulation, we found that offshore CO_2 storage in the United Kingdom is planned for and regulated under a dense mixture of sectoral and fragmented laws and by different public bodies. This regulatory complexity is often characterized as the source of adverse consequences at national, regional, and international scales. Also, we found rules on public participation in the UK's CCS national framework, although this concern has still not been precisely tested because projects have not gone ahead.

Combining the volatility of supportive policies from the UK government with regulatory complexity, we can highlight at least two aspects that will need to be addressed in the near future: (1) inefficient decision-making and (2) ineffective policy.

Over the last decades, inefficient decision-making was observed when the UK government initiatives toward supporting CCS have wavered. This unstable support gives a weak message to investors and to the private sector.

Consequently, there is an ineffective policy. Assuming the right policy would efficiently deliver decarbonization, the opposite could result in higher costs. One example applied to our finding is that the uncertainty regarding future policy support for technology can increase the economic and policy risk, resulting in higher financing costs and less interest from the private sector.

Considering that investors' confidence in new capital-intensive activities (i.e., CO_2 storage) is especially sensitive to regulatory risk types, some measures need to be taken to enable this complex regulatory framework to operate in a coherent, coordinated manner.

Overall, the challenges of making the necessary emission reductions are technical, political, and regulatory. Developed countries, such as the United Kingdom, typically have a strong institutional capacity and are well placed to take the lead and provide supporting mechanisms to deploy critical technologies, such as for CCS and GHG removal. Nevertheless, the level of CCS large-scale facilities in the United Kingdom shows that considerably more work will have to be done.

Acknowledgments

We are grateful to the "Research Center for Gas Innovation—RCGI" (Fapesp Proc. 2014/50279-4), supported by FAPESP and Shell, organized by the University of São Paulo, and the strategic importance of the support granted by the ANP (National Agency of Petroleum, Natural Gas and Biofuels of Brazil) through the R&D clause. We also thank the support from the National Agency for Petroleum, Natural Gas and Biofuels Human Resources Program (PRH-ANP), funded by resources from the investment of oil companies qualified in the R,D&I clauses from ANP Resolution No. 50/2015 (PRH 33.1—Related to Call No. 1/2018/PRH-ANP; Grant FINEP/FUSP/USP Ref. 0443/19).

References

Van Alphen, K., Voorst, Q.V., Hekkert, M.P., Smits, R.E.H.M., 2007. Societal acceptance of carbon capture and storage technologies. Energy Policy 35 (8), 4368–4380. ISSN 0301-4215.

Chalmers, H., et al., 2013. Analysing uncertainties for CCS: from historical analogues to future deployment pathways in the UK. Energy Procedia 37, 7668–7679. Available from: http://doi.org/10.1016/j.egypro.2013.06.712.

Climate Change Committee, 2019. Net-Zero: The UK's Contribution to Stopping Global Warming. CCC. Available online: https://www.theccc.org.uk/publication/land-use-policies-for-a-net-zero-uk.

Curry, T., Reiner, D.M., Ansolabehere, S., Herzog, H.J., 2005. How aware is the public of carbon capture and storage? In: Rubin, E.S., Keith, D.W., Gilboy, C.F., Wilson, M., Morris, T., Gale, J., Thambimuthu, K. (Eds.), Greenhouse Gas Control Technologies, vol. 7. Elsevier Science Ltd., pp. 1001–1009. ISBN 9780080447049.

Department for Business Energy and Industrial Strategy, UK, 2019a. Leading on Clean Growth. The Government Response to the Committee on Climate Change's 2019 Progress Report to Parliament—Reducing UK Emissions.

Department for Business, Energy and Industrial Strategy (BEIS), 2019b. UK Energy in Brief. < www.gov.uk/government/statistics/uk-energy-in-brief-2019 > .

Global CCS Institute, 2018a. Legal and Regulatory Indicator (CCS-LRI). Available online: https://www.globalccsinstitute.com/wp-content/uploads/2020/04/CCS-Legal-and-Regulatory-Indicator-CCS-LRi-DIGITAL-2.pdf. Accessed on 26 December 2020.

Global CCS Institute, 2018b. The Global Status of CCS 2018. < www.globalccsinstitute.com > .

Gross, R., 2016. CCS in the UK: A New Strategy. Advisory Group Report. A Report for the Committee on Climate Change.

Haan-Kamminga, A., Roggenkamp, M., Woerdman, E., 2017. Legal uncertainties of carbon capture and storage in the EU: the Netherlands as an example. Carbon and Climate Law Review 4 (3), 240−249.

Hackett, L.A., Ficheme, C., 2016. Commercialisation of CCS—What Needs to Happen? < https://www.semanticscholar.org/paper/Commercialisation-of-CCS-"-What-needs-to-happen-"-Hackett/8cf6095638ee707827-f44e45de7fc4c481b06958?p2df > .

Hawkes, A., 2019. CCS in the UK: 3rd Time Lucky? Lecture in Energy Transition Research & Innovation Conference, 1−2 October. University of São Paulo, Brazil.

IEA, 2018. IEA Publications World Energy Outlook 2018. < www.iea.org > .

IEA, 2019. Energy Policies of IEA Countries: United Kingdom 2019 Review. pp. 1−220. < www.iea.org/t&c/ > .

IEA, 2020. The role of CCUS in low-carbon power systems. Available online: https://www.iea.org/reports/the-role-of-ccus-in-low-carbon-power-systems. Accessed on 26 December 2020.

IPCC, 2014. Mitigation of Climate Change. Contribution of Working Group III to the Fifth Assessment Report of the Intergovernmental Panel on Climate Change. Cambridge University Press, Cambridge, United Kingdom and New York.

IPCC, 2005. Special Report on Carbon Dioxide Capture and Storage. In: Metz, B., Davidson, H.C.D.C.O., Loos, M., Meyer, L.A. (Eds.), IPCC Special Report on Carbon Dioxide Capture and Storage. B. Prepared by Working Group III of the Intergovernmental Panel on Climate Change. Intergovernmental Panel on Climate Change.

Karimi, F., Toikka, A., 2014. The relation between cultural structures and risk perception: how does social acceptance of carbon capture and storage emerge? Energy Procedia 63, 7087−7095. ISSN 1876-6102.

Lewis, M., Westaway, N., 2020. Public participation in UK CCS planning and consent procedures. In: Havercroft, I., Macrory, R., Stewart, R. (Eds.), Carbon Capture and Storage Emerging Legal and Regulatory Issues (second ed.). Bloomburry Publishing, Hart Publishing.

Malone, E.L., 2005. Finding a way: the potential for adoption and diffusion of carbon dioxide capture and sequestration technologies, Greenhouse Gas Control Technologies, vol. 7. Elsevier Science Ltd, Oxford, pp. 1531−1536, ISBN 9780080447049.

Markusson, N., Kern, F., Watson, J., Arapostathis, S., Chalmers, H., Ghaleigh, N., et al., 2012. A socio-technical framework for assessing the viability of carbon capture and storage technology. Technological Forecasting and Social Change 79 (5), 903−918. ISSN 0040-1625.

Milligan, B., 2014. Planning for offshore CO_2 storage: law and policy in the United Kingdom. Marine Policy 48, 162−171. Available from: http://doi.org/10.1016/j.marpol.2014.03.029.

Musarra, R.M.L.M., Costa, H.K.M., 2019. Comparative international law: the scope and management of public participation rights related to CCS activities. Journal of Public Administration and Governance 9, 93−109.

OECD/IEA, 2005. Legal Aspects of Storing CO_2. OECD/IEA, Paris.

Pearson, P., Watson, J., 2012. Energy Policy UK Energy Policy 1980−2010. A History and Lessons to be Learnt. A Review to Mark 30 Years of the Parliamentary Group for Energy Studies. London.

Reiner, D., Curry, T., De Figueiredo, M., Herzog, H., Ansolabehere, S., Itaoka, K., et al., 2006. An international comparison of public attitudes towards carbon capture and storage technologies. In: Eighth International Conference on Greenhouse Gas Control Technologies (GHGT-8). Trondheim, Norway.

Shackley, S., Reiner, D., Upham, P., Coninck, H., Sigurthorsson, G., Anderson, J., 2009. The acceptability of CO_2 capture and storage (CCS) in Europe: an assessment of the key determining factors: Part 2. The social acceptability of CCS and the wider impacts and repercussions of its implementation. International Journal of Greenhouse Gas Control 3 (3), 344−356. ISSN 1750-5836.

Spence, B., Denise, H., Owain, T., 2014. The peterhead-goldeneye gas post-combustion CCS project. Energy Procedia 63, 6258−6266.

UK Department of Energy and Climate Change, 2012. CCS Roadmap: Supporting Deployment of Carbon Capture and Storage in the UK. London. <http://www.decc.gov.uk/assets/decc/11/cutting-emissions/carbon-capture-storage/4899-the-ccs-roadmap.pdf>.

United Kingdom Energy Act, 2008. United Kingdom, Northern Ireland, Scotland, Wales. <https://www.legislation.gov.uk/ukpga/2008/32/contents>.

United Kingdom Planning Act, 2008. United Kingdom, Northern Ireland, Scotland, Wales. <https://www.legislation.gov.uk/ukpga/2008/29/contents>.

Regulatory framework carbon capture, utilization, and storage in Europe: a regulatory review and specific cases

Israel Araujo Lacerda de*, Vitor Emanoel Siqueira Santos and Hirdan Katarina de Medeiros Costa

Institute of Energy and Environment, University of São Paulo, São Paulo, Brazil

8.1 Introduction

The European Union has been a critical player to implementing carbon capture and storage (CCS) as a role due to it's projects and advances in regulating this activity. In the last decades some countries have shown concerns about climate emissions issues from their industrial hubs, such as Norway's oil industrial clusters and the Amsterdam—Rotterdam—Antwerp fossil fuel cluster, and supporting or entirely dependent industries (steel, chemical, and fertilizer industries, energy sector, or maritime logistics). However, from the high-level political decision to emerge a new business based on decarbonizing those clusters, a complex and detailed tangle of norms needs to be put in place and then reduce the CCS projects' risks. IEA (2019) makes some policy recommendations that will be analyzed in this chapter, such as (1) supporting the development and deployment of the CCS industry's CCS activity in carrying out the technological least-cost portfolio required to achieve climate and energy goals; (2) identifying and prioritizing competitive CCS opportunities in industry to provide learnings and support infrastructure development; and (3) creating or incentivizing the development of CCS hubs in industrial operations with shared transport and storage infrastructure to reduce costs for facilities incorporating carbon capture into production processes.

Even though EU members have approved the main rules under CCS Directive, according to Elkerbout and Bryhn (2019), the new decarbonizing business based on carbon

* http://lattes.cnpq.br/9074391888862438

dioxide storage has not received regulatory attention in order to create certainty for market actors, especially those that operate across multiple countries and for long periods. Regarding EU experience, those scholars understand that the CCS Directive for CCS technology since 2009 is not modernized enough. It should be reviewed to contemplate improving investments in symbiotic sectors, for instance.

Indubitably the European Union is a multifaceted system, and state members may have diverse pathways to achieve climate targets. Perhaps, Germany and Spain have understood that CCS, as an independent industry, may not be feasible in terms of internal return rate (EC, 2018). On the other hand, some members have built infrastructure under rules on setting aside land for future carbon dioxide capture facilities, such as recent power plants in the Czech Republic, Estonia, and Poland (COM, 2017). Other national initiatives have been proven effective, such as the case of Norway and Sweden.

Considering this scenario, this chapter will provide an overview of Directive 2009/31/EC with a description of its content and main features, including a summary discussion on technical assessment. In Section 8.3, we will examine Norway, for example, that new gas-fired power plants must apply carbon dioxide capture and storage from the start of power plant operation (COM, 2013, 2017, 2019). Finally, the Dutch case will be analyzed, mainly through the attempted and concluded projects, and how the policy implementations affected those.

8.2 European Union regulatory framework on CCS

Considering the European Council's goals to reduce the community's overall greenhouse gas (GHG) emissions by at least 20% and 50% below 1990 levels by 2020 and by 2050, respectively, the economic sectors should prepare themselves to incorporate low-carbon technologies (EC, 2009). Thus they approved the Directive 2009/31/EC on April 23, 2009 to build the minimal framework necessary to deal with the geological storage of carbon dioxide technology and support climate change challenge (EC, 2009).

1. The EU-CCS's Directive brings a purpose related to safe geological storage of CO_2 in a permanent way to prevent and eliminate risk to the environment and human health. So, on the selection of storage sites, member states should assess the storage capacity available in parts or the whole of their territory. Also, EU-CO_2 Storage's main rules regarding granting state that any storage exploration site must occur under an exploration permit (article 5). It is possible to consider that the CCS Directive represented an essential milestone for climate change mitigation policies in Europe. It has provided mechanisms to ensure environmentally safe geological storage, rules for transporting, and site choice as adequate liability for damage to health and property. The outlook was for gradual and continuous growth in activity throughout its implementation.

Thinking about critical points, the EU states had to adopt measures to reduce uncertainties in the carbon storage framework, such as:

1. allowing geological sequestration of carbon dioxide when the risk of its leakage is negligible after decommissioning;
2. permission process without discrimination, however, that provides priority to the responsible for the site's operation;
3. the operator must have the technical and financial capacity to support business risks;
4. storage restricted to carbon dioxide;
5. rules for loss of license by the operator in case of leakage;
6. the competent authority must take responsibility only 20 years after decommissioning, or before if necessary, in which case it can draw on financial security made by the operator;
7. monitoring cost will be covered by the operator at least for 50 years (including Competent Authority fee to cover at least 30 years of monitoring costs); and
8. free access to essential facilities under specific conditions.

Previous legal and regulatory frameworks, such as general oil industry's rules, provide a background to EU-CCS's Directive, and countries that want to have storage exploration permits project must use transparent and nondiscriminatory criteria to promote bid rounds for those with the technical, juridical, and financial capacity to participate, and its contractual rules need to be assertive in the area target, over a reasonable period to model geologic storage formation, and allows the business agents involved to have the right to access essential superficial areas and geological formations to develop facilities needed to relevant carbon dioxide storage. Finally, it is relevant to ensure fair rules for user access to a new storage business (Costa, Musarra, and Miranda, 2018).

Chapter 4 of the EU CO_2 Storage Directive states rules on the project life cycle, including operation, closure, and postclosure obligations. For instance, legal frameworks of EU countries have to establish rules for keeping under the operator's responsibility the monitoring process of the whole facilities of the storage complex and underground carbon dioxide plume, and the comparison between carbon dioxide behavior's theoretical model and the resulting plume after injection, noticing important anomalies and leakage of GHG (EC, 2009).

The prevention rules of the EU require that leakage events or irregularities in the project be notified (each country has defined who will be the competent authority to receive it) and be corrected according to the corrective measures plan to be followed by the regulated agents (EC, 2009). Besides that, the CCS complex's decommissioning process could be started by the operator's agent or by the regulator office. Considering that the permitting process has pointed to the main conditions needed to submit a request for the competent authority, once the CCS storage site has met such closure requirements, it will be possible to transfer liability to the government. For instance, the operator shall have to demonstrate the absence of any detectable leakage; and the provision of a minimum period, no shorter than 20 years (EC, 2009).

One of the challenges is ensuring that suitable offshore infrastructure can be maintained while CCS projects are developed instead of prematurely decommissioned and removed (IOGP, 2019). An alternative is to authorize new state-owned companies or the potential operators to ownership of the subsea infrastructure required for CCS, such as wells, pipelines, valves, etc. This would ease the process of changing incumbent operators for a new one responsible for managing essential facilities and decommissioning them at the sites' closure. It is also important that potential CO_2 storage operators access the geological data for the depleted oil and gas reservoirs and other explored areas. This is a challenging subject since there is considerable strategic information to be obtained, and there is a cost associated with the logs, monitoring reports, and production history from previous operators.

The CCS Directive has been implemented for almost all EU members and its guidance is under the European Commission. However, it was observed, since 2011, that the CCS Directive establishes boundary conditions for carrying out CCS projects, mainly to the consumer and government side, and imposed restrictions and limitations to those who will be operating GHG emitting facilities. The consequence is that CCS activity has not happened after more than 10 years, and just a few large-scale projects are being implemented or in operation.

For CO_2 transportation, IOGP (2019) considers that the regulated asset base approach to funding gas infrastructure should be available to carbon dioxide storage managers, and agents responsible for transferring the gas capture to the storage site should be allowed to offer CO_2 transportation as a regulated commercial service, overseen by the national regulatory authority, using equal regulatory conditions applied to natural gas, such as nondiscriminatory third-party access and regulated or negotiated tariffs. This regulated approach works as an alternative to negotiated access to infrastructure. It would also help to predictably fund CO_2 transportation over the long term if users were to pay a tariff to access the infrastructure. Nevertheless, there is a latent demand for other gases to use the pipeline transport system, such as biomethane and hydrogen. Given this reality and the perspectives for carbon dioxide, an expansion of the definitions and the regulatory framework is expected for CO_2.

Regarding CCS rules, the European Commission believed that the legal and regulatory frameworks were well-defined through the CCS Directive, which defined rules addressed to environmental, health, and safety concerns. With procedures in the Member States, there would be certainties needed for investors to develop large-scale projects. Despite that, the Commission did an assessment report based on a survey with stakeholders, where it responded as to why the CCS implementing process was delayed and to investigate possible causes (EU Commission, 2015). They concluded that the rules are adequate for what is proposed. Nonetheless, the absence of practical experience of the CCS regulatory process prevents an assessment of effectiveness in the CCS Directive. Besides that, they thought the revising process of rules could input more uncertainties for investors.

For instance, Norway has developed a carbon dioxide storage atlas covering its North Sea Continental Shelf, and this atlas inputs existing seismic and geologic data from depleted oil and gas fields along the coast. It illustrates the significant potential for large-scale carbon dioxide storage in saline aquifer or mature oil and gas fields. The Netherlands' storage assessment also has been developed in the last decade. It ranked the most attractive options as being P18-2 block for storage using an offshore depleted gas field 20 km far from the Port of Rotterdam, followed by the second option, the P06 block 90 km distant to the coast, and the Q08-A, the K12-B, and Q1 blocks, respectively (Neele et al., 2013).

In fact, a few efforts have been made to make a CCS large-scale project feasible in terms of being legally and regulatory framework-friendly to the business side and profitability.

CLSF (2017) pointed to the main challenges of regulating the current most probable route to large CCS deployment, CO_2 in enhanced oil recovery, and future transition to offshore storage. These barriers include risk assessment and monitoring baseline measurements, overall site characterization, pore space access, ownership and leasing, and postclosure monitoring and liability issues. Many financial instruments used by EU members are dedicated to the research and development (R&D) sector or demonstration projects. On the other hand, financial support for large-scale deployment has been relatively absent. For instance, the UK government proposed a £1 billion ring-fenced capital budget for CCS projects. Still, these incentives were canceled by the government before choosing winning projects, which allowed them to change priorities in the governmental budget to mitigate action plans (Cozier, 2016).

8.3 Norwegian CCS experience

Norway has been pioneering in building an institutional framework for implementing energy changes and CCS technology. In the 1990s it put into place the governmental agenda carbon abatement issue, being responsible for a carbon tax system. A Norwegian carbon tax system was an act concerning sales tax and on the discharge of carbon dioxide in petroleum activity on the continental shelf. Since the Kyoto Protocol, Norway overtaxed final oil goods and the upstream offshore sector. In the beginning, it had used a system where primary fossil fuels are over-taxed, such as gasoline and North Sea Petroleum extraction, being US$ 51 per ton of CO_2 and between US$ 43 and 49 per ton of CO_2, respectively, and an average of US$ 21 per ton of carbon dioxide (Bruvoll and Larsen, 2004). Even though they have been successful in implementing a carbon tax policy, in January 2014 the country increased tax rates to US$ 69 per ton of CO_2.

Above and beyond that, Norwegians have adopted the EU-ETS (Emissions Trading System) to reduce carbon emissions. The Greenhouse Gas Emission Trading Act was enacted in 2005 and wholly harmonized in 2013. In sum, four in five Norwegian emissions have been covered by EU-ETS (exemption for the offshore oil industry) or a carbon tax. Regulatory requirements have also been used as a path to improve CCS activity. To maintain a low GHG emissions level, constraints are being imposed to license oil fields unless technical concerns justified gas ventilation in upstream offshore facilities.

The last two critical insights for Europe's experience were "low hanging fruits," which means a low cost for capturing process or downstream steps, and the enforcement of public sector by State-Owned Enterprises (SOE) as CCS large-scale project operators.

Using an SOE to develop a storage project implies transfer liability to the government once they control its enterprise. It is observed in Sleipner and Snøhvit carbon dioxide storage. There have been successful cases of large-scale CCS projects, and they present the follow characteristics: (1) related to the oil industry (gas processing plants); (2) operated by Equinor; (3) low cost of at least one phase; and (4) supported by carbon tax under Norwegian legislation (Reiner, 2019; Global CCS Institute, 2020).

However, the main reasons to make Sleipner feasible were the compulsory need to separate natural gas from other fluid phases, such as carbon dioxide. The low cost to execute the injection and storage of it (around US$ 17 per ton) means, in other words, the tax penalty for gas ventilation was considerably higher than the marginal cost of storage (Herzog, 2016, 2017).

The Sleipner CO_2 is captured by amine absorption, separated with precombustion technology, and directly injected into the offshore sandstone reservoir in the Utsira aquifer. Over 17 million tons of CO_2 have been injected to date, and the yearly injection capacity is approximately 1 million tons. Besides Equinor (the operator), Vår Energi and Total are participants. This project is the longest-running example of CCS from natural gas, associated with oil production (Rackley, 2017).

Precombustion capture technology is also used in the Snøhvit project. The capture occurs at a liquid natural gas facility through a pipeline and is then injected into the offshore reservoir. Approximately 0.7 million tons of CO_2 are stored per year. Equinor also operates this process, in which it partners with Petoro, Total, Engie, and Hess Norge.

Another important CCS project in Norway is the Northern Lights. This project is expected to start operations in 2022 at the Norwegian Continental Shelf. CO_2 will be captured from onshore cement and waste-to-energy facilities in Eastern Norway, transported by ship to a receiving terminal onshore on the Norwegian west coast, then transferred to intermediate storage tanks. From there, the carbon dioxide is transported through a pipeline to seabed injection wells east of the Troll field. This operation is expected to capture 1.5 million tons of CO_2 per year and results from a partnership between Shell, Equinor, and Total (IOGP, 2020).

It is also important to highlight that both projects started before the CCS Directive. Norway has enhanced rules to incorporate the EU framework but leave the offshore oil sector out of the EU-ETS carbon market. Therefore the Norwegian CCS case would be considered a productive initial legal framework for carbon capture and storage in the European Union.

Lastly, another important contribution to CCS planning is the Nordic Roadmap. The 2013 CCS Roadmap states that Norway cannot meet Nordic climate targets without CCS. This document considers Norway's privileged position of having large-scale storage sites and decades of industry experience as advantages toward a head-start in CCS development. Besides, it provides an overview of the potential to achieve economic viability and the possible combinations with existing and economically viable projects, such as enhanced oil recovery. The document considers the essential policies to drive demonstration and deployment of CCS technology and recommends government support via capital grants, feed-in-tariffs, mechanisms to support the operation of plants, geological carbon storage certificates for agents responsible for GHG emission along the fossil fuel supply chain, the inclusion of biogenic CO_2 sources under the European ETS, and measurement reporting guidelines for CO_2 transport by ship under the ETS (Greig et al., 2016).

8.4 Dutch CCS case

Starting CCS activities in 1988 as one of the first countries globally, the Netherlands soon joined the significant stakeholders in collaboration to develop considerable CCS research advances. As in the Norwegian case, the government progressively integrated efforts to tackle climate change in its agenda. Motivated by an early interest in climate change, scientists, companies, and policymakers contributed to research projects that resulted in the First International Conference on Carbon Dioxide Removal in Amsterdam, 1992 (de Vos, 2014). Some of these Dutch researchers became involved in projects outside the country, such as the North Sea's Sleipner gas field.

The first impactful CO_2 storage project in the Netherlands was proposed in 1999, as a pilot project, as part of the "Green Paper Climate Policy" (Uitvoeringsnota Klimaatbeleid), following two earlier pilot proposals. The project was initially named CRUST (CO_2 re-use through underground storage) and has operated since 2004, storing an annual 20,000 tons of CO_2 captured from locally produced natural gas into the gas field in sector K12-B in the North Sea.

However, perhaps the most impressive example from that decade is the unconcluded Barendrecht CO_2 storage project. Over a period of 25 years, the project planned to store around 10 million tons of CO_2 from the Shell refinery in Pernis in a 2-km-deep depleted gas field under Barendrecht. CO_2 produced would be stored in two small empty gas fields located under a small community in the Netherlands, the city of Barendrecht (van Egmond and Hekkert, 2015). In 2007 this project was the first onshore demonstration to provide evidence of its viability (de Vos, 2014). The same refinery was the source of a pioneer project for CO_2 utilization, as part of the Organic Carbon Dioxide for Assimilation of Plants, projected to bring 300,000 tons of carbon dioxide per year from the refinery through an abandoned oil pipeline to greenhouses.

Barendrecht interested not only its parent companies, Shell and NAM (the Dutch Natural Gas Company), but also the region of Rotterdam, which at the time had ambitions to become a low-carbon industrial area. However, this example shows the importance of accounting for all stakeholders since the local public perception of CO_2 transport and storage safety hindered this project's development.

In 2009, while the environmental licensing procedures for Barendrecht were underway (despite the lateness) and the government had already invested 30 million euros, local involvement in the project heavily increased. The early public information meetings became progressively crowded by critics of the project's safety.

Therefore in 2010 the Dutch government decided to cancel the project altogether after a heavy debate in Parliament (van Egmond and Hekkert, 2015). The formal justification for such a decision was the local opposition combined with a perceived decrease in the demonstration project's necessity due to the accumulated delay. This decision fueled the onshore CO_2 storage debate on a national level and culminated in a decision from the Dutch government in early 2011 to adjourn all onshore CO_2 storage projects. Formally, this moratorium will not be lifted until offshore storage has proven insufficient for achieving Dutch energy and climate goals. Researchers from the CATO2 (Dutch acronym for CO_2 Capture, Transport, and Storage) R&D program were also affected by the lack of a close demonstration project, allowing the operation's testing and modeling.

In the late 2000s the Dutch government's CCS implementation plan was slowly developed despite the urgency to speed up the process toward large-scale deployment. The government created a plan based on the CATO project's research and participated in several stakeholders. It outlines the decisions and actions needed in the short, mid, and longer term to provide the right and sustained conditions for CCS deployment in the country. Also, it defines the roles and responsibilities of stakeholders beyond the demonstration phase of CCS, building a common understanding of who should act and how and when to operate between the stakeholders (Hendriks and Koornneef, 2014). Alongside CATO, the Dutch ROAD project examined various forms of liability applicable to CCS operations, concluding that in several cases the liabilities, to be borne under the Dutch regulatory framework were completely manageable except for the climate liabilities, which exist where a project is required to surrender emissions allowances in the event of leakage (Havercroft, 2019).

One of the most important contributions of the implementation plan was the CCS Roadmap for the Netherlands. This roadmap is not an official document from the government or the collaborating members in CATO, although it offers valuable insight into the deployment planning. It aimed to describe actions needed to accelerate development and deployment to levels that would allow CCS to reach its CO_2 reduction potential until 2050, focusing on short- to medium-term measures. The original idea is that in 10 years CCS technologies will have achieved commercial viability and will not require additional supporting actions from the government. This 2050 CCS vision was built around four main areas: geographical scope of CCS; whether the Netherlands is a leader or follower in CCS technology development; the market share of CCS and its timing toward 2050; and what kind of industry will apply CCS. This vision was considered from five dimensions: legal, economic, public, spatial, and technical. Each of these dimensions can constitute a bottleneck or a driver for the broad implementation of CCS projects in the Netherlands (Hendriks and Koornneef, 2014). Finally, five main issues concerning CCS development were defined and inserted in the roadmap, with suggested actions for each stakeholder: (1) how could CCS collaborate to cut carbon dioxide emissions; (2) the role of R&D to the progress of CCS technology and to build stakeholder's capacity to handle with the challenge to scaled-up CCS; (3) to support economically and accelerate the deployment of CCS; (4) establish short, effective, and transparent procedures to develop and implement CCS projects; and (5) CCS project developers need sufficient certainty about long-term spatial planning, long-term political commitment, and economic viability.

The Barendrecht project exemplifies the importance of public acceptance and the interaction with policy making for CO_2 storage operations. According to de Vos (2014), communication between agents and communities should be improved. The urgency to tackle climate change through CCS must be broadly acknowledged by society. The interactions among politicians, stakeholders, and policymakers at national, regional, and local levels should not be underestimated. Other determinant factors for this project (and possibly others) were the project developer's trust, lack of public engagement, the perceived local impact, how well the technology is known, and the possible benefits for the local population.

According to van Egmond and Hekkert (2015), the Barendrecht project was contested since CCS was a new and relatively unknown technology. There was no systematic debate on either CCS as a climate mitigation technology in Parliament. Momentum was lost because of the project's multiple postponements. Local opposition's arguments' credibility increased. Therefore van Egmond and Hekkert (2015, p. 10) "recommend those projects to prepare for possible local resistance and prepare a strong coalition at the national level by ensuring explicit consensus and support by national stakeholders before local opposition arises." Public engagement, well-designed communication strategies, adequate compensation measures, etc., may reduce the local opposition and decrease its credibility in the national Parliament (van Egmond and Hekkert, 2015; Ashworth et al., 2013).

The repercussions of Barendrecht affected the Dutch government's plans to proceed with investigations about onshore storage: in 2010 a few sites were already preselected candidates, such as Boerakker, Sebaldeburen, and Elevel, but public opposition and the change in government after elections resulted in the 2011 moratorium.

Other early projects that deserve to be mentioned were started in 2006: (1) SEQ's ZEPP—(from the SEQ company) zero-emission power plant with oxyfuel carbon capture technology, combined with CO_2 storage and enhanced gas recovery, with up to 0.2 million ton per year; (2) NUON/Vattenfall (companies)—CO_2 capture pilot at the Buggenum power plant; (3) NAM (from the company Nederlandse Aardolie Maatschappij B.V.)—storage of CO_2 from a Shell refinery to the De Lier gas field. The following year, the Dutch Ministry of Housing Spatial Planning and Environment issued cofunding lines for both capture and storage projects. The former came from the funds that would eventually finance part of the Barendrecht project. Another consortium was awarded 30 million euros to implement the CCS project, capturing CO_2 from an ammonia plant and injectint it into a 1.8-km-deep limestone formation in Limburg.

As part of this incentive program, in 2011 a project pilot was constructed for precombustion CO_2 capture from an integrated gasification combined cycle power plant in Buggenum; nevertheless, the facility was closed in 2013. Another precombustion capture pilot project was implemented at the Netherlands Energy Research Foundation (ECN) facility in Petten, which has extensively been used in the CATO2 program. The postcombustion capture technology also successfully deployed a pilot (Catcher) on a coal-fired power station in Rotterdam in 2008, with a capture capacity of 250 kg of CO_2 per hour. This was one of the first pilot installations in Europe connected to a coal-fired power plant.

CO_2 capture and utilization were also piloted in other ways. Starting in 2011, Twence demonstrated a technology to convert CO_2 into usable sodium bicarbonate. The gas was captured from the flue gases in a waste-to-energy plant. This project has the capacity to capture and convert 6000 tons of CO_2 annually.

There are currently four large-scale Dutch CCS projects: Athos, Magnum, Aramis, and Porthos. Porthos is the third large-scale project in the developing phase in Europe. It is located at Rotterdam, which is a place well-known as a significant world-class industrial cluster. However, it still needs to tackle the transfer of facilities to a public—private consortium responsible for transport and storage phases. Porthos is an industrial capture project for the chemical refining industry with British Petroleum (BP), Shell, Gasunie, the Dutch Port Authority, and Energie Beheer Nederland (EBN). This project is expected to capture approximately 5 million metric tons of CO_2 per year as of 2024 (IOGP, 2020).

Athos is another industrial capture project located at Ijmond and focused on the steel-making industry. This project entails a carbon capture, utilization, and storage network capturing CO_2 from the TATA Steel plant and reusing it or storing it in empty gas fields under the North Sea. It is in the feasibility study phase and is expected to capture 7.5 million metric tons of CO_2 per year as of 2030. Other participants are Gasunie, Port of Amsterdam, and EBN (IOGP, 2020).

The Aramis project is yet another industrial carbon capture operation. Located at Den Helder, it uses CO_2 supplied by third parties from Den Helder and stores it on the North Seafloor. This CO_2 can be brought to Den Helder by boat or by pipeline (IOGP, 2020).

Lastly, the Magnum project at Eemshaven is expected to start operation in 2023, capturing approximately 4 million metric tons of CO_2 per year. Equinor converts Norwegian natural gas into hydrogen for three units of the Magnum gas-power plant in the Eemshaven in Groningen, generating electricity from hydrogen. The process also generates CO_2, gathered and stored in underground facilities on the Norwegian coast. Other participants are Mitsubisi Hitachi Power Systems (MHPS), Gasunie, and Vattenfall (IOGP, 2020). The development of a regulatory framework for CCS activities in the Netherlands took place parallel to the projects and processes starting in the 2000s. The first serious effort to investigate carbon capture details, storage, and utilization regulations came from a legal taskforce within the CRUST project in 2004. Before that, the Mining Law regulated some activities concerning CO_2 underground storage. As mentioned above, the team focused on the environmental effects of CO_2 storage, establishing the framework for future environmental impact assessment reports and procedures on specific onshore locations. After the Directive 2009/31/EC, the Dutch legal framework for CCS was consolidated in 2012 through the Directive's transposition into national law. The most relevant legal framework improvements were related to spatial planning guidance, mining code, and environmental infralegal rules. Moreover, the Dutch mining code was improved to integrate the possibility to develop offshore carbon sequestration under the 2007 Convention for the Protection of the Marine Environment of the North-East Atlantic (OSPAR) decision (de Vos, 2014; Lako et al., 2011).

8.5 Conclusion

Through its CCS Directive, the European Union has played an essential role in creating a regulatory framework for CCS projects. It is necessary to highlight that those rules give legal stability to motivate investments for creating conditions such as permission process without discrimination, loss of license in case of leakage, liability during the first 20 years of decommissioning, monitoring cost for 50 years, and free access to essential facilities.

However, as discussed in this chapter, several factors affect CCS technologies' deployment, even on a small scale. The key players are the national governments, interested in emission reduction goals; local governments, interested in the risk assessment and public perception of long-term safety issues; energy companies and energy-intensive industry (steel, chemicals, refineries, and others), interested in the reduction of emissions; oil and gas companies, interested in maximizing the pore space utilization of its reservoirs; and researchers, whose concern with the enhancement of the processes in the CCS chain is fundamental in reaching the viability of the technology.

The first point to be concerned about is the necessary time and process of technology maturation and its transformation into a real business. From the European Directive and member country initiatives, researchers and business stakeholders began to implement pilot projects, establish incentive rules, and fill in their business guidelines. However, economic conditions, clean energy technologies, and long-term uncertainties have brought the prospect of a dizzying increase in large-scale CCS projects to a halt.

Looking into the various groups of industrial clusters in the countries whose cases are analyzed, it can be seen that there is no solution for net-zero emissions without the use of direct capture of their GHG emissions and their consequent storage via CCS, or compensation for them, also by the geological sequestration mechanism, such as via bioenergy with carbon capture and storage.

The Norwegian case is considered a success because of the convergence between the various agents operating in its territory, the long-term culture rooted in its citizens, and the lock-in of its economy, the oil industry. Thus Norway must remain an essential player in CCS development as a European regional chain.

The Dutch case developed according to the path followed by the Nordic neighbors. It is a country dependent on international trade through its ports. Its industries are considered to be hard-to-reduce in terms of carbon dioxide emissions. Therefore they are highly reliant on the capture in its industrial plants or factories and missing compensation for reaching the emission targets they set out.

Acknowledgments

We are grateful to the "Research Center for Gas Innovation—RCGI" (Fapesp Proc. 2014/50279-4), supported by FAPESP and Shell, organized by the University of São Paulo, and the strategic importance of the support granted by the ANP (National Agency of Petroleum, Natural Gas and Biofuels of Brazil) through the R&D clause. We also thank the support from the National Agency for Petroleum, Natural Gas and Biofuels Human Resources Program (PRH-ANP), funded by resources from the investment of oil companies qualified in the R, D&I clauses from ANP Resolution No. 50/2015 (PRH 33.1—Related to Call No. 1/2018/PRH-ANP; Grant FINEP/FUSP/USP Ref. 0443/19).

References

Ashworth, P., Einsiedel, E., Howell, R., Brunsting, S., Boughen, N., Boyd, A., et al., 2013. Public preferences to CCS: how does it change across countries? Energy Procedia 37, 7410–7418.

Bruvoll, A., Larsen, B.M., 2004. Greenhouse gas emissions in Norway: do carbon taxes work? Energy Policy 32, 493–505.

CLSF, 2017. Enabling Large-scale CCS Using Offshore CO_2, Utilization, and Storage Infrastructure Developments. Online version. <https://www.cslforum.org/cslf/sites/default/files/documents/7thMinUAE2017/Offshore_CO2-EOR_Final_02_Dec_2017.pdf. Access 01 Apr 2020>.

COM, 2013. Communication from the Commission to the European Parliament, the Council, the European Economic and Social Committee and the Committee of the Regions on the Future of Carbon Capture and Storage in Europe. COM (2013) 180, 1–28.

COM, 2017. Report from the Commission to the European Parliament and the Council on the Implementation of Directive 2009/31/EC on the geological storage of carbon dioxide. Brussels. COM (2017) 37, 1–4.

COM, 2019. Report from the Commission to the European Parliament and the Council on the Implementation of Directive 2009/31/EC on the geological storage of carbon dioxide. Brussels. COM (2019) 566, 1–4.

Costa, H.K.M., Musarra, R.M.L.M., Miranda, M.F., 2018. The main environmental permitting requirements on CCS activities in Brazil. In: Sustainability and Development Conference Proceedings. Michigan University, Ann Arbor, MI, p. 123.

Cozier, M., 2016. Reactions to the UK's cut to CCS funding: does CCS have a future in the UK? Society of Chemical Industry and John Wiley & Sons, Ltd | Greenhouse Gas Science and Technology 5, 3–6. Available from: https://doi.org/10.1002/ghg.

de Vos, R. (Ed)., 2014. Linking the Chain. ISBN 978-90-822377-0-2. <https://www.co2-cato.org/publications/linkingthechain (accessed 07.08.20) >.

EC, 2009. European Parliament and Council Directive 2009/31/EC of 23 April 2009 on the geological storage of carbon dioxide and amending Council Directive 85/337/EEC, European Parliament and Council Directives

2000/60/EC, 2001/80/EC, 2004/35/EC, 2006/12/EC, 2008/1/EC and Regulation (EC) No 1013/2006 ('CCS Directive') [2009] OJ L140/114.

EC, 2018. European commission. In-depth analysis in support of the commission communication. COM (2018) 773, 198.

Elkerbout, M., Bryhn, J., 2019. An enabling framework for carbon capture and storage (CCS) in Europe: an overview of key issues. CEPS Policy Brief. No 2019/03, September 23, 2019.

EU Commission, 2015. Report from the Commission to the European Parliament and the Council (Annex).

Global CCS Institute, 2020. Global Status of CCS Report 2020.

Greig, C., Bongers, G., Stott, C., Byrom, S., 2016. Overview of CCS Roadmaps and Projects. The University of Queensland, Brisbane, ISBN 978-1-74272-178-1.

Havercroft, I., 2019. Lessons and Perceptions: Adopting a Commercial Approach to CCS Liability. Global CCS Institute, Thought Leadership Report.

Hendriks, C., Koornneef, J.M., 2014. CCS implementation in the Netherlands. Energy Procedia 63, 6973–6981. Available from: https://doi.org/10.1016/j.egypro.2014.11.730.

Herzog, H., 2016. Lessons Learned from CCS Demonstration and Large Pilot Projects. Massachusetts Institute of Technology (MIT), Massachusetts.

Herzog, H., 2017. Financing CCS demonstration projects: lessons learned from two decades of experience. Energy Procedia 114, 5691–5700. Available from: https://doi.org/10.1016/j.egypro.2017.03.1708.

IEA, 2019. Transforming Industry through CCUS. International Energy Agency.

IOGP, 2019. The Potential for CCS and CCU in Europe. Report to the thirty-second Meeting of the European gas regulatory forum 56 June 2019.

IOGP, 2020. Global CCUS Projects. <https://32zn56499nov99m251h4e9t8-wpengine.netdna-ssl.com/bookstore/wp-content/uploads/sites/2/2020/06/Global-CCS-Projects-Map.pdf (accessed 11.10.20)>.

Lako, P., van der Welle, A.J., Harmelink, M., van der Kuip, M.D.C., Haan-Kamminga, A., Blank, F., et al., 2011. Issues concerning the implementation of the CCS directive in the Netherlands. Energy Procedia 4, 5479–5486. Available from: https://doi.org/10.1016/j.egypro.2011.02.533.

Neele, F., Hofstee, C., Arts, R., Vandeweijer, V., 2013. Offshore storage options for CO_2 in the Netherlands. Energy Procedia 37, 5220–5229. Available from: https://doi.org/10.1016/j.egypro.2013.06.438.

Rackley, S.A., 2017. Carbon Capture and Storage. Butterworth-Heinemann, Boston.

Reiner, D.M., 2019. The political economy of carbon capture and storage. In: Carbon Capture and Storage. Energy & Environment Series. Royal Society of Chemistry, first ed., 576 pp. (Chapter 16).

van Egmond, S., Hekkert, M.P., 2015. Analysis of a prominent carbon storage project failure—the role of the national government as initiator and decision-maker in the Barendrecht case. International Journal of Greenhouse Gas Control 34, 1–11.

9

Australian legislation on new mitigation technologies—the case of carbon capture and storage

Israel Araujo Lacerda de and
Hirdan Katarina de Medeiros Costa

Institute of Energy and Environment, University of São Paulo, São Paulo, Brazil

9.1 Introduction

Australia is one of almost 200 countries that have signed the Paris Agreement, at COP-21,[1] and its target to cut greenhouse gas (GHG) emissions, which reached up to 87% of current emitter nations.

The country was commited to the climate challenge even before the COP-21, and it figured as one of the Kyoto Protocol's country parties, in 1997, but it was ratified only in 2007, when its government has targeted a cap GHG emission's cap of 108% of the 1990 base level by between 2008 and 2012 (in other words, 591.5 Mt of carbon dioxide equivalent per year). In 2015 a second commitment period was endorsed by the Prime Minister, that would cover from 2013 to 2020, setting a limit of 99.5% of the 1990 GHG emissions.[2]

---

[1] The Conference of Parties (COP) is the supreme decision-making body under The United Nations Framework Convention on Climate Change (UNFCCC) Treaty. All States parties of UNFCC are represented on COP, where they review the adopted instruments needed to effectively implement UN Climate Convention. The COP is an annual convention, and the first (COP-1) was held in Berlin, in 1995. COP-21 was in Paris, on November 2015. Available from: https://unfccc.int/process/bodies/supreme-bodies/conference-of-the-parties-cop (accessed 05.09.20).

[2] https://treaties.un.org/doc/Publication/CN/2012/CN.718.2012-Eng.pdf (accessed 05.09.20).

Carbon Capture and Storage in International Energy Policy and Law
DOI: https://doi.org/10.1016/B978-0-323-85250-0.00006-2

167

Despite all international agreements being within its political agenda, Australia's relationship with climate issues occasionally has been a bone of contentious. The main point to be highlighted is the resistance to establishing a commitment to the Kyoto Protocol's target, for instance, in 2002 Prime Minister John Howard had considered refusing to ratify under the guise of causing the the collapse its economy, with effects such as cutting jobs on natural resource industries,[3] and aligning with the US position under the Bush administration (Howarth and Foxall, 2010). At that time, resistance had reflected a lock-in for both the United States and Australia. In the first case, one third of GHG emissions were in US territory, and its economy could be pervasively undermined. On the second case, Australia has not been a big GHG emitter in absolute terms, however, GHG per capita has been increasing in the last decades, and its trade balance became more dependent on coal, natural gas, and iron ore, that is, notable raw materials, with future constraints on a green economy. After the second commitment period, already in the Paris Agreement, the Australian government has been trying to use a controversial system of credit to compute the COP-21 goal.[4] Therefore, from the Kyoto Protocol to the Paris Agreement, a political agenda has swung through the international trend on climate issues, as a symbolic pattern, although economic agenda have inhibited a real engagement toward GHG emission cutting goals (Howarth and Foxall, 2010).

Only in the 2007 federal election did climate change start to enforce the political agenda significantly. From there, the parties elected shifted climate policy rapidly. At the beginning of its government, the Prime Minister of the Labor Party engaged to ratify the Kyoto Protocol. In addition, a cap and trade scheme was approved (carbon pollution reduction scheme), in order to change mainstream GHG emission in the economy, as well as incentives not only for renewable energy sources (solar, biomass, wind, hydro, and geothermal), but also carbon capture and storage (CCS) for natural gas and coal power plants (Betz and Owen, 2010). CCS technology has been a prerequisite to maintaining the pillars of the Australian fossil fuel industry and mining.

First, the chapter shows the Australian energy matrix and its historical changes since the 1990s, when environmental concerns emerged as political priorities, at the time that energy transition arose considering the Asian-Pacific complexity. Then the chapterlooks from a different viewpoint, in which GHG emissions through the production chain could be distributed in order to redistribute losses and gains on the upstream path.

The next section is dedicated to analyzing green technologies under the Paris Agreement, seeing CCS as a key element on those sectors that are hard to decarbonize, such as steel. After that, the chapter considers the legal framework changes on Australian laws, considering national and subnational levels and their particularities. From experience and knowhow, lessons can be inferred in terms of building legal frameworks on a natural resource-dependent economy and how political changes influence environmental issues.

[3] http://news.bbc.co.uk/1/hi/world/asia-pacific/2026446.stm (accessed 25.05.20).

[4] https://www.theguardian.com/environment/2019/oct/22/australia-is-the-only-country-using-carryover-climate-credits-officials-admit (accessed 25.05.20).

9.2 Australian energy sector—framework—energy transition

The primary energy demand in the Australian market steadily increases 1.62% on average since the 1990s, being far supplied by fossil fuels, which reaches currently 92% of total energy consumption (BP, 2019). Demand has been driven by transport and industry, which represent three quarters of the total consumption, and, while transport is closely dependent on oil fuels, other sectors use primarily a share of oil, natural gas, and electricity (International Energy Agency, 2018). Considering the total energy consumption, an increased demand has been provisioned by coal, and, in recent times, also by natural gas (Fig. 9.1).

Looking at power generation capacity, the main drivers are coal (42%) and natural gas (28%), complemented by hydropower plants, wind, solar, and other thermal plants. Despite a relatively diversified capacity installed, in terms of source, electricity demand has been supplied by coal followed by natural gas (Fig. 9.2), which corresponded to up to four fifths in 2018 (BP, 2019; International Energy Agency, 2018).

Those features are linked to energy security and the economics of natural resources available to fulfill commonwealth needs (Howard, 2004). Oil production covers no more than a third of the requirements, and fossil fuel and crude oil have been accessible without much effort from government. The coal sector has been the main agent of the australian energy sector, once 14% of the world's reserves is in its territory, and the internal market consumes 15% of current production, remained the major share for industrial development and international trade. In addition, natural gas reserve accounts a 20 years reserve-to-production ratio, producing 130 billion of cubic meters per year and consuming 31% (BP, 2019). Thus it can be inferred that crude oil is not a central driver for economic development, but there is a high correlation to the development of the coal and natural gas industry as a path to a moderate diversification of sources, as well as a key to support energetic

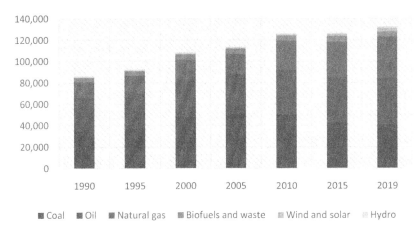

FIGURE 9.1 Total primary energy supply (ktoe) by source in Australia between 1990 and 2019. Coal had a major role in increased demand until the last decade, when it has shared it with natural gas. Renewables generation, exclude hydro, expanded only in the last decade. Source: *Modified from International Energy Agency. Available from: www.iea.org/statistics (accessed 25.05.20).*

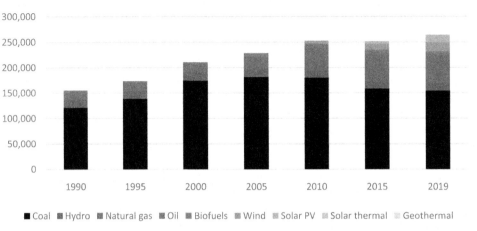

FIGURE 9.2 Historical electricity generation from 1990 to 2019, by source. Clearly, coal has been responsible for supplying the major share of increasing electricity demand over the last decades, and only after the 1990s did natural gas became important as a secondary power supply. *Solar PV*, solar photovoltaic. Source: *International Energy Agency. Available from: www.iea.org/statistics (accessed 25.05.20).*

security in terms of supply constraint (internal production bigger than demanded) and capacity constraint (robust power generation sector adequate to be resilient and able to handle peak load).

In terms of institutions, there is a balanced federative distribution of competences between entities. Since the federative framework was established, the constitution has pointed toward a legislative function of the states to legislate and implement energy policies, such as electricity, transport, and mineral rights, leaving the Commonwealth responsible for general trade issues, climate change policy, foreign affairs, and offshore regulations (Chester and Elliot, 2019; International Energy Agency, 2018).

During the 1990s reform, the Council of the Australian Government settled on a path to an intergovernmental forum of cooperation, such as a national electricity market, in which financial incentives were paid to the States engaged in competition and unbundling on electricity market. A decade after, energy issues were transferred to the Ministerial Council on Energy and two new authorities started to regulate market rules, transmission, and distribution regulations (Chester and Elliot, 2019).

When climate change entered political agenda, the energy sector was faced with political uncertainties, with a shift to renewables and an unprecedented political debate (Chester and Elliot, 2019). Resulting in a new federal framework, it has created several agencies for implementing clean energy, such as Australian Renewable Energy Agency, Clean Energy Finance Corporation, Clean Energy Regulator, and Climate Change Authority. Despite those changes, fossil fuel remained dominant in its economic sector.

Focusing on energy transition in the internal market, it will require displacement of financial capital from coal to natural gas power plants (NGPP). Seen as a transitional fuel to a low-carbon economy, NGPP results in less emission to generate the same unit of electricity (Weisser, 2007). Also, they are quite equivalent in terms of capital needed for deployment and operation, thus expending capital to replace old power plants can be an intermediate path. Besides that, methane is the only national fuel available to deal with coal constraints in the short term, and that is the main reason to choose the natural gas path. As a complementary choice, a transition to a low-carbon economy implies more diversification of input sources of power generation. The potential to increase low GHG energy generation are the solar and wind renewable sources, via local generation and more dynamic retailed trade market (Azzi et al., 2015). However, solutions provided by the political arena have not been effective against the real problems for the transition of the Australian energy market (Chester and Elliot, 2019).

Apart from internal affairs, both natural gas and coal are highly important to international trade commercially for Australia in the Asia-Pacific region. Currently, about 85% of coal is traded overseas, or 29% of the world total, which makes Australia the main exporter followed by Indonesia (25.7%) and Russia (15.9%) (BP, 2019). For natural gas, it figures also as a major player, withs two third of its production allocated to overseas partners as liquid natural gas (LNG). In general, Japan, China, South Korea, and India absorb more than three quarters of the Australian coal and LNG outwardly traded (BP, 2019), which could represent almost a half of its resource exports (Cunningham et al., 2019). Lastly, metal ore and base metals destined to the steel industry have been a major share of exports abroad.

For those reasons, it is inferred that the Australian economy, such as the energy market and overseas commercial businesses, have carried a high carbon footprint with fossil fuel or steel derivate industries, revealing a possible carbon lock-in (Fouquet, 2016; Unruh, 2000) based on its innate abilities and barriers to transmute past choices.

9.3 New technologies to achieve the Paris Agreement

In order to tackle climate change, and after Paris Agreement and its top target. A set of governmental actions must be implemented to reach 2-degree limit (2DS) or beyond (B2DS) determined on COP-21 agreement. Underrating from palpable to intangible, power, transport and industry figure as key targets to effectively reduce GHG emissions on main countries from the Organisation for Economic Co-operation and Development, such as Australia (IEA, 2017). For that there are fewer options that can be highlighted.

Energy efficiency figures as the most important global action in terms of total GHG reduction, and it is projected to contribute up to 40% (IEA, 2017). It is seen as a cross-cut action that includes changes to major emitter sectors even when they figure as a secondary player to climate change. For instance, it is possible to implement public policies on transport, via effective requirements on the urban planning sector, on final uses of energy or even transport sector, such as individual transport cars (Betz and Owen, 2010; IEA, 2017), via inducing deployment of high-efficiency low-emission technologies on power sector (Dodds and Simento, 2017), or via energy market reform (Shaw-Williams and Susilawati, 2020; Skoufa and Tamaschke, 2011).

Electrification of end-uses tends to be a path to remove the carbon footprint from the heating and transport sectors. Regardless of recent advances in electrical vehicle, it seems that the Commonwealth have chosen to establish standards and requirements as a strategy to the transport sector instead of legal enforcements to modernize automotive industry (International Energy Agency, 2018).

Distributed generation also increased markedly in the last decade, and it has been figured as part of the climate change solution and a new trend for electric markets around the world. This concept is intrinsically coupled with new renewables sources and energy efficiency. Once it inputs more controls on grid, next to the main load, it leads to an increase in efficiency on the demand side and also it is more favorable to generate electricity from a variety of renewable sources including wind, solar radiation, biomass, and natural gas. It has an important role at the end of life cycle chain of energy sector (Azzi et al., 2015).

Deployment of renewable sources has contributed expressively to decarbonizing the increased demand of energy, and both wind and solar have increased on the Australian grid (Azzi et al., 2015; Chester and Elliot, 2019; IEA, 2017). However, it is not enough to provide incremental energy demand and replace the old coal-fired plant, or even do not managed high GHG emitting sector's policy.

A key point is transforming the marginal demand of a high-carbon industry as far as technology and cost allows. Meanwhile, one can try to sequester hot spot emissions of other sectors, or compensate for them via BioEnergy with Carbon Capture and Storage (Pee et al., 2018). This refers to CCS as a technological path to the coal, natural gas, and heavy emitters industries. The Commonwealth is driven by investments on clean energy technologies, such as solar power, electricity storage, and CCS. For that, federal agencies are enforced to try a technologically-neutral approach, but include the use of coal in low-emission and efficient power plants, which is not feasible without the storage of GHG emissions (International Energy Agency, 2018). There are a few projects being developed in Australia (Allinson et al., 2017; Morrison et al., 2009; Yoshino et al., 2012), primarily before the commercial phase, and so far just one has been successful as a large-scale project.

9.4 Australian laws on carbon capture and storage

The Australian government has been engaged on climate change and carbon dioxide storage at least since the 2007 election, when the environment entered the political agenda. However, a few legal regimes were being discussed years before and they have been the bases for legislation that would affect CCS activities in the future. After 2007, even before the country signed the Kyoto Protocol and followed international instruments, legal frameworks were approved to support CCS and to allow stability of a new regime dedicated to it. Due to competence distribution between public entities, there are legal models designed to induce the deployment of CCS by the Commonwealth and subnational states (Budinis et al., 2018; Dixon et al., 2015; Walker et al., 2013), both under principles of cooperation instead of predatory concurrence.

The federal government is responsible for offshore activities farther than three nautical miles, and the main activity under its rules is hydrocarbon exploration. The Offshore Petroleum and Greenhouse Gas Storage Act 2006 was approved in 2006, as a regulatory framework mainly related to natural gas extraction.

In late 2007, when the Commonwealth political trends changed, CCS gained prominence. As a group of reforms concerning an environmental framework, the Parliament of Australia enacted in November 2008 the Greenhouse Gas Storage Act, as an amendment of Offshore Petroleum legislation, to enhance a framework in order to regulate CCS. This act covers, under federal jurisdiction, regulation for geological traps that could be dedicated to trapping GHG, as well as the facilities needed (infrastructure), licensing, permitting procedures, a competent authority on Commonwealth Minister's hierarchy, and GHG general regime scope. Several normative regulations have been developed, such as Offshore Petroleum and Greenhouse Gas Safety Regulation, in 2009; Injection and Storage, and Resource Management and Administration, both in 2011. Also, there is pipeline standard under the Petroleum and Geothermal Energy Act, which is applied to GHG infrastructure. In 2012, the Commonwealth established specific authority safety and environmental issues under its jurisdiction. The National Offshore Petroleum Safety and Environmental Management Authority has been responsible for regulating oil infrastructure and related facilities.

At the first proposal, it was argued for a possible new legal framework instead of an amendment of the petroleum legal framework. However, the Federal Parliament saw it as an inefficient system of rule, in theory, decoupled to the oil sector, considering, at that time, that operators were most likely to develop CCS activities on offshore seas under Australian jurisdiction. They treated both activities as synergetic enough to use an amendment as a political path chosen to materialize a new legal and regulatory framework dedicated to carbon sequestration (Ekins et al., 2017). On the other hand, they did not bank on long-term liability as a main issue. For them, carbon dioxide storage could not receive special treatment different from the oil industry.

After enacting the new legislation, Central government has to revisit it, due to possible unfavorable framework for CCS as a nascent industry, and due to the abscence of large-scale projects in the new legislation (Ekins et al., 2017). An amendment was approved in 2015 in order to establish rules that the Commonwealth government will be responsible to cover operator costs for liabilities after the close assurance period, usually defined as 15 years, when coming from the authorization provided by a competent authority (Ekins et al., 2017).

Several States have implemented a specific framework to CCS activities onshore and up to three nautical miles. Based on theory, they presumptively could apply a typical federal framework mirror, even because of needing to homogenize coincident rules.

For onshore CCS activities, Victoria's State approved one of the first regulatory models. In 2008, Parliament approved Greenhouse Geological Sequestration Act 2008 as a legal stand-alone framework for onshore activities. On the other hand, Victoria State do not accept long-term liability transfer like the Central government has done, which makes the CCS activity less atractive. Moreover, we see in the Victorian Onshore Act that a Minister can declare that any land or class of land cannot be used for CO_2 sequestration in order to protect such land for significant environmental reasons, while activities in certain wild lands are prohibited.[5] Carbon dioxide storage activities should not be carried out when

---

[5] Part 1, division 3 of Greenhouse Gas Geological Sequestration Act 2008. Available from: < https://www.east gippsland.vic.gov.au/files/content/public/planning_and_building/planning_reference_and_incorporated_ documents/reference_documents/greenhouse_gas_sequestration_act_2008.pdf > (accessed 05.09.20).

they pose a risk to human health or the environment, and the proposed activities should be subject to the government officials responsible for energy and the environment and the public, with revisions made when appropriate.

For waters under state jurisdiction, it has been a possibility to follow the federal model, by amendment of the offshore oil framework to provide storage activities. The Victorian Offshore Act has environmental protection mechanisms linked mainly to site closure and remediation following the cessation of injection. According to the Act, it is an offense to inject a substance into the seabed or subsoil of the offshore area, or store a substance in such locations, unless authorized under a GHG injection license; or otherwise authorized under the Act or other applicable law or regulation.

Even using two models, Victoria's rules incorporated into its framework were quite close to the Offshore Petroleum Act 2006, at federal level.

Queensland have picked a different path and enacted stand-alone legal frameworks (called Greenhouse Gas Storage Act 2009), as an autonomous act away from the oil industry. Under this Storage Act, exploration is permitted and injection and storage leases for greenhouse gas storage activities cannot be granted or renewed until an environmental authority has been issued for all environmentally relevant activities proposed to be undertaken.

This act predicts that applicants for exploration permits and injection and storage leases must have regard to potential water issues in developing work programs and development plans, which documents cannot be accepted until they have been approved by the minister responsible for administering the Water Act, and also leaseholders cannot take or interfere with water (as defined under the Water Act 2000) unless the taking or interference is authorized under that Act.

In South Australia an enhanced path has been made through an amendment to the Petroleum and Geothermal Energy Act 2000 and its regulations with the aim of stipulating the competent authority on environmental issues as an entity that must be consulted before a project is approved.[6] Also, with regard to the legal system, a Minister and authorized officers have certain powers to give conduct the license process, including directions to take actions to prevent or minimize environmental damage, and to rehabilitate polluted lands. Licensees are liable to compensate the state for costs of environmental rehabilitation that the public sector must carry out as a result of serious environmental damage, or the threat or potential of serious environmental damage, arising from activities carried out under a license.

The most different and important state-run legislation was settled in Western Australia (WA). The Barrow Island Act, enacted in 2003,[7] was approved to deal with only one project, as a mechanism to ratify and authorize an implementing agreement between the State and the Gorgon Joint Venture relating to a proposal to undertake offshore production of

[6] South Australia State innovates on Petroleum and Geothermal Energy Regulations stablishing previous and mandatory consultation process before approving a CCS project. Available from: < https://legislation.sa.gov.au/LZ/C/R/PETROLEUM%20AND%20GEOTHERMAL%20ENERGY%20REGULATIONS%202013/CURRENT/2013.25.AUTH.PDF> (accessed 05.09.20).

[7] < https://www.legislation.wa.gov.au/legislation/statutes.nsf/main_mrtitle_76_homepage.html > (accessed 27.05.20).

natural gas and other petroleum and a gas processing and infrastructure project on Barrow Island. To homogenize long-term liability clauses, The WA Parliament made an amendment adjusting its framework to new terms corresponding to federal postclosure requirements.[8] Congressmen argued, in sum, that it was an unprecedented world-class example of what government and industry can achieve to reduce carbon footprint, and it could deal with a high percentage of CO_2 in the Gorgon gas field.[9]

It can be highlighted that long-term liability was a key issue for stakeholders of the industries involved, and parliament's composition allowed them to infer the support of main opposition parties of Labor government. During the discussion of the bill in the Senate, the government gave in and accepted an agreement with the opposition to set a 20-year minimum of liability after site closure, instead of an indefinite period (Gibbs, 2018). Therefore the debate was focused on feasible economic issues rather than environmental concerns, presumably due to the high investment required for a single project and consequent rent which WA could takes as royalties and taxes (Gibbs, 2018).

Gongon project has its own legal framework and, as a *sui generis* case, WA is responsible to cover one fifth while the Commonwealth would be in charge of the remaining eventual costs in terms of postclosure long-term liability, and both entities remained silent about the ownership of pore space (Swayne and Phillips, 2012).

The Australian system has at least a double jurisdiction layer, at State and at Commonwealth level, which covers numerous models, such as stand-alone (Victoria and Queensland), amendments to petroleum framework (federal level, Victoria's offshore, and South Australia) and a "tailor-made" framework (Borrow Island), which cover the only one current large-scale project in Australia.

9.5 Lessons to be learned

Australia has been highly engaged in the development of CCS as a technology to be used for the avoidance of GHG emission, and CCS is considered one of the most important options to reduce GHG emissions on its own territory. CCS technology emerged onto the political agenda concomitant to the Kyoto Protocol promulgation. In fact, the Commonwealth government was a pioneer of promoting efforts to develop industrial processes of clean energy technologies to be used in the synergy of the fossil energy sources available, since the main systems have not been able to internalize fully the environmental costs involved (Ekins et al., 2017).

Despite initial resistance to Kyoto Protocol, this agreement has been signed. Nevertheless, this support can be overshadowed by political swings regarding environmental agreement issues and potential conflicts against national economic interests and natural resource policies (Reiner, 2020). This controversy could be caused by the nondefinition of better solutions for the Commonwealth economy under the Kyoto Protocol and a

---

[8] < https://www.parliament.wa.gov.au/publications/tabledpapers.nsf/displaypaper/ 3912728a9d54fcf4c071806048257e0c003d6ddb/$file/2728.pdf > (accessed 27.05.20).

[9] < https://www.parliament.wa.gov.au/Parliament/Bills.nsf/9D4676B11F1471BC48257E0C001AA55F/$File/ Bill87-1BSR.pdf > (accessed 27.05.20).

possible adverse consequence on key export drivers. At that time, fossil fuels did not figure as a participant of green economies, and they would generate resistance from the Australian establishment. Then, CCS emerged not only as a GHG mitigation path, but also as vital to maintain the health of the natural resource sector and to respond to the climate change agenda on international arena.

Governmental behavior swings occurred after political changes in 2008, with a Labor Prime Minister, and CCS moved to a proactive position. It can be highlighted that the Global CCS Institute, which has headquartered currently in Melbourne, and the National Low Emission Coal Initiative[10] aimed to keep coal as a player in the future green economy. After a half decade of efforts, another political wave changed coalitions and a Liberal became Prime Minister. In its first year, they cut the CCS budget by some 70%, cutting A $460 million out of A$650 million, after having campaigned against Labor's climate-friendly agenda (Reiner, 2020). Accordingly, it is possible to infer that, due to political oscillation, institutions involved with CCS or any climate agenda that could be confounded with political debate tend to be deteriorated.

On the one hand, observing the demand side, such as International Energy Agency view, dealing with Australian GHG emission has been characterized as a complex challenge. On the other hand, life cycle analysis shows an economic trap due to the dependence of high-carbon emission goods for the exports sector.

International stakeholders have pointed to resolutions based on carbon markets, carbon pricing processes, and regulation to promote efficiency through supply chains (Azzi et al., 2015; Brink et al., 2016; Doda and Fankhauser, 2020; Finon, 2019; International Energy Agency, 2009; Ramstein et al., 2019). If the global market decided to overburden fossil fuel and steel industry as an incentive to reduce GHG emission, the early consequences would be an economic deterioration of major destination countries of Australian commodity exports (China, Japan, South Korea, Taiwan, and India). Secondly, they could negotiate to transfer impairments to the commodity sector, in other words, reduce the natural resource rents of Australian exporter goods. Considering the fair scenarios, the CCS technology remained as a keysolution to targetto low carbon emission from hard-to-reduce sector.

As an internal feature, national and subnational levels are responsible for almost the same activity, with the possibility of having antagonistic interests. The Commonwealth started to assume relevant rule on energy sector in the 1990s, despite remaining as a State competence on its territory. Latent conflicts between both levels can also work as a mechanism to deteriorate confidence in institutions. For instance, when the federal level enhanced legislation in order to allow better conditions of long-term liabilities, subnational levels had no duties to follow same path (Gibbs, 2018). A handful of states implicitly remained preceding legal framework, therefore, reducing the long-term business friendly environment. Moreover, just emphasized a strict inevitability to improve harmonious coordination systems between federal entities.

An exemption is the Gorgon project, which is a joint venture of major operators of the oil industry and represent a world-class natural gas project. Also, it can be qualified as low-hanging fruit in terms of carbon dioxide capture and storage, considering that GHG is

---

[10] < https://www.industry.gov.au/funding-and-incentives/national-low-emission-coal-initiative > (accessed 03.06.20).

easily captured after natural gas processing plant, and costs were considered adequate under a specific legal framework approved only for it, representing a nascent industry.

In sum, (1) the coordination between national and subnational levels can figure as a main challenge; (2) an excess of legislation can obfuscate the legal framework and input uncertainties without cause and consequence factors; and (3) for a nascent industry, wasteful regulation means implicit decisions not to encourage CCS technology.

Finally, not only in Australia, but also in any country that would like to transform CCS into a powerful tool for developing green economy, without a well defined and stable process for monetizing, which means transforming carbon dioxide storage into revenue to cover expenses along the project life cycle as well as to pay business risks, it will not be feasible to move forward in the carbon storage industry.

9.6 Final remarks

The Australian energy matrix is based mainly on fossil fuels. Electricity consumption has been mostly supplied by coal followed by natural gas. At the same time, Australia's economy is very dependent on natural resources exports. Thus considering that Australia have been also one of almost 200 that have signed the Paris Agreement, the challenge to reach its target to cut GHG emissions is costly and high.

So, technological options to tackle climate change must be considered to help Australia's targets. Energy efficiency, distributed generation, and CCS are linked as a technologic path for the coal, natural gas, and heavy emitters industries. In fact, CCS fits as a perfect solution to avoid a low carbon fossil fuel sector.

Particularly, CCS may be a key element on those sectors that are hard to be decarbonized, such as steel. Because of that, one may say that a legal framework was designed into Australian laws, considering national and subnational levels and its particularities.

Basically, the Australian system covers a number of models, such as stand-alone (Victoria and Queensland), amendments to petroleum framework (federal level, Victoria's offshore, and South Australia), and "tailor-made" framework (Barrow Island), covering the Gorgon project.

However, it seems that an existence of a legal framework is not only a key factor to stimulate this technology. Excess of legislation can be a problem if it creates uncertainties or wasteful regulation. In fact, there are a few projects being developed in Australia, primarily before commercial phase, and just one has been successful as a large-scale project.

Therefore, we may ask which obstacles exist to create an environment to stimulate CCS projects. Of course, we have seen that legal framework is important in order to induce deployment of CCS, mainly considering principles of cooperation instead of predatory concurrence. However, due to political oscillation, institutions involved in CCS or any climate agenda that could be confounded with political debate tend to be deteriorated.

In addition, there are two kinds of different approaches: the federal level enhanced legislation in order to allow better conditions for long-term liabilities, while at the same time subnational levels had no duties to follow the same path. This scenario creates a risky and an insecure environment. Harmonious coordination systems between federal entities are needed.

Thus, countries that desire to increase CCS projects must create coordination between national and subnational levels, adequate legislation that clearly defines rights and duties, and transform carbon dioxide storage into revenue to cover expenses across the project life cycle as well as to pay business risks.

Acknowledgments

We are grateful to the "Research Center for Gas Innovation—RCGI" (Fapesp Proc. 2014/50279-4), supported by FAPESP and Shell, organized by the University of São Paulo, and the strategic importance of the support granted by the ANP (National Agency of Petroleum, Natural Gas and Biofuels of Brazil) through the R&D clause. We also thank the support from the National Agency for Petroleum, Natural Gas and Biofuels Human Resources Program (PRH-ANP), funded by resources from the investment of oil companies qualified in the R,D&I clauses from ANP Resolution No. 50/2015 (PRH 33.1−Related to Call No. 1/2018/PRH-ANP; Grant FINEP/FUSP/USP Ref. 0443/19).

References

Allinson, K., Burt, D., Campbell, L., Constable, L., Crombie, M., Lee, A., et al., 2017. Best practice for transitioning from carbon dioxide (CO_2) enhanced oil recovery EOR to CO_2 storage. Energy Procedia . Available from: https://doi.org/10.1016/j.egypro.2017.03.1837.

Azzi, M., Duc, H., Ha, Q.P., 2015. Toward sustainable energy usage in the power generation and construction sectors—a case study of Australia. Automation in Construction 59, 122−127. Available from: https://doi.org/10.1016/j.autcon.2015.08.001.

Betz, R., Owen, A.D., 2010. The implications of Australia's carbon pollution reduction scheme for its National Electricity Market. Energy Policy 38 (9), 4966−4977. Available from: https://doi.org/10.1016/j.enpol.2010.03.084.

BP, 2019. Statistical Review. In BP Statistical Review of World Energy 2019. <https://doi.org/10.1001/jama.1973.03220300055017>.

Brink, C., Vollebergh, H.R.J., van der Werf, E., 2016. Carbon pricing in the EU: evaluation of different EU ETS reform options. Energy Policy 97, 603−617. Available from: https://doi.org/10.1016/j.enpol.2016.07.023.

Budinis, S., Krevor, S., Dowell, N., Mac, Brandon, N., Hawkes, A., 2018. An assessment of CCS costs, barriers and potential. Energy Strategy Reviews 22 (May), 61−81. Available from: https://doi.org/10.1016/j.esr.2018.08.003.

Chester, L., Elliot, A., 2019. Energy problem representation: the historical and contemporary framing of Australian electricity policy. Energy Policy 128 (January), 102−113. Available from: https://doi.org/10.1016/j.enpol.2018.12.052.

Cunningham, M., Uffelen, L., Van Chambers, M., 2019. The Changing Global Market for Australian Coal. Reserve Bank of Australia. Available from: https://www.rba.gov.au/publications/bulletin/2019/sep/pdf/the-changing-global-market-for-australian-coal.pdf.

Dixon, T., McCoy, S.T., Havercroft, I., 2015. Legal and regulatory developments on CCS. International Journal of Greenhouse Gas Control 40, 431−448. Available from: https://doi.org/10.1016/j.ijggc.2015.05.024.

Doda, B., Fankhauser, S., 2020. Climate policy and power producers: the distribution of pain and gain. Energy Policy 138 (October 2019), 111205. Available from: https://doi.org/10.1016/j.enpol.2019.111205.

Dodds, K., Simento, N., 2017. Accelerating Australian demonstration projects through focused research and development. Energy Procedia 114 (November 2016), 5888−5896. Available from: https://doi.org/10.1016/j.egypro.2017.03.1726.

Ekins, P., Hughes, N., Pye, S., Macrory, R., Milligan, B., Haszeldine, S., et al., 2017. The role of CCS in meeting climate policy targets. Understanding the potential contribution of CCS to a low carbon world, and the policies that may support that contribution.

Finon, D., 2019. Carbon policy in developing countries: giving priority to non-price instruments. Energy Policy 132 (September 2018), 38−43. Available from: https://doi.org/10.1016/j.enpol.2019.04.046.

Fouquet, R., 2016. Path dependence in energy systems and economic development. Nature Energy 1 (8). Available from: https://doi.org/10.1038/nenergy.2016.98.

Gibbs, M., 2018. The regulation of underground storage of greenhouse gases in Australia. In: Havercroft, I., Macrory, R., Stewart, R. (Eds.), Carbon Capture and Storage : Emerging Legal and Regulatory Issues, second ed. Hart Publishing, pp. 213−230. Available from: http://www.bloomsburycollections.com/book/carbon-capture-and-storage-emerging-legal-and-regulatory-issues-1/ch11-the-regulation-of-underground-storage-of-greenhouse-gases-in-australia-the-views-in-this-chapter-are-those-of-the-author-and-do-not-necessaril.

Howard, J., 2004. Securing Australia's Energy Future. Australian Government. Department of the Prime Minister and Cabinet.

Howarth, N.A.A., Foxall, A., 2010. The Veil of Kyoto and the politics of greenhouse gas mitigation in Australia. Political Geography 29 (3), 167−176. Available from: https://doi.org/10.1016/j.polgeo.2010.03.001.

IEA, 2017. Energy Technology Perspectives 2017. International Energy Agency (IEA) Publications, p. 371. < https://webstore.iea.org/download/summary/237?fileName = English-ETP-2017-ES.pdf > .

International Energy Agency, 2009. Technology roadmap: carbon capture and storage. Current 1−52. Available from: https://doi.org/10.1007/SpringerReference_7300.

International Energy Agency, 2018. Energy Policies of IEA Countries: Australia 2018 Review. Data & Publications, p. 244.

Morrison, H., Schwander, M., Bradshaw, J., 2009. A vision of a CCS business—the ZeroGen experience. Energy Procedia 1 (1), 1751−1758. Available from: https://doi.org/10.1016/j.egypro.2009.01.229 > .

Pee, A. de, Pinner, D., Roelofsen, O., Somers, K., Speelman, E., Witteveen, M., 2018. Decarbonization of Industrial Sectors: The Next Frontier. McKinsey & Company, p. 68. < https://www.mckinsey.com/ ~ /media/McKinsey/BusinessFunctions/Sustainability and ResourceProductivity/Our Insights/How industrycan move toward a low carbon future/Decarbonization-of-industrial-sectors-The-next-frontier.ashx > .

Ramstein, C., Dominioni, G., Ettehad, S., Lam, L., Quant, M., Zhang, J., et al., 2019. State and Trends of Carbon Pricing 2019. < https://doi.org/10.1596/978-1-4648-1435-8 > .

Reiner, D.M., 2020. Political economy of carbon capture and storage. RSC Energy and Environment Series, 26, 536−558 (Chapter 16). Available from: https://doi.org/10.1039/9781788012744-00536.

Shaw-Williams, D., Susilawati, C., 2020. A techno-economic evaluation of Virtual Net Metering for the Australian community housing sector. Applied Energy 261 (December 2019), 114271. Available from: https://doi.org/10.1016/j.apenergy.2019.114271.

Skoufa, L., Tamaschke, R., 2011. Carbon prices, institutions, technology and electricity generation firms in two Australian states. Energy Policy 39 (5), 2606−2614. Available from: https://doi.org/10.1016/j.enpol.2011.02.029.

Swayne, N., Phillips, A., 2012. Legal liability for carbon capture and storage in Australia: where should the losses fall? Environmental and Planning Law Journal 29 (3), 189−216.

Unruh, G.C., 2000. Understanding carbon lock-in. Energy Policy 28 (12), 817−830. Available from: https://doi.org/10.1016/S0301-4215(00)00070-7.

Walker, I., Tantala, S., Senanayake, W., Leamon, G., 2013. Regulating carbon dioxide storage operations near oil and gas field, Australia's approach. Energy Procedia 37, 7766−7773. Available from: https://doi.org/10.1016/j.egypro.2013.06.723.

Weisser, D., 2007. A guide to life-cycle greenhouse gas (GHG) emissions from electric supply technologies. Energy 32 (9), 1543−1559. Available from: https://doi.org/10.1016/j.energy.2007.01.008.

Yoshino, Y., Harada, E., Inoue, K., Yoshimura, K., Yamashita, S., Hakamada, K., 2012. Feasibility study of "CO_2 free hydrogen chain" utilizing Australian brown coal linked with CCS. Energy Procedia 29, 701−709. Available from: https://doi.org/10.1016/j.egypro.2012.09.082.

CHAPTER
10

Carbon capture and storage: Intellectual property, innovation policy, and climate change

Matthew Rimmer

Faculty of Business and Law, Queensland University of Technology (QUT), Brisbane, Australia

10.1 Introduction

In 2020 World Intellectual Property Day focused on the theme, "Innovate for a Green Future." Francis Gurry, the Director-General of the World Intellectual Property Organization, discussed the relationship between intellectual property and green energy:

> Innovation provides options and is the key to unlocking the solutions and approaches we need to create a sustainable and green future. But to achieve that goal, we need to act. We need to make environmental sustainability a priority—it is the bedrock of our future well-being. We need to invest in national innovation ecosystems and enable broad access to effective national IP systems that support the development and deployment of the technologies, products and services we need for the transition to a brighter, greener, low-carbon future. *Gurry (2020)*

Gurry (2020) suggested: "Let us reflect on the role that innovation and the IP system can play in driving global efforts to create a green future that supports the well-being of humanity."

There is a growing academic and scholarly literature about the topic of intellectual property, clean technologies, and climate change. Barton (2007) provided an early analysis of patent landscapes in respect of renewable technologies. Lane (2011) has focused upon green patents, ecomarks, and other forms of clean tech innovation. In my work, I have focused on intellectual property and climate change (Rimmer, 2011); and charted some of the issues brought into relief by the *Paris Agreement* 2015 (Rimmer, 2018a). In a number of works, Brown (2013, 2019) has explored the intersections between intellectual property, environmental technologies, and climate change. Sarnoff (2015) has collected a range of works in the field on intellectual property and climate change. Zhuang (2017) has discussed intellectual property rights and climate change against the background of international trade law. Baker et al. (2017) have highlighted the need to reform intellectual property laws in light of the United Nations Sustainable Development Goals and the need for substantive climate action.

However, there have been a set of technologies, which have had a somewhat anomalous status within this policy discussion about intellectual property and climate change. There has been discussion as to whether biofuels should be classed as clean technologies (Rimmer et al., 2015). There has also been controversy over the status of geoengineering technologies (Rimmer, 2018b). There has been debate as to whether carbon capture and storage (CCS) technologies should be classified as clean technologies.

Accordingly, this chapter will focus upon the peculiar and strange position of CCS as part of the larger debate about intellectual property and climate change. Section 10.2 of this survey looks at patent law and its treatment of CCS. There is an investigation of patent landscapes, patent litigation, and patent flexibilities in relation to CCS. Section 10.3 explores government schemes to support research and development and deployment of CCS. There is a discussion of Australia's CCS Flagships program; as well as some of the collaborations between the United States and China on CCS. There is also a conversation about whether clean technology accelerators like the United Nations Framework Convention on Climate Change (UNFCCC) Clean Technology Centre and Network should provide support for CCS. Section 10.4 focuses upon innovation prizes and challenges related to CCS. In particular, it examines the Mission Innovation focus on CCS; the proposed *Carbon Capture Prize Act*; and the Carbon XPRIZE. The conclusion questions whether further incentives should be provided for research and development of CCS, given the lack of substantive progress in the field. It also highlights Elon Musk's support for an XPRIZE for Carbon Removal in 2021.

10.2 Patent law

10.2.1 A patent landscapes

In respect of the academic literature, there have been a growing collection of studies which have sought to evaluate technological developments in CCS by reference to an analysis of patent data. There have been a series of patent landscapes conducted and developed in respect of CCS technologies.

A 2009 Chatham House study (Lee et al., 2009) considered the position of carbon capture patents. This report observed that "A number of technology families underpin the different stages in CCS systems." The study noted: "Despite significant industrial experience in individual areas, there remain great opportunities for advancing the integration of the CCS components, scaling up the process and decosting the various processes." The study highlighted that key patent portfolio holders included "Oil and gas companies (ExxonMobil, Shell, Texaco, Chevron, BP), equipment manufacturers (Alstom, General Electric, Mitsubishi), chemicals industry players (Air Liquide, Dow Chemical, BASF) and specialists service providers (such as Honeywell)." In terms of geographical distribution, "More patents were filed in the United States than in all other countries taken together." The study commented: "There is a relatively high IP ownership concentration in this field, with the top 20 assignees accounting for 32.4% of all patents." The study also warns that emerging IP issues may be related to vertical integration, and the embedding of technology in the design of equipment: "Many oil and gas companies may be in the position of holding large patent portfolios (and underlying technology capabilities) across the full chain of CCS."

In early patent landscapes for clean technologies, carbon capture was featured as the sixth highest ranking technology field, with 616 patents from 1988 to 2007 (Rimmer, 2011, p. 180). Carbon storage was lower down the scale, with 54 patents from 1988 to 2007 (Rimmer, 2011, p. 180).

A series of Intellectual Property Offices around the world—including the US Patent and Trademark Office, IP Australia, and the Canadian Intellectual Property Office—have introduced green fast-tracks for expedited patent examination of clean technologies. Dechezleprêtre (2013) conducted an empirical study of fast-tracking green patent applications in various jurisdictions. The majority of patent-applications related to renewable energy—followed by other environmental technologies, such as recycling and pollution control. The main technology to use the green fast-tracks in Australia and Canada has been CCS. Dechezleprêtre (2013, p. 10) suggested: "This can be linked to Australia's dependence on coal-based electricity production and to Canada's booming tar sand mining industry." The participation rates in respect of green patent fast-track examination systems have remained relatively low (Dechezleprêtre and Lane, 2013). Lane (2012) has argued that there should be a global green patent highway—with an expedited examination of green technologies.

The green patent fast-track systems have largely taken an inclusive and broad view of what constitutes a "green patent." In my view, it is debatable whether CCS deserves to be classified as "green technologies," "environmentally sound technologies," or "clean technologies"—to the use the various nomenclature which has been employed. There has been a range of research—for instance the work of (Jacobson, 2019)—which has questioned whether CCS has made a substantive and meaningful contribution to addressing climate change.

A 2013 paper conducted a patent review of advances in CO_2 capture technology (Li et al., 2013). This study noted that there had been a steady rise in patents in respect of carbon capture: "More than 60% of these patents were published since the year 2000, and a sharp increase in patent numbers was seen in the last several years; $\sim 25\%$ patents were published in the last 2 years" (Li et al., 2013). The study also highlighted the geographical distribution of the patents—"The top four major types of patents, which consist of more than 2/3 of these patents, were patents granted by Japan (JP), United States (US), World Intellectual Property Organization (WO), and China (CN), and approximately half of the patents were JP and US patents" (Li et al., 2013). The study concluded that there remained major innovation gaps in the area of carbon capture: "Unfortunately, no current technologies for removing CO_2 from large sources like coal-based power plants exist which satisfy the needs of safety, efficiency, and economy; further enhancement and innovation are much needed" (Li et al., 2013).

There have been concerns expressed about the corporate domination of CCS. Raiser et al. (2017) have noted the rise in patents for carbon capture technologies. They observed: "Considering previously reviewed literature on the need for technology transfer, this unequal distribution of patents globally may have a large effect on a global societies 'access' to mitigation technologies." Raiser et al. (2017) elaborated upon their concerns:

> Not all but many applications of renewable energy (RE) and CCS still require follow-on research and development to lower the costs of implementation or to be optimized for a specific implementation setting... Exclusive rights would obstruct access to such research and development know-how. Similar effects are true for carbon capture technologies, which also rely on incremental advances in efficiency in order to lower costs global patent and licensing trends in mitigation technologies focus predominantly on a few developed nations. Whilst innovation in mitigation has grown in developed countries, and especially in Europe, the study relates these trends to policy initiatives incentivizing such research.

Raiser et al. (2017) argued: "Given the examples of RE and CCS, the importance of promoting follow-on research to overcome barriers to the implementation of mitigation technologies is clear, independent of what technological portfolio is politically targeted."

It is also worthwhile adding that some fossil fuel companies have faced scrutiny under consumer law for alleged misleading and deceptive representations about the nature of their investments in new technologies (including CCS) (Drugman, 2020).

10.2.2 Patent litigation

While the patent landscapes provide a useful indication of the ownership and distribution of patents in a field, patent litigation is often revealing about larger questions in respect of patent quality in a field. There has been increasing patent litigation over clean technologies across a range of fields—including in the areas of solar power, wind power, renewable energy, energy efficiency, electric vehicles, smart grids, and carbon-ready crops. There have been skirmishes so far in respect of patent law and CCS, according to legal records.

CO_2 Solutions was a Canadian company, which sought to lower the cost barrier to carbon capture, sequestration and utilization. The company built an extensive patent portfolio covering the use of carbonic anhydrase, or analogues thereof, for the efficient postcombustion capture of carbon dioxide with low-energy aqueous solvents.

In the 2016 dispute of *CO$_2$ Solutions Inc.* versus *Akermin Inc.* E.D.Mo._416-cv-00520_1 (Missouri Eastern District Court 2016) Reference 4:16-cv-00520-RLW, the Canadian company CO_2 Solutions Inc. bought an action for patent infringement against the rival United States company Akermin Inc. The Canadian company noted: "The asserted CO_2 Solutions Patents relate generally to systems, methods, apparatus, and materials for capturing carbon dioxide (CO_2) from gas mixtures containing carbon dioxide, using biocatalysts such as carbonic anhydrase." The Canadian company alleged: "Defendant Akermin has been and is engaged in research and development of systems, methods, apparatus, and/or materials for capturing carbon dioxide from gas mixtures containing carbon dioxide, using biocatalysts such as carbonic anhydrase." The Canadian company argued: "Defendant Akermin's activities for capturing carbon dioxide from gas mixtures, using carbonic anhydrase as a biocatalyst, directly infringe one or more claims of each of the asserted CO_2 Solutions Patents, including claim 1 of the '377 patent, claim 1 of the '709 patent, claim 10 of the '769 patent, and claim 1 of the '796 patent." The Canadian company observed of the action of Akermin: "Its ongoing infringement of the asserted CO_2 Solutions Patents has been objectively reckless and, therefore, wilful and in deliberate disregard of CO_2 Solutions' rights in the patents."

In response, Akermin challenged the validity of a number of the patent claims of CO_2 Solutions in *Akermin, Inc.* versus *CO_2 Solutions Inc.* bpai_IPR2015-00880_25 (PTAB—other actions 2016). The administrative judge held: "We determine that Petitioner has shown by a preponderance of the evidence that claims 1, 2, 15, 16, 22–26, and 40–43 of the '458 patent are unpatentable, but has not shown that claims 3, 4, 17–19, 27, and 28 of the '458 patent are unpatentable."

CO_2 Solutions (2016) put out a press release, claiming to have won a further patent challenge in the European Union. In the reexamination decision, the Danish Patent Office maintained the issued claims in the Corporation's registered Utility Model #BR 201400144 entitled "System for CO_2 Capture Using Packed Reactor and Absorption Mixture with Micro-Particles Including BioCatalysts." The decision came after a challenge filed by Akermin Inc., a US company that had intended to utilize similar CO_2 capture technology for a biogas-related project in Denmark, known as ENZUP. President and CEO of CO_2 Solutions, Evan Price stated: "The decision clearly confirms our ownership of the IP related to enzyme-enabled carbon capture, the most economical and cleanest commercial technology available to date for this purpose" (CO_2 Solutions, 2016). He commented: "While we offered the ENZUP partners a single commercial license for use of the CO_2 Solutions' IP in their project, it appears that, pursuant to this decision, Akermin has indefinitely ceased operations and the ENZUP project itself, in Denmark, has been cancelled." He observed: "CO_2 Solutions welcomes initiatives to implement carbon capture technology, but shall defend the Corporation's IP wherever we observe actual or imminent infringement on our rights, as we have done successfully in Denmark."

In the end, CO_2 Solutions (2020) finished up in bankruptcy. The company negotiated the sale of the entirety of the Corporation's assets in two separate transactions. In December 2019 the Corporation sold its portfolio of intellectual property, including approximatively 90 patents granted or pending and the Corporation's trademarks, the CO_2 capture unit located at the pulp mill of Resolute Forest Products in Saint-Félicien, Québec, and certain contracts related thereto, to Saipem S.P.A and Saipem Canada Inc. In the second transaction, approved on January 17, 2020, the Corporation sold its CO_2 capture unit located in Montreal-East, Québec, to Chimie Parachem.

In the 2017 case of *Ex parte Li* us-ptab-2015000389-04-212017 (PTAB—Administrative Hearings 2017), the appellants' claimed subject matter related to "Power plants with integrated CO_2 capture as well as CO_2 capture ready power plants." Claim 1 of the patent provided for:

> A power plant, with a CO_2 capture system, comprising: at least one of a steam power plant 1 or a combined cycle power plant 2, wherein a water steam cycle of the power plant 1, comprises two steam turbine arrangements, the first steam turbine arrangement comprising steam turbines with at least two pressure levels, and a second steam turbine arrangement comprising at least one back pressure turbine configured to expand steam to a supply pressure of the CO_2 capture system, wherein the second steam turbine arrangement further comprises a low-pressure steam turbine, which is designed for a supply pressure that matches an outlet pressure of the at least one back pressure turbine, and the at least one back pressure steam turbine and the low-pressure steam turbine are both configured for a steam mass flow of the CO_2

capture system in order to convert thermal energy of an outlet steam of the back pressure steam turbine into mechanical energy when the CO_2 capture system is not operating.

In examination, a number of claims were rejected on various grounds—including that they were unpatentable; anticipated by prior art; and indefinite.

The administrative judge held: "We find that one having ordinary skill in the art, upon reading the Specification and being familiar with the use of the term 'space' in this field as shown in the art, would understand 'space' to refer to a footprint (area) needed to house the CO_2 capture system." The administrative judge concluded: "For these reasons, we do not sustain the first ground of rejection of claims 1, 2, 4, 5, 12, and 16 under 35 U.S.C. § 112, second paragraph, as indefinite."

The administrative judge, though, upheld the examiner's position in respect of anticipation: "We agree with the Examiner's determination that Hegerland discloses the subject matter of claim 1." The judge found: "Accordingly, we sustain the rejection of claim 1, and claim 12 which falls with claim 1, under 35 U.S.C. § 102(b) as anticipated by Hegerland." The administrative judge added: "Because we find no deficiencies in the anticipation rejection of claim 1 based on Hegerland, we likewise sustain the third through fifth grounds of rejection of claims 2, 4, 5, and 8–11." Moreover, the administrative judge held: "The rejections of claims 1, 2, and 4–18 under 35 U.S.C. § 103(a) are affirmed."

In the United States matter of *Ex Parte Paul J. Berlowitz* us-ptab-2018009128-12-30–2019 (PTAB—Administrative Hearings 2019), the invention at stake related to the use of molten carbonate fuel cells in the processing or production of cement. The appellant appealed against the examiner's finds that a number of claims were invalid, pointing to research on carbon capture:

> According to Appellant, Hendriks is referenced numerous times in a subsequent 2008 report by the International Energy Agency Greenhouse Gas R&D Programme, titled "CO_2 Capture in the Cement Industry" ("2008 Report"), which allegedly provides "the current state of the art (in 2008) for mitigating CO_2 emissions during cement manufacture." Appellant points to Table 1 of the 2008 Report to establish that the two primary methods of carbon capture depicted allow for avoidance of 77% and 52% of CO_2 emissions. Appellant compares these CO_2 emissions statistics with the 90% or greater capture in Table A of Appellant's Specification, but acknowledges that Table A is not a direct comparison with Table 1 of the 2008 Report.

Donna Praiss—the administrative judge—affirmed the rejection of a number of claims (1–4, 6–21) on the basis that they were obvious.

In the case of *Celgard, LLC* versus *Shenzhen Senior Technology Material Co. Ltd. (United States) Research Institute*, etc. (California Northern District Court 2019), there was an action for patent infringement and trade secrets theft by Celgard LLC in respect of battery technology. There was a discussion in passing about CCS during the dispute.

In terms of utility, the US patent system requires specific, credible, and concrete uses of inventions (Rimmer, 2007). CCS technologies which do not work according to specifications will struggle to meet such requirements. That could be a further area of litigation in respect of patent validity related to CCS.

10.2.3 Patent flexibilities

There have been perennial debates in international climate talks and negotiations about whether there should be text on the topic of intellectual property and climate change

(Rimmer, 2011, 2014). There has been argument over the treatment of patentable subject matter; public sector licensing; patent pools; compulsory licensing; and technology transfer. However, such discussions have often ended in stalemate between nation states—without any lasting agreement.

During the negotiations of the *Paris Agreement* 2015, there was a discussion about a range of patent flexibilities (Rimmer, 2018a, 2019). Options included Green Climate Fund financing for use of intellectual property; an intellectual property mechanism for clean technologies; collaborative approaches to intellectual property rights; intellectual property enforcement measures; and public goods dedications. However, given the lack of consensus between the parties, text about patent law was not included in the final text of the *Paris Agreement* 2015. There was instead some language relating to technology transfer under the *Paris Agreement* 2015. At the sidelines of the Paris negotiations, there was also an announcement of a range of innovation initiatives (Rimmer, 2019).

Given the experimental state of CCS, there is still going to be a need for researchers rely upon the defense of experimental use and the research exemption under patent law.

In light of the proliferation of patents in respect of CCS, there could be problems with patent thickets—perhaps even a "tragedy of the anticommons." Raiser et al. (2017) observed: "Carbon Capture processes often include multiple patents, all of which will need to be licensed if an actor wants to access a 'state of the art' technology." They maintained: "Such complicated licensing arrangements accompanying the diffusion and transfer of patented technology therefore often dissuade firms from investing in certain, heavily patented, technological research area."

Raiser et al. (2017) commented: "Although technological progress can be observed for RE and CCS development, these important mitigation technologies today still face many technological and economic barriers to become viable alternatives to the simple and cheap combustion of fossil fuels." They observed: "These barriers might be most readily overcome if the global research community collaborates on prioritizing the incremental improvements in cost and efficiency, rather than obstructing follow-on research and the subsequent dissemination of technologies through patenting."

Maskus and Okediji (2014) have considered a range of mechanisms to facilitate technology transfer of environmentally sound technologies. There remains a debate, though, as to whether CCS are truly environmentally sound technologies.

Taubman (2009) from the World Trade Organization observes that there are a range of technology sharing models, which could facilitate the sharing of climate technologies—including CCS. A patent pool would enable the sharing of technologies—either in closed form (amongst members) or in an open form. A patent commons would allow technology holders to pledge patents for widespread use for no royalty payment. A patent pledge or covenant would enable an owner to agree not to use patent rights against anyone using their technology in certain circumstances. Given the massive public investment in CCS technologies, there should also be a conversation over the need for flexible licensing in respect of public sector-funded work. Humanitarian or preferential licensing would provide for accessible use of patented inventions. Inventions can also be dedicated to the public domain. There is also scope for open innovation, commons-based peer production, and distributed innovation.

Compulsory licensing and crown use provisions could be used to provide access to patented inventions on public interest grounds—such as competition policy, or the need to address an emergency (including conceivably, a climate emergency).

Carrier (2011) discusses the treatment of competition and monopoly issues, such as refusals to license intellectual property in the United States and European Union. He considers the application of this law to the treatment of patents that assist in the removal of carbon dioxide from the atmosphere. Carrier (2011, p. 532) concludes that "A party's control of patented carbon-capture technologies would lead to different outcomes depending on the jurisdiction." He suggests: "The United States most likely would not require a monopolist to license the patents" (Carrier, 2011, p. 532). Carrier (2011, p. 532) argues: "In contrast, the EU would be more willing to require a firm with a dominant position to share its patented climate-change technology." Teachout (2020) has argued that there is a need for a more rigorous application of competition law in respect of energy markets in the United States—and the monopolies need to be broken up.

10.3 Government funding

Given concerns about the reluctance of the market to invest in CCS, various governments have considered grants and other forms of funding to provide incentives for research, development, and deployment of carbon capture technologies. Australia established carbon capture storage flagships and has supported CCS with a range of other policy initiatives. The United States has conducted some joint collaborative work on CCS with China. However, that cooperation has broken down—as there have been increasing clashes over intellectual property and trade between the superpowers. There has been debate that the UNFCCC Climate Technology Centre and Network should provide technical assistance in respect of CCS. There has been concerted criticism from Al Gore and others about the wisdom of governments making public investments in CCS—given their uninspiring performance thus far.

10.3.1 Australia's carbon capture storage flagships

In 2008 Ross Garnaut conducted a wide-ranging policy review in respect of climate change. On the subject of carbon capture, he contended: "There is a compelling case for Australia to play a major role in accelerating the international research effort on CCS across the range of technological change" (Garnaut, 2008, p. 500).

Subsequently, the Australian governments under Kevin Rudd and Julia Gillard established the carbon capture storage flagships. The CCS Flagships program supported a small number of demonstration projects, which sought to "capture carbon dioxide emissions from industrial processes," "provide transport infrastructure (generally pipelines)," and "safely store carbon dioxide underground in stable geological formations" (Department of Industry, Science, Energy and Resources, 2020).

There has also been funding provided for CCS through various other programs—including the Low Emission Technology Demonstration Fund, the Asia-Pacific Partnership on Clean Development and Climate, the National Low Emissions Coal Initiative, CCS RD&D, and Geoscience Australia's National CO_2 Infrastructure Plan.

Despite various forms of government funding, there has not been much progress with CCS projects in Australia. Browne and Swann (2017) noted: "Despite the promises and spending, there has never been an operational large-scale deployment of coal with CCS in Australia." Browne and Swann (2017) commented: "Attempts to develop coal with CCS, such as the ZeroGen project, have been expensive failures." Browne and Swann (2017) stressed: "Australia's only close-to-operational CCS project is connected to gas extraction." Browne and Swann (2017) noted: "Apart from three carbon storage projects, not expected to be operational until the 2020s, there are no other large-scale CCS projects at any stage of development." Browne and Swann (2017) conclude: "CCS is so uncommercial that even the enormous subsidies of the last decade have mostly resulted in cancelled, failed and bankrupt projects." Browne and Swann (2017) added: "Moreover, even if the technology could be demonstrated reliably at scale, the proponents of these projects consistently identify a high carbon price as being necessary for their commercial viability." Browne and Swann (2017) warned: "Without such a price, any new projects will need an even greater subsidy from government."

Lipski (2018) from Environmental Justice Australia maintained that the coal industry's carbon capture dream is a dangerous fantasy: "CCS fails from a technological perspective, an economic perspective, and a pollution reduction perspective."

There was controversy over the *Clean Energy Finance Corporation Amendment (CCS) Bill 2017 (Cth)*—especially in respect of its intent to provide funding for CCS. The Minister for the Environment and Energy Josh Frydenberg (2017) argued that the legislative "Change will provide direct support for CCS technologies, encourage greater private sector investment and reduce risk for potential investors."

The Coalition government—under Prime Ministers Tony Abbott, Malcolm Turnbull, and Scott Morrison—have supported CCS technologies. In 2015 the Minister for Industry and Science Ian Macfarlane maintained CCS technologies were worthwhile investing in:

> Just as we are using science to boost our key economic sectors, investment in research for CCS technologies will be important as the coal and gas industries continue to develop both for our domestic use and for export. As Australia and our major trading partners continue to use our valuable resources responsibly, further research and development in low emissions energy sources will further strengthen Australia's role as an energy superpower. Industry has a critical role to play in developing CCS technologies and investing in its own future, through the application of science and research in this field. *Macfarlane (2015)*

The Coalition government established the $25 million CCS Research Development and Demonstration Fund to focus on transport and storage projects.

Nonetheless, Prime Minister Malcolm Turnbull had his doubts as to whether CCS technology had been effective. He lamented at a 2017 National Press Club event: "We've invested $590 million since 2009 in clean coal technology research and demonstration and yet we do not have one modern high-efficiency low-emissions coal-fired power station, let alone one with CCS" (Turnbull, 2017).

The Coalition government persisted with its support for CCS under the leadership of Scott Morrison. The fossil fuel industry welcomed this support (Morton, 2020). Andrew McConville, the head of the Australian Petroleum Production and Exploration Association, commented: "Importantly, the report underlines the key role the oil and gas industry can play in cutting emissions" (Morton, 2020). Environmentalists questioned the wisdom of supporting CCS. Suzanne Harter, a campaigner with the Australian Conservation Foundation, said: "Australia has virtually no effective climate policy and no pathway to achieve net zero emissions by 2050" (Morton, 2020).

Opposition leader of the Australian Labor Party—Anthony Albanese—has argued: "Labor is willing to support CCS technologies being able to generate carbon offsets as long as the usual quality safeguards are met" (Mazengarb, 2020). He has promised: "We would also support the government if it reinstates the CCS Flagship program that was established by Labor with $177 million of funding and abolished by the Abbott government" (Mazengarb, 2020). However, Albanese maintained that renewable energy agencies should not be involved in the support of CCS: "We won't agree that renewable energy agencies like Australian Renewable Energy Agency (ARENA) and the Clean Energy Finance Corporation should have their funds for renewables raided in order to invest in carbon capture technology" (Mazengarb, 2020).

The leader of the Australian Greens leader Adam Bandt had questioned the government investment in carbon capture technology. He observed: "The only energy sources in Australia that have cut pollution are wind and solar" (SBS News, 2020). Bandt was of the view that support for renewable energy was a priority: "Now is the time to be backing them in and supporting investment, not cutting support" (SBS News, 2020). He commented: "CCS won't save coal and it won't save people from the climate crisis" (SBS News, 2020). Bandt warned: "For over a decade this has been hailed as the miracle cure, but it's just snake oil" (SBS News, 2020). He was of the view that fossil fuel companies engaged in "greenwashing" in respect of carbon capture technology. Bandt concluded: "Minister Taylor's huffing and puffing about CCS is the same failed fossil fuel industry pipe dream they have been promoting for years" (Morton, 2020).

In retrospect, Garnaut (2019) observed: "Despite the commitment of large financial support from the Australian government, the Australian coalmining industry hardly invested at all in the research, development, and commercialisation of CCS." He contended that carbon capture storage might still have a role in bioenergy combustion but "Not in capturing and storing emissions from coal-fired generation."

Reflecting on the government's proposals, Professor Frank Jotzo observed that CCS "Tends to be technically difficult and costly per unit of tonne of emissions saved, and usually does not capture all of the emissions" (Jotzo, 2020). He commented: "The obvious criticism is that extending government support to CCS locks in some fossil fuel use, when Australia has great opportunities to put our energy system on a zero-emissions footing using cheap renewable energy" (Jotzo, 2020).

In his book *Windfall*, Joshi (2020) recalls the long history of failure of public investment in respect of CCS technologies in Australia. He observed: "In 2018 CCS failed to meet every international target set for it, and in Australia in mid-2017, $1.3 billion had been invested with no operational projects at the time" (Joshi, 2020, p. 141). Joshi reflected: "Perhaps most tellingly, an organisation named COAL21, set up in 2005 to research CCS, has been spending its kitty on large-scale fossil fuel advertising campaigns" (Joshi, 2020, p. 141).

Grandia (2013) has contended: "Unfortunately, the mythical distraction of 'clean coal' and still unrealized CCS commercialization remain a shiny penny for the technocentric crowd."

For all the discussion of "clean coal" over the past couple of decades, Australia's coal plants remain rather dirty and deadly (Perkins, 2020).

10.3.2 United States—China collaboration on carbon capture and storage

In the Obama administration, Energy Secretary Steven Chu was an advocate of collaboration and cooperation in respect of CCS technologies (Rimmer, 2011, pp. 264–6). Steven

Chu supported joint projects between the United States and China in respect of CCS technologies. A flagship example of the collaboration was the United States–China Clean Energy Research Center. The United States Department of Energy (2020) maintained: "Working with counterparts in China, Department of Energy (DOE) has made advancing CCS technologies from large fossil energy facilities a top action agenda item for coal utilization and climate mitigation."

Drahos (2009) has argued that China and the United States should develop an integrated approach to the climate change, energy and intellectual property regimes if both countries are to respond to the problem of climate change in time. He questioned whether there was sufficient investment in basic research and development of CCS for the technologies to arrive in time:

> The crucial question is how much of the global funding effort going into CCS is going into basic R&D that CCS requires if it is to play a significant role in enabling the world to keep to the 2°C guardrail. Unfortunately we do not have an answer to this question, because the funding packages being announced by governments around the world do not provide this level of detail. *Drahos (2009, p. 127)*

Drahos (2009, p. 131) argued that China and the United States "Can recognise that climate change is a matter of survival governance in which they must create strong regimes that will bind them together in cooperation."

Far from strengthening scientific collaboration and cooperation, there has been increasing conflict over intellectual property between the United States and China. The Trump Administration has alleged that China has engaged in the theft of trade secrets from American companies. The Trump Administration has also brought an action in the World Trade Organization, alleging breaches of the *TRIPS Agreement* 1994.

Gerrard and Dernbach (2019) have sought to develop a comprehensive collection of Legal Pathways to Deep Decarbonization in the United States. A couple of chapters consider the topic of carbon capture and sequestration. In the collection, Jacobs and Craig (2019) make the case for the adoption of CCS technologies. They argue that "Significant legal reforms that include a combination of financial incentives, mandates, and other forms of government support are needed to drive full-scale diffusion of (CCS) technology in the United States" (Jacobs and Craig, 2019, p. 714). They suggest: "While a national program would be most effective in providing uniformity and consistency, there are many ways in which states can band together to create markets for the purchase of electricity from plants equipped with (CCS), offer financial incentives such as tax credits and other forms of tax relief, absorb some of the potential long-term liability for sequestration sites, and impose stricter standards on carbon dioxide emissions from fossil fuel-fired power plants" (Jacobs and Craig, 2019, p. 748).

Beck (2020) contends that the United States can play a particular role in innovation leadership in the commercialization of CCS.

10.3.3 United Kingdom

The United Kingdom government had announced a £1 billion competition to design CCS—but there has been little progress with the initiative. Baxter (2017) has argued that

"It is time for governments to stop wasting time and money on technologies like CCS that aren't working."

10.3.4 UNFCCC Climate Technology Centre and Network

The UNFCCC Climate Technology Centre and Network (CTCN) was established as part of the outcome of international climate talks in order to help accelerate the research, development, and deployment of clean technologies.

There has been debate as to whether it is appropriate for the CTCN to provide support for CCS technologies. There have also been concerns that the CTCN has provided technical assistance related to CCS (Kelly, 2018).

The UNFCCC Climate Technology Centre and Network (2020) provides an overview of CO_2 technologies. In terms of its overview of carbon storage, the CTCN discuss technical feasibility; regulatory perception and public perception; environmental impact and risks; the status of the technology and its future market potential; financial requirements and costs; and the impact on the climate.

10.3.5 Critical analysis

Former US Vice President, and leader of the Climate Reality program, Al Gore has questioned whether carbon capture technologies are viable:

> The fact none of the existing technologies are considered ready for primetime, in the sense that nobody knows how to execute at scale, makes it a daunting challenge for sure. I'm mindful of that. I just think it's an extremely improbable solution right now, but maybe they will come up with some breakthrough. There are so many fossil-fuel burning installations now that they are just shutting down well before their useful lifetime because it's just simply cheaper to move to renewable energy. *Harder (2018)*

Al Gore's position is that fossil fuels must be cut back in order to reduce carbon emissions.

In his book, *Our Choice*, Gore (2009) devotes a chapter to carbon capture and sequestration. He concludes: "Most experts who have studied the CCS option have concluded that it is probably impracticable for many years to come, because the technology for capturing CO_2 would either require a dramatic increase in the overall use of coal and gas for the same amount of electricity, or sharply reduce the amount of electricity obtained from burning the same amount of fuel as at present—and because every one of the potential geological repositories presents a unique and extremely difficult challenge in characterizing its geology deep underground and estimating both storage capacity and the safety of storing CO_2 there" (Gore, 2009, p. 148). Gore comments that a carbon price would help determinate whether CCS is economically plausible. He contends: "When the reality of the need to sharply reduce CO_2 emissions is integrated into all market calculations—including the decisions by utilities and their investors—market forces will drive us quickly toward the answers we need" (Gore, 2009, p. 148).

In his book, *The Future*, Gore (2013, p. 353) discusses carbon capture and sequestration in terms of being "false solutions." He observed: "I have long supported research and development of CCS technologies, but have been skeptical that they will play more than a minor role" (Gore, 2013, p. 353). Gore commented upon the cost of CCS:

> Barring breakthroughs, however, the cost of the CCS technology presently available—both in money and energy—is so high that utilities and others are unlikely to use it. A utility operating a coal-fired generating plant and selling electricity to its customers would have to divert approximately 35% of all the electricity it produces just to provide power for the capture, compression, and storage of the CO_2 that would otherwise be released into atmosphere. While that might be interpreted as a bargain if it saved civilization's future, the utility could not afford to do it and still stay in business. And the volumes of CO_2 emissions involved are so enormous that taxpayers do not have much appetite for shouldering the expense. *Gore (2013, p. 354)*

Gore (2013, p. 354) noted: "There has been notable public opposition to the siting of such underground storage facilities near populated areas." Gore (2013, p. 354) concluded that "The overall expense of CCS has prevented its adoption by large carbon polluters." Gore (2013, p. 355) suggests that, in spite of its problems, the psychological appeal of CCS is that it provides a possibility that a single technological strategy might lead to a relatively quick fix for the complex problems of climate change.

10.4 Innovation prizes and grand challenges

Innovation prizes and challenges have been a popular alternative means of promoting research and development. The Nobel-Prize winning economist Stiglitz (2019) has advocated the use of innovation prizes in the field of medicine. There has been extensive discussion about the use of innovation prizes to encourage research, development, and deployment of clean technologies (Rimmer, 2011).

There have been a number of specific innovation prizes, which have focused on CCS. Richard Branson established a prize—the Virgin Earth Challenge—for the removal of carbon dioxide from the atmosphere (Rimmer, 2011). However, that prize initiative did not result in the technological progress, which had been hoped for. More recently, Mission Innovation—launched in 2015 during the *Paris Agreement* 2015 negotiations—has a specific innovation focus in relation to CCS. In 2017 Representative Grace Meng introduced the *Carbon Capture Prize Act* 2017 (United States) into the US Congress. Furthermore, in 2019 the XPRIZE organization has launched a Carbon XPRIZE, which is particularly interested in CCS.

Nonetheless, there have been concerns about whether such innovation prizes and grand challenges are worthwhile, given the history of failures in respect of carbon capture. In her book *On Fire*, Klein (2019, p. 24) contends that carbon capture technologies are "unproven and expensive."

10.4.1 The Virgin Earth Challenge

In 2007 Richard Branson established the Virgin Earth Challenge as an innovation prize for the removal of carbon dioxide from the atmosphere (Rimmer, 2011). The prize established a kitty of US$25 million for "Whoever can demonstrate to the judges' satisfaction a commercially viable design which results in the removal of anthropogenic, atmospheric greenhouse gases (GHGs) so as to contribute materially to the stability of Earth's climate" (Virgin Earth Challenge, 2020).

Rather presciently, Adler (2010, p. 42) observed of Richard Branson's prize and others like it: "These prize awards could be important, but they are unlikely to produce the degree of technological innovation necessary to achieve current climate policy goals in a cost-effective manner."

In her book *This Changes Everything*, Klein (2014) is critical of Richard Branson's initiative with the Virgin Earth Challenge Prize. She noted that a shortlist of contenders were presented at a conference, but there was no final award of a prize:

> In November 2011, at an energy conference in Calgary, Alberta, … Branson announced the 11 most promising entries. Four were machines that directly sucked carbon out of the air (though none at anywhere near the scale needed); three were companies using the biochar process, which turns carbon-sequestering plant matter or manure into charcoal and then buries it in the soil and is controversial on a mass scale; and among the miscellaneous ideas was a surprisingly low-tech one involving revamping livestock grazing to boost the carbon-sequestering potential of soil. According to Branson, none of these finalists was ready yet to win the $25 million prize.

Klein (2014) noted that a number of the finalists have positioned themselves as startups in the oil industry. She comments upon the perverse nature of the prize: "Richard Branson has gone from promising to help us get off oil to championing technologies aimed at extracting and burning much more of it." Klein (2014) contends that the Earth Challenge could be seen as a regulatory avoidance strategy.

The Virgin Earth Challenge (2020) has become dormant. Virgin maintained: "The good news is it's becoming increasingly obvious that many carbon removal solutions have real potential; and are starting to prove themselves in the real world." Nonetheless, Virgin insisted that, "As with many other climate solutions, a better enabling environment is desperately needed" including "Stronger climate policies (in line with the climate science) and higher prices on carbon." The Virgin Earth Challenge (2020) insisted: "Though the Earth Challenge is no longer active, Virgin remains committed to carbon removal as part of our wider efforts to address the climate emergency."

10.4.2 Mission innovation

Mission Innovation has established a number of innovation challenges, which have been aimed at accelerating research, development, and demonstration in a range of technology areas (Mission Innovation, 2015). There was a carbon capture innovation challenge—named IC3.

The coleads of the Mission Innovation Project in respect of carbon capture are storage are Mexico, Saudi Arabia, and the United Kingdom. The participants included Australia, Canada, China, Denmark, the European Commission, Finland, France, Germany, India, Indonesia, Italy, Japan, the Netherlands, Norway, the Republic of Korea, Sweden, the United Arab Emirates, and the United States. US$103 million in funding for carbon capture had been allocated from the European Commission, the US Department of Energy, and the Accelerating Carbon Capture and Storage Consortium.

In terms of its pitch, Mission Innovation maintains that carbon capture, utilization, and storage technologies could be useful and helpful in addressing a reduction in carbon emissions:

> Globally, power and industry account for about 50% of all GHG emissions. Carbon Capture, Utilisation and Storage (CCUS) can achieve significant CO_2 reductions from power plants (fuelled by coal, natural gas, and biomass) and industrial applications (Figure 1). Industrial applications of CCUS include upstream oil and gas production, cement production, iron and steel production, and fertilizer manufacturing. These large ($> 100,000$ ton CO_2/year) point sources of CO_2 emissions have few alternative options for significant reductions. Efforts to integrate bioenergy with CCUS also represent a pathway to negative emission technologies, which models suggest will become increasingly important in achieving deep decarbonisation.
> *Mission Innovation (2020)*

Mission Innovation (2020) contends: "Coordinated decarbonisation efforts must include the development of additional technologies that (1) prevent and curtail emissions of CO_2, (2) result in carbon negative solutions, and (3) lead to safe and secure carbon storage."

In terms of its framing of the issue, Mission Innovation (2020) maintains that "Achieving *Paris Agreement* targets will require a significant acceleration of the development and deployment of technologies that dramatically reduce the output of CO_2." Mission Innovation (2020) observed: "CCUS developments to date are noteworthy, but additional extensive and far-reaching efforts are required to combat climate change." Mission Innovation (2020) acknowledged: "Globally, the total CO_2 capture capacity of the 22 current projects (in operation or construction) is about 40 million tonnes per annum." Mission Innovation (2020) recognizes that "Overall costs need to be reduced for the technology to be adopted at a sufficient scale to meet the challenges of climate change."

In respect of the opportunity, Mission Innovation (2020) argues: "CCUS is one of the only technologies able to achieve significant decarbonisation of our fossil fuel based economies, particularly in carbon-intensive industries such as cement, iron and steel production." Mission Innovation (2020) concludes: "Operating CCUS projects offer important insight into the technical capabilities, policy and financing mechanisms, and permitting frameworks that could enable the successful deployment of CCUS." Mission Innovation (2020) insists: "The goal of the Carbon Capture Innovation Challenge is twofold: first, to identify and prioritize breakthrough technologies; and second, to recommend research, development, and demonstration (RD&D) pathways and collaboration mechanisms."

In respect of implementation, Mission Innovation (2020) argues that "Further efforts must be focused on research and development to enable new and novel carbon capture technologies, aimed at driving down costs and facilitating broader deployment." In its view, "Fundamental research should be directed in areas that could result in revolutionary, not just incremental, advances in gas separation and geologic storage of CO_2" (Mission Innovation, 2020). Mission Innovation (2020) also suggests: "Parallel efforts to utilize CO_2 must also be pursued, exploring the use of captured CO_2 to create plastics or algal biofuels, carbonate materials, or other uses yet-to-be-discovered." Mission Innovation (2020) emphasizes that there is a need to test and evaluate emerging technologies in the field: "Technologies will need to be developed, tested, and vetted in collaborative forums, building on past experiences and improving on current efforts to further reduce costs."

In mid-2017 a technical Mission Innovation CCUS workshop will be hosted by the United States. The workshop will convene top experts to discuss breakthrough opportunities and find international RD&D synergies in carbon capture, geologic storage, and CO_2 utilization.

In 2017 there was a report on the work of Mission Innovation in respect of carbon capture, utilization, and storage. In June 2019 there was a workshop held in respect of the implementation and commercialization of carbon capture, utilization and storage. There was a report issued from the workshop, with a range of recommendations. In December 2019 the IC3 Action Plan was released. Thus far, it seems that Mission Innovation has played a dialogical role in encouraging discussion of the development of CCS technologies.

10.4.3 Carbon Capture Prize Act

The US government has previously legislated for innovation prizes—such as the L-Prize and the H-Prize (Rimmer, 2011). This past experience has interested some legislators in the option of an incentive prize for research and development in respect of CCS.

In October 2017 New York Democrat Representative Meng (2017) introduced legislation into the US Congress to create a prize competition to reduce carbon in the atmosphere. The legislation was entitled the *Carbon Capture Prize Act* 2017 (United States) H.R. 4906. Representative Meng (2017) commented: "Presence of carbon in the atmosphere contributes to rising sea levels and heat waves around the globe, and we are already seeing the devastating consequences." Meng added (2017): "We must take action to reduce the amount of carbon in our atmosphere." Meng concluded (2017): "Prize competitions have long been an effective tool to find cost-effective solutions for expensive problems, and establishing a prize competition to decrease the quantity of carbon in the atmosphere is the next logical step in tackling this issue." There were a number of cosponsors of the bill—including Reps. Nydia Velázquez (D-NY) and Jamie Raskin (D-MD).

The legislation directed the DOE to conduct a prize competition to incentivize the development, research, or commercialization of technology that reduces the amount of carbon dioxide in the atmosphere. The measure authorized an aggregate prize amount of US $5,000,000. The DOE may run the competition individually or with other agencies. The legislation stipulated that any competition must comply with the *Stevenson-Wydler Technology Innovation Act* 1980 (United States).

The legislation is described as a "Bill to authorize the Secretary of Energy to establish a prize competition for the research, development, or commercialization of technology that would reduce the amount of carbon in the atmosphere, including by capturing or sequestering carbon dioxide or reducing the emission of carbon dioxide."

Section 2(a) provides for the establishment of a prize competition: "The Secretary of Energy, individually or in cooperation with the heads of other Federal agencies, shall carry out a program pursuant to section 24 of the *Stevenson-Wydler Technology Innovation Act* of 1980 (15 U.S.C. 3719) to award prizes competitively to incentivize the research, development, or commercialization of technology that reduces the amount of carbon dioxide in the atmosphere, including by capturing or sequestering carbon dioxide or reducing the emission of carbon dioxide."

Section 2(b) stipulates a prize amount: "In carrying out the program established under subsection (a), the Secretary may award not more than $5,000,000, in the aggregate, to the winner or winners of the prize competition."

Section 2(c) provides that "The term 'Federal agency' has the meaning given the term under section 24(a)(3) of the *Stevenson-Wydler Technology Innovation Act* of 1980 (15 U.S.C. 3719 (a)(3))."

The bill was referred to House Science, Space, and Technology Subcommittee on Energy—but did not progress further in the US Congress.

The legislation was reintroduced in 2019 as the *Carbon Capture Prize Act* 2019 (United States) H.R. 3282, sponsored again by the Hon. Meng, with 18 cosponsors. In June 2019 the Hon. Meng (2019) called on her colleagues to support the bill. Meng (2019) argued: "While it is critical that all nations transition to a low-carbon future, we must also explore technologies that remove and sequester carbon pollution to keep global average temperatures from rising above 1.5 degrees." Meng (2019) made a case for the adoption of CCS technologies:

> Technologies, like direct air capture, can provide nations the tools needed to reduce carbon pollution in the atmosphere. The benefit of this technology is that it can be located anywhere, making its potential scale of deployment enormous. A major challenge facing direct air capture technology, however, is cost, which can range between $800 and $250 per metric ton of CO_2 removal from the atmosphere.

Meng (2019) discussed the suitability of innovation prizes: "Prize competitions have long been an effective tool to find cost-effective solutions for expensive problems." Meng (2019) commented: "It is undeniable that the fate of our children and future generations rests on the decisions we make today about fighting climate change."

The failure of the legislation to progress further perhaps reflects a flagging interest in CCS on the part of the United States.

10.4.4 The Carbon XPRIZE

The XPRIZE Foundation is a nonprofit 501(c)(3) company, which has operated since 1994. The XPRIZE Foundation has designed and operated seventeen competitions in the areas of Space, Oceans, Learning, Health, Energy, Environment, Transportation, Safety, and Robotics.

Peter Diamandis has been a key organizer of XPRIZE competitions. He is a technooptimist. His book *The Future if Faster Than You Think* (Diamandis and Kotler, 2020) is concerned about technological solutions to global challenges—such as climate change. Diamandis and Kotler (2020, p. 266) explain the approach of the XPRIZE Foundation:

> The XPRIZE Foundation uses large-scale global incentive competitions to crowdsource solutions to the world's grand challenges. XPRIZE believes that solutions can come from anyone, anywhere. Scientists, engineers, academics, entrepreneurs, and other innovators with new ideas from all over the world are invited to form teams and compete to win the prize. Rather than throw money at a problem, we incentivize the solution and challenge the world to solve it. *Diamandis and Kotler (2020, p. 266)*

Diamandis and Kotler (2020, p. 212) identify a range of interlocking ecological threats—including water crises, biodiversity loss, extreme weather, climate change, and pollution. Considering technological means to respond to such problems, Diamandis and Kotler (2020, p. 212) reflect: "Ours is not a techno-utopian argument." They maintained: "Solving our planet's ecological woes requires technology, for certain, but it also demands one of the largest cooperative efforts in history" (Diamandis and Kotler, 2020, p. 212).

In 2018 the XPRIZE Foundation established the Carbon XPRIZE to provide an incentive for the development of carbon capture and utilization technologies, which converted carbon dioxide (CO_2) into useful products, while mitigating climate change. As the FAQ explains, the focus of the Carbon XPRIZE is a little bit different from traditional carbon capture and utilization projects:

> CCS can also be a critical part of the portfolio for addressing global CO_2 emissions. Investment in CCS has succeeded in advancing some technologies to a stage where there are now at least a dozen commercial-scale CCS projects worldwide, with approximately 50 additional CCS projects in various stages of development around the world. However, there may be many regions where subsurface storage is not the optimal solution. In addition, no matter how advanced, CCS technologies will always fundamentally treat CO_2 as a waste product that requires disposal rather than as an asset with value. CO_2 conversion has the potential to be a complementary solution with additional investment and technological development. *(XPRIZE Foundation, 2020)*

Thus, the Carbon XPRIZE is particularly focused upon carbon conversion technologies. The Carbon XPRIZE has a prize of US$20 million dollars. The Carbon XPRIZE is sponsored by NRG (an American energy company), and COSIA (Canada's Oil Sands Innovation Alliance). The Carbon XPRIZE has a number of prize partners—including Canadian Natural Resources Limited; Cenovus Energy Inc.; ConocoPhillips Canada; Devon; Suncor Energy; ANA; Wyoming Integrated Test Center; InnoTech Alberta; 350 Solutions; AFARA; Calgary Economic Development; CMC Research Institutes; Enviro Innovate; Innovate Calgary; Kinetica Ventures; and Google. It is notable that the Carbon XPRIZE sponsors and partners are dominated by fossil fuel companies, and related innovation groups.

There are 10 finalists for the Carbon XPRIZE. There are several competitors from the United States—including AIR Co., C2CNT, CO_2 Concrete, and Dimensional Energy. From Canada, there is Carbon Upcycling-NLT, CarbonCure, and CERT. There is Carbon Capture Machine from the United Kingdom. There is Breathe from India; and C4X from China. According to the XPRIZE Foundation (2020), "The winning team will convert the most CO_2 into products with the highest value as determined by: (1) How much CO_2 they convert; and (2) The net value of their products."

There has been discussion as to whether an innovation prize system works best as an alternative or a complement to the intellectual property system. Overett (2018) emphasizes that "The private XPRIZE Foundation does not limit the IP rights of contestants, and most rely on patent protection as part of their commercialisation strategy." He observed: "The largest company in the competition, Carbon Cure from Canada, has at least 11 patent families underpinning its commercialised retrofit technology for CO_2 curing of Portland cement" (Overett, 2018). He commented: "At the opposite end of the scalability spectrum, the US-based start-up C2CNT has a small patent portfolio related to the synthesis of tailored carbon nanotubes, via solar-powered electrolysis of CO_2 in molten carbonate salts" (Overett, 2018).

It remains to be seen whether the Carbon XPRIZE will result in substantial technological developments.

10.5 Conclusion

In terms of intellectual property law, policy, and practice, CCS pose some challenging and awkward questions. There is a debate about whether CCS should be classified as part of the family of clean technologies. It has been argued that, at present, the technology field is not best described as a green technology. The patent landscape in respect of CCS has become increasingly crowded—but there remain questions about the quality of patent applications in the field. There is concern about the domination of patent filings in respect of CCS by fossil fuel companies and high-carbon emitting countries. There are a range of patent flexibilities which could be used to provide access to patented inventions—if need be. Given market concerns about the viability of CCS, governments have often provided extensive support for such technologies. However, there has been a record of poor outcomes in respect of public investment in CCS. Indeed, there has been criticism that many CCS projects have been boondoggles and "white elephants." There have been a range of innovation prizes and grand challenges designed to promote research, development, and deployment of CCS. The Virgin Earth Challenge has turned out to be fruitless. The Mission Innovation has focused on CCS but this far its contribution has largely dialogical. The US Congress has not passed the *Carbon Capture Prize Act* 2017 (United States) into the US Congress. The XPRIZE organization has launched a Carbon XPRIZE—but it is too early to determine whether it will result in lasting innovation.

Despite the disappointments of these past innovation prizes, billionaire Elon Musk—associated with Tesla Inc., Space X, Neuralink, The Boring Company, and Open AI—has been enthusiastic about providing new incentives for research into carbon capture and storage. Elon Musk and the Musk Foundation have established the XPRIZE Carbon Removal, with a prize purse of $100 million (Neate, 2021). Musk has explained that he wanted scientists to make a 'truly meaningful impact' and achieve 'carbon negativity, not neutrality' (Neate, 2021). He commented, 'This is not a theoretical competition; we want teams that will build real systems that can make a measurable impact and scale to a gigaton level' (Neate, 2021). Musk declared: 'Whatever it takes. Time is of the essence' (Neate, 2021). The XPRIZE Foundation stressed that teams must create 'a solution that can pull carbon dioxide directly from the atmosphere or oceans and lock it away permanently in an environmentally benign way' (Neate, 2021). The XPRIZE Foundation declared, 'this $100M competition is the largest incentive prize in history, an extraordinary milestone' (XPRIZE Foundation (2021)). Likewise, billionaire Bill Gates (2021) has been enthusiastic about carbon capture and storage technologies. While recognising that such technologies are currently very expensive and quite ineffective, he argues, 'smart public policies could create incentives to use carbon capture' (Gates, 2021, 95).

This chapter considered the field of CCS in terms of innovation law and policy. After much hype about the potential of the technology, there has been a trough of disillusionment about CCS, for a variety of reasons. To use the language of Gardiner (2011), CCS has been a "shadow solution" to the larger problem of climate change. Research from Stanford University (2019) by Jacobson (2019) suggests that current approaches to carbon capture can increase air pollution and are not efficient at reducing carbon in the atmosphere. Jacobson (2019) reaches the conclusion.

> Synthetic direct air carbon capture and use (SDACCS/U) and carbon capture and use (CCS/U) are opportunity costs, not close to zero-carbon technologies. For the same energy cost, wind turbines and solar

panels reduce much more CO_2 while also reducing fossil air pollution and mining, pipelines, refineries, gas stations, tanker trucks, oil tankers, coal trains, oil spills, oil fires, gas leaks, gas explosions, and international conflicts over energy. CCS/U and SDACCS increase these by increasing energy use and always increase total social costs relative to using renewables to eliminate fossil fuel and bioenergy power generation directly.

Jacobson argues: "There is a lot of reliance on carbon capture in theoretical modelling, and by focusing on that as even a possibility, that diverts resources away from real solutions" (Stanford, 2019). Jacobson argues that carbon capture "gives people hope that you can keep fossil fuel power plants alive" and "delays action" (Stanford, 2019). In his view, "Carbon capture and direct air capture are always opportunity costs" (Stanford, 2019).

Ultimately, it is currently doubtful as to whether it is worthwhile providing incentives for further research in respect of CCS.

References

Adler, J., 2010. Eyes on a climate prize: rewarding energy innovation to achieve climate stabilization. Harvard Environmental Law Review 35 (1), 1–46.

Baker, D., Jayadev, A., Stiglitz, J., 2017. Innovation, Intellectual Property, and Development: A Better Set of Approaches for the 21st Century. AccessIBSA. <http://ip-unit.org/wp-content/uploads/2017/07/IP-for-21st-Century-EN.pdf>.

Barton, J., 2007. Intellectual Property and Access to Clean Energy Technologies in Developing Countries: An Analysis of Solar Photovoltaic, Biofuel and Wind Technologies. International Centre for Trade and Sustainable Development, Geneva. <http://www.ictsd.org/downloads/2008/11/intellectual-property-and-access-to-clean-energy-technologies-in-developing-countries_barton_ictsd-2007.pdf>.

Baxter, T., 2017. It's time to accept carbon capture has failed—here's what we should do instead. The Conversation, 24 August. <https://theconversation.com/its-time-to-accept-carbon-capture-has-failed-heres-what-we-should-do-instead-82929>.

Beck, L., 2020. Carbon capture and storage in the USA: the role of US innovation leadership in climate-technology commercialization. Clean Energy 4 (1), 2–11. Available from: https://doi.org/10.1093/ce/zkz031.

Brown, A. (Ed.), 2013. Environmental Technologies, Intellectual Property and Climate Change: Accessing, Obtaining and Protecting. Edward Elgar, Cheltenham and Northampton, MA.

Brown, A., 2019. Intellectual Property, Climate Change and Technology: Managing National Legal Intersections, Relationships and Conflicts. Edward Elgar, Cheltenham and Northampton, MA.

Browne, B., Swann, T., 2017. Money for Nothing: A Submission on the Clean Energy Finance Corporation Amendment (Carbon Capture and Storage) Bill 2017 (Cth). Australian Parliament. <https://www.aph.gov.au/DocumentStore.ashx?id=f906d1ec-e24d-43ab-aa4a-a494c5c34c9a&subId=564349>.

Carrier, M., 2011. An antitrust framework for climate change. Northwestern Journal of Technology and Intellectual Property 9 (8), 513–532.

CO$_2$ Solutions, 2016. CO$_2$ Solutions Wins Patent Challenge, August 10. <https://www.prnewswire.com/news-releases/co2-solutions-wins-patent-challenge-589737431.htmlr>.

CO$_2$ Solutions, 2020. CO$_2$ solutions sells its CO$_2$ capture technology assets, including unit at Resolute pulp mill in Quebec. Press Release, January 23. <https://technology.risiinfo.com/mills/north-america/co2-solutions-sells-its-co2-capture-technology-assets-including-unit-resolute-pulp-mill-quebec>.

Dechezleprêtre, A., 2013. Fast-Tracking Green Patent Applications: An Empirical Analysis, ICTSD Global Platform on Climate Change. Trade, and Sustainable Energy. Available from: https://www.files.ethz.ch/isn/161230/fast-tracking-green-patent-applications-an-empirical-analysis.pdf.

Dechezleprêtre, A., Lane, E., 2013. Fast-tracking green patent applications. WIPO Magazine 3, <https://www.wipo.int/wipo_magazine/en/2013/03/article_0002.html>.

Department of Industry, Science, Energy and Resources, 2020. Carbon Capture Storage Flagships, Australian Government. <https://www.industry.gov.au/funding-and-incentives/carbon-capture-storage-flagships>.

Diamandis, P., Kotler, S., 2020. The Future Is Faster Than You Think: How Converging Technologies Are Transforming Business, Industries, and Our Lives. Simon & Schuster, New York.

Drahos, P., 2009. The China—US relationship on climate change, intellectual property and CCS: requiem for a species? The WIPO Journal 1, 124—131.

Drugman, D., 2020. Exxon Sued again for "misleading" advertising. DeSmog Blog, May 20. <https://www.desmogblog.com/2020/05/20/exxon-sued-misleading-advertising-beyond-pesticides>.

Frydenberg, J., 2017. Second Reading Speech on the Clean Energy Finance Corporation Amendment (Carbon Capture and Storage) Bill 2017 (Cth), Hansard, Australian Parliament, May 31, p. 5476.

Gardiner, S., 2011. A Perfect Moral Storm: The Ethical Tragedy of Climate Change. Oxford University Press, Oxford.

Garnaut, R., 2008. The Garnaut Climate Change Review: Final Report. Cambridge University Press, Cambridge.

Garnaut, R., 2019. Superpower: Australia's Low-Carbon Opportunity. La Trobe University Press, Melbourne.

Gates, B, 2021. *How To Avoid A Climate Disaster: The Solutions We Have and the Breakthroughs We Need*. Allen Lane, Penguin Random House, London.

Gerrard, M., Dernbach, J. (Eds.), 2019. Legal Pathways to Deep Decarbonization in the United States. Environmental Law Institute, Washington, DC.

Gore, A., 2009. Our Choice: A Plan to Solve the Climate Crisis. Bloomsbury, London.

Gore, A., 2013. The Future: Six Drivers of Global Change. Random House, New York.

Grandia, K., 2013. Coal summit's pipe dream of carbon capture and storage, Renew Economy, November 19. <https://reneweconomy.com.au/coal-summits-pipe-dream-of-carbon-capture-and-storage-91782/>.

Gurry, F., 2020. World Intellectual Property Day 2020—Innovate for a Green Future. World Intellectual Property Organization. <https://www.wipo.int/ip-outreach/en/ipday/2020/dg_message.html>.

Harder, A., 2018. Al Gore: Technology Capturing CO_2 Emissions Is "Nonsense." Axios, December 12. <https://www.axios.com/al-gore-technology-capturing-co2-emissions-nonsense-c1e5e230-0ffb-4b5c-ba24-8bcdfd78e7d3.html>.

Jacobs, W., Craig, M., 2019. Carbon capture and sequestration. In: Gerrard, M., Dernbach, J. (Eds.), Legal Pathways to Deep Decarbonization in the United States. Environmental Law Institute, Washington, DC, pp. 713—748.

Jacobson, M.Z., 2019. The health and climate impacts of carbon capture and direct air capture. Energy and Environmental Science 12, 3567—3574.

Joshi, K., 2020. Windfall: Unlocking a Fossil-Free Future. New South Books, Sydney.

Jotzo, F., 2020. Morrison government dangles new carrots for industry but fails to fix bigger climate policy problem. The Conversation, May 20. <https://theconversation.com/morrison-government-dangles-new-carrots-for-industry-but-fails-to-fix-bigger-climate-policy-problem-138940>.

Kelly, S., 2018. How America's clean coal dream unravelled. The Guardian, March 2.

Klein, N., 2014. This Changes Everything, Capitalism vs the Climate. Penguin Books, London.

Klein, N., 2019. On Fire: The (Burning) Case for a Green New Deal. Simon & Schuster, New York.

Lane, E., 2011. Clean Tech Intellectual Property: Eco-Marks, Green Patents, and Green Innovation. Oxford University Press, Oxford.

Lane, E., 2012. Building the global green patent highway: a proposal for international harmonization of green technology fast track programs. Berkeley Technology Law Journal 27 (3), 1119—1170.

Lee, B., Iliev, I., Preston, F., 2009. Who Owns Our Low Carbon Future? Intellectual Property and Energy Technologies. London: A Chatham House Report. <https://www.chathamhouse.org/sites/default/files/public/Research/Energy,%20Environment%20and%20Development/r0909_lowcarbonfuture.pdf>.

Li, B., Duana, Y., Luebkea, D., Morreale, B., 2013. Advances in CO_2 capture technology: a patent review. Applied Energy 102, 1439—1447.

Lipski, B., 2018. Coal industry's carbon capture dream. Renew Economy, March 23. <https://reneweconomy.com.au/coal-industrys-carbon-capture-dream-dangerous-fantasy-41399/>.

Macfarlane, I., 2015. New support for carbon capture and storage R&D. Press Release, Australian Government, August 31. <https://www.minister.industry.gov.au/ministers/macfarlane/media-releases/new-support-carbon-capture-and-storage-rd>.

Maskus, K., Okediji, R., 2014. Legal and economic perspectives on international technology transfer in environmentally sound technologies. In: Cimoli, M., Dosi, G., Maskus, K., Okediji, R., Reichman, J., Stiglitz, J. (Eds.), Intellectual Property Rights: Legal and Economic Challenges for Development. Oxford University Press, Oxford, pp. 392—414.

II. Case studies on CCS and related policies, and their consequences for climate change

Mazengarb, M., 2020. Albanese set to deal on carbon capture if renewables funding protected. Renew Economy, June 24. <https://reneweconomy.com.au/albanese-set-to-deal-on-carbon-capture-if-renewables-funding-protected-66989/>.

Meng, G., 2017. Meng introduces legislation to create a prize competition to reduce carbon in atmosphere. Press Release, October 25. <https://meng.house.gov/media-center/press-releases/meng-introduces-legislation-to-create-a-prize-competition-to-reduce>.

Meng, G., 2019. Carbon Capture Prize Act, House of Representatives. United States Congress, June 13. <https://www.congress.gov/116/crec/2019/06/13/modified/CREC-2019-06-13-pt1-PgE765-5.htm>.

Mission Innovation, 2015. Innovation Challenges. <http://mission-innovation.net/our-work/innovation-challenges/>.

Mission Innovation, 2020. Our Work—Innovation Challenges—Carbon Capture. <http://mission-innovation.net/our-work/innovation-challenges/carbon-capture/>.

Morton, A., 2020. Fossil fuel industry applauds coalition climate measures that support carbon capture and storage. The Guardian, May 20. <https://www.theguardian.com/australia-news/2020/may/20/fossil-fuel-industry-applauds-coalition-climate-measures-that-support-carbon-capture-and-storage>.

Overett, M., 2018. The Carbon XPRIZE and CSIRO: CO_2 mitigation technology in Australia and beyond. Phillips Ormonde Fitzpatrick, June 20. <https://www.pof.com.au/the-carbon-xprize-and-csiro-co2-mitigation-technology-in-australia-and-beyond/>.

Perkins, M., 2020. Coal-fired pollution killing 800 Australians a year: report. The Sydney Morning Herald, August 26. <https://www.smh.com.au/environment/climate-change/coal-fired-pollution-killing-800-australians-a-year-report-20200825-p55ozz.html>.

Neate, R, 2021. Elon Musk Pledges $100m to Carbon Capture Contest. The Guardian. Available from: https://www.theguardian.com/environment/2021/feb/08/elon-musk-pledges-100m-to-carbon-capture-contest.

Raiser, K., Naims, H., Bruhn, T., 2017. Corporatization of the climate? Innovation, intellectual property rights, and patents for climate change mitigation. Energy Research and Social Science 27, 1–8.

Rimmer, M., 2007. The new conquistadors: patent law and expressed sequence tags. Journal of Law, Information, and Science 16, 10–50.

Rimmer, M., 2011. Intellectual Property and Climate Change: Inventing Clean Technologies. Edward Elgar, Cheltenham and Northampton.

Rimmer, M., 2014. Intellectual property and global warming: fossil fuels and climate justice. In: David, M., Halbert, D. (Eds.), The Sage Handbook of Intellectual Property. Sage Publications, London, pp. 727–753.

Rimmer, M., et al., 2015. Intellectual property and biofuels: the energy crisis, food security, and climate change. Journal of World Intellectual Property 18 (6), 271–297.

Rimmer, M. (Ed.), 2018a. Intellectual Property and Clean Energy: The Paris Agreement and Climate Justice. Springer, Singapore.

Rimmer, M., 2018b. Intellectual ventures: patent law, climate change, and geoengineering. In: Rimmer, M. (Ed.), Intellectual Property and Clean Energy: The Paris Agreement and Climate Justice. Springer, Singapore, pp. 235–271.

Rimmer, M., 2019. Beyond the Paris Agreement: intellectual property, innovation policy, and climate justice. Laws 8 (1), 7. Available from: https://doi.org/10.3390/laws8010007.

Sarnoff, J. (Ed.), 2015. Research Handbook on Intellectual Property and Climate Change. Edward Elgar, Cheltenham and Northampton, MA.

SBS News, 2020. Government flags hydrogen and carbon capture investment, may pave the way for new coal fired power. SBS News, February 28. <https://www.sbs.com.au/news/government-flags-hydrogen-and-carbon-capture-investment-may-pave-the-way-for-new-coal-fired-power>.

Stanford University, 2019. Study casts doubt on carbon capture. Stanford News, October 25. <https://news.stanford.edu/2019/10/25/study-casts-doubt-carbon-capture/>.

Stiglitz, J., 2019. People, Power, and Profits: Progressive Capitalism for an Age of Discontent. W.W. Norton & Company, New York.

Taubman, A., 2009. Sharing technology to meet a common challenge. WIPO Magazine 2, <https://www.wipo.int/wipo_magazine/en/2009/02/article_0002.html>.

Teachout, Z., 2020. Break 'Em Up: Recovering Our Freedom From Big Ag, Big Tech, and Big Money. Macmillan Publishers, New York.

Turnbull, M., 2017. Address at the National Press Club and Q&A—Canberra. <https://www.malcolmturnbull.com.au/media/address-at-the-national-press-club-and-qa-canberra>.

The UNFCCC Climate Technology Centre and Network, 2020. CO_2 Capture Technologies. <https://www.ctc-n.org/technologies/co2-capture-technologies>.

United States Department of Energy, 2020. US–China Clean Energy Research Center—CERC. <https://www.energy.gov/ia/initiatives/us-china-clean-energy-research-center-cerc>.

Virgin Earth Challenge, 2020. <https://www.virgin.com/content/virgin-earth-challenge-0>.

XPRIZE Foundation, 2020. Carbon XPRIZE: Frequently Asked Questions. <https://carbon.xprize.org/prizes/carbon/faq>.

XPRIZE Foundation, 2021. $100M Gigaton Scale Carbon Removal. XPRIZE Foundation. Available from: https://www.xprize.org/prizes/elonmusk.

Zhuang, W., 2017. Intellectual Property Rights and Climate Change: Interpreting the TRIPS Agreement for Environmentally Sound Technologies. Cambridge University Press, Cambridge.

Legislation

Carbon Capture Prize Act 2017 (United States) H.R. 4906.
Carbon Capture Prize Act 2019 (United States) H.R. 3282.
Clean Energy Finance Corporation Amendment (Carbon Capture and Storage) Bill 2017 (Cth).
Stevenson-Wydler Technology Innovation Act 1980 (United States) (15 U.S.C. 3719 (a)(3)).

Cases

Akermin, Inc. v. *CO_2 Solutions Inc.* bpai_IPR2015-00880_25 (PTAB—other actions 2016).

Celgard, LLC v. *Shenzhen Senior Technology Material Co. Ltd. (United States)* Research Institute etc. N.D.Cal._519-cv-05784_1 (California Northern District Court 2019) Reference 5:19-cv-05784-SVK.

CO_2 Solutions Inc. v. *Akermin Inc.* E.D.Mo._416-cv-00520_1 (Missouri Eastern District Court 2016) Reference 4:16-cv-00520-RLW.

Ex Parte Paul J. Berlowitz us-ptab-2018009128-12-302019 (PTAB—Administrative Hearings 2019).

Ex parte Li us-ptab-2015000389-04-212017 (PTAB—Administrative Hearings 2017).

International treaties

Marrakesh Agreement Establishing the World Trade Organization, opened for signature 15 April 1994, 1867 UNTS 3 (entered into force 1 January 1995) annex 1C ("*Agreement on Trade-related Aspects of Intellectual Property Rights*"—*TRIPS Agreement* 1994).

Paris Agreement to the United Nations Framework Convention on Climate Change, opened for signature 12 December 2015 (entered into force 4 November 2016) (in UNFCCC, Report of the Conference of the Parties on its Twenty-First Session, Addendum, UN Doc FCCC/CP/2015/10/Add.1, 29 January 2016).

United Nations Framework Convention on Climate Change 1992, Opened for signature 9 May 1992, 1771 UNTS 107 (entered into force 21 March 1994).

CHAPTER

11

Negative-emission technologies and patent rights after COVID-19*

Joshua D. Sarnoff

Depaul University College of Law, Chicago, IL, United States

11.1 Introduction

Globally, concerns continue to mount over the accelerating increase in worldwide levels of carbon dioxide (CO_2) and other greenhouse gas (GHG) emissions, and over the consequent rise in temperatures and associated adverse effects. Regulatory efforts and market incentive mechanisms to develop and deploy mitigation (GHG emission reducing) measures—from carbon taxes and emission-trading schemes to research and development (R&D) and production-tax subsidies and intellectual property (IP) rights—have so far proven inadequate to control and reduce emissions. Humans are resistant to behavioral changes, and research and development and deployment (RD&D, with deployment also including demonstration) of emission-reducing technologies is costly, time-consuming, and may strand valuable assets (Livesay, 2020; Juma, 2016). Even with the dramatic contraction of economic activity and associated reductions of GHG emissions following lockdowns to address the COVID-19 pandemic, "no one should think that the climate crisis is therefore over—far from it" (UN Environment Programme, 2020; Betts et al., 2020).

Nevertheless, the pandemic has shown us that the private, philanthropic, and public sectors can attempt big, collaborative innovation projects and can leverage massive amounts of funding if the need is apparent (World Health Organization, 2020b; Sagonowsky, 2020; "World leaders donate," 2020). In order to address COVID-19 innovation and production needs, some private companies volunteered to donate or not to assert their IP rights, were encouraged to share their data, and began to collaborate with individuals, entities, and governments in myriad and unexpected ways (Open Covid Pledge, b; MIT Technology Licensing Office, 2020; U.S. National Institutes of Health, 2020a; U.S.

* This chapter is based in large part upon Negative Emissions Technologies and Patent Rights After COVID-19, 10 Climate Law 225−265 (Brill Nijhoff 2020). Thanks to Sydney Warda for research assistance.

National Institutes of Health, 2020b). Yet, deep philosophical disputes over the nature of IP rights, over compulsory licensing and mandated production, over price controls, and over free or regulated markets continue to confront governments and intergovernmental organizations seeking to balance innovation incentives with assuring affordable and widespread access (Silverman, 2020b; Liu et al., 2020). In particular, some private companies may oppose any measures to address COVID-19 that would not be voluntary, based on concerns with recoupment, uniform licensing requirements, and lack of control over downstream access, distribution, and uses of future products. These companies assert that such measures will deter voluntary collaboration or the required high-levels of investment (Brachman, 2020; Silverman, 2020a). For example, Pfizer's Chief Executive Albert Bourla responded to a voluntary COVID-19 patent pool created by the World Health Organization (WHO) by stating that companies are "investing billions to find a solution and, keep in mind, if you have a discovery, we are going to take your [intellectual property], I think, is dangerous" (Silverman, 2020c). And the COVID-19 Vaccines Global Access (COVAX) facility developed by the WHO to deliver vaccines equitably around the world has been underfunded and lacks sufficient access to supplies, as richer countries (particularly those that have developed the vaccines) have contracted for prioritized access for their own populations, resulting in so-called "vaccine nationalism" (Nature, 2021; Serhan, 2020).

It is increasingly apparent that we need additional collaboration and additional funding for innovation to minimize and to respond to climate change as part of the transition to a post-COVID-19 world (Gronewold, 2020; Sheffi, 2020). Conversely, it is also increasingly apparent that we need to address climate change to reduce the incidence of future pandemic disease (Stern, 2020). But unlike for pandemic disease, the world has yet to respond to climate change with the same sense of urgency, or to take the same kinds of far-reaching and extensive emergency measures to act in the public's interest (Bronin, 2020). I believe it is well past time to do so. Perhaps we are starting to see such a change with the new US administration of President Biden, whose transition plan for climate change included hopes for making "far reaching investments" (Temple, 2020), and which has now supported a waiver of intellectual property rights during the COVID-19 pandemic (Tai, 2021).

I believe that we are already in a serious climate emergency that requires urgent action to reduce GHGs and to minimize the adverse effects of existing GHG emissions and concentrations. I also believe that the nature of negative emissions technologies (NETs) and of the IP rights that will be obtained in them make the analogy to COVID-19 vaccine and therapeutic developments apt. Although there may be many technological methods developed soon to reduce atmospheric GHG concentrations (just as there will be many different vaccines and therapeutics developed to treat COVID-19), the earliest cost-effective NETs will likely be viewed as breakthrough technologies (like the first approved, effective vaccines and therapeutics) and will likely lead to technological lock-in (Nelson and Winter, 1982) that will be less like for vaccines and therapeutics, for which substitution will depend on survival and on the duration of any immunity. Depending on the breadth of IP rights and the licensing behaviors of the rights-holders, there also will likely be manufacturing and supply limits for NETs and inadequate cost-effective substitutes (as opposed to regulatorily approved COVID-19 vaccine and therapeutic substitutes) developed in the relevant time frames. If left to normal market mechanisms, those early breakthrough NETs will therefore most likely be priced to maximize private returns to investors by charging high prices to purchasers and users. This will reduce affordable access to and limit uptake of NETs in many countries around the world that otherwise could respond to the climate crisis sooner, and at lower overall worldwide economic and social costs. These additional expenditures (even in developed countries) will come at the opportunity cost of greater spending on adaptation and other needed climate-related and social measures. In other words, as with vaccines and therapeutics, NETs will need to be deployed broadly and at low cost around the world, and not just in a few jurisdictions, but unlike for vaccines governments may not fully subsidize their purchase.

Further, and notwithstanding the global public good nature of NETs (Bodansky, 2012), IP rights in NETs may result in "climate nationalism," similar to the current vaccine and therapeutics nationalism (Bollyky and Brown, 2020; Lexchin, 2020). IP rights holders also may be unwilling or unable to license their rights and to share their know-how with third-parties to rapidly expand manufacturing and reduce costs of global NET supplies or processes (Price et al., 2020; Black, 2020). Not only will this increase the overall time frame for and the economic and social costs of responding to the climate crisis, but also it will require payments from poorer jurisdictions to wealthier jurisdictions where the NETs are developed and the IP rights are owned (Bosetti et al., 2013; Miguez et al., 2018). Those wealthier, developed countries are most responsible for GHG emissions, and are obligated by international climate treaties to support and to finance technology transfer to the poorer and less developed countries (United Nations Framework Convention on Climate Change—UNFCCC). Thus the IP rights will direct worldwide finances in direct conflict with the UNFCCC's underlying principle of "common but differentiated responsibilities and respective capabilities."

Prior voluntary efforts to pool technologies to address climate change to prevent our current climate crisis and to develop other eco-innovations have so far failed to increase diffusion of patents and patented technologies or to generate technological solutions adequate to address these problems (Contreras et al., 2019). It is long past time to increase worldwide efforts to address climate change, particularly with regard to technological collaboration, innovation funding, IP rights, and equitable access to developed technologies (Sarnoff and Chon, 2018). Focusing on patent rights, this chapter argues for adopting such emergency measures now, to avoid the increasingly foreseeable problems that will otherwise occur later, including further pandemics (Wyns, 2020; Shope, 1991).

11.2 Failure to meet climate change goals, the need for NETs, and current funding

Collectively to date, the world's nations have failed to reduce emissions adequately to achieve the1992 UNFCCC's goal of "stabiliz[ing] GHG concentrations in the atmosphere at a level that would prevent dangerous anthropogenic interference with the climate system" (UNFCCC). The 1997 Kyoto Protocol, with its focus on developed countries only, has been widely seen as a failure in achieving the needed emission reductions and transitioning of economies. Further efforts under the UNFCCC to achieve more voluntary reductions in the 2015 Paris Agreement have proved unavailing to adequately limit GHGs. Accordingly, countries, industry, academics, and others are looking to accelerate alternative approaches to the traditional means of limiting GHG emissions, so as to reduce or prevent adverse effects.

The two most basic forms of these alternative approaches are NETs and geoengineering (or solar climate engineering, SCE). As a US democratic legislative plan to address climate change recommended, the United States (and other countries) should "develop, manufacture, and deploy cutting-edge carbon removal technology … [and to] jumpstart a direct air capture industry in the United States, Congress should dramatically increase federal investment in carbon removal R&D" (U.S. House of Representatives, 2020). The proposal also recommended that the United State "consider alternative approaches to intervene in the atmospheric climate system" using SCE (U.S. House of Representatives, 2020).

Many current R&D efforts seek to develop large-scale NETs and to begin field testing of SCE approaches (US National Academies of Sciences, Engineering and Medicine (USNASEM), 2018; USNASEM, 2010; Robock, 2020; Jones, 2020; Fialka, 2020; Smedley, 2019). But the degree to which NETs and SCE can substitute for conventional emission reduction approaches is uncertain (European Commission; Victor, 2020; van Vuuren et al., 2018).

Performing R&D on NETs, and particularly on SCE technologies is controversial. This is due in part to the so-called moral hazard that such R&D may generate, by diminishing efforts in, and diverting public attention from, the need to reduce GHG emissions directly through conventional approaches (Wagner and Zizzamia, 2020; Gertner, 2017). Such alternative R&D is also controversial due to the lack of clear governance mechanisms over the R&D itself (Mace et al., 2018; Carnegie Council for Ethics in International Affairs, 2019; Reynolds, 2019). Further, such R&D efforts will affect the developmental paths and costs of any ultimate procurement and deployment decisions that may be made in the future, including as a result of technology lock-in effects (Gertner, 2017; Jaffe et al., 2001).

NETs include a wide range of technologies, which in theory could substitute for each other or for mitigation technologies (although they would then shift the burdens and costs of GHG emission reductions and result in cross-subsidies for polluting behaviors). The most prominent areas of NET R&D currently fall into two basic groupings: (1) carbon capture and sequestration (CCS, or with utilization CCUS), and more particularly biomass energy with carbon capture and sequestration (BECCS); and (2) direct air capture technologies (DACs, sometimes called direct air carbon capture and sequestration) (Haszeldine et al., 2018; Honneger and Reiner, 2018; USNASEM, 2019; Jacobs and Craig, 2018; Hester, 2018; McGlashan et al., 2012). Agriculturally based NETs (such as afforestation) that in theory could readily scale up face practical problems, including competition for land uses and other social concerns. DAC technologies, unlike BECCS and some other NETs, provide no side-benefits (such as pollution control) that attract funding, and thus may be principally reliant on government funding for their development (USNAMES, 2019; but see Fuss, 2017). Other approaches, such as biochar and ocean-based carbon dioxide absorption are also being explored.

Recently, attention has focused more on DAC than on BECCS, as the amount of land needed for BECCS to achieve needed reductions would likely be prohibitive (Gerard, 2020; Forster et al., 2020). Nevertheless, funding for DAC, as for BECCS, has been limited to date. As one prominent academic put it, R&D for DAC "technologies has been severely underfunded," notwithstanding the 45Q tax credit program, and widespread DAC adoption might require an "economy-wide carbon tax with credits for DAC" (Gerard, 2020).

Although US and other governments' funding has increased for both BECCS and DAC in recent years, it is nowhere near the levels of funding currently flowing to develop treatments and vaccines for COVID-19. For example, the USNASEM in 2018 recommended dramatically increasing the amounts of funding for RD&D for a wide range of CCS and DAC technologies, in the billions of dollars over 10- or 20-year time frames (Hezir et al., 2019). More recently, the FY2020 budget for the Office of Fossil Energy of the US Department of Energy (which directs much of the US government's NET funding efforts) requested Congress to provide $218 million for CCS programs, which is $19 million above 2019 levels (American Institute of Physics, 2020). In contrast, for COVID-19, the US government in 2020 had allocated $4.5 billion to R&D by just one agency, the Biomedical Advanced Research and Development Authority—out of $55 billion allocated to the Department of Health and Human Services to purchase tests, treatments, and vaccines (some of which funding also may be used for COVID-related R&D) (California Legislative Analyst's Office, 2020).

Lack of adequate funding is not the only impediment to NETs, as other concerns have discouraged NET RD&D funding. As a 2018 analysis explained, concerns over moral hazards, liability, governance, implementation of planetary-scale interventions, and disinterest in NETs and SCE have limited governmental investments in research, noting the need for a "significant boost [in] the funding available to support climate engineering research proposals" to achieve removal at the necessary scale in any relevant time frame (Hester, 2018). Accordingly, analysts have recommended that Congress and states in the United States "should significantly boost" NET RD&D funding so as to "achieve CO_2 removal at the necessary scale within a relevant time frame" (Gerard and Dernbach, 2018).

Although US government expenditures on NETs have yet to scale up to anything close to the levels needed, private investments in RD&D have been increasing in the last few years. One study observed increasing capture capacity of operating large-scale CCS facilities from 2017 to 2019, but noted a need for an "estimated 2000-plus large scale CCS facilities" to be deployed by 2040 to meet mitigation targets, "requiring hundreds of billions of dollars in investment" (Global CCS Institute, 2019). For DAC, "[h]istorical [US] federal investment for carbon removal-related direct air capture activities amounted to $10.9 million across 15 RD&D projects from 2009 to 2019, which was less than 1% of the total historical funding for carbon removal-related RD&D projects identified in the analysis" (Hezir et al., 2019). More recently, the US Congress approved $35 million for DAC R&D (Lebling, 2020). In contrast, private investments reached about $200 million in 2019, particularly in light of the US government's increased tax credits provided for each ton of CO_2 removed (Jacobson, 2019). Venture capital also is flowing to DAC. As reported in June 2020, Climeworks (one of the most prominent private DAC projects) raised $75 million in private investment capital (Climeworks, 2020). Finally, large petrochemical and other traditional energy companies devoted similar amounts to DAC in 2019 (Hook, 2019).

In summary, many billions of dollars soon will be spent on NETs, by private industry and increasingly by governments. These amounts still may be far less than needed to address the adverse effects of climate change over the next few decades, and are to be contrasted with the massive spending in the course of only a few months to address COVID-19. In developing NETs, companies and venture capitalists that invest substantial financial resources will of course seek to obtain IP rights to better assure returns on their investments. The IP rights gained may permit (indeed, will be designed to assure) supra-competitive pricing, as an ex-ante incentive to make those investments. Similarly, governments may look to those IP rights to induce private investments, in preference to having to impose unpopular taxes to generate the needed revenues to pay for NET RD&D.

Such supracompetitive pricing from patent rights may raise the overall costs of addressing both the COVID-19 pandemic and climate change through NETs. Controversial disputes over patent rights in NETs are all too likely to occur, because raising the costs of NETs and potentially limiting access to them may have dramatic local effects on economies and on national decisions to deploy NETs or to further control GHG emissions through (even more costly) mitigation measures. IP rights controversies have already occurred with regard to climate mitigation technologies (and other "environmental sound technologies"). Various measures (such as prohibiting or revoking "climate friendly" technology patents, reducing duration of such patents, and mandating patent pools and compulsory licensing of GHG emission-reducing technologies) have been proposed, but so far have not been adopted (largely due to developed country objections) in the context of UNFCCC technology transfer deliberations (Barooah, 2008; Santamauro et al., 2009; Chavez, 2015; Parthasarathy et al., 2010).

I have explained elsewhere why I think mandatory measures to prohibit patent rights in climate mitigation technologies on environmental grounds *might* be found to conflict with requirements of the World Trade Organization (WTO) Agreement on Trade Related Aspects of Intellectual Property Rights (WTO TRIPS Agreement, 1994). But such prohibitions are not likely to achieve political acceptance even if they were normatively justified (Sarnoff, 2011; WTO TRIPS Agreement, 1994; Maskus and Okediji, 2010). Instead, I suggested various less controversial alternatives to address patent concerns that could more feasibly be adopted, without international agreement (Sarnoff, 2011). There is every reason to think that patent disputes will be even more severe in the context of NETs, which need to be adopted precisely because mitigation technologies and practices have to date proved too expensive for or resistant to widespread adoption.

11.3 Intellectual property rights, invention incentives, and affordable-access concerns

As has been frequently discussed, IP rights are one mechanism for providing incentives for private investment in R&D. But they are only one of many measures that governments can take to fund innovation. Other measures include direct or indirect subsidization (including through various forms of tax subsidies, deductions, or credits), government procurement, direct development by government (including feasibility demonstrations leading to reduction of private risk), creation of various forms of commons (which reduce development costs), and regulatory adjustments that expand market returns (including product, process, and antitrust regulatory measures) (Sarnoff, 2013; Hemel and Ouellette, 2019). A recent World Bank study in the climate change context explained various supply side and demand side measures that could expand R&D for climate innovation, including subsidies and tax credits, IP, procurement, and prizes (World Bank Group, 2020). What that study fails to note is that these measures may potentially increase the overall costs of supplying the desired innovation outputs. Further, as has been explored in detail in the access-to-medicines context, measures such as granting patent rights may make access to the desired technologies unaffordable for low-income individuals or in low-income juris-dictions, or otherwise may reduce diffusion to achieve the desired goals (Burci and Gostin, 2017). And as more recently demonstrated in regard to COVID-19, trade secrecy rights may have limited production of needed worldwide supplies of vaccine technologies (Levine, 2020).

As discussed further below, these competing concerns for incentivizing innovation through the grant of patent rights while ensuring affordable access have already led to international conflicts in the context of COVID-19. To date, the mechanisms developed to minimize those conflicts have yet to prove successful. Given that extensive patent rights will also be granted with regard to NETs, particularly for complex products or processes having multiple inputs, similar concerns are likely to arise with regard to NETs.

11.3.1 COVID-19, patent rights, ex-ante incentives for RD&D, and ex-post conflicts over affordable access

Shortly after COVID-19 was recognized as a pandemic, the WHO (and others) launched two initiatives to incentivize innovation to develop vaccines and treatments and to assure equitable worldwide access to any such products that would be developed. The first initia-tive, the Access to COVID-19 Tools (ACT) Accelerator, was launched at the end of April 2020 by the WHO and by "governments, scientists, businesses, civil society, and philanthro-pists ... to speed up an end to the pandemic by supporting the development and equitable distribution of the tests, treatments and vaccines" (World Health Organization, 2020f). By the end of June 2020, the ACT Accelerator had developed costed plans running to $31.3 billion for distribution of diagnostics, therapeutics, and vaccines to lower- and middle-income countries (LMICs). As the WHO noted, "the total cost of the ACT Accelerator's work is less than a tenth of what the International Monetary Fund (IMF) estimates the global economy is losing every month due to the pandemic" (World Health Organization, 2020a). A key principle of the ACT Accelerator is to assure "equitable distribution of COVID-19 tools to those most in need. Whether a person can access tests, treatment and a vaccine for COVID-19 should not be determined by where they live" (World Health Organization, 2020g).

How that will be achieved has yet to be fully determined. Pledged funding to the ACT Accelerator by June 2020 reflected a $27.9 billion shortfall (World Health Organization, 2020a). Assuming that the contemplated funding materializes, the goals of the ACT Accelerator may be met. But the overall costs of attaining the goals may be much higher than they might have been if measures were adopted to restrict the IP rights in, and the market prices for, the developed products. The ACT Accelerator does not itself require open scientific sharing of data or trade secrets, nor voluntary or low-cost contributions of patented technologies. Nor does it impose specific access require-ments or price constraints on any technologies developed under its auspices, although the parties taking the lead to develop those products could choose to impose them, or political or publicity concerns may push toward similar results.

But if such access measures were adopted more explicitly, the (unanswerable) question would arise of whether the needed products would be developed in as timely a fashion (or at all), given the reduction of private investment incentives from lower anticipated market returns. Delays in developing the needed products also may impose substantially higher social costs. For example, although the US National Institutes of Health in 1989 had inserted a "fair pricing" clause in all of its cooperative research and development agreements (CRADAs), it removed the clause from its standard agreement after pharmaceutical companies became reluctant to sign CRADAs containing the clause, "reasoning that this would better 'promote research that can enhance the health of the American people'" (Contreras, 2020). As the Director of the World Intellectual Property Organization (WIPO) put the issue in April 2020 with regard to COVID-19, "[t]he main challenge *at the present time* is not access to vaccines, treatments or cures for COVID-19, but the absence of any approved vaccines, treatments or cures to have access to" (Gurry, 2020). Nevertheless, Harvard University, MIT, Stanford University, and others have chosen to temporarily provide their IP for free for COVID-19-related product developments, while imposing downstream commitments for affordable and widespread public access. "In return for these royalty-free licenses, we are asking the licensees for a commitment to distribute the resulting products as widely as possible and at a low cost that allows broad accessibility during the term of the license" (Office of Technology Development, Harvard University, 2020).

The second initiative of the WHO was to adopt a patent pool for COVID-19-related inventions, originally suggested by Costa Rica and colaunched by that country. Like the ACT Accelerator, the COVID-19 Technology Access Pool (C-TAP) is "an initiative aimed at making vaccines, tests, treatments and other health technologies to fight COVID-19 accessible to all" (World Health Organization, 2020c). The C-TAP pool has five "key elements": (1) public disclosure of gene sequences and data (rather than patenting them or keeping them as trade secrets); (2) transparency in publishing all clinical-trial results; (3) governments and other funders "are encouraged" to impose contractual funding agreement clauses with pharmaceutical and other recipients on "equitable distribution, affordability, and publication of trial data"; (4) licensing any treatments, diagnostics, vaccines, or other health technology to the Medicines Patent Pool (MPP) for low-cost provision of final products to LMICs; and (5) promoting "open innovation models and technology transfer that increase local manufacturing and supply capacity, including through joining the Open Covid Pledge and the Technology Access Partnership (TAP[artnership])" (World Health Organization, 2020c).

Participation in C-TAP, however, is voluntary and the terms of licensing of any pooled patents and developed products have yet to be determined. (Nevertheless, pooling patent rights would permit generic production, and therefore is likely to result in lower-cost products than would voluntarily licensed production outside of such pooling arrangements.) Prior MPP licenses have authorized sublicensing for low-cost production (typically by generic-pharmaceutical manufacturers), but do not appear to contain any mandatory final product access or pricing requirements, although they may contain royalty terms and revocation rights for not obtaining regulatory approvals in a timely fashion (License Agreement). The WHO nevertheless seeks as a matter of "social solidarity" to "promote" publicly funded and donor-funded research outcomes that are "affordable, available and accessible to all on a global scale" (World Health Organization 2020d).

In contrast to the C-TAP, the TAPartnership does not include therapeutics and vaccines, and for diagnostics, medical devices, and personal protective equipment provides only an information-sharing platform. That platform is intended to facilitate information sharing about products; provide technical support (including in relation to market dynamics and regulatory hurdles); and facilitate partnership development (which would include licensing) by identifying expertise, needs, and capacity (World Health Organization, 2020e). In this sense it is similar to the WIPO-Green platform, which facilitates identification of potential voluntary licensing partnerships for patented (and unpatented) eco-friendly innovations (WIPO, 2020b).

Finally, the Open COVID Pledge encourages companies to provide a free, open license of their patents (or copyrights and research data) relating to COVID-19 during the duration of the COVID-19 pandemic (Open Covid Pledge, a). However, the Open COVID Pledge does not require that any final products created with any such freely licensed IP be made available on a widespread or nondiscriminatory basis, nor does it require that they be provided at affordable prices. In this respect, it is just like many technical standards adopted by standard-setting organizations (SSOs) that incorporate patented technologies or employ patented processes and require "fair, reasonable, and non-discriminatory" (FRAND) licensing of those patents, but not of access to or pricing for the final products (Open Covid Pledge, a; Contreras, 2015).

Although these voluntary measures are to be strongly encouraged, concerns remain that they will not be sufficient to improve access or affordability to final COVID-19 products, such as vaccines. For example, the African Union and the Africa Centres for Disease Control and Prevention (Africa CDC) issued a communiqué noting "the barriers that IP, including patents, trade secrets and other technological knowhow, has posed to timely introduction of affordable vaccines in developing countries in the past" (African Union and Africa CDC, 2020). Further, they advocated for African countries "to receive a sufficient global allocation for timely access to a potential COVID-19 vaccine, including partnership with the COVAX facility within" WHO's ACT Accelerator; and they called upon all countries to "remove all obstacles" to using TRIPS Agreement flexibilities to assure that technologies, IP, data, and knowhow "are openly and immediately made available and rapidly scaled-up" (African Union and Africa CDC, 2020). Clearly, these African countries believe that additional, nonvoluntary measures will need to be adopted by governments to prevent IP rights from becoming a barrier to needed innovation and to affordable access.

With regard to vaccine nationalism, "American and British exceptionalism" resulted in prioritized access to the University of Oxford's vaccine, given significant financial investments from those countries. "While questions have been raised about why two of the wealthiest countries should receive priority access to the vaccine, little attention has been paid to the role of the university in reinforcing ... 'vaccine sovereignty'" (Ahmed, 2020; Rutschman, 2020; Adepoju, 2020). Similarly, the United States contracted to purchase Gilead's entire production capacity for 3 months of remdesivir, which was the first therapeutic found to have significant efficacy in shortening COVID-19 hospitalizations and likely in reducing deaths (Lee, 2020). As the old expression goes, "money talks."

The costs of providing access to the treatment in the United States also were substantial. "The ... cost to the US government will be US$3,200 for a 6-day treatment. In contrast, production costs for remdesivir are estimated at 93 US cents for 1 day's treatment, or less than US$6 for an entire course" (Mintzes and 't Hoen, 2020). Other developed countries paid 25% less (Denham et al., 2020), and "[i]n 127 developing countries, Gilead has allowed generics companies to manufacture the drug, pricing a 5-day course at about US $600" (Ren, 2020). This likely still makes the drug unaffordable to billions of people in those developing countries (assuming the generic pharmaceutical companies can make the drug available in a timely fashion), as well as in other countries that were not included for authorized generic distribution. Given its pricing policy, Gilead could turn a profit of US $2.3 billion in 1 year alone (Ren, 2020). The concerns over pricing are particularly cogent given that remdesivir was originally developed using substantial government funding (Mintzes and 't Hoen, 2020; Hughes and Rai, 2020).

As with the WHO's efforts at fundraising to assure affordable access, other measures have been suggested that would help to decouple innovation incentives from product provision and pricing (although it still requires the volunteering of industry to perform the R&D given the anticipated controlled pricing). In particular, the use of an "advanced market commitment" (AMC), which essentially is a form of prize, can in theory induce needed innovation without risking subsequent price controls or insurance reimbursement policies that might make industry hesitant to engage in the needed RD&D (Sachs et al., 2020; Hemel and Ouellette, 2020; Athey et al., 2020). AMCs have been proposed for climate-change-related technologies. For example, the US House of Representatives recently proposed using bulk government procurement (which is an AMC, particularly if the purchased products are intended for further resale or distribution to the general public) for a post-COVID-19 green economic recovery, in addition to "grants, rebates, and tax incentives." "Bulk procurement orders would provide manufacturers the demand certainty needed to make long-term investments and hiring decisions to bring their factories back to full capacity ... [and] help drive down the costs of production, reducing the costs of clean energy and decarbonization efforts and further increasing demand in a virtuous cycle" (U.S. House of Representatives, 2020).

To be credible, AMCs must be able to assure the amount of overall funding that government or private usage would entail, and must be at prices sufficient to induce the desired RD&D (even when grants, rebates, and tax incentives already have proven insufficient to do so). That incentive may need to be substantial overall, given the alternative investments to which capital (and human capital) could be put and the substantial research, production, regulatory, and liability risks and profit motives for the companies that such AMCs (or other funding) seek to induce to perform the needed RD&D (Hemel and Ouellette, 2020). Returning to COVID-19, whether the WHO's cost estimates and fundraising for the C-TAP will be sufficient to induce the desired R&D has yet to be seen (or may never be known as compared to what higher amounts and prices might have induced) (Newey, 2020). And even if the funding is considered sufficient to generate the needed products, "vaccine nationalism [may cause] a misuse of these (prepurchase, advanced market) agreements. Contracts should not trump equitable access to global public health goods" (Rutschman, 2020).

In summary, the COVID-19 experience reiterates earlier concerns in the access-to-medicines context that patent rights can lead to increased overall costs in obtaining and deploying technologies needed to address emergencies, even if such rights may be useful (or even necessary) to incentivize the private sector to invent, develop, and disseminate the needed technologies in the relevant time frames. However, the access-to-medicines context typically focuses on products subject to relatively few patents, unlike complex products involved in standard setting, which raise additional concerns over rights to use patents and over the pricing of patent licenses and end products. Further, although voluntary donation or humanitarian licensing of patent rights by well-meaning rights holders may require licensees to assure widespread access at affordable prices (Sarnoff, 2011), it is unclear whether the products developed with or using patented technologies will be sufficient to address the needs in the relevant time frames and across all relevant jurisdictions. Instead, it may lead to preferential sovereign access to the needed technologies, generating further conflicts that may deter international cooperation, as with the European Union's threat to prohibit vaccine exports (Boffey, 2021). In that case, resort to compulsory measures adopted by national governments to assure access and affordable pricing are more likely.

11.3.2 More controversial, ex-post compulsory measures to control prices and to assure access to patented technologies

Under the US "Bayh Dole Act" (Patent and Trademark Law Amendments Act), the US government retains so-called "march-in rights" that allow it to authorize a third-party license for a particular field of use (effectively a compulsory license) when one of four conditions is met. The two most relevant of the conditions are (1) the contractor (or its assignee) patent holder "has not taken, or is not expected to take within a reasonable time, effective steps to achieve practical application of the subject invention in such field of use"; and (2) "to alleviate health or safety needs which are not [being] reasonably satisfied" by the patent holder or its licensees (35 U.S.C. § 203(a)(1)and(2)).

To date, the US government has never exercised its march-in rights. Rather, in various cases seeking to authorize generic production of pharmaceuticals, the US government has held that the "practical application" and "available to the public on reasonable terms" requirements are satisfied by placing drugs on the market, even if they are not affordable to a significant segment of the public that needs them (Thomas, 2016). Even more surprisingly, with regard to some COVID-19-related innovation funding, the US government purported to contractually limit its own ability to exercise its statutory march-in rights (Herman, 2020). However, the US government has extensively used or threatened to use its inherent sovereign authority to employ patented technologies for the public benefit without being subject to judicial injunction (28 U.S.C. § 1498(a); Knowledge Ecology International). That authority provides immunity to government contractors who are directed to infringe when satisfying government contractual obligations, but then is subject to a statutory takings remedy against the government that effectively provides a compulsory license (28 U.S.C. § 1498(a); Thomas, 2016).

More explicit authorities for compulsory licensing for third-party or government use also exist in the United States and elsewhere, as well as judicial authority to adopt explicit or effective compulsory licensing remedies (Correa, 2020). Similarly, a royalty-free compulsory license is sometimes granted as a remedy for antitrust violations (Gilbert, 2019; Delrahim, 2004; Schlam, 1998), although the United States does not generally follow practices in much of the rest of the world, as other nations more routinely find competition violations from excessive pricing or from effective denial of access to so-called "essential facilities" (Gilo and Spiegel, 2018; Waller and Tasch, 2010; First, 2019; Frischmann and Waller, 2008). Finally, a court may refuse to grant the "extraordinary" equitable remedy of injunctive relief prohibiting continuing infringement (or, prior to a trial, potential infringement), particularly when focusing on the public interest in accessing a patented technology, and may instead grant an ongoing royalty award that is effectively a compulsory license. Such judicial decisions to deny prohibitory injunctions while awarding reasonable compensation are fully consistent with the requirements of the WTO's TRIPS Agreement (Sarnoff, 2010, 2020). Similarly, low-cost (and possibly even royalty-free) compulsory licenses to remedy competition violations are also fully consistent with the TRIPS Agreement, which would apply to excessive pricing and essential facilities competition violations (Correa, 1999; Schlam, 1998).

Nevertheless, explicit compulsory licensing and the other measures to address and to assure affordable access and reasonable pricing of patented goods are highly contentious, and are often the subject of retaliatory trade sanctions (Peets et al., 2007; Correa, 2020). Resort to such measures is therefore often avoided, due to concerns over their effects on investments and RD&D incentives, or due to trade pressures. But the consequence of failures to impose these measures may increase overall costs to and adverse consequences for the public from the higher prices and reduced access, in both developing and developed countries (Bond and Saggi, 2014).

In theory, such ex-post measures to control prices and to assure access may be preferable to the ex-ante measures, where the potential inadequacies of promised pricing in pooling arrangements or of advance market commitments may fail to induce the needed RD&D. Inadequate expectations on pricing may result in failures to create, or in delays to obtain, the needed technologies, which may sometimes be the greater concern. But as a political matter, ex-ante measures (such as those suggested in the final section of this article) may be much easier to adopt given the fraught politics of competing views on innovation policy, compulsory licensing, reliance on and imperfections of economic markets, and government-imposed price controls. Further, ex-ante measures may be easier to adjust by increasing prices, should the pricing prove inadequate to induce the desired RD&D within the relevant time frames.

11.3.3 NETs, the creation of patent rights, and prior unsuccessful international efforts to pool climate-change-related patents and otherwise address price and access concerns ex-ante to innovation

In light of the substantial amounts of funding and extensive efforts that will flow toward NETs and R&D for them, NETs researchers and their funders will seek to patent many of their inventions. Perhaps the most extensive patent landscaping of CCS technologies was published in 2018, addressing patent and scientific documents from 2007 to 2017 found that "[m]ost patent generation activity is concentrated in capture technologies that use absorption and adsorption chemical processes (within 35% and 30% of the collected documents, respectively)," with large companies from different countries dominating the CCS portfolios and with "most of the patent activity … occurring in the United States, Japan, China, and Korea" (Miguez et al., 2018, 2020).

A more European-focused analysis, from 2017, of CCUS R&D investment measured by patenting showed that funding "strongly increased after 2008," due in large part "to a larger investment from the public sector" and showed a positive association with corporate R&D expenditures. Similar trends were observed for China and South Korea, while Japanese inventions remained constant from 2000 to 2012 (Fiorini et al., 2017).

A 2018 patent landscaping report on utilization technologies for CCS concluded that "Overall, based on the patent trends, the patents published are increasing every year, especially during the last 5 years (2013–17) whereby a sharp increase was observed" (Norhasyima and Mahlia, 2018). Similar trends were identified in another 2018 CCS patent landscaping report focusing on the United States and China, which also looked at global assignees and found that the United States led in applications, followed by China (which had no top-10 companies in applications) and then Europe, Canada, and Japan (Qui and Yang, 2018). A 2017 patent-and-literature landscaping for "carbon capture" methods performed by the Canadian Intellectual Property Office found that France led in the number of researchers (followed by Germany, Canada, the United Kindom, and Korea), whereas Canada led in patents in relation to reducing the carbon-emission intensity of hydrocarbon fuels (Canadian IPO, 2017).

No similar, recent, comprehensive patent landscaping report was identified for DAC. In 2014, Oldham and colleagues found only four families of originally filed patent applications on DAC (and only 42 families of technologies indirectly related to DAC, in the sense that they may need to be licensed to perform DAC) (Oldham et al., 2014). Nevertheless, many companies have filed for, or obtained patents on, their particular approaches to DAC (Sandalow et al., 2018; Keith et al., 2018), and many more patents are likely in the future given the increasing commercial investments noted above. For example, one of the most publicized DAC approaches, using filter material and moisture in the air to bind carbon dioxide, was developed by the Swiss Company Climeworks AG, which has many patents on its technology (WIPO, 2020a), and which recently entered into a joint venture with a Canadian company, Svante Inc., to employ its patented solid-absorbent technology (Orland, 2020).

Many patents, not just those specific to CCS or DAC, may need to be licensed to be able to effectively and efficiently develop, manufacture, and operate NETs. The technology and IP terrain for BECCS and DAC is highly heterogeneous. Accordingly, both licensing platforms such as WIPO-Green and patent pools such as the EcoPatent Commons (EcoPC) are likely to be needed, in order to provide assurance that the patented technologies can be used without giving rise to patent-infringement litigation. For more complex NETs that incorporate many patented inputs or use many patented processes, additional licensing arrangements may be needed, such as by developing standards subject to FRAND licensing commitments (Biddle et al., 2019).

To date, however, efforts to create such ex-ante pooling arrangements, as with the WIPO C-TAP for COVID-19 with regard to pharmaceuticals, have not been successful. For example, as some prominent analysts have noted, earlier efforts to create a "Global

Technology Pool for Climate Change" failed, due to international opposition (Baker et al., 2017). Similarly, the EcoPC was ultimately disbanded, both for lack of adequate maintenance efforts and because the patented inputs included mostly low-value technologies. As a review of EcoPC noted, the contributed patents "were, on average, less cited than comparable patents, ... the contribution of these patents ... did not increase their rates of citation," and there was no evidence that the availability of the patents "increased the diffusion of the pledged inventions" (Contreras, 2019).

Patent rights will increase the technology transfer costs, which will flow opposite to the UNFCCC's "common but differentiated responsibilities" and technology transfer obligations (UNFCCC). This is because patents on such technologies will mostly be owned by companies from developed countries, diminishing the ability and willingness of those countries to adopt tough emission-control measures (Baker et al., 2017; Sarnoff, 2011). To address those concerns, various measures were unsuccessfully suggested in the UNFCCC for international adoption, such as prohibiting patents on environmentally sound technologies, compulsory licensing, creating a declaration similar to the Doha Declaration (which also encouraged compulsory licensing for export where local capacity for production was inadequate), patent buyouts and payments of licensing fees, and banning patents on such technologies outright (Baker et al., 2017; Barooah, 2008; Santamauro et al., 2009; Chavez, 2015; Parthasarathy et al., 2010).

Some analyses suggest that there is enough competition in mitigation-technology markets—such as for photovoltaics, wind power, biofuels, and hybrid vehicles—"to keep prices low and limit the monopoly impact of patents in these fields" (Baker et al., 2017). Nevertheless, "[t]here is a tendency for patent owners in developed countries to refuse to license technologies at the cutting edge of research, for fear of enabling developing country competitors" (Baker et al., 2017). Further, as I have previously noted, the "assumptions of price constraints on patented climate change technologies assume ready substitutes for existing technologies, or development of incremental, rather than breakthrough, technologies" (Sarnoff, 2011). I believe that NETs are precisely the kind of technologies that are unlikely to have ready, cost-effective substitutes and that will be viewed as breakthroughs, at least in the short term, and then lock-in, switching costs, and network effects will further restrict both substitution behaviors and consequent market incentives for further innovation (Farrell and Klemperer, 2007). Accordingly, controlling prices ex-post to the development of those technologies will be highly controversial, and obtaining ex-ante agreements to limit pricing of future products through pooling and other measures will likely lead to the same kinds of international impasses that have so far developed both with regard to past patent pools for climate change and to present COVID-19 patent pools. In turn, we will likely have to look to national measures to assure affordable access while continuing to incentivize the needed innovations.

11.3.4 Minimizing ex-post conflict over patents in NETs by national and private adoption of ex-ante measures

Because ex-ante *international* mechanisms to limit the scope of patent rights or the prices of patent licenses to assure affordable access to NETs will likely encounter substantial resistance and may be unsuccessful, *national* measures will likely be adopted ex-post to NET innovation. Those measures will seek to control market prices for NETs incorporating or using patented technologies and to authorize the unlicensed making and use of those patented technologies so as to permit third-party production. Such measures are consistent with TRIPS Agreement obligations. In most countries, price controls are a routine part of the supply and purchase of patented medical technologies (Dearment, 2020), even if this is not as common for patented products in other social fields. In theory, there is no reason why countries cannot impose price controls on other important social goods, such as NETs or other climate-change-related technologies. Further, imposing price controls will still result in compensation to patent holders (just as the compulsory licensing with "reasonable compensation" authorized by the TRIPS Agreement), providing an additional reason why it may not conflict with obligations under the TRIPS Agreement (World Trade Organization, 2020e; Sarnoff, 2007).

The TRIPS Agreement did not seek to harmonize the substantive grounds on which compulsory licenses could be granted or on which prohibitory injunctive relief could be denied. The WTO's Doha Declaration on the TRIPS Agreement and Public Health expressly recognized the ability of member states to unilaterally determine both the grounds for granting compulsory licenses and what qualifies as an "emergency or other circumstances of extreme urgency," avoiding the need to negotiate with rights holders (World Trade Organization, 2001). Climate change certainly is a circumstance of extreme urgency, and should be considered an emergency. Further, the TRIPS Agreement does not harmonize the substantive grounds for countries to determine what conduct constitutes an antitrust violation that would justify granting compulsory licensing without providing compensation (or providing only reduced compensation) to rights holders. Such conduct can clearly encompass excessive pricing and unilateral refusals to deal.

Accordingly, countries may adopt such measures or may refuse to impose prohibitory injunctive remedies for infringement of patent rights (and may do so and also deny compensation for antitrust violations) for essentially whatever substantive legal reasons that they choose. The only limitation is that they must follow the procedures required by Article 31 of the TRIPS Agreement (but need not consult in emergencies) for granting compulsory licenses. Any WTO disputes challenging such actions, moreover, would likely face the problem of inconsistency of the positions of the complaining parties. Further, as has already occurred with regard to COVID-19 in the United States under the Defense Production Act (50 U.S.C. §§4501 et seq), governments can simply prohibit excessive pricing or can compel production and take over distribution of needed, patented technologies in order to assure affordable access to relevant markets, requiring patent holders to sue the government for (diminished) compensation (Congressional Research Service, 2020). Issuing threats to impose such measures could lead patent owners to supply markets that they otherwise might avoid, and if they could not obtain sufficient profits in those jurisdictions to warrant market entry (while charging above marginal costs of production) to voluntarily license others to supply those markets when generic producers would be willing to do so (Sarnoff, 2011; Ghosh and Calboli, 2018).

Such measures to control prices, to authorize third-party production without payment or at reduced licensing fees, or to threaten such actions may be adopted by national governments even if they were not consistent with the TRIPS Agreement. Countries can avoid retrospective liability if they are later found to be in conflict with TRIPS obligations, simply by prospectively changing their laws (World Trade Organization, g). Challenges in the WTO to such measures in the context of international emergencies such as climate change are particularly unlikely to result, given the adverse publicity and diplomatic conflicts likely to ensue, as well as the precedents that would be established for intruding on national sovereignty to regulate domestic markets and public health (World Trade Organization, 2018; World Trade Organization, 2020d; Tobacco Control Laws Global Legal Center).

Moreover, controlling the price of NETs or assuring alternative, competitive supplies of NETs that make them freely available or available at affordable prices may not result in substantial reductions of innovation—at least not to the extent frequently claimed by pharmaceutical companies as a reason for not imposing any constraints on market prices. Thus, concerns over the inherent but uncertain tradeoff in lost innovation incentives by assuring affordable access ex-ante to producing any innovations may be overstated. This is especially true if investment levels are already relatively high and the needed innovative outputs remain low (Frank and Ginsburg, 2017). Investment levels are likely to remain high if comparatively higher prices continue to be charged in wealthier jurisdictions, so that the reduction of profits in developing countries from price controls on or competitive supply of NETs would not substantially affect worldwide profits. However, imposing price controls that deter NET patent holders from market entry, or having to issue compulsory licenses to generate needed, affordable supplies of NETS, may lead to delays in providing access. This is particularly likely if trade secrets and knowhow are needed for production. Such failures to supply or delays would further exacerbate growing concerns over preferential sovereign access to NETs (similar to what has happened with COVID-19), and would create delays in adopting those technologies, which might increase the costs of other mitigation measures as well as worldwide social and economic costs.

In summary, given the political, ideological, and empirical disputes over patent rights, intergovernmental organizations are unlikely to agree on the needed ex-ante measures to assure widespread and affordable licensing of patent rights and downstream controls over pricing of NETs that will ensure affordable access to them, particularly in developing countries that would have to pay the costs of access to patent holders from developed countries (Sarnoff, 2011). But all countries may unilaterally adopt ex-post measures such as price controls and compulsory licensing, generating intense political conflicts regarding NETs. One way to minimize such conflicts would be for countries to announce in advance the conditions on which they will employ such pricing and access measures (including compulsory licensing or the exercise of march-in rights), before investments in RD&D efforts to supply NETs occur (Sarnoff, 2011). This should eliminate any concerns ex-post regarding prejudicial reliance, unfair treatment, or inadequate returns in such countries. Those conditions would simply be the new conditions of access in the relevant product markets, fixing structural problems in those markets that would otherwise result in unaffordability of the NETs or in inadequate production and access. Those conditions, moreover, can be modified and additional compensation offered or limits on the conditions for exercising such domestic measures adopted, if the desired worldwide NET innovations are not forthcoming in the anticipated time frames.

As noted above, ex-post controls are in theory preferable to ex-ante controls, as they avoid diminishing ex-ante innovation incentives before the needed NET technologies are developed and before conditions of unaffordability or of limited supply materialize (Bond and Saggi, 2014; Miguez et al., 2018, 2020). Of course, the mere potential for governments to impose such controls may already diminish investment and RD&D incentives. But adopting such measures ex-post without announcing the conditions for their exercise in advance has continuously proven to be politically difficult, and it generates both conflicts and retaliatory trade sanctions.

Similarly, any prize mechanisms developed by intergovernmental organizations or by private funding entities could condition the receipt of the prize on commitments to supply NETs to, or voluntarily license production for, all jurisdictions at prices that are affordable in those jurisdictions. (Note that granting the prize could then potentially require pricing below marginal costs, or licensing rates that would enable low-cost production, with the differentials being fully or partially offset, without the NET patent holder or licensees *being required* to produce at a loss or below marginal costs.) Of course, the patent holders themselves, such as universities, could impose such conditions on licensing to further NET research or product development and deployment (Sarnoff, 2011). As with other funding conditions, patent holders can then decide whether or not they are willing to accept the prizes or licenses on the terms offered, without incurring any prejudicial reliance. Alternatively, such prizes could require public dedication or transfer to the prize offeror of any patent rights, which again would be voluntary as part of the decision on whether to accept the prize. As has been previously noted, prizes and patents can be complementary (or substitutionary) policies, both having deadweight losses that can be compensated for with corresponding adjustments such as "subsidies, tax credits, and price controls" (Sichelman, 2017; Roin, 2014; Hemel and Ouellette, 2013; Sarnoff, 2013). And as with other funding commitments, the prize amounts can be raised if the desired innovation (or prize acceptance) does not occur within reasonable time frames. But such holdout also poses risks, because if prize acceptance does not occur after the NETs are developed, compulsory licensing in other jurisdictions may result. In contrast, compelling trade secrets and know-how to be disclosed for public benefit is more difficult and raises potential ethical concerns (Vermeij, 2020).

Nevertheless, other ex-ante measures may be adopted by individual governments (or a group of governments) or by private or public–private entities (including standard-setting bodies) for the development of NETs, that also may help to avoid the forthcoming ex-post conflicts. For example, government funders of NET R&D may make their funding more expressly conditional on commitments to supply on affordable terms and to particular markets. Unlike the growing movement towards vaccine sovereignty with regard to COVID-19, where assuring local access may be more critical to avoiding harms, assuring worldwide uptake of NETs will provide local benefits from reducing overall GHG levels.

Such funding conditions need not be contractually negotiated with the direct funding recipients, thereby reducing overall administrative costs, if general principles for their application (and potentially some enforcement mechanism) can be developed. (An enforcement mechanism may not even be needed, given that such measures should be a conflict-reducing alternative to ex-post measures in the jurisdictions intended as beneficiaries of the conditions.) Such principles could establish automatic public licenses under specified conditions, authorizing third-party production without any governmental or private liability should the conditions occur. Such restrictions and authorizations might be attached to measures such as current lobbying efforts in the United States seeking to expand the ability to expense R&D costs in the year incurred, which would induce R&D expenditures and improve US short-term and long-term economic competitiveness (Hood and Bseiso, 2020). As with conditions on negotiated funding agreements, and clarifying conditions for march-in rights, parties will have the grounds to know in advance whether they wish to take the tax deductions, and thus may not later complain about any unfairness that may result if the conditions they agree to are not met. As one seminal US case determined, providing conditions on the receipt of government benefits (in that case, prospectively assuring that trade-secret information submitted for regulatory marketing approvals of pesticides could be used for generic competitor marketing approvals) is not subject to any takings liability or other concerns, as private entities voluntarily decide whether they are willing to accept the benefits that are tied up with those conditions (Ruckelshaus v. Monsanto Co., 1984). As there is no constitutional right to tax deductions in the United States, and as imposing such conditions would not require relinquishing constitutional rights in exchange for legislated benefits, the "unconstitutional conditions" doctrine would not apply to any such conditions placed on tax or other government funding benefits (Sullivan, 1989).

Further, as noted above, antitrust-law measures to compel price reductions and to assure that patents are broadly licensed have a strong basis in the laws of most jurisdictions (but not in the United States) (Gilo and Spiegel, 2018; Waller and Tasch, 2010; First, 2019; Frischmann and Waller, 2008; Correa, 1999; Schlam, 1998). In contrast, competition law tends to view unfavorably collective efforts to set prices or to allocate markets geographically, particularly within a single geographic jurisdiction (U.S. Department of Justice, 2015). But in the context of providing NETs (or other needed goods) at differential prices around the world, ex-ante agreements to set prices (including pricing of patent licenses) and to allocate markets may make more sense. This is particularly true where patent holders may choose not to enter particular jurisdictions on the grounds of lack of profitability, but also may refuse to license third parties to produce in or for those jurisdictions, given concerns about product or price arbitrage across jurisdictions. In such circumstances, ex-ante agreements to allocate international markets and to set prices may make sense, as promoting both consumer and international welfare (Day and Schuster, 2021). This approach would likely make even more sense in the context of SSOs for complex technologies, where patent holders need to coordinate their patent-licensing efforts, and typically do so by adopting FRAND obligations under SSO requirements (Contreras, 2019). But FRAND obligations can lead to significant licensing costs and litigation, and may not result in development of the needed NETs in or for many jurisdictions. Thus, ex-ante agreements on relative patented contributions to, and of the overall costs of, downstream products in different jurisdictions (perhaps supplied by different producers) could result in greater overall accessibility at affordable prices. Of course, such agreements would likely need to be scrutinized carefully by multiple competition agencies, and could be subject to preclearance mechanisms similar to merger approvals to assure that the price-fixing and market allocations are in fact socially beneficial (U.S. Federal Trade Commission, a).

Public-interest-regarding countries (or potential licensees) also could collectively adopt "humanitarian" approaches to NET deployment, by indicating to SSOs or to individual patent holders that they will not permit sales in high-cost jurisdictions (or will not license the technologies) unless affordable access is provided to jurisdictions that otherwise would not be supplied. Such "group boycott" dynamics are usually viewed unfavorably by competition laws (U.S. Federal Trade Commission, b). The TRIPS Agreement likely does not prohibit individual or collective government restrictions of this type, as it does not regulate patented-product-market access per se, but only requires that patents not be denied based on regulatory product prohibitions (TRIPS Agreement). Arguably, such measures are needed to avoid serious prejudice to the environment, although, for NETs, that would result from the failure to supply the technologies rather than from utilization of the patented products. Accordingly, such measures might still violate the WTO's Technical Barriers to Trade Agreement or other WTO treaty provisions (World Trade Organization, 2020f). But when used to counter patents conveying monopoly market power, such private or governmental restrictions might be viewed more favorably (Day and Schuster, 2021). And in the context of worldwide supply of NETs for worldwide public benefit, such measures *should* be viewed more favorably.

Finally, similar measures to condition funding for NETs development should support open-science and open-data approaches, to assure that competitive RD&D is possible in the most rapid time frames (USNASEM, 2018; Reynolds, 2019). As others have argued, although it remains the subject of vigorous theoretical and empirical debates, competitive development tends to lead to greater and more rapid innovation (Merges and Nelson, 2003; Kitch, 1977). Accordingly, most intergovernmental organizations and national governments encourage the open sharing of data developed with governmental funding, and many foundations do so as well (U.S. National Institutes of Health; Bill and Melinda Gates Foundation; Cogsgriff et al., 2020). Similar policies should be adopted for NET RD&D efforts.

11.4 Conclusion

Funding will and should continue to grow for RD&D of NETs. Patents in NETs also will develop, but if left unaddressed by international agreements will generate conflicts over affordable access to patented NETs. Although international cooperative measures may develop, it is more likely that national governments will impose controversial measures such as price controls and compulsory licensing to assure affordable access. However, alternatives exist for both national governments and private funders to impose measures ex-ante to the needed innovation that could minimize the impending conflicts. They should do so, rather than failing to serve the publics that will need but cannot afford those NETs.

References

28 U.S.C. § 1498(a), 2020.
35 U.S.C. § 203(a)(1), (2), 2020.
50 U.S.C. §§4501 et seq, 2020.
Adepoju, P., July 11, 2020. COVID-19 vaccine nationalism limits Africa's options. HealthPolicy Watch Blog.

African Union & Africa CDC, June 25, 2020. Communiqué From Africa's Leadership in COVID-19 Vaccine Development and Access Virtual Conference. https://africacdc.org/news-item/covid-19-vaccine-development-and-access-virtual-conference/.

Ahmed, A.K., June 4, 2020. Oxford, AstraZeneca COVID-19 deal reinforces "vaccine sovereignty." We need a people's vaccine instead. STAT.

American Institute of Physics, January 30, 2020. Final FY20 appropriations: DOE Applied Energy R&D. https://www.aip.org/fyi/2020/final-fy20-appropriations-doe-applied-energy-rd.

Athey, S., Kremer, M., Snyder, C., Tabarrok, A., May 4, 2020. In the race for the coronavirus vaccine, we must go big. Really, really big. New York Times. https://www.nytimes.com/2020/05/04/opinion/coronavirus-vaccine.html.

Baker, D., Jayadev, A., Stiglitz, J., 2017. Innovation, intellectual property, and development: a better set of approaches for the 21st century, 50. https://cepr.net/images/stories/reports/baker-jayadev-stiglitz-innovation-ip-development-2017-07.pdf.

Barooah, S.P., June 23, 2008. UNFCCC requiring a new patent regime? SpicyIP. https://spicyip.com/2008/06/unfccc-requiring-new-patent-regime.html.

Betts, R., Jones, C., Jin, Y., Keeling, R., Kennedy, J., Knight, J., et al., May 7, 2020. Analysis: What impact will the coronavirus pandemic have on atmospheric CO_2? Carbon Brief. https://www.carbonbrief.org/analysis-what-impact-will-the-coronavirus-pandemic-have-on-atmospheric-co2.

Biddle, C.B., Contreras, J.L., Love, B.J., Siebrasse, N.V. (Eds.), 2019. Patent Remedies and Complex Products: Toward A Global Consensus. Cambridge University Press.

Bill & Melinda Gates Foundation. How we work: information sharing approach. https://www.gatesfoundation.org/How-We-Work/General-Information/Information-Sharing-Approach.

Black, S., 2020, August 13. Sharing is caring: why COVID-19 vaccine manufacturers must collaborate. https://www.scienceboard.net/index.aspx?sec=sup&sub=Drug&pag=dis&ItemID=1179.

Bodansky, D., 2012. What's in a concept? Global public goods, international law, and legitimacy. European Journal of International Law 23 (3), 651. Available from: https://doi.org/10.1093/ejil/chs035.

Boffey, D., January 25, 2021. EU threatens to block Covid vaccine exports amid AstraZeneca shortfall. The Guardian. https://www.theguardian.com/world/2021/jan/25/eu-threatens-to-block-covid-vaccine-exports-amid-astrazeneca-shortfall.

Bollyky, T.J., Brown, C.P., September 2020. The tragedy of vaccine nationalism: only cooperation can end the pandemic. Foreign Affairs. https://www.foreignaffairs.com/articles/united-states/2020-07-27/vaccine-nationalism-pandemic.

Bond, E., Saggi, K., 2014. Compulsory licensing, price controls, and access to patented foreign products. Journal of Development Economics 109, 219.

Bosetti, V., Carraro, C., De Cian, E., Massetti, E., Tavoni, M., 2013. Incentives and stability of international climate coalitions: an integrated assessment. Energy Policy 55, 47. Available from: https://doi.org/10.1016/j.enpol.2012.12.035.

Brachman, S., May 31, 2020. WHO's C-TAP initiative pushes for non-exclusive global licensing amid pharma industry concerns. IP Watchdog. https://www.ipwatchdog.com/2020/05/31/whos-c-tap-initiative-pushes-non-exclusive-global-licensing-amid-pharmaceutical-industry-concerns/id = 122041/.

Bronin, S.C., 2020. What the pandemic can teach climate attorneys. Stanford Law Review Online 72, 159. https://www.stanfordlawreview.org/online/what-the-pandemic-can-teach-climate-attorneys/.

Burci, G.L., Gostin, L.O., 2017. Privatized pharmaceutical innovation vs access to essential medicines. Journal of the American Medical Association 317.

California Legislative Analyst's Office, May 13, 2020. Overview of federal COVID-19 research funding. https://lao.ca.gov/Publications/Report/4230.

Canadian IPO, Ministry of Innovation, Science and Economic Development Canada, 2017. Patented inventions in climate change mitigation technologies. 68 Figure 67, 69 Figure 68. https://www.ic.gc.ca/eic/site/cipointernet-internetopic.nsf/eng/h_wr04289.html.

Carnegie Council for Ethics in International Affairs. November 26, 2019. C2G2 Policy brief: governing solar radiation management. https://www.c2g2.net/publications/.

Chavez, A.E., 2015. Exclusive rights to saving the planet: the patenting of geoengineering inventions. Northwestern Journal of Technology & Intellectual Property 13, 17–34.

Climeworks, June 2, 2020. Climeworks raises CHF 73M (USD 75M)—the largest ever private investment into direct air capture. https://www.climeworks.com/news/climeworks-raises-chf-73m-usd-75m.

Cogsgriff, C.V., Ebner, D.K., Celi, L.A., 2020. Data sharing in the era of COVID-19. The Lancet Digital Health 2, E224. Available from: https://doi.org/10.1016/S2589-7500(20)30082-0.

Congressional Research Service, May 15, 2020. Insight: COVID-19: Defense Production Act (DPA) developments and issues for Congress, 2, 3. https://www.everycrsreport.com/reports/IN11387.html.

Contreras, J.L., 2015. A brief history of FRAND: analyzing current debates in standard setting and antitrust through a historical lens. Antitrust Law Journal 80.

Contreras, J.L., 2019. Global rate setting: a solution for standards-essential patents? Washington Law Review 94.

Contreras, J.L., 2020. Association for molecular pathology v. myriad genetics: a critical reassessmentforthcoming Michigan Technology Law Review 27, 53.

Contreras, J.L., Hall, B.H., Helmers, C., 2019. Pledging patents for the public good: rise and fall of the eco-patent commons. Houston Law Review 57. https://houstonlawreview.org/article/10854-pledging-patents-for-the-public-good-rise-and-fall-of-the-eco-patent-commons.

Correa, C.M., 1999. Intellectual property rights and the use of compulsory licenses: options for developing countries. In: South Centre Trade Related Agenda, Development and Equity (T.R.A.D.E.), Working Paper 5, p. 9.

Correa, C.M., 2020. Special Section 301: US Interference with the Design and Implementation of National Patent Laws. South Centre Research Paper 115, pp. 3–4, 11–26.

David S. Levine, 2020. Editorial: Trade secrets and the battle against Covid, 15 J. Intellectual Property L. & Practice 849.

Day, G., Schuster, M. 2021. Available from: https://papers.ssrn.com/sol3/papers.cfm?abstract_id = 3799477.

Dearment, A., February 19, 2020. How much will drug price controls harm innovation? It depends. MedCity News. https://medcitynews.com/2020/02/how-much-will-drug-price-controls-harm-innovation-it-depends/.

Delrahim, M., May 10, 2004. Deputy Assistant Attorney General, Antitrust Division, U.S. Dept. of Justice. Forcing firms to share the sandbox: compulsory licensing of intellectual property rights and antitrust.

Denham, H., Abutaleb, Y., Rowland, C., June 29, 2020. Gilead sets price of coronavirus drug remdesivir at $3,120 as Trump administration secures supply for 500,000 patients. Washington Post. https://www.washingtonpost.com/business/2020/06/29/gilead-sciences-remdesivir-cost-coronavirus/.

Editorial, 2021. The world's vaccine plan must succeed. Nature 589.

European Commission. European Climate Change Program. https://ec.europa.eu/clima/policies/eccp_en#:~:text=The%20European%20Commission%20established%20the,to%20cut%20greenhouse%20gas%20emissions.

Farrell, J., Klemperer, P., 2007. Coordination and lock-in: competition with switching costs and network effects. In: Armstrong, M., Porter, R. (Eds.), Handbook of Industrial Organizations, third ed. p. 1967.

Fialka, J., January 23, 2020. U.S. geoengineering research gets a lift with $4 million from Congress. Science. https://www.sciencemag.org/news/2020/01/us-geoengineering-research-gets-lift-4-million-congress.

Fiorini, A., Pasimeni, F., Georgakaki, A., Tzimas, E., 2017. Analysis of the European CCS research and innovation landscape. Energy Procedia 114. Available from: https://doi.org/10.1016/j.egypro.2017.03.1897.

First, H., 2019. Excessive drug pricing as an antitrust violation. Antitrust Law Journal 82, 705. n.18, 719 and n.103. https://www.antitrustinstitute.org/wp-content/uploads/2019/04/First-ALJ-82-2-FINAL.pdf.

Forster, J., Vaughan, N., Gough, C., Lorenzoni, I., Chilvers, J., 2020. Mapping feasibilities of greenhouse gas removal: key issues, gaps and opening up assessments. Global Environmental Change 63. Available from: https://doi.org/10.1016/j.gloenvcha.2020.102073.

Frank, R.G., Ginsburg, P.B., November 13, 2017. Pharmaceutical industry profits and research and development. Health Affairs Blog. https://www.healthaffairs.org/do/10.1377/hblog20171113.880918/full/.

Frischmann, B., Waller, S.W., 2008. Revitalizing essential facilities. Antitrust Law Journal 75, 3. Available from: https://papers.ssrn.com/sol3/papers.cfm?abstract_id=961609.

Fuss, S., 2017. The 1.5°C target, political implications, and the role of BECCS. In Oxford Research Encyclopedia, Climate Science. Available from: http://doi.org/10.1093/acrefore/9780190228620.013.585.

Gerard, M.B., Dernbach, J.C. (Eds.), 2018. Legal Pathways to Deep Decarbonization in the United States. Environmental Law Institute.

Gerard, M., 2020. Direct air capture: an emerging necessity to fight climate change. Trends ABA Section of Environment, Energy, and Resources 51.

Gertner, J., April 18, 2017. Is it ok to tinker with the environment to fight climate change? New York Times. https://www.nytimes.com/2017/04/18/magazine/is-it-ok-to-engineer-the-environment-to-fight-climate-change.html.

Ghosh, S., Calboli, I. (Eds.), 2018. Exhausting Intellectual Property Rights: A Comparative Law & Policy Analysis. Cambridge University Press.

Gilbert, R., October 8, 2019. Compulsory licensing: an underrated antitrust remedy. Antitrust Chronicle. https://www.competitionpolicyinternational.com/compulsory-licensing-an-underrated-antitrust-remedy/.

Gilo, D., Spiegel, Y., 2018. The antitrust prohibition of excessive pricing. International Journal of Industrial Organization 61, 503. Available from: https://doi.org/10.1016/j.ijindorg.2018.05.003.

Global CCS Institute, 2019. Global status of CCS 2019, pp. 22–23. https://www.globalccsinstitute.com/resources/global-status-report/.

Gronewold, N., May 28, 2020. E.U.'s coronavirus recovery plan also aims to fight climate change. Scientific American. https://www.scientificamerican.com/article/e-u-s-coronavirus-recovery-plan-also-aims-to-fight-climate-change/.

Gurry, F., April 24, 2020. Some considerations on intellectual property, innovation, access and COVID-19. WIPO. https://www.wipo.int/wipo_magazine/en/2020/02/article_0002.html.

Haszeldine, R.S., Flude, S., Johnson, G., Scott, V., 2018. Negative emissions technologies and carbon capture and storage to achieve the Paris Agreement commitments. Philosophical Transactions of the Royal Society A 376. Available from: https://doi.org/10.1098/rsta.2016.0447.

Hemel, D.J., Ouellette, L.L., 2013. Beyond the patents-prizes debate. Texas Law Review 92 (310–325), 351–352. https://texaslawreview.org/wp-content/uploads/2015/08/HemelOuellette.pdf.

Hemel, D.J., Ouellette, L.L., 2019. Innovation policy pluralism. Yale Law Journal 128. https://www.yalelawjournal.org/article/innovation-policy-pluralism.

Hemel, D.J., Ouellette, L.L., May 4, 2020. Want a coronavirus vaccine fast? Here's a solution. Time. https://time.com/5795013/coronavirus-vaccine-prize-challenge/.

Herman, B., July 1, 2020. Federal government weakened its march-in rights for coronavirus drugs. AXIOS. https://www.axios.com/federal-government-barda-contracts-moderna-regeneron-aaf9fde2-2ee1-46fb-8465-0d573e6af1ed.html.

Hester, T., 2018. Legal pathways to negative emissions technologies and direct air capture of greenhouse gases. In: Gerard, M.B., Dernbac, J.C. (Eds.), Legal Pathways to Deep Decarbonization in the United States. Environmental Law Institute.

Hezir, J.S., Bushman, T., Stark, A.K., Smith, E., 2019. Carbon removal: comparing historical federal research investments with the National Academies' recommended future funding levels. Bipartisan Policy Center. https://bipartisanpolicy.org/wp-content/uploads/2019/06/Carbon-Removal-Comparing-Historical-Investments-with-the-National-Academies-Recommendations.pdf.

Honneger, M., Reiner, D., 2018. The political economy of negative emissions technologies: consequences for international policy design. Climate Policy 18. Available from: https://doi.org/10.1080/14693062.2017.1413322.

Hood, D., Bseiso, F., July 14, 2020. Manufacturers hope R&D tax break is revived in next relief bill. Bloomberg Daily Tax Report: State.

Hook, L., March 24, 2019. Climate change fears spur investment in carbon capture technology. Financial Times. https://www.bizjournals.com/albany/news/2019/03/25/climate-change-fears-spur-investment-in-carbon.html.

Hughes, J., Rai, A.K., May 8, 2020. Acknowledging the public role in private drug development: lessons from remdesivir. STAT. https://www.statnews.com/2020/05/08/acknowledging-public-role-drug-development-lessons-remdesivir/.

Jacobs, W.B., Craig, M., 2018. Carbon capture & sequestration. In: Gerard, M.B., Dernbac, J.C. (Eds.), Legal Pathways to Deep Decarbonization in the United States. Environmental Law Institute.

Jacobson, R., May 28, 2019. The case for investing in direct air capture just got clearer. GreenBiz. https://www.greenbiz.com/article/case-investing-direct-air-capture-just-got-clearer.

Jaffe, A.B., Newell, R.G., Stavins, R.N., 2001. Technological change and the environment. In: Resources for the Future. Discussion Paper 00-47REV. Available from: https://doi.org/10.22004/ag.econ.298440.

Jones, L., May 12, 2020. Have we entered the era of climate engineering? Cloud brightening technology trialed over Great Barrier Reef. Reset. https://en.reset.org/blog/have-we-entered-era-climate-engineering-cloud-brightening-technology-trialled-over-great-barrie.

Juma, C., July 6, 2016. Why do people resist new technologies? History might provide the answer. World Economic Forum. https://www.weforum.org/agenda/2016/07/why-do-people-resist-new-technologies-history-has-answer/.

Keith, D.W., St. Holmes, G., Angelo, D., Heidel, K., 2018. A process for capturing CO_2 from the atmosphere. Joule 2, 1594. Available from: https://doi.org/10.1016/j.joule.2018.05.006.

Kitch, E.W., 1977. The nature and function of the patent system. The Journal of Law and Economics 20.

Knowledge Ecology International. Compulsory licensing as limitations on remedies for infringement under 28 U.S.C. §1498: Patent and copyright. https://www.keionline.org/cl/28usc1498.

Lebling, K., March 9, 2020. To unlock the potential of direct air capture, we must invest now. World Resources Institute Blog. https://www.wri.org/blog/2020/03/to-unlock-the-potential-of-direct-air-capture-we-must-invest-now.

Lee, J., July 10, 2020. A new Gilead analysis indicates remdesivir can reduce death but company says finding "requires confirmation." MarketWatch. https://www.marketwatch.com/story/a-new-gilead-analysis-indicates-remdesivir-can-reduce-death-but-company-says-finding-requires-confirmation-2020-07-10.

Lexchin, J., July 5, 2020. As U.S. buys up redesivir, "vaccine nationalism" threatens access to COVID-19 treatments. The Conversation. https://theconversation.com/as-u-s-buys-up-remdesivir-vaccine-nationalism-threatens-access-to-covid-19-treatments-141952.

License Agreement [AbbVie & MPP]; [Appendix 6-A for existing licensees in India] [first/second/third] amended and restated license agreement.

Liu, M., Feldman, W.B., Avorn, J., Kesselheim, A.S., May 6, 2020. March-in rights and compulsory licensing— safety nets for access to a COVID-19 vaccine. Health Affairs Blog. https://ramaonhealthcare.com/march-in-rights-and-compulsory-licensing-safety-nets-for-access-to-a-covid-19-vaccine/.

Livesay, A., February 4, 2020. Lex in depth: the $900bn cost of "stranded energy assets." Financial Times. https://www.ft.com/content/95efca74-4299-11ea-a43a-c4b328d9061c.

Mace, M.J., Fyson, C.L., Schaeffer, M., Hare, W.L., 2018. Governing large-scale carbon dioxide removal: are we ready? Carnegie Climate Geoengineering Governance Initiative. https://www.c2g2.net/wp-content/uploads/C2G2-2018-CDR-Governance-1.pdf.

Maskus, K.E., Okediji, R., 2010. Intellectual property rights and international technology transfer to address climate change: risks, opportunities, and policy option. International Centre for Trade and Sustainable Development 35. Available from: http://citeseerx.ist.psu.edu/viewdoc/download?doi=10.1.1.471.6938&rep=rep1&type=pdf.

McGlashan, N.R., Workman, M.H.W., Caldecott, B., Shah, N., 2012. Negative Emissions Technologies, Briefing Paper No 8. Grantham Institute for Climate Change, Imperial College London. https://www.imperial.ac.uk/media/imperial-college/grantham-institute/public/publications/briefing-papers/Negative-Emissions-Technologies---Grantham-BP-8.pdf.

Merges, R.P., Nelson, R.R., 2003. On the complex economics of patent scope. Columbia Law Review. Available from: https://doi.org/10.2307/1122920.

Miguez, J.L., Porteiro, J., Perez-Orozco, R., Patino, D., 2020. Biological systems for CCS: patent review as a criterion for technological development. Applied Energy 257. Available from: https://doi.org/10.1016/j.apenergy.2019.114032.

Miguez, J.L., Porteiro, J., Perez-Orozco, R., Patino, D., Rodriguez, S., 2018. Evolution of CO_2 capture technology between 2007 and 2017 through the study of patent activity. Applied Energy 211. Available from: https://doi.org/10.1016/j.apenergy.2017.11.107.

Mintzes, B., 't Hoen, E., July 3, 2020. The US has bought most of the world's remdesivir. Here's what it means for the rest of us. The Conversation. https://theconversation.com/the-us-has-bought-most-of-the-worlds-remdesivir-heres-what-it-means-for-the-rest-of-us-141791.

MIT Technology Licensing Office, April 7, 2020. Covid-19 Technology Access Framework. https://tlo.mit.edu/engage-tlo/covid-19/covid-19-technology-access-framework.

Nelson, R.R., Winter, S.G., 1982. An Evolutionary Theory of Economic Change. Belknap Press.

Newey, S., May 29, 2020. WHO patent pool for potential Covid-19 products is "nonsense," pharma leaders claim. The Telegraph. https://www.telegraph.co.uk/global-health/science-and-disease/patent-pool-potential-covid-19-products-nonsense-pharma-leaders/.

Norhasyima, R.S., Mahlia, T.M.I., 2018. Advances in CO_2 utilization technology: a patent landscape review. Journal of CO_2 Utilization 26, 333. Available from: http://dspace.uniten.edu.my/bitstream/123456789/11596/1/Advances%20in%20CO2%20utilization%20technology%20A%20patent%20landscape%20review.pdf.

Office of Technology Development, Harvard University, April 6, 2020. Licensing during COVID-19: COVID-19 technology access framework. https://otd.harvard.edu/about-otd/our-values/licensing-during-covid-19.

Oldham, P., Szerszynski, B., Stilgoe, J., Brown, C., Eacott, B., Yuille, A., 2014. Mapping the landscape of climate engineering. Philosophical Transactions of the Royal Society of London A 372, 10 Table 1. Available from: https://doi.org/10.1098/rsta.2014.0065.

Open Covid Pledge (a). Frequently asked questions. https://opencovidpledge.org/faqs/.

Open Covid Pledge (b). https://opencovidpledge.org/.

Orland, K., January 27, 2020. Bid to cut carbon-capture cost 80% gets nudge in joint venture. Bloomberg Environment & Energy Report. https://www.bloomberg.com/news/articles/2020-01-27/bid-to-cut-carbon-capture-cost-80-gets-nudge-in-joint-venture.

Parthasarathy, S., Avery, C., Hedberg, N., Mannisto, J., Maguire, M., September 9, 2010. A Public Good? Science, Technology, & Public Policy Program, Working Paper No. 10-1. http://www.umt.edu/ethics/ethicsgeoengineering/Workshop/articles1/Chris%20Avery.pdf.

Patent and Trademark Law Amendments Act, Pub. L. 96−517, December 12, 1980. (codified at 35 U.S.C. §§ 200−212 (2019)).

Peets, L., Young, M., Cheek, M., 2007. Special 301: China and Russia top the annual U.S. list of IPR offenders. Trademark World #199. https://www.cov.com/en/news-and-insights/insights/2007/07/special-301-china-and-russia-top-the-annual-u-s--list-of-ipr-offenders.

Price, W.N., Rai, A.K., Minssen, T., 2020. Knowledge transfer for large-scale vaccine manufacturing. Science 369, 912. Available from: https://doi.org/10.1126/science.abc9588.

Qui, H.H., Yang, J., 2018. An assessment of technological innovation capabilities of carbon capture and storage technology based on patent analysis: a comparative study between China and the United States. Sustainability 10, 3−7. Available from: https://doi.org/10.3390/su10030877.

Ren, G., June 29, 2020. World surpasses 500,000 COVID-19 deaths; medicines access experts challenge US $2340 per remdesivir treatment course price set by Gilead for developed countries. Health Policy Watch. https://healthpolicy-watch.news/gilead-to-charge-us-320-per-vial-of-remdesivir-in-developed-countries/.

Reynolds, J.L., 2019. The Governance of Solar Geoengineering: Managing Climate Change in the Anthropocene. Cambridge University Press.

Robock, A., 2020. Benefits and Risks of Stratospheric Solar Radiation Management for Climate Intervention (Geoengineering). The Bridge 50. Available from: https://www.nae.edu/228936/Benefits-and-Risks-of-Stratospheric-Solar-Radiation-Management-for-Climate-Intervention-Geoengineering.

Roin, B.N., 2014. Intellectual property versus prizes: reframing the debate. University of Chicago Law Review 81, 999. https://chicagounbound.uchicago.edu/uclrev/vol81/iss3/3.

Ruckelshaus v. Monsanto Co., 467 U.S. 986, 1006, 1984.

Rutschman, A.S., June 4, 2020. How "vaccine nationalism" could block vulnerable populations' access to COVID-19 vaccines. Utica College of Public Affairs and Election Research. https://www.ucpublicaffairs.com/home/2020/6/24/how-vaccine-nationalism-could-block-vulnerable-populations-access-to-covid-19-vaccines-by-ana-santos-rutschman.

Sachs, R., Sherkow, J.S., Ouellette, L.L., Price, N., May 12, 2020. Remdesivir part I: incentivizing antiviral innovation. Written Description Blog. https://writtendescription.blogspot.com/2020/05/remdesivir-part-i-incentivizing.html.

Sagonowsky, E., May 11, 2020. Novavax scores $384 M deal, CEPI's largest ever, to fund coronavirus vaccine work. Fierce Pharma. https://www.fiercepharma.com/vaccines/novavax-scores-cepi-s-largest-award-up-to-384m-to-support-covid-19-vaccine-work.

Sandalow, D., Friedmann, J., McCormick, C., McCoy, S., 2018. Direct air capture of carbon dioxide. ICEF Roadmap 12. Available from: https://www.globalccsinstitute.com/wp-content/uploads/2020/06/JF_ICEF_DAC_Roadmap-20181207-1.pdf.

Santamauro, J.P., Martella, R.R., Mendenhall, J., Sachse, T., Shoyer, A.W., 2009. The international climate change negotiations and intellectual property: another climate change risk? International Finance &

Treasury 35 (1). Available from: https://www.sidley.com/-/media/files/publications/2009/09/the-international-climate-change-negotiations-an__/files/view-article/fileattachment/sidleyaustin_ft081509.pdf?la=en.

Sarnoff, J.D., Chon, M., 2018. Innovation law and policy choices for climate change-related public—private partnerships. In: Roffe, P. (Ed.), Global Intellectual Property, Public—Private Partnerships, and Sustainable Development Goals, 245. Cambridge University Press.

Sarnoff, J.D., 2007. BIO v. DC and the new need to eliminate federal patent law preemption of state and local price and product regulation. Patently-O Patent Law Journal 2007, 30—35. https://patentlyo.com/lawjournal/2007/08/bio-v-dc-and-th.html.

Sarnoff, J.D., 2010. Lessons from the United States in regard to the recent, more flexible application of injunctive relief. In: Correa, C.M. (Ed.), Research Handbook on the Interpretation and Enforcement of Intellectual Property Under WTO Rules, second ed. Edward Elgar Publishers, p. 48.

Sarnoff, J.D., 2011. The patent system and climate change. Virginia Journal of Law & Technology 16, 339. https://www.law.uh.edu/eenrcenter/North-American/2020/Sarnoff%20VJOLT%20patent%20system%20and%20climate%20change.pdf.

Sarnoff, J.D., 2013. Government choices in innovation funding (with reference to climate change). Emory Law Journal 62. https://scholarlycommons.law.emory.edu/cgi/viewcontent.cgi?article=1272&context=elj.

Sarnoff, J.D., 2020. forthcoming Comparative Permanent and Preliminary Injunctive Relief After COVID-19, in the Context of Patented Pharmaceuticals in the U.S. and Canada. South Center Research Paper.

Schlam, L., 1998. Compulsory royalty-free licensing as an antitrust remedy for patent fraud: law, policy and the patent-antitrust interface revisited. Cornell Journal of Law & Public Policy 7. https://scholarship.law.cornell.edu/cgi/viewcontent.cgi?article=1234&context=cjlpp.

Serhan, Y., December 8, 2020. Vaccine nationalism is doomed to fail. The Atlantic. https://www.theatlantic.com/international/archive/2020/12/vaccine-nationalism-doomed-fail/617323/.

Sheffi, Y., August 10, 2020. From COVID to climate: four pandemic lessons that may mitigate global warming. https://www.linkedin.com/pulse/from-covid-climate-four-pandemic-lessons-may-mitigate-yossi-sheffi/.

Shope, R., 1991. Global climate change and infectious disease. Environmental Health Perspectives 96, 174. http://www.ciesin.org/docs/001-366/001-366.html.

Sichelman, T., 2017. Patents, prizes, and property. Harvard Journal of Law and Technology 30, 281. https://jolt.law.harvard.edu/assets/digestImages/a13-Sichelman_Round5Complete.pdf.

Silverman, E., April 24, 2020a. WHO launches ambitious global project to develop COVID-19 medical products. STAT. https://www.statnews.com/pharmalot/2020/04/24/covid19-coronavirus-who-vaccines-medicines-access/.

Silverman, E., May 15, 2020b. WHO embraces plan for COVID-19 intellectual property pool. STAT. https://www.statnews.com/pharmalot/2020/05/15/who-covid19-coronavirus-patents-intellectual-property/.

Silverman, E., May 28, 2020c. Pharma leaders shoot down WHO voluntary pool for patent rights on COVID-19 products. STAT. https://www.statnews.com/pharmalot/2020/05/28/who-voluntary-pool-patents-pfizer/.

Smedley, T., February 25, 2019. How artificially brightened clouds could stop climate change. BBC Future. https://www.bbc.com/future/article/20190220-how-artificially-brightened-clouds-could-stop-climate-change.

Stern, N., May 4, 2020. Financing climate ambition in the context of COVID-19. Grantham Research Institute on Climate Change and the Environment. https://www.lse.ac.uk/granthaminstitute/news/financing-climate-ambition-in-the-context-of-covid-19/.

Sullivan, K.M., 1989. Unconstitutional conditions and the distribution of liberty. San Diego Law Review 96. https://digital.sandiego.edu/sdlr/vol26/iss2/10.

Tai, K.C., 2021. Statement from Ambassador Katherine Tai on the Covid-19 Trips Waiver, 5 May 2021, https://ustr.gov/about-us/policy-offices/press-office/press-releases/2021/may/statement-ambassador-katherine-tai-covid-19-trips-waiver.

Temple, J., 2020. Biden calls for major investments into carbon removal tech. MIT Technology Review. Available from: https://www.technologyreview.com/2020/11/09/1011859/biden-calls-for-major-investments-into-carbon-removal-tech/.

Thomas, J.R., 2016. March-in rights under the Bayh-Dole Act. Congressional Research Service Report R44597 8—9. https://fas.org/sgp/crs/misc/R44597.pdf.

Tobacco Control Laws Global Legal Center. Australia—Tobacco Plain Packaging TRIPS Agreement, Articles 27, 30, 31(k), 44.1.

U.S. Department of Justice, June 25, 2015. Price fixing, bid rigging, and market allocation schemes. https://www.justice.gov/atr/price-fixing-bid-rigging-and-market-allocation-schemes.

U.S. Federal Trade Commission (a), Premerger notification and the merger review process. https://www.ftc.gov/tips-advice/competition-guidance/guide-antitrust-laws/mergers/premerger-notification-merger-review.

U.S. Federal Trade Commission (b). Group boycotts. https://www.ftc.gov/tips-advice/competition-guidance/guide-antitrust-laws/dealings-competitors/group-boycotts.

U.S. House of Representatives, 2020. Select Committee on the Climate Crisis, Majority Staff Report. Solving the Climate Crisis: The Congressional Action Plan for a Clean Energy Economy and a Healthy, Resilient, and Just America, 276.

U.S. National Academies of Sciences, Engineering & Medicine, 2010. Consensus Study Report: Advancing the Science of Climate Change (Chapter 15).

U.S. National Academies of Sciences, Engineering & Medicine, 2018. Data matters: Ethics, Data, and International Research Collaboration in a Changing World: Proceedings of a Workshop, 1, 4, 9.

U.S. National Academies of Sciences, Engineering & Medicine, 2019. Consensus Study Report Negative Emission Technologies and Reliable Sequestration: A Research Agenda.

U.S. National Institutes of Health, Office of Data Science Strategy, 2020a, May 13. Open-access data and computational resources to address Covid-19. https://datascience.nih.gov/covid-19-open-access-resources.

U.S. National Institutes of Health, 2020b. 3D Print Exchange. https://3dprint.nih.gov/.

U.S. National Institute of Health, '3D Print Exchange'; Herbert Smith Freehills, 'Covid-19: Pressure Points: A Catalyst for Collaborations (Global)', 8 April 2020.

UN Environment Programme, May 11, 2020. Record global carbon dioxide concentrations despite COVID-19 crisis. https://www.unenvironment.org/news-and-stories/story/record-global-carbon-dioxide-concentrations-despite-covid-19-crisis.

United Nations Framework Convention on Climate Change, Paris Agreement, Article 2, 3.1, 4.3, and 10.6. https://unfccc.int/files/meetings/paris_nov_2015/application/pdf/paris_agreement_english_.pdf.

van Vuuren, D.P., Stehfest, E., Gernaat, D.E.H.J., et al., 2018. Alternative pathways to the 1.5°C target reduce the need for negative emission technologies. Nature Climate Change 8. Available from: https://doi.org/10.1038/s41558-018-0119-8.

Vermeij, J., April 4, 2020. A compulsory license on trade secrets? In times of corona, much is possible. Verj. https://solv.nl/en/blog/a-compulsory-licence-on-trade-secrets-in-times-of-corona-much-is-possible/.

Victor, D.C., 2020. Deep decarbonization: a realistic way forward on climate change. Yale Environment 360. https://e360.yale.edu/features/deep-decarbonization-a-realistic-way-forward-on-climate-change.

Wagner, G., Zizzamia, D., January 9, 2020. Working draft. Green moral hazards. https://gwagner.com/wp-content/uploads/Wagner-Zizzamia-Green-Moral-Hazards-DRAFT-200708.pdf.

Waller, S.W., Tasch, W., 2019–2010. Harmonizing essential facilities. Antitrust Law Journal 76, 741.

WIPO, 2020a. Climeworks: A Technology to Reverse Climate Change. https://www.wipo.int/ip-outreach/en/ipday/2020/case-studies/climeworks.html.

WIPO, 2020b. WIPO-GREEN -The Marketplace for Sustainable Technology. https://www3.wipo.int/wipogreen/en/.

World Bank Group, June 17, 2020. Transformative climate finance: a new approach for climate finance to achieve low-carbon resilient development in developing countries, 40. https://openknowledge.worldbank.org/handle/10986/33917.

World Health Organization, 26 June, 2020a. Act Accelerator Update. https://www.who.int/news/item/26-06-2020-act-accelerator-update.

World Health Organization, April 24, 2020b. Access to COVID-19 (ACT) Accelerator: a global collaboration to accelerate the development, production, and equitable access to new COVID-19 diagnostics, therapeutics, and vaccines. https://www.who.int/publications/m/item/access-to-covid-19-tools-(act)-accelerator.

World Health Organization, May 29, 2020c. International community rallies to support open research and science to fight COVID-19. https://www.who.int/news/item/29-05-2020-international-community-rallies-to-support-open-research-and-science-to-fight-covid-19.

World Health Organization, 29 May 2020d. Solidarity Call to Action, https://www.who.int/initiatives/covid-19-technology-access-pool/solidarity-call-to-action.

World Health Organization, 2020e. Tech Access Partnership: A COVID-19 Technology Sharing Platform, https://techaccesspartnership.net/.

World Health Organization, 2020f. The Access to COVID-19 Tools (ACT) Accelerator, 2020, https://www.who.int/initiatives/act-accelerator.

World Health Organization, 2020g. The Access to COVID-19 Tools (ACT) Frequently Asked Questions at https://www.who.int/initiatives/act-accelerator/faq.

World leaders donate to COVID-19 vaccine funding drive, May 5, 2020. PMLive. http://www.pmlive.com/pharma_news/world_leaders_donate_to_covid-19_vaccine_funding_drive_1339756.

World Trade Organization, Doha Declaration, November 14, 2001. Declaration on the TRIPS Agreement and public health, para. 5.c.

World Trade Organization, June 28, 2018. Reports of the Panels, Australia—Certain Measures Concerning Trademarks, Geographical Indications and Plain Packaging Requirements Applicable to Tobacco Products, WT/DS435/R, WT/DS441/R, WT/DS458/R, WT/DS467/R.

World Trade Organization, April 1994. 2020d. Agreement on Trade Related Aspects of Intellectual Property Rights, 33 I.L.M. 81, 15 April 2020d: WTO, Reports of the Panels, Australia—Certain Measures Concerning Trademarks, Geographical Indications and Plain Packaging Requirements Applicable to Tobacco Products, WT/DS435/R, WT/DS441/RWT/DS458/R, WT/DS467/R (28 June 2018), aff'd, WTO, Appellate Body Reports WT/DS435/AB/R & WT/DS441/AB/R (9 June 2020); Tobacco Control Laws Global Legal Center, 'Australia – Tobacco Plain Packaging'.2020d: WTO, Panel Report, Canada—Patent Protection of Pharmaceutical Products, paras. 7.54–7.57, 7.69, WT/DS114/R (Mar. 17, 2000).

Wyns, A., April 9, 2020. Climate change and infectious diseases. Scientific American Blog. https://blogs.scientificamerican.com/observations/climate-change-and-infectious-diseases/.

PART III

Comparative experiences around the world

Who is taking climate change seriously? Evidence based on a comparative analysis of the carbon capture and storage national legal framework in Brazil, Canada, the European Union, and the United States

Carolina Arlota[1,2] and Hirdan Katarina de Medeiros Costa[3]

[1]Visiting Assistant Professor of Law at the University of Oklahoma, College of Law, Norman, OK, United States [2]Law and Economics Research Fellow, Capitalism and Rule of Law Project (Cap Law) at Antonin Scalia Law School, George Mason University, Fairfax, VA, United States [3]Institute of Energy and Environment, University of São Paulo, São Paulo, Brazil

12.1 Introduction

Scientific consensus relates climate change to global warming (Tol, 2014). The leading factor contributing to the increase in the Earth's temperature is the accumulation of greenhouse gases (GHGs; carbon dioxide being the most famous) in the atmosphere. As global GHGs emissions continue increasing, it becomes less likely that the emissions gap will be closed or considerably narrowed by 2020. Previous research by the Nobel laureate in economics William Nordhaus concludes that, with the limited global action on climate change so far, the reduction established in the Paris Agreement on Climate Change, namely, limitation of 2°C of average global temperature increase is unattainable (Nordhaus, 2018). Therefore the necessity of policies to reduce climate change is urgent.

Due to such urgency, countries will have to rely on more complex, expensive, and risky choices to limit the increase in the global average temperature below 2°C. This amplifies the need to deploy more energy efficiency technologies and bioenergy with carbon capture and storage to reach the temperature target (United Nations Environment Programme UNEP, 2013). Recent studies have recognized the establishment of a carbon capture and storage (CCS) regulatory framework as a key issue for the large deployment of CCS globally (UNEP, 2013).

In this scenario, this article surveys the different CCS experiences of the European Union, Canada, the United States, and Brazil, using the comparative methodology. The countries were chosen due to federal systems (except the EU supranational system, which in some respects resembles federative experiences). The countries include diverse interpretations of the role of the government, and all have been based on a market-based economy. Gathering knowledge on the various aspects of the CCS federal (or supranational, in the European Union's case) legislation may offer relevant insights. Learning from different experiences may accelerate the timely progress of the safe deployment of CCS in all jurisdictions, and in developing countries.

The focus is on the federal level, because their actions are determinative for each country. Moreover, subnational unities' actions on climate change have their own limitations, and may require significant legal expertise to be effective (Coglianese and Starobin, 2019). In order to foster comparisons of legislation and replicable research material in the future, this article focuses on the national level—except on the European Union's paradigmatic example, where it considers the supranational sphere.

Hence, this article is premised on the notion that climate change action—to which CCS belongs—is better taken on the highest sphere of governance as possible to avoid greater harm, such as when unconstitutionality or illegality challenges are successful (Wiener, 2007; Arlota, 2020). Gathering knowledge on the different federal legislative experiences on CCS is relevant, because different countries face similar challenges on CCS, including its unique technology and costly price. Countries also have (on their national level) global incentives to pursue carbon emissions reductions under the Paris Agreement.

Despite the international arena per se not being the focus of this article, recent efforts on CCS deserve notice. First, the London Protocol, which was amended in 2006 to include in its Annex 1 a provision on carbon dioxide streaming from carbon dioxide storage. Subsequent amendments aimed at the transboundary movement of carbon dioxide. Second, the Convention for the Protection of the Marine on the North-East Atlantic (OSPAR) was amended in 2007 to authorize storage of carbon dioxide in subseabed geological formations. Third, the Intergovernmental Panel on Climate Change (IPCC) Guidelines for GHG inventories designated CCS as a zero- or low-emission mechanism. Finally, there was the inclusion of CCS as a clean development mechanism in the Kyoto Protocol (Dixon et al., 2015).

As this research targets CCS, which could have a significant impact on the reduction on carbon emissions, it addresses the concerns of the United Nations Security Council. This is the case as the Council recently recognized climate change as a "threat multiplier," due to climate-related risks and conflicts no longer being mere hypotheticals for significant populations as it affects their security and peace (Climate Change Recognized, 2019). This research also fills a void in the literature as comparative studies on federal (and supranational, in the case of the EU) level legislation of CCS concerning the countries referenced are yet to be published.

The article is organized as follows. Section 12.2 defines CCS and its main policy-related challenges. Section 12.3 surveys the CCS federal (or technically, supranational, in the case of the EU) experiences of Brazil, Canada, the European Union, and the United States. Section 12.4 discusses those findings, and section 12.5 concludes.

12.2 Definition of CCS and its main challenges

Climate change is complex. Although the scientific knowledge regarding the broad scientific principles on the topic are unimpeachable (Millner et al., 2010), there may exist discrepancies considering its impact (Posner and Sykes, 2013). Nonetheless, countries and world leaders have not taken action that is remotely significant to answer the dangers posed by climate change (Gerrard and Hester, 2018). Private actors are already concerned with the impact of climate change on the infrastructure and its systemic risks to the financial system as a whole (CIEL, 2019). The IPCC Report of 2018 finds, with high confidence, that all scenarios to limit global warming to 1.5°c require carbon dioxide removal (IPCC, 2018). Hence, CCS as a mitigation mechanism is crucial for meeting international climate change targets (Global CCS Institute, 2017). It also must be global and rapid (Consoli et al., 2017). Moreover, recent empirical studies estimate that a persistent annual increase in average global temperature by 0.04°c in the absence of mitigation policies reduces world real gross domestic product per capita by 7.22% by 2100, whereas limiting the temperature increase to 0.01°c per year, as determined in the Paris Agreement, significantly reduces to loss to 1.07% (Kahn et al., 2019).

In such a context, it is worth emphasizing that adaptation is not the best approach to curb emissions, despite climate change being a dynamic event that scientists are still understanding (Sovacool et al., 2016). Accordingly, countries shall invest on mitigation of their carbon emissions (Turn Down the Heat, 2012). This is not merely from a normative perspective, because mitigation is not only necessary but indispensable, as it is the sole alternative to effectively reduce carbon emissions (Sovacool et al., 2016; Turn Down the Heat, 2012). Among the main incentives for CCS are the low- or zero-emissions it permits, which would enable countries to include fossil fuels in their energy sources (Budinis et al., 2018). Globally, this is also relevant for the so-called "unburnable carbon," that is, known fossil fuels reserves cannot all be used if the world does not want to face the perils of climate change (Budinis et al., 2018). Illustrating the relevance of the topic, without carbon capture utilization and storage, mitigation cost is estimated to be 138% higher (Wang and Oko, 2017).

The literature review for this article looks at the following common challenges faced for CCS adoption: (1) the uncertainty on the permanence of ocean storage; (2) cross-border disagreements; (3) long-term liability provisions; (4) difficulties on GHGs reporting; (5) lack of an international CCS regulation; and (6) the potential infringement of international law, because the relevant international treaties were not designed to accommodate CCS issues (IPCC Special Report, 2005). In addition, CCS has cost as its main barrier, followed by location and capacity of storage (Budinis, 2018). Risks of storage and lack of information regarding global, regional, and national information have also appeared frequently in the specialized literature (Havercroft, 2019).

In such a scenario, predictability regarding environmental policy is of crucial importance as it reassures foreign investors (Havercroft, 2019). Related concerns include social license issues associated with CCS in the context of environmental licenses and community participation (Havercroft, 2019). Additional research and development are also cited as key factors in establishing a pro-CCS framework (Havercroft, 2019).

12.3 Comparative experience

Energy policy is informed by several factors ranging from environmental protection, national sovereignty, and energy security concerns (Heffron et al., 2018). This part surveys the federal legislation on CCS across the following jurisdictions: the United States, Canada, Brazil, and the supranational energy framework of the European Union.

12.3.1 The United States

In the United States, carbon capture and emissions have been occurring at the state level, but there is no national legislation on this subject (Folger, 2017). CCS appears in the Environmental Protection Agency (EPA) guidance on Underground Injection Control Program for Carbon Dioxide Geological Storage (EPA UIC, 2018). Despite not having federal legislation specific to the CCS topic, the EPA collects information for its GHG reporting program (EPA GHGRP, 2018). Importantly, EPA has issued rules for CCS that arise out of the congressional mandate to do so based on the Safe Drinking Water Act of 1974, and, more recently, on the Energy Policy Act of 2005 (Dixon et al., 2015).

There have been attempts to implement CCS legislation in the House of Representatives as well as in the Senate (Folger, 2017). Carbon capture and sequestration is coherent with the Trump administration's main goals to invest in infrastructure projects and to create jobs (Rebuilding America's Infrastructure, 2019). CCS is also a welcomed technology as per the terms of President Biden's recent Executive Order on the Environment (Executive Order on Tackling the Climate Crisis at Home and Abroad: Exec. Order N. 14008 of January 27, 202, 86 FR 7619).

Regarding recommendations to foster the adoption of CCS policies in the United States, the following are noteworthy: (1) state-imposed restrictions on carbon dioxide emissions, such as carbon emissions standards in new sources that require full CCS on new coal-powered units, coupled with partial and full CCS on new natural gas combined cycle (NGCC) units; (2) states should mandate the purchase of renewable energy as well as of energy originated from coal-fired and NGCC plants which are equipped with carbon dioxide capture; and (3) state governors should use executive orders to make agencies buy a minimum amount of CCS-produced energy (Jacobs and Craig, 2018).

It remains to be seen if these recommendations will be considered. Meanwhile, the federal guidance is on the deregulatory trend, as President Trump directed agencies to review (suspend, modify, or rescind) regulations which may "unduly burden" energy development—including those aimed at reducing GHGs emissions (Exec. Order No. 13,783, 2017).

12.3.2 Brazil

Following the international trend, Brazil ratified in 2016, through the National Congress, a document to define the strategy for implementing the country's commitments from 2020, according to the Nationally Determined Contribution (Brazilian Environmental Minister, NDC, 2020).

Brazil expects to reduce GHG emissions by 37% below 2005 levels by 2025 with a subsequent indicative contribution of reducing GHG emissions to 43% below 2005 levels by 2030 (Brazilian Environmental Minister, Paris Agreement, 2020). In order to achieve that, the country committed itself to increasing the share of sustainable bioenergy in its energy matrix to approximately 18% by 2030, restoring and reforesting 12 million hectares of forests, as well as achieving an estimated 45% share of renewable energy in the composition of the energy matrix in 2030 (Brazilian Environmental Minister, Paris Agreement, 2020).

According to the International Energy Agency (IEA), energy generation is one of the main carbon-emitting sources in Brazil, accounting for 43.6% of the CO_2 emitted by stationary sources, which shows the intensification of new approaches, discussions, technological development, and incentives for CO_2 capture and storage (Musarra and Costa, 2018). However, there is a lack of CCS regulation matters in Brazil (Romeiro-Conturbia, 2014). Therefore it is necessary to develop legal scenarios to ensure the safe and effective implementation of CCS, considering the role of relevant regulatory authorities, the main environmental licensing requirements, definition of CO_2 ownership and the allocation of long-term liabilities (Costa, 2019). On the other hand, Petrobras has undertaken CCS injection under enhanced oil recovery activities in order to improve its operations in the presalt area (Kury, 2019). Thus one may say that under oil and gas production CCS injection has be applied as a recovery methodology, however, Brazil is behind under a regulation that promotes a large CCS implementation projects.

12.3.3 Canada

According IEA, to strengthen Canada's position as a responsible energy supplier and user, federal mechanisms were created for provincial and territorial governments' cooperation to collectively meet climate targets through energy efficiency improvements, interconnections, and the development of renewable energy and other low-carbon energy technologies (IEA, Canada).

Thus Canada has availed itself as having world-class expertise in CCS, due to its leadership in developing related technologies (Switzer, 2018). A database from IEA has pointed out rules on this subject (IEA, CCS).

Alberta, for instance, has enacted Mines and Minerals Act, R.S.A. 2000, c. M-17 (Alberta M&M Act), that states a general prohibition on carbon injection without authorization (Section 54), and requires lessees to obtain a well license and approval of the Energy Resources Conservation Board prior to drilling or using a well (Section 116) (Costa et al., 2018).

Also, in Alberta we have found a Carbon Sequestration Tenure Regulation (Alberta Tenure Regulation). Under this act, for the injection and storage of CO_2 (in accordance with Section 116 of the Alberta M&M Act) the rules are (1) a carbon sequestration lease entered into by the Minister (as lessor) to grant CO_2 injection and storage rights may permit the drilling of wells, evaluation and testing, and the injection of CO_2 into deep subsurface reservoirs within the location covered by the lease (regulation 9); (2) the term of a carbon sequestration lease is 15 years (regulation 10), extending for a further 15 years in accordance with regulation 11; (3) the area covered by a carbon sequestration lease must not exceed 73,728 hectares (regulation 12); and (4) the rent for a carbon sequestration lease is as prescribed in the Mines and Minerals Administration Regulation (regulation 13) (Costa et al., 2018).

A carbon sequestration lease does not grant the holder the right to win, work, or recover any minerals found within the land covered by the lease. A party wishing to obtain a carbon sequestration lease must: (1) submit an application for a lease in a form satisfactory to the Minister; (2) pay the application fee prescribed in the Mines and Minerals Administration Regulation; (3) pay the rent applicable for the first year of the lease; (4) submit evidence that the area covered by the application is suitable for CO_2 sequestration; and (5) submit a monitoring, measurement, and verification plan for approval; and submit a closure plan.

12.3.4 The European Union

The European Union CCS Directive (2009/31/EC) coupled with additional regulations establish a comprehensive legal regime for CCS within all its member states (Condor et al., 2011). The European Union is concerned with sustainability, as enunciated in article 191 Treaty on the Functioning of the European Union, and Article 3(3) Treaty on the European Union, both determining the EU to act considering the overall goal of sustainable development (Terhechte, 2019). The European Union has been implementing a holistic approach toward CCS, which includes actions within the EU as well as its cooperation with other countries. Recently, the EU has been pursuing cooperation with China, including energy security among the main areas of collaboration (Espa, 2018).

12.4 Relevant findings of the comparative survey

The previous session presented the federal legal framework regarding CCS in Brazil, Canada, the United States, and the supranational experience of the European Union.

The use of CCS is particularly important for developing countries due to CCS being the only proven technology to disassociate carbon dioxide emissions from fossil fuel use at a large scale (Forbes et al., 2011). Based on energy security considerations, CCS enables the exploitation of oil or coal while not increasing GHGs emissions significantly (Sloss, 2019).

The findings described are particular relevant now, in a post Paris Agreement global scenario. This is the case as the Paris Agreement (Articles 4, 9, 10, and 13) establishes a new system of responsibilities for developed and developing countries, albeit not defining which countries belong to the former or the latter (Ari and Sari, 2017). Developed countries should remain leading the way through absolute emission reduction targets, with developing countries also reducing their emissions in accordance with different national circumstances. Hence, economic growth and preserving the environment are not conflicting aims, as the Paris Agreement (Articles 4 and 6) exemplifies.

It is noteworthy that CCS is no panacea. The technology is also viewed as environmentally risky, complex, and may enable the extraction of fossil fuels while also competing with investments in renewable resources (Boyd, 2017).

In such a context, our findings are overall consistent with previous research on the CCS Index, which found that nations who are high consumers and producers of fossil fuels have long recognized CCS to be an effective mechanism in reducing emissions and therefore achieving climate change goals (Consoli et al., 2017).

Our findings support the understanding that innovation in the energy sector relies on government intervention through different policies encompassing the research and development of early technology to its demonstration and commercialization (Havercroft and Consoli, 2018).

Our findings are also consistent with previous studies showing that perceptions of trust and confidence in key institutions to develop and monitor CCS projects are dependent on the track record of the organization and their past experiences with different technologies (Gough et al., 2018).

Additional research on the public perception of CCS in the countries analyzed is needed and may corroborate our analysis (Kamari and Toikka, 2018). On a related note, the United Nations has been fostering environmental protection and the reduction of the impacts of climate change. They have been instrumental regarding global action on such matters, specifically. As mentioned throughout this book, the United Nations recently enacted the Sustainable Development Goals, as part of the urgent call for action to all countries in the 2030 agenda for the Sustainable Development for the planet. That call for immediate action has 17 principles, and there are three goals of particular relevance to CCS and to this research, namely: the affordable and clean energy (goal 7), sustainable cities and communities (goal 11), and climate action (goal 13). These goals demonstrate the necessity of all countries—developed and developing nations alike—to commit to effective and responsible actions to protect the environment and to combat climate change, specifically.

12.5 Conclusion

This article surveys the regulatory framework existing in the countries referenced, comparing and contrasting their federal (or supranational) experiences. The incentives for stakeholders to act fostering CCS implementation were addressed.

This article concludes that the European Union has been the leading actor in taking climate change seriously, and that additional comparative research targeting the jurisdictions studied in this article is needed. Our counterintuitive findings, such as Brazilian efforts being potentially more effective than those made by the United States are of interest today, and would benefit from follow-up studies. Finally, our findings recommend all the countries to take global warming seriously, that is, to actively invest in CCS policies immediately, and to consider comparative regulatory experiences when designing their own legal framework.

Annex 1

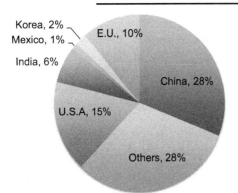

CHART 1 Global carbon dioxide emissions in 2016. *Chart from Arlota, C., 2020. Does the United States' withdrawal from the Paris Agreement on climate change pass the cost-benefit analysis test? University of Pennsylvania Journal of International Law 41, 4. For additional data: see International Energy Agency, IEA Atlas of Energy (2016). http://energyatlas.iea.org/#!/tellmap/1378539487. For per capita emissions, consult Table 1, Annex 1.*

Korea, 2%
Mexico, 1%
India, 6%
E.U., 10%
China, 28%
U.S.A, 15%
Others, 28%

TABLE 1 States and emissions per capita.[a]

Country	GDP per capita (K)	Emissions per capita (rounded metric tons per capita)
Australia	62	17
Austria	51	8
Belgium	48	9
Bulgaria	8	7
Canada	50	14
Chile	15	5
Colombia	8	2
Cyprus	27	7
Czech Republic	20	10
Denmark	61	7
Estonia	20	14
Finland	50	10
France	43	5
Germany	48	9
Hungary	14	5
Iceland	52	6
Indonesia	4	2
Ireland	53	8
Italy	35	7
Japan	36	9

(Continued)

TABLE 1 (Continued)

Country	GDP per capita (K)	Emissions per capita (rounded metric tons per capita)
Latvia	16	4
Liechtenstein	135	1
Lithuania	16	5
Luxembourg	111	21
Malta	23	6
Mexico	10	4
Monaco	163	–
Mongolia	4	7
Netherlands	52	10
New Zealand	42	7
Norway	97	9
Panama	12	3
Peru	7	2
Poland	14	8
Portugal	22	5
Rep. of Korea	28	12
Romania	10	4
Spain	30	6
Sweden	59	6
Switzerland	85	5
United Kingdom	46	7
United States	55	17
Vietnam	2	2

[a]*This table was built by the authors based on data from the Green Climate Fund. See Green Climate Fund, Status of Pledges and Contributions made to the Green Climate Fund (May 8, 2018). https://www.greenclimate.fund/how-we-work/resource-mobilization.*
GDP, Gross domestic product.

Acknowledgments

Hirdan Katarina de Medeiros Costa is grateful to the "Research Center for Gas Innovation—RCGI" (Fapesp Proc. 2014/50279−4), supported by FAPESP and Shell, organized by the University of São Paulo, and the strategic importance of the support granted by the ANP (National Agency of Petroleum, Natural Gas and Biofuels of Brazil) through the R&D clause. She also thanks to the support from the National Agency for Petroleum, Natural Gas and Biofuels Human Resources Program (PRH-ANP), funded by resources from the investment of oil companies qualified in the R,D&I clauses from ANP Resolution No. 50/2015 (PRH 33.1−Related to Call No. 1/2018/PRH-ANP; Grant FINEP/FUSP/USP Ref. 0443/19). Carolina Arlota acknowledges the excellent research assistance of Natosha D. Greene.

References

Ari, I., Sari, R., 2017. Differentiation of developed and developing countries for the Paris Agreement. Energy Strategy Reviews 18, 175.

Arlota, C., 2020. Does the United States' withdrawal from the Paris Agreement on climate change pass the cost-benefit analysis test? University of Pennsylvania Journal of International Law 41, 4.

Boyd, A., 2017. Communicating About Carbon Capture and Storage. The Oxford Research Encyclopedia of Climate Science.

Brazilian Environmental Minister, 2020. MMA. NDC. Retrieved from: https://www.mma.gov.br/component/k2/item/15137-discuss%C3%B5es-para-implementa%C3%A7%C3%A3o-da-ndc-do-brasil.html.

Brazilian Environmental Minister, 2020. Paris Agreement. https://www.mma.gov.br/clima/convencao-das-nacoes-unidas/acordo-de-paris. (accessed 03.02.20).

Budinis, et al., 2018. An assessment of CCS costs, barriers and potential. Energy Strategy Review 22, 61.

Center for International Environmental Law (CIEL), 2019. Trillion Dollar Transformation: Fiduciary Duty, Divestment, and Fossil Fuels in an Era of Climate Risk: CCS Institute. IPCC Report on Global Warming, 7.

Climate Change Recognized as 'Threat Multiplier', UN Security Council Debates its Impact on Peace, U N News, Jan. 25, 2019, https://news.un.org/en/story/2019/01/1031322.

Coglianese, C., Starobin, S., 2019. The legal risks of regulating climate change at the subnational level. The Regulatory Review (Sep. 18, 2019), https://www.theregreview.org/2017/09/18/coglianese-starobin-legal-risks-climate-change-subnational/.

Condor, J., et al., 2011. CA comparative analysis of regulations for the geologic storage of carbon dioxide. Energy Procedia 4, 5895.

Consoli, C., Havercroft, I., Irlam, L., 2017. Carbon capture and storage readiness index: comparative review of global progress towards wide-scale deployment. Energy Procedia 114, 7348.

Costa, H.K.M., et al., 2018. Environmental license for carbon capture and storage (CCS) projects in Brazil. Journal of Public Administration and Governance 8, 163–185.

Costa, H.K.M., et al., 2019. Legal aspects of offshore CCS: case study–salt cavern. Polytechnica 1.

Costa, H.K.M., Musarra, R.M.L.M., Miranda, M.F., 2018. The main environmental permitting requirements on CCS activities in Brazil. Sustainability and Development Conference. Michigan University, p. 1.

Dixon, T., McCoy, S., Havercroft, I., 2015. Legal and regulatory developments on CCS. International Journal of Greenhouse Gas Control 40, 431.

Environmental Protection Agency—EPA, 2018. Green House Gas Reporting Program (GHGRP): Capture, Supply and Underground Injections of Carbon Dioxide. Retrieved from: https://www.epa.gov/ghgreporting/capture-supply-and-underground-injection-carbon-dioxide.

Environmental Protection Agency—EPA, 2018. Underground Injection Control (UIC): Final Class VI Guidance Documents. Received from: https://www.epa.gov/uic/final-class-vi-guidance-documents.

Espa, I., 2018. Climate, energy and trade in EU–China relations: synergy or conflict? China–EU Law Journal 6, 57.

Executive Order on Promoting Energy Independence and Economic Growth (Exec. Order No. 13,783, §1, C, 82 Fed. Reg, 2017).

Forbes, S., et al., 2011. CCS Demonstration in Developing Countries: Priorities for a Financing Mechanism for Carbon Dioxide Capture and Storage. World Resources Institute.

Folger, P., 2017. Carbon Capture and Sequestration (CCS) in the United States, vol. 1. Congressional Research Service.

Gerrard, M., Hester, T., 2018. Climate Engineering and the Law, vol. IX. Cambridge University Press.

Global CCS Institute, 2017. The Global Statistics of CCS. Retrieved from: https://www.globalccsinstitute.com/wp-content/uploads/2018/12/2017-Global-Status-Report.pdf.

Gough, C., Cunningham, R., Mander, S., 2018. Understanding key elements in establishing a social license for CCS: an empirical approach. International Journal of Greenhouse Gas Control 68, 16.

Havercroft, I., 2019. Lessons and Perceptions: Adopting a Commercial Approach to CCS Liability, Global CCS Institute, vol. 6.

Havercroft, I., Consoli, C., 2018. Is the World Ready for Carbon Capture and Storage? Global CCS Institute.

Heffron, et al., 2018. A treatise for energy law. Journal of World Energy Law and Business 11, 34.

Intergovernmental Panel on Climate Change, 2005. IPCC Special Report on Carbon Dioxide Capture and Storage.

Intergovernmental Panel on Climate Change, 2018. IPCC Report on Global Warming of 1.5°C: Summary for Policy Makers. Retrieved from: https://www.ipcc.ch/site/assets/uploads/sites/2/2019/05/SR15_SPM_version_report_LR. pdf.

Jacobs, W., Craig, M., 2018. Carbon capture and sequestration. In: Gerrard, M., Dernbach, J. (Eds.), Legal Pathways to Deep Decarbonization in the United States: Summary and Key Recommendations, vol. 71. Environmental Law Institute.

Kahn, M.E., et al., 2019. Long term macroeconomics effects of climate change: a cross-county analysis. National Bureau of Economic Research 44.

Karimi, F., Toikka, A., 2018. General public reactions to carbon capture and storage: does culture matter? International Journal of Greenhouse Gas Control 70, 193.

Kury, F., 2019. Unlocking the Opportunities in the Brazilian Oil Industry. Retrieved from: http://www.anp.gov. br/arquivos/palestras/unlocking-opportunities-brazilian-oil-industry.pdf.

Millner, A., et al., 2010. Ambiguity and Climate Policy. National Bureau of Economic Research, NBER Working Paper Series 3.

Musarra, R.M.L.M., Costa, H.K.M., 2018. Elements of public action and governance in capture, stocking and carbon transportation activities. International Journal of Humanities and Social Science Invention (IJHSSI) 7, 46−53.

Nordhaus, W., 2018. Projections and uncertainties about climate change in an era of minimal climate policies. American Economic Journal: Economic Policy 10, 333.

Posner, E., Sykes, A.O., 2013. Economic Foundations of International Law, Harvard University Press.

Rebuilding America's Infrastructure: The 2019 Budget Fact Sheet, 2019. The White House, Received from: https:// www.whitehouse.gov/wp-content/uploads/2018/02/FY19-Budget-Fact-Sheet_Infrastructure-Initiative.pdf.

Romeiro-Conturbia, V.R.D.S., 2014. Carbon Capture and Storage Legal and Regulatory Framework in Developing Countries: Proposals for Brazil. Institute of Energy and Environment, University of São Paulo.

Sloss, L., 2019. Current estimations of the international energy agency suggest that coal-fired power plants could achieve an almost zero carbon dioxide emission index. Technology Readiness of Advanced Coal-Based Power Generation Systems. Retrieved from: https://www.iea-coal.org/technology-readiness-of-advanced-coal-based-power-generation-systems-ccc-292/.

Sovacool, B.K., Brown, M.A., Valentine, S.V., 2016. Fact and Fiction in Global Energy Policy. Johns Hopkins University Press.

Switzer, J., 2018. CO_2 Capture, Utilization and Storage: A Canadian Snapshot. Pembina Institute.

Terhecte, J.P., 2019. Sustainability, thus, entails a positive objective of European integration towards a European Natural Resources Law? European Yearbook of International Economic Law 203.

Tol, R.S.J., 2014. Quantifying the consensus on anthropogenic global warming in the literature: a re-analysis. Energy Policy 73, 701.

Turn Down the Heat: Why a 4°C Warmer World Must Be Avoided 18, 2012. The World Bank. Retrieved from: http://documents.worldbank.org/curated/en/865571468149107611/pdf/NonAsciiFileName0.pdf.

United Nations Environment Programme (UNEP): Annual Report., 2013. Available at: https://wedocs.unep.org/ bitstream/handle/20.500.11822/8607/-UNEP%202013%20Annual%20Report-2014UNEP%20AR%202013-LR.pdf? sequence = 8&isAllowed = y.

Wang, M., Eni Oko, E., 2017. Special issue on carbon capture in the context of carbon capture, utilization and storage (CCUS). International Journal of Coal Science and Technology 4, 1.

Wiener, J.B., 2007. Think globally, act globally: the limits if local climate policies. University of Pennsylvania Law Review 155, 1961.

Further reading

Bradbrook, A.J., 1996. Energy law as an academic discipline. Journal of Energy and Natural Resources Law 14 (2), 193−217.

International Energy Agency. Carbon Capture and Storage. Retrieved from: https://www.iea.org/topics/carbon-capture-and-storage.

International Energy Agency. Canada. Retrieved from: https://www.iea.org/countries/canada.

Musarra, R.M.L.M., Costa, H.K.M., 2019. Liability in civil and environmental subjects for carbon capture and storage (CCS) activities in Brazil. International Journal for Innovation Education and Research 7.

Sustainable Development Goals, 2015. The United Nations. Retrieved from: https://www.un.org/sustainabledevelopment/sustainable-development-goals/.

Reducing CO_2 emissions through carbon capture use and storage and carbon capture and storage in Mexico and Alberta, Canada: addressing the legal and regulatory barriers

Allan Ingelson[1] and Teresa Castillo Quevedo[2]

[1]Canadian Institute of Resources Law, University of Calgary, Faculty of Law at the University of Calgary, Calgary, Canada [2]COFECE, Offentlicher, Dienst, Benito Juarez, Ciudad de Mexico

13.1 Introduction

Fossil fuel combustion is a major source of CO_2 emissions in both Mexico and Canada. For instance, in 2010 the combustion of fossil fuels accounted for approximately 80% of Mexico's CO_2 emissions (Fifth National Communications, 2012, p. 179). As a member of the G20 and world's 15th largest economy, Mexico depends heavily on its oil and gas resources for energy security; 30% of federal government budget revenues are secured from Pemex (Wood and Martin, 2018). Numerous scientists agree that a significant reduction in CO_2 emissions is an important element in mitigating the effects of global warming and climate change. In 2015 the governments of Mexico and Canada committed to significantly reduce their national greenhouse gas (GHG) emissions (Paris Agreement, 2015). In 2016 Mexico ratified the Paris Agreement and committed to reduce its national GHG emissions by 22% by 2030 and by 50% from 2000 levels by 2050 (Altamirano et al., 2016). The Government of Mexico's GHG emissions reduction goals are furthered under its Transition Strategy to promote the use of Cleaner Technologies and Fuels, which is one of the guiding instruments of the Mexican national energy policy (Mexico Secretariat of Energy [SENER], 2018a).

There are two types of CO_2 emissions reduction projects that have been deployed and are operating in Canada, carbon capture, use and storage (CCUS) and carbon capture and storage (CCS) (Navarro et al., 2016). These types of projects are to be constructed and tested in Mexico, by the Mexican Government. Currently, there are two pilot projects under construction in the state of Veracruz, Mexico: "Campo Brillante" in Coatzacoalcos for enhanced oil recovery (EOR), and "Poza Rica" for CCS (Navarro et al., 2016). Pemex is interested in developing EOR operations due to the potential to capture 50 million of tons of CO_2 annually (Navarro et al., 2016).

CCUS includes EOR in which CO_2 is injected into pore spaces in depleted oil and natural gas reservoirs, that were previously filled with oil and gas that has been pumped out of the ground (Metz et al., 2005). In light of the additional revenue from EOR, these projects, which have been developed and operated in Canada for decades, are more economically attractive to oil and gas companies than CCS projects that have the sole objective of long-term CO_2 storage to mitigate the effects of climate change. Due to the proximity of CO_2 emitting industrial facilities to Pemex's oil fields, EOR projects are one element of the Mexican government's plans to reduce the level of national GHG emissions. In Mexico there is significant potential for CCUS projects at industrial installations that include both upstream and downstream processes, like coal-fired power and natural gas processing plants, such as those situated near the Gulf of Mexico (Lacy et al., 2013).

CCS refers to a variety of technologies used to permanently capture, inject, and store CO_2 emissions from industrial processes and fossil fuel electrical generation facilities into subsurface geological reservoirs for centuries, to prevent additional emissions entering the atmosphere (Ingelson et al., 2010; Woodford, 2017). Globally, there are 37 major CCS projects, operating or in development, most of which have an approximate CO_2 storage capacity of 37 million tons annually (Global CCS Institute, 2017a).

In the last 10 years the Government of Mexico has adopted policies and strategies to promote CCS and CCUS to reduce its CO_2 emissions (SENER, 2018a). In 2008 Mexico joined the Global CCS Institute and the Carbon Sequestration Leadership Forum (SENER, 2018a). The government in 2010 adopted a National Energy Strategy, in which sustainable development is a critical element (SENER, 2018a). The Secretariat of Energy in 2011 carried out a study called "State that holds the technologies of CCS in Mexico," which stated that in light of the economic and investment considerations, Mexico should incorporate both CCS and CCUS projects in its plans to reduce GHG emissions (SENER, 2018a). The "Mexican Atlas of CO_2 Geological Storage" was published in 2012, as part of a trilateral project in the North American Carbon Sequestration Atlas. The General Law of Climate Change came into force in 2012, a significant milestone in Mexican environmental policy.

In 2013 a carbon tax on fossil fuels was approved and the National Climate Change Strategy (Vision 10−20−40) published. To facilitate CCUS test projects and coordinate research activities on CO_2 capture and regulation, in 2014 the Minister of Energy (SENER) established a working group comprising representatives from the Minister of the Secretariat of Environment and Natural Resources (SENVNR), Pemex, the Federal Commission of Electricity, the Autonomous National University of Mexico, the National Technologic Polytechnic, and the Mario Molina Center (Mexico. Secretariat of Environment and Natural Resources [SEMENR], 2014). A Technological Roadmap for CCUS projects in Mexico was created in 2014 and is currently being used as a planning mechanism to assist in the development of CCUS technology (Mexico. Secretariat of Environment and Natural Resources [SEMENR], 2014). A National Emissions Registry has been established, under which all companies that generate GHG emissions higher than 25,000 tCO_2e must prepare and submit an annual report on those emissions. The Special Climate Change Program 2014−18 was launched, and for the first time in Mexican history the Electricity Industry Law classified "geological capture and storage or carbon dioxide bio sequestration" as Clean Energy, and CCS to be a low-carbon source.

In 2015 three studies were undertaken in collaboration with the World Bank, aiming to facilitate deployment of CCUS in EOR projects and long-term CO_2 storage (Navarro et al., 2016). In addition, the Energy Transition Law came into force. In 2016 a National Transition Strategy to promote the use of cleaner technologies and fuels was released (Mexico, 2016). and the following year an inventory of fixed emissions sources and CO_2 storage sites in Mexico and the Mexican CCUS Center was created (SENER, 2018b). At the end of 2017 SENER created a Mexican CCUS Center to further support CCUS development by promoting collaboration among academics in the research community with industry representatives that support the development and completion of CO_2 capture pilot plants (Global CCS Institute, 2017b). In 2018 Wood and Martin (2018) noted that privatization of the Mexican energy sector and "the Mexican government's desire to build an energy sector that meets the need to shift to a low-carbon growth model" (p. 35). In 2018 the CCUS National Strategy and CCS-READY Strategy were under development, which are Mexican public policy documents intended to establish a strategy to implement CCUS and CCS technologies through the integration of criteria to promote, plan, and execute projects at different scales in the country (SENER, 2018a).

13.2 Alberta's regulatory experience with enhanced oil recovery and acid gas disposal projects

Since the 1970s CO_2 has been successfully injected and sequestered for the purpose of EOR in Alberta to increase the amount of petroleum that can be recovered from subsurface oil and gas reservoirs. Injecting CO_2 into some oil and gas fields can increase the reservoir pressure to facilitate additional production of hydrocarbons from the fields (Metz et al., 2005). In light of the economic value of the hydrocarbons produced from EOR projects, there have been a much higher number of EOR projects in Alberta (at least 48 projects) compared to only two CCS projects.

With regard to regulating these projects, under section 39 of the *Oil and Gas Conservation Act* (*OGCA*) no company may commence an EOR scheme or the injection, disposal, or storage of any fluid into a subsurface geological formation through a well without the approval of the Alberta Energy Regulator (AER) (OGCA, 2000). As CCS projects employ similar gas injection technology and processes, the AER with experience in regulating EOR well injections and monitoring, evaluates and approves CCS projects that satisfy the regulator's criteria (OGCA, 2000). An application is referred to the Minister of the Environment and Parks, the primary government department responsible for environmental protection, and AER approval is subject to the conditions imposed by the relevant Minister (OGCA, 2000). The OGCA and the associated regulations and AER directives classify different types of wells for injection of CO$_2$ or other gases used in EOR or disposal or storage. These wells are subject to increased levels of monitoring and higher standards for well cementing, casing, completions, and testing. EOR project applicants must notify companies which hold adjacent subsurface interests that may be affected by their activities. The plugging and abandonment of oil and gas wells, including EOR wells and those for acid gas disposal operations, to protect the environment and public health and safety is regulated by the AER to ensure that downhole and subsurface infrastructure is left in a stable and safe condition, including the isolation and protection of aquifers and groundwater (AER, 2018). Well completion reports and plug logs must be submitted to the AER. Much of this regulatory scheme can be applied directly to the abandonment phase of a CCUS project, with one exception; it does not provide for ongoing monitoring and verification of the well after abandonment (AER, 2018). This monitoring and verification, over a long period, is required for CCS projects.

Regulation of EOR projects is aimed at optimizing worker and public safety, environmental protection, and equitable treatment of adjacent well licensees (AER, 2020, Section 2.1.2). EOR projects have similar geographical scale to likely CCS projects, which makes them a useful analogy in terms of regulation, but on the other hand their objectives are very different; the injection of CO$_2$ is merely a means to the end of increased hydrocarbon recovery.

Acid gas disposal projects are another type of gas injection project that, similar to CCS, are designed to permanently inject and dispose of another gas via wells. However, CCS project are designed to permanently store a much larger volume of CO$_2$ than most of the acid gas disposal projects approved to date in Alberta. The AER considers several analogous issues with CCS to be equitable treatment of existing hydrocarbon rights holders and safety with regard to potential above-ground gas releases. When considering gas containment, reservoir characteristics, and hydraulic isolation, the AER requires a significant amount of information to demonstrate that the injected fluids will be contained "within a defined area and geologic horizon," to protect the environment and public health and safety (AER, 2020, Section 4.2.2). The above environmental protection and health and safety issues are analogous considerations in AER review and of proposed CO$_2$ well injection and storage CCS projects.

13.3 Injecting CO$_2$ for long-term storage

As in Mexico, CO$_2$ storage sites in Alberta include empty pore spaces in depleted oil and natural gas reservoirs from which the hydrocarbons have been extracted, active oil and natural gas fields, and deep saline geological formations (Orr, 2009). Initially oil, natural gas, helium, and CO$_2$ have been stored naturally in subsurface geological reservoirs around the world for hundreds of millions of years. In Canada, where natural gas is abundant, after the gas is withdrawn and in some cases treated to remove impurities, the energy source has been reinjected (pumped) back into empty pore spaces in subsurface geological reservoirs for storage and future withdrawal at peak demand times such as winter when more gas is required for heating homes and buildings. Decades of Canadian petroleum reservoir engineering experience is now being applied to the injection and storage of another gas, CO_2, to mitigate the impacts of climate change. Scientists and engineers have estimated that the fraction of CO_2 retained in a well-scrutinized and well-managed CCS storage reservoir is very likely to exceed 99% over 100 years and is likely to exceed 99% over 1000 years (Metz et al., 2005).[1] However, it is important to identify storage reservoirs with adequate seals that will prevent CO_2. In Alberta, decades of gas injection and storage regulatory experience with EOR injection projects was applied to the review and approval of the first dedicated long-term CCS project approved in 2012, the Quest Carbon Capture and Storage Project (Alberta. Energy Resources Conservation Board [ERCB], 2012). The Quest project is designed to capture and store more than 1 million tons of CO_2 annually in deep subsurface saline reservoirs (ERCB, 2012).

There are four stages in the CCS process: capture, transport, injection, and storage or disposal. In the first stage, CO_2 is captured and separated from the industrial production source, usually a facility where there is significant fossil fuel combustion of CO_2. A variety of technologies may be employed to separate CO_2 from other gases. In Alberta to transport CO_2 for long-term injection and storage, the proposal to develop a CO_2 pipeline must include a detailed plan. Efficient and reliable CO_2 pipelines have been approved by the provincial regulator (Aydin et al., 2010). For instance a 240-km pipeline called the Alberta Carbon Trunk Line was proposed in 2009 to transport CO_2 throughout the region (Government of Alberta, 2009). The Canadian experience has revealed there are three fundamental legal issues that must be carefully considered in the development of CCS projects:

1. the subsurface and surface property rights required for the projects;
2. the required changes in existing oil and gas and environmental legislation and/or new regulations; and
3. the importance of a meaningful consultation process with area residents and other stakeholders.

13.4 Clear and certain property rights to enable carbon capture and storage projects

Under the Alberta *Mines and Minerals Act, 2000* the Department of Energy grants subsurface petroleum rights, such as petroleum and natural gas licenses and leases, and the rights for CCS projects. Operators that inject gases into subsurface geological reservoirs may impact other types of operations. In the context of an extended history of Alberta hydrocarbon development and proposed CCS projects, to facilitate CCS development in light of prior subsurface oil and gas property interests, in 2008 Bankes, Poschwatta, and Shier noted the importance of a legal system that clearly allocates legal rights and regulations covering the injection of different types of gases into different geological formations to prevent conflicts among project operators to minimize disputes and lawsuits that could delay CCS project development. The legal system needs to balance competing commercial interests to provide fairness and certainty to encourage investment in new CCS projects

---

[1] "Very likely" is a probability between 90% and 99%, "likely" is one between 66% and 99%.

(Bankes et al., 2008). In light of the Alberta CCS development experience there is a strong argument that Mexico should develop similar legal requirements to provide security of tenure for project investors and developers. Alberta has developed such a legal system. The Alberta government added a ninth part to the *Mines and Minerals Act* (MMA) to clarify the rights of CCS project developers. Section 116(1) provides:

> "the Minister may enter into an agreement with a person that grants that person the right to inject captured CO$_2$ into a subsurface reservoir for sequestration," on the condition that "the lessee of an agreement referred to in subsection (1) shall obtain a well licence and the approval of the Regulator under the *OGCA* prior to drilling or using a well for the purposes of this section." (MMA, section 116(2))

In addition, under the MMA the Alberta government created a new regulation called the Carbon Sequestration Tenure Regulation (2011). that contains clearly definitions as follows, that provided increased certainty to CCS project developers:
Definitions:

> 1. In this Regulation,
> (b) "carbon sequestration lease" means an agreement under section 116 of the Mines and Minerals Act issued in the form of a lease specified under section 9;
> (c) "deep subsurface reservoir," in respect of a permit or lease, means the pore space within an underground formation that is deeper than 1000 m below the surface of the land within the location of that permit or lease.

A second important consideration is the establishment of a clear and certain legal framework for surface rights, including access to the land for CCS project infrastructure, and CO$_2$ pipelines that may be required to transport CO$_2$ from the capture site to the well injection site. Just as the Surface Rights Act (2000) has for more than six decades provided oil and gas developers with the right to enter and use the surface of the land for energy development which is considered to be in the public interest, a similar law should be considered in Mexico. Two key elements of the surface rights legislation are the requirement for mineral developers to make a genuine effort to negotiate reasonable compensation with surface rights owners for surface access and land disturbances or nuisance (such as noise) that may arise. In Alberta, if the surface rights owner and the project developer are unable to agree on the amount of compensation and sign a written surface lease agreement, a government tribunal called the Surface Rights Board will hold a hearing at the request of either party, where either party can make submissions and the tribunal will decide the amount of compensation. The legal certainty and fairness provided under this surface rights compensation system has encouraged oil and gas development in Alberta and will facilitate the development of new CCS projects as well.

13.5 Environmental protection considerations

In Alberta regulation is a completely different process than the allocation of the tenure rights by the Department of Energy discussed in the previous section, as it involves granting of approvals to drill wells and to inject substances. As such, proposed well drilling and operations are reviewed and considered by the independent AER, including the potential environment and public safety impacts of the proposed project (OGCA, 2000). The provincial oil and gas regulator has adapted its comprehensive and sophisticated oil and gas well regulatory framework for CCS projects. The regulation of petroleum operations in Alberta is under the jurisdiction of the AER, under the Responsible Energy Development Act (2012) and the Alberta *EPEA*. Under EPEA, no person may discharge any contaminant from any industrial or trade premises onto or into land, or into water, unless the discharge is expressly allowed by a rule in a regional plan, by a resource consent, or by regulations.

Project review and approval regulation by the AER is separate from ownership and the granting of dispositions by the Alberta Department of Energy. The main instrument employed by the AER to regulate drilling and the operation and CCS injection wells, is the approval process for well licenses under the *OGCA* to protect the environment and the health and safety of industry workers, residents, and the general public. Even if it holds a petroleum and natural gas license, a company cannot drill an oil or gas well without a license from the AER pursuant to the *OGCA*. The above discussed environmental and health and safety considerations also apply to CCS operations that may take place near oil and gas operations, especially if the target storage formation is a depleted oil or gas reservoir. One main issue is that the CO_2 that is injected in CCUS operations can migrate into adjoining formations, damaging petroleum resources, and making petroleum production more difficult. A second important issue is that CO_2 storage formations can be compromised by surrounding oil and gas wells. In cases when plugging and abandonment of an injection well has been inadequate, a CO_2 leak is possible. CO_2 is corrosive, and when combined with water forms acid that can attack steel and cement. In light of the potential impacts on oil and gas production from injecting CO_2 as part of CCS projects, oil and gas developers may be concerned about new restrictions on their operations and that is why regulations are important to respect the rights of oil and gas operators.

A significant addition was made to the *Alberta Mines and Minerals Act* (Part 9) that contains new requirements specifically tailored to CCS projects that must be satisfied to drill CO_2 test wells into a potential subsurface reservoir and for long-term storage to reduce GHG emissions in response to climate change. An agreement with the provincial government is required to inject CO_2 and the following new terms were added to provide clear direction to CCS project developers:
Definitions:

(d) "Directive 65" means AER Directive 65, "Resources Applications for Conventional Oil and Gas Reservoirs," published by the Regulator;
(e) "evaluation permit" means an agreement under section 115 of the Act issued in the form of an evaluation permit under section 3;
(f) "lessee" means the holder of a carbon sequestration lease;
(g) "minerals" means minerals as defined in the Act;
(h) "permittee" means the holder of an evaluation permit;
(i) "pore space" means the pores contained in, occupied by or formerly occupied by minerals or water below the surface of land;
(j) "Regulator" means the AER.

In section 114 of the MMA,
(b) "facility" is defined as any building, structure, installation, equipment or appurtenance over which the Regulator has jurisdiction and that is connected to or associated with the injection or sequestration of captured carbon dioxide pursuant to an agreement under this Part (Mines and Minerals Act, 2000, section 114(b)).

The Alberta government has adopted the following process, information, reporting, and requirements to drill CCS injection test wells for the evaluation of potential CO_2 storage reservoirs (Mines and Minerals Act, 2000, section 115):

> Rights to drill CO_2 storage evaluation wells
> Section 115(1)... Minister may enter into an agreement with a company that grants that company the right to evaluate the geological or geophysical properties of a subsurface reservoir in a location to determine its suitability for use for the sequestration of captured carbon dioxide.

The lessee of an agreement referred to in subsection (1) shall obtain a well license and approval of the Regulator under the *OGCA* prior to drilling or using a well for the purposes of this section.

> (2) A lessee of an agreement under this section shall in accordance with the regulations
> (a) submit a monitoring, measurement and verification plan for approval;
> (b) comply with the monitoring, measurement and verification plan that has been approved;
> (c) provide reports with respect to the lessee's compliance with the monitoring, measurement and verification plan;
> (d) fulfil the work requirements with respect to the location of the agreement.

As a condition of evaluating whether a subsurface reservoir would be suitable to store CO_2 for the long-term by drilling a CO_2 test well, as with oil and gas wells drilled on provincial lands in the province, the Alberta government requires the lessee who has entered into an agreement to "obtain a well license and approval of the Regulator under the OGCA prior to drilling or using a well" for testing the suitability of a reservoir to hold CO_2 as part of a CCS project (Mines and Minerals Act, 2000, section 116(2)). In addition to review the proposed site testing program, the AER requires the following information requirements for project review and consideration:

A lessee of an agreement under this section shall be in accordance with the regulations:

1. submit a monitoring, measurement and verification plan for approval;
2. comply with a monitoring, measurement and verification plan that has been approved;
3. provide reports with respect to the lessee's compliance with the monitoring, measurement and verification plan;
4. fulfill the work requirements with respect to the location of the agreement;
5. submit a closure plan for approval;
6. comply with a closure plan that has been approved...(Mines and Minerals Act, 2000, section 116(3)).

13.6 Liability

Potential liability for environmental and property damage from CO_2 leaks and human injuries was a major issue that had to be addressed in Alberta's legal framework before CCS site developers were prepared to invest in the development of CCS projects. Different types of liability that might arise include civil liability (e.g., under the torts of negligence, nuisance, trespass and occupiers' liability); criminal liability for acts or omissions causing serious harm to workers, residents, and members of the public; environmental liability for failing to comply with the Alberta *EPEA* and the associated regulations including the unauthorized release of substances; and gas emissions exceeding those allowed under the regulations. Consistent with its GHG emissions reduction policies and laws and to encourage private investment to fund the desired CCS projects the Alberta government decided to assume liability for damage arising from long-term CO_2 storage projects by allowing the facility operator to transfer liability to the government subject to certain conditions (Mines and Minerals Act, 2000, section 121). The Alberta government recognized that corporations simply do not continue in perpetuity and de facto the government would inherit the risk of damage from CO_2 leakage over an extended period of time. However, an important condition of the government assuming long-term liability is that the operator remains liable for negligence. CO_2 storage developers will only be liable if they fail to meet a duty of care.

The Alberta legal approach to long-term liability parallels the following approach adopted in the European Union (EU) as outlined in EU Directive 2009 that also provides for a transfer of responsibility to a competent authority for monitoring, corrective measures, and liability, but only under the following stringent conditions:

1. after a minimum of 20 years has passed since the closure of the long-term CO_2 injection and storage facility;
2. if all of the available evidence indicates that the stored CO_2 will be completely and permanently contained; and
3. after providing a financial contribution for 30 years of monitoring (European Parliament, 2009, Articles 18–20).

In both the EU and Alberta, there are strict conditions that must be satisfied to complete a transfer, including no negligence on the part of the facility developer and operator. The strict transfer conditions reflected in the Alberta and EU approaches to manage long-term liability are consistent with the "Polluter pays" principle that is embedded into the Alberta Environmental Protection and Enhancement Act, which provides that the CCS developer and operator must pay for the total costs of the industrial project.

13.7 Meaningful consultation—an essential element of an effective and efficient regulatory framework

In July 2012, the provincial oil and gas regulator considered and approved the first application for a CCS project in Alberta (2012 AERCB 008). The CCS project proponent Shell Canada Limited stated at the hearing that "one of the principles of its sustainable development policy is its commitment to engage with stakeholders in a meaningful way, to identify issues, and to search for mutually agreed-upon solutions" (2012 AERCB 008, paragraph 1). At the hearing, Shell reported that it had carried out an extensive and comprehensive stakeholder consultation program for the CCS Project (Shell International, 2015). The company indicated that it provided information about the proposed project to landowners, occupants, and residents within 5 km of the CO_2 capture site and within the CO_2 pipeline emergency planning zone, and to indigenous peoples and their organizations in the project area (ERCB, 2012). In addition the project proponent also notified leaseholders, petroleum and natural gas rights holders, government agencies, regional and municipal governments, special interest groups, nongovernmental organizations, industry participants, and industry associations about the proposed CCS project (Shell

International, 2015). As part of meaningful consultation, the CCS project proponent distributed project information packages, organized and offered community information and project feedback sessions, prepared and distributed a CCS project site-specific newsletter, and create a dedicated toll-free telephone number to answer questions from residents and the general public about the proposed project. The company also carried out several information workshops in communities located in the project area to provide information, answer questions, and discuss potential concerns with residents and members of the general public (Shell International, 2015). The proponent met face-to-face with landowners along the pipeline right-of-way and within the emergency planning zone to provide project information, discuss questions and concerns, and get access to properties for seismic surveys, water well testing, and other environmental data collection. Shell stated that these meetings resulted in about 30 pipeline reroutes and confirmation of nonobjection from 109 of 111 landowners (ERCB, 2012, paragraph 84). The proponent also established a website where stakeholders could find information, ask questions, and express concerns. The proponent had representatives at various community events to provide local residents with the opportunity to ask questions and get information about the project (ERCB, 2012, paragraph 368).

The proponent reported that it learned several things from its CCS project stakeholder consultation program. First, the importance of early consultation. The proponent reported that it had initiated community consultations long before reaching a final decision on whether or not to proceed with the project and long before design details were established (ERCB, 2012, paragraph 370). Second, the importance of meaningful consultation. The proponent submitted that its engagement showed a meaningful concern for how people felt about the project, and it identified problems and potential solutions. Based on feedback from community people, project details were modified, as illustrated by the 30 reroutes of the project pipeline. The proponent indicated that it was committed to maintaining open lines of communication with municipal authorities and government agencies (ERCB, 2012, paragraph 374).

13.8 Project approval

The regulator noted that in assessing the CCS project, it had to consider whether the project application was in the public interest generally and analyze the social, economic, and environmental impacts of the project (ERCB, 2012, paragraph 390). The regulator also noted that it had the power, where necessary, to apply conditions to mitigate site-specific or local impacts. At the hearing the regulator concluded that the CCS project proponent's public consultation program initiatives were effective. The regulator noted the number of project modifications to the CCS project made by the company to address public concerns, and that the proponent was able to secure 109 nonobjections to the project with only two landowners continuing to object. The small number of parties appearing at the hearing in opposition to the project supports the regulator's assessment that participant involvement programs had been effective. The Board noted the apparent strong foundation of communication and engagement with area landowners and communities. The regulator concluded that the communication and public consultation program initiated exceeded the minimum oil and gas industry Participant Involvement Program requirements in AER *Directive 056*. The regulator commended the proponent on its meaningful communication and consultation to date. The Board strongly supported the proponent's plan to consider forming community advisory panels to help with the communication of complex monitoring data and developments.

The Board noted that in determining whether or not there was a need for the project and its assessment of whether the project is in the public interest, the regulator must weigh the benefits against the risk factors that are present, given the nature of the development, the proposed location, and other factors associated with the specific situation. In addition, the regulator had to ensure that any site-specific or local impacts are mitigated to an appropriate and acceptable level. Should the regulator determine that the risks cannot be sufficiently mitigated and that the risk exceeds the potential benefit, the project would not be in the public interest and therefore the project would not be approved (ERCB, 2012, paragraph 393).

The regulator noted that in evaluating whether or not to approve the project it considered the GHG emission reduction policies of the governments of Canada and Alberta. The regulator notes that both the national and regional governments had adopted strategies and targets to reduce GHG emissions as part of their respective climate change strategies. Developing effective CCS technologies is an important part of both the Canadian and Alberta government GHG reduction strategies. To this end, the Alberta government enacted legislation to provide for the funding of CCS projects, to establish a framework to address long-term liability for stored or sequestered CO_2, and to address access and ownership of pore space for CO_2 storage. The government of Canada also supports the reduction of GHG emissions through its Clean Energy Fund and is aiming to reduce CO_2 emissions by 325 Mt/a by 2050 through a number of approaches, including CCS (ERCB, 2012, paragraph 60). The Canadian federal government has committed to supporting research into clean energy technologies and will also invest public funds into the CCS project.

The regulator concluded that direct capture and storage of CO_2 emissions has the potential to be one of the leading available processes capable of significantly reducing the amount of GHGs released to the atmosphere in Alberta and elsewhere. The regulator also concluded that the project is at a suitable surface location to minimize impacts to existing land-use activities in the area, and that the project design and monitoring program were adequate to mitigate potential impacts of construction and operation of the project. The regulator also concluded that impacts to individual residents or landowners in the area will be minimal. The regulator, after reviewing the evidence presented at the hearing, concluded that the project was in the public interest as there was a need for the project and after considering the environmental, social, and economic impacts of the project as discussed in this decision (ERCB, 2012, paragraph 409). The regulator concluded that the potential for a CO_2 release was "very low" and that the overall public safety, environmental, and social risks and impacts associated with the capture of CO_2, and its subsequent transmission and injection were also very low.

After approving the application for the project in light of the novelty of the CCS project and the uncertainty as to potential impacts from the CCS operations, the regulator imposed the following conditions:

1. before construction of the pipeline, submit additional detailed information on the technical, operational, cost, and public safety considerations regarding the CO_2 stream;

2. submit a complete prebaseline plan by the specified date;
3. submit an annual report of operational performance, that describes how the operational performance of the scheme conforms with the modeling and predictions, and discuss whether there is a need for changes in the program;
4. submit plan updates as required by the regulator; at a minimum, the proponent must submit updates at critical milestones such as commencement of injection, closure, and postclosure;
5. evaluate the need for additional deep monitoring wells, and provide analysis by the specified date;
6. immediately report any anomalies that indicate fracturing out-of-zone;
7. complete and submit the final results of the geomechanical testing;
8. immediately report evidence of loss of CO$_2$ containment;
9. submit a more comprehensive project model using site-specific parameters to reevaluate the issue of deformations caused by pressure changes, if monitoring shows loss of containment;
10. address the potential need for microseismic arrays at other injection well pads by the specified date to monitor for seismic events;
11. two years before commencing injection, provide a preliminary report on baseline data that addresses the suitability of the data for geomechanical modeling and analysis recommended in its plan;
12. allow additional water well owners to participate in the landowner water well portion of its program at any time. The proponent was required to include such wells in the plan and associated reports;
13. in its prebaseline report on the project plan, address a phased assessment of natural variability of the geochemistry of the water in the domestic water wells included in its baseline study, including the need for more frequent sampling during both the baseline data collection and early operational monitoring periods; and
14. address the potential need for installing additional monitoring wells during the project life. This information requires the information to be included in each annual report and presentations to the regulator (ERCB, 2012, Appendix 2).

After reviewing important issues of Alberta's regulatory experience and how they have been addressed in relation to specific aspects of clear property rights, environmental protection considerations, responsibility, meaningful consultations, and public interest, the current Mexican regulatory landscape of the CCS and CCUS projects is addressed below to identify specific gaps subject to attention for a suitable regulation of the CCS and CCUS projects development.

13.9 Current gaps in Mexican regulations

Sources of law in Mexico include legislation, jurisprudence, customs, individualized provisions, and general principles of law (Mexico. Supreme Court of Justice of the Nation, 2006, p. 8). Unlike in Alberta there is no standardized specific approval process or requirements for CCS projects. There is a general type of authorization from the SENVNR to operate industrial facilities that include Federal stationary sources of gas emissions that may be revised and tailored to CCS sites (Mexico. General Law of Ecological Balance and Environmental Protection, 1988, Article 111). There is no specific definition of the CO_2 stream in Mexican regulations. However, Article 5, section XIX, in the National Agency of Industrial Safety and Protection Law of the Environment of the Hydrocarbons Sector, could be tailored to CCS projects. There are existing regulations that may be revised and tailored to CCS projects regarding the classification of CO_2, air emissions (Regulations on the General Law of Ecological Equilibrium and Environmental Protection, 1988, Article 3(V, XIV)), waste (Mexico. General Law for the Prevention and Integral Management of Wastes, 2004, Article 5(XXIX)), hazardous waste (Regulations of the LGPGIR, 2006, Articles 2(VIII) and 91(II)),[2] and GHGs. There are no specific regulations to manage CO_2 from CCS facilities as waste, however, there is a General Law on Prevention and Integral Management of Waste that may be tailored to CCS operations. The Secretary of Health and Commission has the power to prevent and control hazardous effects of environmental factors, to assess health risks and order security measures under the General Law on Health. In regard to monitoring, reporting and enforcing environmental regulations for CCUS projects, the SENVNR could issue approvals, like the AER does in Alberta. There are no specific regulations regarding CCS project monitoring, reporting and verification requirements, however, there are general provisions under the *Ecological Balance and Environmental Protection Act* applicable to industrial projects. There are no specific regulations covering habitat and wildlife protection, however, the *General Wildlife Act*, Land-Use Programs and Ecological Management Programs could be revised to address CCS site-specific issues. In regard to potential impacts on vegetation, the General Law for Sustainable Forestry Development could be revised as well. There are no specific regulatory requirements or processes in Mexico to approve CO_2 pipelines and monitor their operation. As CO_2 is a gas like natural gas, in light of the experience of the Regulatory Energy Commission regulating natural gas pipelines, the Commission would be a logical regulator for CO_2 pipelines along with SEMARNAT. Unlike in Alberta, there are no specific regulations in Mexico to address post-closure liability for CCS projects. A fund has been created in Alberta to address some of the potential liability issues that may arise in the future. As a starting point, Article 147 of the General Law of Ecological Equilibrium and Environmental Protection and Article 8 of the Federal Law on Environmental Responsibility that requires providing financial security to cover the cost of environmental, property or health damage from a CCS project should be considered. Analysis of the legal and regulatory issues that have been considered and addressed in Canada should assist Mexican regulators in evaluating proposed CCS projects to determine whether or not they are in the public interest.

CCS has the potential to contribute to a suite of GHG mitigation measures. Due to the novelty of CCS projects in Mexico and the arguments discussed above, the Mexican legal framework remains incipient; nevertheless, Mexico can benefit significantly from the Canadian CCS experience, specifically, regarding the inclusion of clear property rights, environmental protection considerations, liability, meaningful consultation, and public interest.

---

[2] If it is mixed with any other hazardous waste can be considered as it, and it is subject to the legal framework that allows materials to be contained in stable geological formations.

References

Alberta Energy Regulator, 2018. AER directive 020: well abandonment guide. Surface Land Reclamation is Dealt With by Alberta Environment. Retrieved from: https://www.aer.ca/documents/directives/Directive020.pdf. (Accessed 13 May 2018).

Alberta Energy Regulator, 2020. AER directive 065: resources applications for oil and gas reservoirs. Retrieved from: https://www.aer.ca/documents/directives/Directive065.pdf.

Alberta Energy Resources and Conservation Board, 2012. Shell Canada Limited, Application for the Quest Carbon Capture and Storage Project, Radway Field, July 10, 2012, 2012 AERCB 008. Retrieved from: https://www.aer.ca/documents/decisions/2012/2012-ABERCB-008.pdf. (Accessed 16 March 2019).

Altamirano, J.-C., Sanchez, E.O., Rissman, J., Ross, K., Fransen, T., Sola, C.B., et al., 2016. Achieving Mexico's Climate Goals: An Eight-Point Action Plan. World Resource Institute, Washington, DC. Retrieved from: https://energyinnovation.org/wp-content/uploads/2016/11/WRI_OCN_Mexico_final.pdf. (Accessed 13 January 2019).

Aydin, G., Karakurt, I., Aydiner, K., 2010. Evaluation of geologic storage options of CO$_2$: applicability, cost, storage capacity and safety. Energy Policy 38 (9), 5072–5080.

Bankes, N., Poschwatta, J., Shier, E.M., 2008. The legal framework for carbon capture and storage in Alberta. Alberta Law Review 45 (3), 585–630.

Carbon Sequestration Tenure Regulation, Alta Reg 68/2011. Retrieved from: http://canlii.ca/t/52q6b. (Accessed 21 January 2019).

Environmental Protection and Enhancement Act, RSA 2000, c. E-12. Retrieved from: http://canlii.ca/t/54qb9. (Accessed 12 March 2019).

European Parliament. Council of 23 April 2009, 2009. Directive 2009/31/EC on the Geological Storage of Carbon Dioxide, retrieved from: https://eur-lex.europa.eu/LexUriServ/LexUriServ.do?uri = OJ:L:2009:140:0114:0135:EN:PDF. (Accessed 20 October 2018).

Global Carbon Capture and Storage Institute, 2017a. The global status of CCS: 2017. Retrieved from: https://www.globalccsinstitute.com/wp-content/uploads/2018/12/2017-Global-Status-Report.pdf. (Accessed 12 December 2018).

Global Carbon Capture and Storage Institute, 2017b. Global CCS Institute engages mexico's decision makers on carbon capture. Retrieved from: https://www.globalccsinstitute.com/news-media/insights/global-ccs-institute-engages-mexicos-decision-makers-on-carbon-capture/.(Accessed 5 January 2019).

Government of Alberta. News release, 2009, November 24. New Pipeline will Enhance Carbon Capture and Storage, retrieved from Government of Alberta: https://www.alberta.ca/release.cfm?xID = 27386278A12C1-C3D7-722E-E591EC672F9FC009. (Accessed May 28, 2018).

Ingelson, A.E., Nielson, N., Kleffner, A., 2010. Long-Term Liability for CCS in Depleted North American Oil and Gas Reservoirs-A Comparative Study. Energy Law Journal 31 (2), 431–469.

Lacy, R., Serralde, C., Climent, M., Vaca, M., 2013. Initial assessment of the potential for future CCUS with EOR projects in Mexico using CO$_2$ captured from fossil fuel industrial plants. International Journal of Greenhouse Gas Control 19, 212–219. Available from: https://www.sciencedirect.com/science/article/pii/S1750583613002958 (Accessed 10 February 2019).

Metz, B., Davidson, O., de Coninck, H., Loos, M., Meyer, L. (Eds.), 2005. IPCC Special Report on Carbon Dioxide Capture and Storage. Cambridge University PressRetrieved from Intergovernmental Panel on Climate Change:. Available from: https://www.ipcc.ch/site/assets/uploads/2018/03/srccs_wholereport-1.pdf (Accessed 14 February 2019).

Mexico. General Law of Ecological Balance and Environmental Protection, 1988. Retrieved from: https://www.wipo.int/edocs/lexdocs/laws/en/mx/mx028en.pdf. (Accessed 16 October 2018).

Mexico. Regulations on the general law of ecological equilibrium and environmental protection regarding the registry of emissions and transfer of pollutants, 1988.

Mexico. General law for the prevention and integral management of wastes, 2004.

Mexico. Supreme Court of Justice of the Nation, 2006. The Mexican Legal System, Retrieved from: www.scjn.gob.mx/sites/default/files/material_didactico/2016-11/Sistema-Juridico-Mexicano.pdf. (Accessed 4 April 2018).

Mexico. Advisory Council for the Energy Transition, 2016. Transition strategy to promote the use of cleaner technologies and fuels. Retrieved from the Grantham Research Institute on Climate Change and the Environment: https://climate-laws.org/cclow/geographies/mexico/policies/transition-strategy-to-promote-the-use-of-cleaner-technologies-and-fuels. (Accessed 12 January 2019).

México. Secretariat of Energy, 2018a. Advances in the implementation of CO_2 capture, use and storage in Mexico. Retrieved from: https://www.gob.mx/sener/en/documentos/advances-in-the-implementation-of-carbon-capture-use-and-storage-in-mexico. (Accessed 23 December 2018).

Mexico. Secretariat of Energy (SENER), 2018b. National Inventory of Emission Sources and sites for the use and storage of CO_2 in Mexico. Online: https://www.gob.mx/sener/articulos/inventario-nacional-de-fuentes-de-emision-y-sitios-para-el-uso-y-almacenamiento-de-co2-en-mexico. (Accessed 23 December 2018).

Mexico. Secretariat of Environment and Natural Resources [SEMENR], 2014. March. CCUS Technology Roadmap in Mexico. Retrieved from: www.gob.mx/publicaciones/articulos/mapa-de-rutatecnologica-de-ccus-en-mexico?idiom = es. (Accessed 18 March 19).

Mines and Minerals Act, RSA, 2000. c M-17, Retrived from CanLii: http://canlii.ca/t/53pcm. (Accessed 2 January 2019).

Navarro, F.A., Gallego, E.H., Arronte, L.J.A., Campos, Q.R., Ballesteros, M., Jurkiewicz, K., et al., 2016. Development of a regulatory framework for carbon capture, utilization and storage in Mexico (Report No. AUS8579-1). World Bank, Washington, DC.

Oil and Gas Conservation Act, RSA, 2000, c O-6. Retrieved from: http://canlii.ca/t/54c23. (Accessed 2 January 2019).

Orr, F.M., 2009. Onshore geologic storage of CO_2. Science 325 (5948), 1656−1658.

Paris Agreement, 2015. December 12. United National Framework Convention on Climate Change (Report No. FCCC/CP/2015/L.9/Rev.1.).

Responsible Energy Development Act, SA, 2012. c R-17.3. Retrieved from: http://canlii.ca/t/54cw1. (Accessed 19 January 2019).

Shell International BV, 2015. The Quest for Less CO2: Learning from CCS Implementation in Canada. A Case Study on Shell's Quest CCS Project. Retrieved from: https://www.globalccsinstitute.com/archive/hub/publications/196788/quest-less-co2-learning-ccs-implementation-canada.pdf. (Accessed 6 January 2019).

Surface Rights Act, RSA, 2000. c S-24. Retrieved from: http://canlii.ca/t/54qhz. (Accessed 21 January 2019).

United National Framework Convention on Climate Change, 2012. December 6. Fifth National Communication, Annex I-Mexico to the UNFCCC. Retrieved from: http://unfcc.int/essential_background/library/items/3599.php_?rec = j&priref = 7675#beg. (Accessed 18 December 2018).

Wood, D., Martin, J., 2018. October. Mexico's new energy model of paradigm shifts and political conflict: "The History of Mexico's Second Energy Revolution" in Mexico's New Energy Reform, Mexico Institute Wilson Center, Retrieved August 17, 2020 from Mexico Institute Wilson Center. https://www.wilsoncenter.org/sites/default/files/media/documents/publication/mexicos_new_energy_reform.pdf.

Woodford C., 2017. Explain that stuff: carbon capture and storage. Retrieved from: http://www.explainthatstuff.com/carbon-capture-and-storage.html. (Accessed 26 March 2019).

14

Legal and regulatory barriers to CO_2 geological storage in Brazil: Lessons from the European Union

Haline Rocha and Hirdan Katarina de Medeiros Costa

Institute of Energy and Environment, University of São Paulo, São Paulo, Brazil

14.1 Introduction

Carbon capture, utilization, and storage (CCUS) is a key suite of technologies for reducing carbon dioxide emissions. It is essential to decarbonize the power and industrial sectors worldwide (IPCC, 2005, 2014) and already has resulted in 260 million tonnes of CO_2 being stored to date (GCCSI, 2019). Besides its significant role in the Intergovernmental Panel on Climate Change (IPCC) 1.5-degree target, CCUS is necessary to meet net-zero goals in developing nations, such as Brazil, where it works as a transitional technology to a more sustainable energy mix and provides CO_2 mitigation to its continued use of fossil fuels. Accordingly, Brazil has substantial opportunities for large-scale CCUS deployment.

Although CCUS technologies only gained recognition and became essential in 2005 by the IPCC report, CCUS has been in operation for 45 years and currently is applied at 21 large-scale facilities operating at a commercial scale worldwide, that capture approximately 500,000 tonnes of CO_2 per year (Global CCS Institute, 2019). Overall, there are 51 large-scale CCUS facilities globally, and hundreds that operate at different stages of development (IPCC, 2005, 2018). However, existing commercial and legal barriers still limit the expansion of large-scale CCUS projects at the required scale.

Besides CO_2 abatement, there are also economic issues associated with CCUS deployment. The CCUS strategy is significantly cheaper than climate change consequences and other strategy costs (Adelman and Duncan, 2011; IPCC, 2014; Pop, 2015). According to the International Energy Agency, there is an estimated cost increase of 71% for meeting 2050 targets without carbon capture and storage (CCS) (IEA, 2015). Moreover, the development and deployment of CCUS technologies can bring revenue through CO_2 utilization and by exporting and improving this worldwide demanded strategy and emerging global CCUS industry (Pop, 2015). The combined potential for mitigating CO_2 emissions and for supporting a sustainable energy transition, coupled with economic development opportunities, is what brings efforts worldwide to support CCUS projects and research.

CO_2 geological storage is the segment within the CCUS chain that plays a major role in mitigating the worst impacts of climate change. This role arises from the ability to injecting large volumes of CO_2 into adequate geologic formations. For instance, global targets are estimated at 10 $GtCO_2$ emission abatement per year by 2050 and this can be achieved with geological CO_2 storage (Krevor et al., 2019). Despite so far there being few large-scale facilities, already 97.5 million tonnes of CO_2 are annually (GCCSI, 2020).

Overall, CO_2 geological storage has become so important in the net-zero emissions scenario that global efforts are taking place to support such technology, such as the European Commission with the Directive 2009/21/EC—here referred as "CCS Directive." The CCS Directive regulates CO_2 geological storage within the European Union, and it is an important first step toward a legal and regulatory regime for CCUS technologies worldwide. The geological storage segment of the CCUS chain is the focus of the EU CCS Directive and will be the focus of this chapter.

Brazil has a high potential for CO_2 geological storage due to the large occurrence of sedimentary basins within its territory—a total of 12 prospective basins that cover an area of approximately 12,000,000 km^2 (Ketzer et al., 2015), with an estimated storage capacity of 2030 $GtCO_2$ (GCCSI, 2019). In addition to geological feasibility, most of the CO_2-emitting stationary sources, especially in the southeast region of Brazil, lay on top of these sedimentary basins (Fig. 14.1). Estimates show that CO_2 emissions in the surroundings of these basins (within a 300 km radius) reach approximately 368 Mt/year, from which 49% are from the Brazilian power generation sector (Ketzer et al., 2015).

Brazil stands out among the developing countries for its comparatively low CO_2 equivalent footprint in the power generation sector, with 80.2% of its electricity mix and 43.5% of primary energy mix derived from renewable sources (MME, 2019). However, similarly to most developing countries, fossil fuels tend to play a continuously increasing share in the Brazilian energy and electricity mix, especially with the growing exploration of the Brazilian Presalt—one of the largest hydrocarbons reserves worldwide. For instance, oil production in Brazil will account for 23% of the total increase in oil production worldwide until 2030 and will double its natural gas production capacity, from 112 to 220 million m^3/day (MME, 2019).

Brazil also has a national energy security establishment dependent on fossil fuels. With regard to the oil and gas expansion supply, the energy expansion plan (from the Portuguese, *Plano Decenal de Expansão de Energia*—PDE), which reports expansion prospects for the Brazilian energy sector over a 10-year time frame, states natural gas power plants to be essential to energy capacity supply in the current scenario of hydraulic potentials exhaustion (PDE, 2020). Consequently, CO_2 emissions in the Brazilian energy generation sector have an upward tendency (EPE, 2020).

In this context, Brazil voluntarily committed in the 2015 Paris Agreement to reduce 37% of its CO_2 emission by 2025, and to reduce 43% by 2030, based on 2005 levels. This accounts for mitigating 6.2 tonnes of CO_2 (NDC Brazil, 2015; Brazil, 2016). However, Brazil is facing challenges to meet this ambitious target: deep economic recessions and political crisis make difficult the future of climate action in Brazil. Since 2018 the flexibilization of existing national

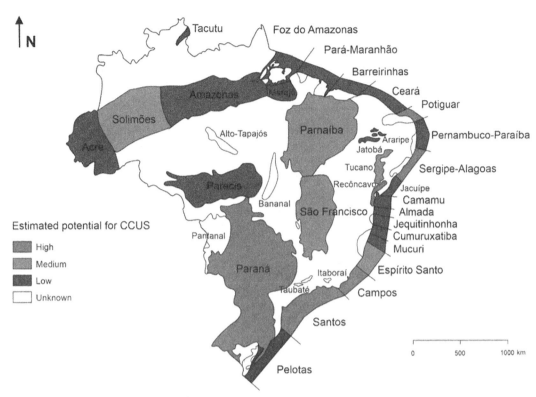

FIGURE 14.1 Map of Brazil and sedimentary basins with prospects for CO_2 geological storage. *Source: Adapted from Ketzer, J.M., Machado, C.X., Rockett, G.C., Iglesias, R.S., 2015. Brazilian Atlas of CO_2 Capture and Geological Storage.*

climate and environmental policies has resulted in increasing deforestation and, consequently, increasing CO_2 emissions (Nobre, 2016)

Aiming to surpass these challenges and fulfill the nationally determined contribution (NDC) Brazil commitment toward net-zero emissions, CCUS technologies represent a promising solution: large-scale CO_2 geological sequestration could lead to a major CO_2 abatement in power and industrial sectors and could make national greenhouse gas (GHG) reduction efforts achievable. Additionally, CCUS could work as transitional technology toward a more sustainable energy future in Brazil.

In this chapter, we identify and discuss the main legal and regulatory barriers to CCUS deployment in Brazil, with a focus on CO_2 geological storage. To do this, we start by describing the CCUS and CO_2 geological storage technologies, followed by an analysis of the existing regulation in Brazil that can be applied to CCUS. Further, we identify and discuss gaps and barriers in the Brazilian regulatory regime toward CCUS. Subsequently, we analyze the legal and regulatory regime for CCUS in Europe—the "CCS Directive" (i.e., Directive 2009/21/EC—which regulates CCUS projects with CO_2 quantities above 10^5 tons within the European Union), and carry out a comparative analysis with the Brazilian study case. We conclude with the lessons learned from the EU CCS Directive and proposals for a regulatory regime in Brazil that is supportive of CCUS.

14.2 CCUS and CO_2 geological storage

Carbon capture, utilization, and storage (CCUS) refers to a suite of technological processes. These are (1) CO_2 capture and separation from gaseous effluents; (2) CO_2 transport through pipelines or ships from the capture facility to a storage site or industrial facility; (3) CO_2 utilization into service and/or product with economic value; and/or (4) CO_2 storage into suitable geological formations (Lal, 2005; Bradshaw et al., 2007; IPCC, 2014).

CO_2 capture consists of the separation of CO_2 from other effluent gases, compression for volume reduction and fluid accommodation (Ramírez, 2020). It can be categorized into three different CO_2 capture processes: (1) precombustion—fuel conversion into CO_2 and H_2 gaseous mixture, followed by H_2 separation and combustion without producing CO_2, usually applied to natural gas processing; (2) postcombustion—separation of CO_2 from combustion exhaust gases through solvents or adsorption processes, being a highly applied method within food and beverage industries; and (3) oxyfuel with postcombustion—utilization of oxygen for the combustion of fuels, which produces H_2O and CO_2 that can easily be captured afterwards (GCCSI, 2014, 2019). CO_2 capture can be efficiently applied to industrial and energy generation facilities, such as thermoelectric power plants or cement, steel, paper, chemical, and natural gas processing industries, through different methods, such as CO_2 absorption, desorption, membrane separation, and cryogenic separation (IEA, 2009, 2015).

CO_2 transport includes its compression through pipelines, or adequate ships and trucks, from the capture unit to the storage site (IEA, 2019).

CO_2 utilization is a broad term that refers either to the direct usage of CO_2 in physical processes, such as enhanced oil recovery (EOR) or to its indirect usage as a feedstock supply to industrial processes, such as in the production of chemicals, materials, and fuels (Ramírez, 2020). CO_2 subsurface utilization accounts for the exploitation of energy resources, such as oil (EOR), natural gas (enhanced gas recovery; enhanced shale gas recovery; and enhanced coalbed methane), and geothermal power and heat (enhanced geothermal systems). Among these technologies, CO_2-EOR is the most used and mature technology, routinely used by the petroleum industry—worldwide CO_2-EOR can potentially produce 470 billion barrels of oil and could facilitate the storage of 140 billion tons of CO_2 (GCCSI, 2014, 2019). Consequently, CO_2-EOR represents a cost-effective pathway within the CCUS chain. However, there are concerns and legal uncertainties associated to CO_2-EOR being used for increasing oil production and, understandably, a debate around it being addressed as a CO_2 abatement technology (GCCSI, 2020)

In contrast, CO_2 storage is the segment within the CCUS chain that has a major contribution to net-zero goals (Krevor et al., 2019). It stands out among a diverse portfolio of CO_2 mitigation strategies, together with hydrogen production, energy efficiency, and renewable energy generation, and it is required as a component for other CO_2 abatement technologies such as bioenergy with carbon capture and storage (BECCS) and direct air capture (IPCC, 2018), which also can result in negative emissions.

CO_2 storage can occur through two different technological segments: biological fixation (biotic sequestration) and geological storage (abiotic sequestration) (Herzog and Colomb, 2004; Rodrigues et al., 2015). Regarding economic feasibility, CO_2 geological sequestration is far more advanced than the biotic sequestration pathway, being already widely implemented at commercial scale worldwide in subsurface geological formations such as depleted oil and gas reservoirs, deep saline aquifers, coal seams, organic-rich shale formations, and basalts (Rodrigues et al., 2015).

CO_2 geological storage consists of CO_2 injection and storage into adequate reservoirs for a geologically significant period (Bachu, 2002). Suitable reservoirs must present suitable porosity and permeability, a satisfactory seal, and a stable tectonic environment to avoid compromising the integrity of the storage site (Bachu et al., 2007). Reservoirs should be permanently monitored to ensure that the CO_2 remains stored within the geological formation (Bachu et al., 2017). These combinations of features are found in sedimentary geological systems around the world, that include both conventional (i.e., sandstones and carbonates) and unconventional reservoirs (i.e., coal seams and organic-rich shales). However, it is important to highlight that CO_2 storage sites are not simply associated with the occurrence of a sedimentary basin. It relies on geological, geochemical, and petrophysical processes and properties such as porosity, permeability, caprock integrity, injectivity, and fluid dynamics (Krevor et al., 2019; Haszeldine, 2019).

Conventional geological reservoirs considered favorable to carbon storage must have the following characteristics (Bachu et al., 2007; Bachu, 2008): (1) sufficient capacity to store large volumes of CO_2; (2) adequate "injectivity" to allow the injection of carbon into the geological formation; and (3) reservoir confinement or integrity, (i.e., geological configuration with traps and sealing that retain the upward CO_2 leakage for the desired period).

The effectiveness of CO_2 geological storage depends on the combination of physical factors that define the trapping system, or a geological trap, which consist of three-dimensional arrangements of rocks in the subsurface that prevent the buoyant CO_2 from migrating upwards (Krevor et al., 2019). In general, reservoir—seal—trapping systems are well-known by the oil industry, since they are analogous to the petroleum systems exploited by them. Petroleum systems' elements and processes configure a trapping mechanism that store hydrocarbons for millions or even tens of millions of years (Bachu, 2008), which supports the statement that these geological mechanisms can be used effectively for CO_2 storage.

However, similarly to petroleum production, there are environmental risks associated, mainly represented by fluid leakage. Leakage can be caused by improper sealing of the reservoir, integrity/confinement loss due to drilling and abandoned wells, and by geological faults and fractures (Barros et al, 2016).

14.2.1 Geologic risk and environmental impacts of CO_2 storage

The environmental impacts of CO_2 geological storage can be divided into global and local environmental risks. Global risks arise by the lack of efficiency of the CO_2 storage process and the consequent CO_2 release to the atmosphere. Local risks are usually associated with CO_2 leakage, which may result from high CO_2 concentrations in the gas phase near the surface, CO_2 dissolution in groundwater (aquifers), and effects induced by displacement of fluids with the injection of CO_2 (seismic activity) (Zhang and Bachu, 2011; Rodrigues et al., 2015; Almeida et al., 2017).

Local large-scale onshore leakages of CO_2 can result in direct risks to human health, both through immediate deaths resulting from asphyxiation or health impacts of prolonged exposure to high concentrations of carbon dioxide (Graus et al., 2011). The effects of CO_2 leakage on the local fauna and flora, however, are less certain and could disrupt the local ecology and agriculture (Gerard and Wilson, 2009). The main concern relates to the contamination of drinking water and surface water through the displacement of saline water and/or CO_2 into the aquifers. Moreover, subsurface CO_2 injection can trigger small seismic events. Although there is a low probability of risk occurrence, it is essential to safely manage CO_2 injection to guarantee human and ecological safety (Gerard and Wilson, 2009).

The consequences associated with CO_2 leakage are highly dependent on the location of the storage site and on how the country deals with such incidents. For instance, leakage from onshore storage sites is likely to affect a much more significant number of people than in the case of offshore sites. However, CO_2 leakage from offshore can trigger ocean acidification and, consequently, have adverse effects on marine ecosystems and the livelihoods of coastal populations (van der Zwaan and Gerlagh, 2016).

On the other hand, CO_2 leakage from a CCUS site most probably will be a gradual seepage rather than a sudden and rapid leakage, which lowers CO_2 concentrations and associated consequences and can be more easily predicted and remediated than natural events (Pop, 2015). While this means that the likelihood of catastrophic accidents leading to extensive and uncontrollable leakage is extremely low, a degree of uncertainty remains as to the possibility of CO_2 escaping from transport pipelines, through operational or abandoned wells on the site, as well as through fault lines, depending on the characteristics of the reservoir (Lako et al., 2011). Importantly, this risk is dictated to a significant extent by the level of maintenance of both active and abandoned infrastructure and the standard of quality applied in well design and construction (IEAGHG, 2019). Overall, the risk of CO_2 leakage from CCUS facilities relies significantly on human action and is mainly mitigated with storage site control, through appropriate subsurface monitoring and remediation measures.

Monitoring is essential to the safe deployment of CO_2 geological storage. It is required from the permitting process and should be carried out during all stages of CO_2 injection, from planning to completion. Monitoring includes a characterization of the reservoir and caprock properties, to select appropriate storage sites with adequate capacity, injectivity, and integrity to handle high-pressure CO_2 injection (Dino and Gallo, 2009). Appropriate monitoring include: (1) tracking the location of the CO_2 plume; (2) ensuring that injection and abandoned wells are not leaking; and (3) verifying the quantity of CO_2 that has been injected underground. Additionally, depending on site-specific considerations, monitoring may be required to (4) ensure that natural resources, such as groundwater and ecosystems, are protected and that the local population is not exposed to unsafe concentrations of CO_2 (Bachu, 2017). Together with monitoring requirements, it is important to quantify the volume of CO_2 injected underground to estimate storage effectiveness. In addition to monitoring activities, a detailed risk assessment is required to ensure that any CO_2 injection associated impact can be appropriately identified, monitored, and mitigated (EC, 2015).

Evidence indicates that CO_2 leakage is not likely to occur if site selection, characterization, and storage project design are undertaken correctly. However, there is a minor risk that inevitably remains associated with the up-to-date insufficient deployment of CCUS projects and the consequent lack of practical experience within the emerging learning curve. This risk should be addressed in the legal framework to be minimized.

Nevertheless, addressing the geologic risk adds complexity to the regulatory regime, since the geological sequestration exceeds the life span of conventional commercial-scale projects and entities. Accordingly, large-scale and long-term environmental safety of CO_2 storage requires a well-designed and comprehensive policy and regulatory framework, that can address specific barriers to this technology deployment, such as the geologic risk and long-term liabilities.

14.3 Brazil as a study case

The Brazilian context is characterized by the lack of a legal and/or regulatory framework specific for CCUS challenging the large-scale CO_2 storage deployment. In this sense, this section brings an overview on the prospects for CCUS in Brazil, followed by an analysis of the existing regulatory regime and what can be applied to CCUS and, finally, the regulatory barriers to rapid CO_2 storage deployment are addressed.

The first steps toward CCUS development in Brazil are to provide a legal and regulatory regime that is supportive to CCUS, to evaluate the country's geological storage capacity; and to identify potential storage sites that are geographically close to CO_2-emitting sources and storage infrastructure (Ketzer et al., 2015).[1]

Based on preliminary geological studies, numerous opportunities for CO_2 geological storage Brazil were identified, due to its geology and the occurrence of favorable storage sites. These opportunities include CO_2 storage in saline aquifers; CO_2 storage in depleted oil and gas fields; CO_2 storage through mineral carbonatation in basalts; CO_2 storage in organic-rich formations such as coal seams and shales—potentially applied to unconventional hydrocarbon production (i.e., coalbed methane and shale gas); and CO_2 storage in salt caverns within the Brazilian Presalt complex (Maia da Costa et al., 2019). Additionally, there is a clear potential for CO_2 utilization, such as enhanced oil recovery (CO_2-EOR), which can bring economic feasibility to CCUS projects (Beck et al., 2011; Ketzer et al., 2015).

CO_2 utilization and geological storage technologies are already being deployed in Brazil. However, CCUS projects have been mainly restricted to the oil industry, through EOR. The CO_2-EOR technology has been applied by the Brazilian national oil and gas company—Petrobras—since 1982. The company has accumulated 25 years of experience on onshore CO_2-EOR in the northeast of Brazil, in Bahia State, where 130,000 tonnes of CO_2 were injected per year into the Buracica oil field and the Rio Pojuca aquifer, both at Recôncavo Basin (Ketzer et al., 2015). Additionally, CCUS demonstration projects have been applied at saline aquifers and depleted reservoirs, such as the Petrobras Miranga

---

[1] In 2015 was published the Brazilian Atlas of CCS: a guide to CO_2 geological storage in Brazil through theoretical storage capacity assessment. Overall, CCUS Atlases are important first steps toward CCUS deployment in a country, since they address general aspects and fill the knowledge gap on geological data and storage site characterization.

Project, and coal seams with enhanced coalbed methane production in the CEPAC Carbometano Porto Batista Project (Beck et al., 2011).

Petrobras' expertise with CO_2 injection has advanced toward CO_2-EOR research and development in offshore environments. In 2015, 3 million tonnes of CO_2 were reinjected to deepwater producing oil fields in the Santos Basin, southeast of Brazil: Libra and Sapinhoá (Hatimondi et al., 2011). This CO_2 injection can be crucial to oil production in certain presalt reservoirs, due to elevated CO_2 concentrations, which can vary from 8% to 20% (Petrobras, 2019). Accordingly, Petrobras has invested in operating and planned CCUS demonstration projects, with partners, to promote CO_2 geological sequestration from the Santos and Campos basin's presalt oil fields cluster (Hatimondi et al., 2011; Beck et al., 2011). Additional onshore storage sites will be targeted for carbon-intensive industrial facilities, fossil-fueled power plants, and biofuel production (BECCS).

Within the energy generation sector, there are also various opportunities for CO_2 capture in Brazil, such as retrofitting of existing coal and diesel-fueled power plants and incorporating postcombustion CO_2 capture technologies, and CO_2 capture implementation on new natural gas-fueled power plants—with precombustion CO_2 capture integrated with combined cycle technologies (Rochedo et al., 2016).

Regarding CO_2 transportation, Brazil has a limited natural gas pipeline network. This network is geographically located in the proximities of potential CO_2 capture facilities and storage sites (e.g., the southeast region of Brazil, where most of the industrialization within the country is concentrated), and these existing pipelines can be retrofitted to CO_2 transportation. In contrast, CO_2 transportation in Brazil can rely on trucks and/or ships, which are already consolidated as a primary means of products transportation and distribution within the country.

CO_2-EOR concentrates financial investments and is the most deployed technology among the CCUS chain—to date, 84% of stored CO_2 worldwide is used for CO_2-EOR operations (Zahasky and Krevor, 2020). This is mainly due to economic viability, overlapped by technical similarities, such as the injection infrastructure, and to petroleum companies being familiar with the subsurface geology complexities and research and development (R&D).

In addition to economic and technical aspects, there are legal and regulatory mechanisms to the oil sector that are relatable to CO_2 geological storage, such as fluid injection into geological formations and associated geologic risks, environmental impacts, and liabilities. This applicability makes large-scale CO_2 geologic sequestration most likely to the deployed within the petroleum sector, especially in countries that have no regulation specific to CCUS, such as Brazil.

14.3.1 Existing regulation in Brazil and what can be applied to CCUS

CCUS technologies, especially CO_2 geological storage, are well-known and established in Brazil through Petrobras and partners, and have been mostly associated with CO_2-EOR. Therefore the regulatory framework for the oil and gas industry is usually applied.

The legal regime for the Brazilian oil and gas industry includes Law 9.478/97—here called "Oil Law," which establishes rules and conditions for the petroleum industry in Brazil. It includes provisions on exploration and production, imports and exports, refinement, processing, transport and other associated activities (Costa et al., 2018). The Brazilian Oil Law created the National Council for Energy Policy, an entity responsible for national policy development toward energy resources and for supporting the National Energy Policy implementation and effectiveness. Additionally, the Brazilian Oil Law established the sector's regulatory agency: National Agency for Oil, Natural Gas and Biofuels (ANP) as the competent authority to regulate and supervise contracts and activities related to the oil and gas exploration, development and production. Besides the Oil Law, there is Law 11.909—the so-called "Gas Law," the main guidance specifically to the natural gas sector. The Gas Law established rules for natural gas treatment, processing, storage, liquefaction, reliquefaction, commercialization, and other associated activities. Additionally, other states and national-level laws regulate the broader aspect of the CCUS activity, such as permission and gas property (Romeiro-Conturbia, 2014).

In this context, ANP has been the competent authority of CCUS activities in Brazil since most of it is connected to the oil and gas industry. Besides ANP, other governmental agencies should be considered for regulating different segments of the CCUS chain. For instance, CO_2 capture could be regulated by ANP and the National Electric Energy Agency; CO_2 transport by ANP, the National Land Transportation Agency, and the National Waterway Transportation Agency; and CO_2 geological storage could be regulated by ANP, the National Mining Agency, and National Water Agency (Romeiro-Conturbia, 2014).

In existing Brazilian legislation that addresses environmental risks, the Brazilian Civil Code (Law 10.406/2002) mentions that "Industrial activities that generate risks to the environment or property" are under the risk-generating/polluter-payer theory, which attributes to the activity itself the harmful potential of this activity. Additionally, as Musarra, Cupertino and Costa (2019) pointed out "Federal Constitution of 1988[2] welcomed Article 14 of Law No. 6.938/81, addressing objective liability by damage to the environment, where the conduct of the agent causing the damage has nothing to do with its legality and is obliged to repair it." Consequently, the Brazilian current legal regime applies the theory of total risk and strict liability to avoid possible leakage of CO_2 from the geological storage (Almeida et al., 2017; Costa et al., 2018).

14.3.2 Challenges and regulatory barriers to CCUS in Brazil

CCUS legislation and CO_2 geological storage regulation provide social-environmental protection against the potential leakage of CO_2 from storage sites. However, legislative instruments for CCUS activities are as yet nonexistent in Brazil.

In this sense, the Brazilian study case is in accordance to the international experience: the lack of carbon trading or taxation policies, CCUS not yet being mandatory to new oil production installments, or other initiatives that could encourage CCUS deployment without the revenue of the CO_2 utilization, initially constrain CCUS to CO_2-EOR and the oil industry. Additionally, specificities such as the elevated CO_2 content in the offshore presalt reservoirs,

---

[2] Article 225 from the Brazilian Federal Constitution of 1988 states that "Everyone has the right to an ecologically balanced environment, a good for the common use of the people and essential to a healthy quality of life, imposing on the Public Power and the community the duty to defend and preserve it for present and future generations."

and petroleum regulatory mechanisms being relatable to CCUS, make CCUS mostly likely to the deployed by the oil and gas industry. In this scenario, to propose a supportive regulatory regime to CCUS in Brazil, it is essential to analyze the existing legislation and competent authorities of the Brazilian petroleum sector. Additionally, CO_2 geological storage in Brazil includes the analysis of the Brazilian Civil Code and the Civil Liability concept, the petroleum sector regulation with the National Oil and Gas Laws, and the environmental legal and regulatory regime, which were raised in this chapter.

Risks regarding CO_2 geological storage are primarily associated with CO_2 leakage, which goes beyond generations, ongoing to hundreds to thousands of years—geological timescale rather than human timescale. Such a timescale goes way beyond the life span of conventional entities and/or operators that may be managing CCUS projects and storage sites and has regulatory implications on the property on the subsoil as well as on the direct ownership over the storage fluid. Consequently, to date, risk allocation and liability uncertainties of CO_2 geological storage projects are the main regulatory barriers to CCUS large-scale deployment in Brazil.

Additionally, heterogeneities on national and subnational (state-level) jurisdictions can barrier transboundary aspects of CCUS projects, such as interstate tax regulation, CO_2 transportation (i.e., gas pipelines) and storage (CO_2 plume and its possible migration) that might go across several Brazilian states and municipalities. As the fifth largest country in the world, Brazil has a very heterogeneous territory. This contributes to different regulatory demands from each state, due to the divergent nature of its natural resources and even the lack of access to these, adding complexity to the regulatory systems.

Another barrier is the lack of jurisdiction regarding the classification of CO_2, which also presents heterogeneities: CO_2 can be classified as polluting or as a commodity, depending on its origin and end-usage (Costa et al., 2018).

To surpass these regulatory barriers and promote safe and economic CO_2 storage projects in Brazil, a specific regulatory framework for CCUS is necessary. Accordingly, this new legal/regulatory regime for CCUS in Brazil should address the mentioned issues unresolved by the existing regulation such as long-term liability ownership transfer, which is a barrier to large industrial-scale CCUS programs. The CCUS regulatory regime should also include mandatory monitoring requirements and risk assessments, as well as provisions on regulatory agents, licensing, and Environmental Impact Assessment (EIA) procedures. Also, it is important to create a competent national authority on CCUS—or attribute regulatory capacity to an existing national entity (Costa et al., 2018).

So far, R&D programs have been created in Brazil to research CCUS regulatory barriers and solutions. Regarding national and state-level provisions, Câmara et al. (2011) presented a multiple scenario approach in which the Brazilian states would be responsible for regulating CO_2 emissions and conceding exploration permits to private entities to perform CO_2 capture, transport, injection, and monitoring processes. In this scenario, a private company would be responsible for the project execution and the CO_2 would be the state's property. Consequently, CO_2 capture processes would be regulated by the state-level authority and all the processes going forward (transportation, utilization, and storage) would be regulated by the national authority (Câmara et al., 2011). Although most authors agree that there should be some level of state regulation (Câmara et al., 2011; Costa, 2018; Romeiro-Conturbia, 2014), this imposes a challenge for investors, since each Brazilian state might have different regulatory regimes, and some might even lack the regulation by the time the opportunity arises.

As previously mentioned, large-scale deployment of CCUS is still subject to political, economic, environmental, and social challenges. In Brazil, regarding CCUS activities, it is necessary to increase regulatory capacity and build support for government authorities to develop a deeper understanding of how this technology and the liabilities should be applied (Romeiro-Conturbia, 2014).

14.4 European Union CCS Directive

In this concern, the Europe Commission created the Storage Directive (i.e., Directive 2009/31/EC), which provides a legislative framework for the CO_2 geological storage component of the CCUS chain. The Directive guides European Commission Member States by requirements of appropriate project design to ensure the storage of CO_2 is permanent and safe (EC, 2009).

The European Parliament together with the Council on the Geological Storage of Carbon Dioxide agreed on April 23, 2009 and implemented on June 25, 2009, the Directive 2009/31/EC. The so-called Carbon Capture and Storage Directive—here referred to as "CCS Directive" establishes a regulatory regime for an environmentally secure geological storage of CO_2 (Billson and Pourkashanian, 2017; Fiorini et al., 2017)

As previously mentioned, the CCS Directive focuses on the geological storage component of the CCUS chain, more especially on the prevention of CO_2 leakage and on geologic risk. The CCS Directive addresses the exploration of potential CO_2 storage sites, storage operations, closure and postclosure obligations, transfer of responsibility, financial security, and financial mechanisms (EC, 2009).

To ensure the coherent implementation of the CCS Directive throughout the European Union (EU), State Members are required to present a report every 4 years on the implantation of the Directive (EC, 2009). Subsequently, the EC reports to the European Parliament and the Council on the CCS Directive implementation. These include reporting on the implementation, facilitating exchanges between the competent authorities, publishing guidance documents, and adopting Commission Opinions on draft storage permits. The first implementation report was released in 2013, and it was based on Member States' reports delivered between 2011 and 2013, such as the report "CCS in Europe"—published by the EC in partnership with the European Academies Science Advisory Council in 2013—on how to secure CCS as a viable strategy towards net-zero emissions (EASAC, 2013). A second report was released in 2017, based on practical applications of the Directive. The third and last implementation report was released in 2019, based on national reports produced from 2016 to 2019 focused on the progress toward the CCS Directive implementation. These guidance documents assist the insurance industry and regulators to gain a better understanding of the risks involved and their significance in terms of likelihood of occurrence and their potential to cause harm to the environment.

14.4.1 EU CCS Directive on geologic risk and long-liability issues

The CCS European Directive 2009/31/EC is an important first step toward a legal and regulatory regime for CCUS technologies worldwide. It addresses the main challenges and difficulties of regulating this complex chain of CCUS, and especially the questions regarding the geologic risk and long-term liability associated with the CO_2 geological storage (EC, 2015, 2018).

Concerning geologic risk, the CCS Directive determines that the absence of CO_2 leakage risk, as well as the absence of significant risks to health and the environment, are fundamental first steps to storage site selection (EC, 2009).

To identify and assess the potential hazards associated with a storage location, a thorough characterization of the site and surrounding areas should be performed in accordance with the best practices adopted by Costa et al. (2018, 2019) and Mussara et al. (2019). These consist of (1) data collection, to construct a 3D static model of the reservoir, sealing rock/trap and the surrounding area, including hydraulically connected areas; (2) construction of the 3D static terrestrial geological model, by developing a series of scenarios for each parameter and calculating the appropriate confidence limits and associated uncertainty; and (3) characterization of the dynamic behavior of the storage complex, sensitivity characterization, and risk assessment, through computerized simulations of CO_2 injection at the storage site (Costa et al., 2018, 2019; Musarra et al., 2019; Musarra and Costa, 2020).

Furthermore, the EU CCS Directive demands a constantly updated monitoring plan of the storage complex. According to the monitoring requirements of the CCS Directive, the plan should consider (1) detection of CO_2 migration; (2) detection of CO_2 leakage; and (3) effects on the surrounding environment, including the biosphere and all its resources, and particularly human populations (EC, 2009). The monitoring plan should be detailed for each of the main stages of the project, including (1) baseline; (2) operational; and (3) post-closure monitoring. In addition to the monitoring plan, a plan for mitigation and corrective measures should be addressed. This would update the safety evaluation of the storage complex integrity both in the short and long term, assuring that the CO_2 storage would be environmentally safe (Barros et al., 2016; Oliveira, 2016).

The safety evaluation of the CO_2 storage site integrity should be based on an EIA and/or by a leakage risk assessment, carried out at various stages of the CCUS operation. In this sense, large-scale CCUS projects—that deal with more than 100 kilotons of CO_2 being stored, are regulated by Directive 2009/31/EC and by Directive EIA 2011/92/EU (EC, 2009). The latter being an update of the Directive 85/337/EEC, regarding EIA, capture and transport of CO_2 for geological storage, as well as storage sites (EC, 2015). Accordingly, the Directive 2009/31/EC added amendments related to CCUS and projects phases, in addition to EIA obligation (Annexes I and II of the former EIA Directive) (Barros et al., 2016; Oliveira, 2016). CCUS projects phases and obligations are presented in Table 14.1.

According to Barros et al. (2016) and Oliveira (2016), the EIA of a CCUS project should consider the entire life cycle of the projected unit, analyzing not only environmental issues but also the social and economic effects of the project and the risk assessment. Thus the purpose of the EIA is to identify the possible origins of the problems, propose alternatives, and define measures to avoid, reduce, and, if possible, remedy significant adverse effects.

TABLE 14.1 The life cycle of carbon capture and storage (CCS) projects considering the obligations established in the CCS and EIA Directives, according to the analysis of Oliveira (2016).

CCS project/activity phase	CCS Directive (2009/31/EC)	EIA Directive (2011/92/EU)
	Detailed project description; Storage complex characterization	
Phase 0 Site selection; Planning; Project feasibility evaluation; Local storage risks assessment	Explorations permit request; Development of site-specific monitoring, control and corrective measures plan	The selected site and surroundings environmental characterization regarding natural resources and human population; Environmental Impact Assessment (EIA)
Phase 1 Construction and substructures for on-site testing and operation	EIA full report submitted and approved: required for the storage permit; Site-specific monitoring and control, and development of a corrective action plan;	
Phase 2 Testing: CO_2 injection tests	Consider possible environmental impacts as criteria in strategic decision making; Mitigation measures incorporated into the installation project and project review.	
Phase 3 Exploration: CO_2 storage at commercial scale		Monitoring of significant impacts to the environment, and the effectiveness of corrective measures; Quantitative risk assessment and re-evaluation of the EIA report;
Phase 4 Deactivation: site closure	Maintenance of the specific Plan of Monitoring, Control Plan and Corrective Measures; Records and quantitative assessment of risks and associated impacts.	

CO_2 geological storage projects introduce to the EIA the concept of permanent storage, which means the CO_2 is stored for thousands of years. Considering this perpetual underground storage, some authors tend to establish comparisons between the geological sequestration of CO_2 and the disposal of radioactive waste. However, in addition to the permanent storage required in underground rock formations, there are no other similarities between these two situations (Barros et al., 2016). In this sense, there is no reference in the legal system for dealing with the permanent storage concept—which far exceeds the conventional life span of commercial projects and entities. Overall, the permanent storage concept brings legal uncertainties to the liability regime. And these associated uncertainties restrain stakeholder investment and technology development, which are essential to commercial-scale CCUS deployment.

The CCS Directive states that "all available evidence indicates that the stored CO_2 will be completely and permanently contained" (EC, 2009). According to Article 18 of the CCS Directive, most of the liabilities associated with storage sites are to be transferred to the national competent authority after postclosure stability—which can be determined after 20 years minimum (EC, 2009). These requirements are very general and could be arbitrary—and possibly will take much longer than 20 years.

The long timescale and associated uncertainties add complexity to the legal and regulatory framework. Consequently, risk allocation becomes imbalanced and mostly carried by the operator. Considering that risk allocation overlaps with financial obligations, the liability issue associated costs can be prohibitively expensive for CCUS projects operators. Such a regime does not incentivize the competent authority to accept the responsibility transfer and to move toward equal risk allocation which could also undermine investment in CCUS projects (Pop, 2015).

Since the liability regime and the permanent storage concept place unnecessary barriers on investment in CCUS due to an, arguably, inappropriate allocation of the burden of uncertainty, the legal liability regime proposed by the CCS Directive needs improvement (Elkerbout and Bryhn, 2019). The legal liability regime should address a clear division of responsibilities between CCUS operators and Member States and should address the nature, extent, and duration of liabilities very clearly to reach a balance between legal certainty, security, and investment. Additionally, considering the significance of geological storage of CO_2 toward net-zero ambitions, it seems reasonable that States and respective national competent authorities assume part of the risks until CCUS reaches significant deployment scale and progress on the learning curve.

14.4.2 EU CCS Directive effectiveness and lessons learned

The CCS Directive contributed to the knowledge and development of CCUS in Europe, especially through requiring assessments for available storage capacity and technical and economic feasibility for CCUS projects. The CCS Directive promoted national research programs, demonstration projects, and funds, together with transboundary collaborative projects toward CCUS deployment (Shogenova et al., 2014).

With regard to CCUS retrofitting, the CCS Directive requires that when applying for a licence, operators assess the technical and economic feasibility of carbon capture, transport, and storage. If the assessment is positive, "space on the installation site must be set aside for the equipment necessary to capture and compress CO_2" (EC, 2009, 2015, 2018). However, technical assessments carried out in Belgium, the Czech Republic, Germany, Romania, Poland, Slovenia, and Spain, found that CCS was not economically feasible (EC, 2018). The unfeasibility was mainly associated with lack of access to suitable geological storage sites (e.g., the case of Belgium and Estonia) or associated with technical incompatibility for power plants retrofitting.

In contrast, newly built power plants have been going beyond the established legal requirements and setting aside land for future CO_2 capture installations, and are being designed in such way that is suitable for further CCUS installations (e.g., the Czech Republic, Estonia, Germany and Poland) (COM, 2017). Other national initiatives have been proven successful, such as the case of Norway, Sweden, and the United Kingdom.[3] In

---

[3] The United Kingdom represents a specific case, since its legislation goes beyond the EU CCS Directive requirments and UK legilslation only grants permisison to power plants that can assure technical and economic feasibility during all its life cycle (COM, 2017).

Norway, for instance, new gas-fired power plants must apply CO_2 capture and storage from the start of power plant operation. In Sweden, power plants installations that surpass 300 MW are conducting preliminary studies on CCUS retrofitting (COM, 2013, 2017, 2019).

With regard to CO_2 transport and storage networks, the CCS Directive promoted international collaborations to develop common and transboundary solutions to access CO_2 transport networks and geological storage sites (e.g., the North Sea and the Baltic Sea Basins case—cooperation between the Member States bordering the North Sea and the Baltic Sea, which has proven geological suitability and ongoing CO_2 storage—involving the United Kingdom, the Netherlands, Norway, Germany, Belgium, and the Baltic Sea Region CCS network with Estonia, Germany, Finland, Norway, and Sweden). These cross-chain integration initiatives ensure that each capture facility will have full access to transport and storage infrastructure (COM, 2017, 2019).

Storage capacity assessments are the first step toward CCUS large-scale deployment and have been highly promoted through the CCS directive: Article 4(2) of the CCS Directive requires that the Member States that intend to allow geological sequestration on their territory must carry out assessments of the available storage capacity (EC, 2009). In this sense, many European countries are analyzing theoretical storage capacities of saline aquifers, depleted oil and gas fields, and hydrocarbon production suitable for CO_2-EOR. Once carried out, these assessments are used for explorations and storage permit applications, also addressed in the CCS Directive.

Overall, the CCS Directive effectively created a general legal framework to storage site selection, permitting exploration and storage, monitoring, reporting, and management of storage sites. It established a legislative framework for CO_2 geological storage by requiring appropriate project design to ensure the storage of CO_2 is permanent and safe. Based on studies undertaken by EC, the following conclusions were drawn (Lions et al., 2014; Patil, 2012):

1. Impacts from CO_2 leakage are expected to be small compared to impacts caused by other stressors. These additional stressors include, but are not limited to, changes in land use, extreme onshore weather events, periods of abnormal weather, and activities such as bottom trawler fishing, as well as the impacts that CCS seeks to mitigate such as climate change and ocean acidification.
2. It is recommended that storage operators, and relevant Competent Authorities, demonstrate that an appropriate level of understanding has been developed with regard to the potential impacts that might arise if a leak did occur from the specific site being considered for CO_2 storage.
3. Evaluation of risks of leakage and potential impacts should be undertaken at each site since each will have specific characteristics which will influence the nature and scale of the environmental response. The context of what specific impacts mean for a particular storage site (e.g., selection of crops) is fundamental and should be explained where relevant.
4. Potential impacts will be further reduced by careful site selection and appropriate monitoring and mitigation plans.
5. All monitoring programs should use ecosystem evaluation techniques. Monitoring technologies and assessment methodologies have been developed and tested that allow the impacts of CO_2 in terrestrial and marine environments to be assessed.

6. Indicator species that occur within specific onshore sites have been identified that can be monitored in conjunction with other environmental factors to assess the scale of an impact and the efficacy of any remediation.
7. Regarding public participation, the CCS Directive states that "Member States shall make available to the public environmental information relating to the geological storage of CO_2 following applicable Community legislation." Thus the requirement to advertise information about storage is clear (Musarra and Costa, 2020). This provision should inspire countries like Brazil to make all information available for public consultation.

Nevertheless, the CCS Directive left regulatory uncertainties toward the allocation of responsibility between the operator and the competent national authority in the case of geologic risk in a CCUS project (i.e., CO_2 leakage). This general approach toward long-term liability created an unbalance risk-sharing: the operator is left with a high degree of uncertainty in estimating the financial magnitudes of potential leakages. Consequently, undefined liability provisions became a significant barrier to CCUS investment and deployment.

In summary, the lessons learned after more than 10 years since the CCS Directive was implemented include the (1) review on long-term liability regime, adopting a more balanced risk-sharing approach to CO_2 leakage and revising the concept of permanent storage by attributing a more tangible timescale; (2) continued mandatory assessments regarding geological storage capacity and technical and economic feasibility of CCUS; (3) continued support on R&D through funding and grants; (4) continued support on transboundary collaborations for developing a CO_2 transport and storage network; (5) make CCUS mandatory for the power generation and industrial sectors, through CCUS retrofitting and new installation requirements; (6) consider the CO_2 utilization as part of the CCS chain, since it brings economic feasibility to the process and could accelerate CCS deployment; and finally (7) promote supportive perspectives from policymakers and the general public toward CCUS and its significance to 2050 climate goals. These lessons could help to build regulatory support to CCUS in emerging economies, such as Brazil.

14.5 Final remarks and conclusions

The role of CCUS in Brazil, and in most developing countries and industrial nations, is to provide CO_2 mitigation to its continued usage of fossil fuels and to be a transitional technology toward a more sustainable energy future. Opportunities for CCUS in Brazil rely on carbon and energy-intensive industries, such as cellulose-based products, cement, steel, aluminum, and fertilizer production, and especially on the power generation sector.

Regarding the power sector, Brazil is characterized by emerging demand for energy resources and by an increase of natural gas-fueled power-plants for baseload energy supply—complementary to intermittent hydropower-generation. Consequently, fossil fuels tend to play a continuous role in the Brazilian energy mix. In addition, fossil fuels' exploration and production are expanding within the country, especially due to the offshore exploration of the Brazilian Presalt and to the high concentration of CO_2 within these reservoirs.

Facing this scenario of an increasing share of fossil fuels in the Brazilian energy mix and consequently the CO_2 emissions upward trend, CCUS technologies become an evident strategy for the decarbonization of the Brazilian energy generation sector (Román, 2011). In this sense, CCUS can enable the continuous participation of fossil fuels in the Brazilian energy mix and still meet national (and international) climate change targets. However, to meet the Paris Agreements and 2050 net-zero goals, a regulatory regime, that is, supportive of CCUS is necessary.

In this chapter, we identified and discussed the main legal and regulatory barriers to CO_2 storage in Brazil, as an initial step toward a CCUS regulatory framework proposal. The analysis was conducted on the existing national oil and gas legal and regulatory structure, and the environmental law and regulation, in addition to references to the Brazilian Civil Code with regard to civil liability and risk theory.

CO_2 geological storage, as well as the injection of fluids into the subsurface, such as to EOR, is generally a mature technology, already extensively deployed in Brazil. Although there is no specific regulation for CO_2 geological storage, the Brazilian regulatory regime for the oil and gas industry can be adopted and/or adapted to CCUS. However, the existing regulation is yet insufficient since it poorly addresses the postoperational phase to the injection of fluids, that is, the process of "permanent" CO_2 geological storage that mostly requires specific regulation and public policies worldwide.

Based on the EU experience, most of the regulatory challenges are attributed to this geologic CO_2 storage component of the CCUS chain, especially on how to manage CO_2 geological storage risks and how to predict and quantify the costs of potential CO_2 leakage in this long-term liability scenario.

Estimating the CCUS cost is yet challenging due to the lack of enough empirical data, since these are emerging technologies, and to heterogeneity in the existing projects, which operate in different currencies and with different infrastructure, adding complexity to a comparative analysis (Herzog, 2017; Budinis et al., 2018). Consequently, It is important to ensure that potential associated costs and liabilities of the operator are predictable and quantifiable.

Additional barriers to CCUS deployment rely on undefined and unlimited provisions on liabilities in the CCS Directive, such as "evidence of CO_2 being completely and permanently contained." Such a general approach represent unreasonable guarantees that no entity, private or public, would be able to assure. Therefore it is important to aim for a balanced distribution of the burden of uncertainty.

This is the main barrier to CCUS, and especially to CO_2 geological storage, which is the same in Brazil and worldwide: the lack of a specific regulation to this suite of technologies and overlapping economic and regulatory uncertainties. Although there are jurisdictions that present regulatory elements, such as the aforementioned Brazilian oil and gas regulation, these are usually incomplete and insufficient to address juridical safety and financial risk prevention to the Brazilian State and operating company.

Additional steps toward CCUS development in Brazil include (1) assessments of Brazilian geological storage capacity, prioritizing storage sites that are geographically close to CO_2 emitting sources and storage infrastructure; (2) technical and economic assessments of potential CCUS projects; (3) continued search for financial and political support of early-phase CCUS projects to be able to latter meet the long-term goals for net-zero GHG emissions; (4) investments in R&D and on the international partnership; and (5) continuously investigating the mechanisms for carbon prices and how these could promote CO_2 abatement technologies.

Ensuring an environmentally safe deployment of CCUS must remain a priority in every regulatory regime. A very permissive regulation can lead to socioenvironmental risks, which go beyond generations, maybe for hundreds to thousands of years—geologic risks rely on the geological timescale rather than the human timescale. However, long-term liability uncertainties can, instead of incentivizing CCUS deployment, become a significant barrier for investors.

It is important to highlight that CCUS are a suite of climate change mitigation technologies and that with CO_2 geological storage climate goals can be met until 2050. And that CCUS investment relies on the improvement on liability and cost allocation issues, yet not resolved by the CCS Directive. Therefore there should be a balance of obligations between stakeholders to prevent binding and establish legal certainty regarding the obligations of each involved entity. This balanced approach could provide legal certainly until sufficient CCUS projects deploy and the financial costs associated with the potential risks of CO_2 leakage become quantifiable.

Acknowledgments

We are grateful to the "Research Center for Gas Innovation—RCGI" (FAPESP Proc. 2014/50279−4), supported by FAPESP and Shell, organized by the University of São Paulo, and the strategic importance of the support granted by the ANP (National Agency of Petroleum, Natural Gas and Biofuels of Brazil) through the R&D clause. We also thank the support from the National Agency for Petroleum, Natural Gas and Biofuels Human Resources Program (PRH-ANP), funded by resources from the investment of oil companies qualified in the R,D&I clauses from ANP Resolution No. 50/2015 (PRH 33.1—Related to Call No. 1/2018/PRH-ANP; Grant FINEP/FUSP/USP Ref. 0443/19).

References

Adelman, D.E., Duncan, I.J., 2011. The Limits of Liability in Promoting Safe Geologic Sequestration of CO_2. Duke Environmental Law and Policy Forum 22 (1).

Almeida, J.R.L., Rocha, H.V., Costa, H.K.M., Santos, E.M., Rodrigues, C.F.A., Lemos de Sousa, M.J., 2017. Analysis of civil liability regarding CCS: the Brazilian case. Modern Environmental Science and Engineering 3, 382−395.

Bachu, S., 2002. Sequestration of CO_2 in geological media in response to climate change: road map for site selection using the transform of the geological space into the CO_2 phase space. Energy Conversion and Management 43 (1), 87−102. Available from: https://doi.org/10.1016/S0196-8904(01)00009-7.

Bachu, S., 2008. Legal and regulatory challenges in the implementation of CO_2 geological storage: an Alberta and Canadian perspective. International Journal of Greenhouse Gas Control 2 (2), 259−273. Available from: https://doi.org/10.1016/j.ijggc.2007.12.003.

Bachu, S., 2017. Analysis of gas leakage occurrence along with wells in Alberta, Canada, from a GHG perspective—gas migration outside well casing. International Journal of Greenhouse Gas Control 61, 146−154. Available from: https://doi.org/10.1016/j.ijggc.2017.04.003.

Bachu, S., Bonijoly, D., Bradshaw, J., Burruss, R., Holloway, S., Christensen, N.P., et al., 2007. CO_2 storage capacity estimation: methodology and gaps. International Journal of Greenhouse Gas Control 1 (4), 430−443. Available from: https://doi.org/10.1016/S1750-5836(07)00086-2.

Billson, M., Pourkashanian, M., 2017. The evolution of European CCS policy. Energy Procedia 114, 5659−5662. Available from: https://doi.org/10.1016/j.egypro.2017.03.1704 (accessed 00.11.16).

Barros, N., Oliveira, G.M., Lemos de Sousa, M.J., 2016. Environmental impact assessment of carbon capture and sequestration: general overview. In: 32nd Annual Conference of the International Association for Impact Assessment.

Beck, B., Cunha, P., Ketzer, M., Machado, H., Rocha, P.S., Zancan, F., et al., 2011. The current status of CCS development in Brazil. Energy Procedia 4, 6148–6151. Available from: https://doi.org/10.1016/j.egypro.2011.02.623.

Bradshaw, J., Bachu, S., Bonijoly, D., Burruss, R., Holloway, S., Christensen, N.P., et al., 2007. CO_2 storage capacity estimation: issues and development of standards. International Journal of Greenhouse Gas Control 1 (1), 62–68. Available from: https://doi.org/10.1016/S1750-5836(07)00027-8.

Brazil, 2016. Ministry of Science, Technology and Innovation. Third National Communication of Brazil to the United Nations Framework Convention on Climate Change–Executive Summary. Tech. Rep., Ministry of Science, Technology and Innovation, Brasília.

Budinis, S., Krevor, S., Dowell, N., Mac Brandon, N., Hawkes, A., 2018. An assessment of CCS costs, barriers and potential. Energy Strategy Reviews 22, 61–81. Available from: https://doi.org/10.1016/j.esr.2018.08.003.

Câmara, G.A.B., Andrade, J.C.S., Ferreira, L.E.A., Rocha, P.S., 2011. Regulatory framework for geological storage of CO_2 in Brazil—analyses and proposal. International Journal of Greenhouse Gas Control 5 (4), 966–974. Available from: https://doi.org/10.1016/j.ijggc.2010.12.001.

COM, 2013. Communication from the Commission to the European Parliament, the Council, the European Economic and Social Committee and the Committee of the Regions on the future of carbon capture and storage in Europe. COM (2013) 180, 1–28.

COM, 2017. Report from the Commission to the European Parliament and the Council on the Implementation of Directive 2009/31/EC on the geological storage of carbon dioxide. Brussels. COM (2017) 37, 1–4.

COM, 2019. Report from the Commission to the European Parliament and the Council on the Implementation of Directive 2009/31/EC on the geological storage of carbon dioxide. Brussels. COM (2019) 566, 1–4.

Costa, H.K.M., Musarra, R.M.L.M., E Silva, I.M.M., De Carvalho Nunes, R., Cavalcante, I.L., Cupertino, S.A., 2019. Legal aspects of offshore CCS: case study–salt cavern. Polytechnica 1, 1–10.

Costa, H.K.M., Musarra, R.M.L.M., Miranda, M.F., Moutinho Dos Santos, E., 2018. Environmental license for carbon capture and storage (CCS) projects in Brazil. Journal of Public Administration and Governance 8, 163–185.

Dino, R., Gallo, Y. Le, 2009. CCS project in Recôncavo Basin. Energy Procedia. 1 (1), 2005–2011. Available from: https://doi.org/10.1016/j.egypro.2009.01.261.

(EASAC), 2013. Carbon Capture and Storage in Europe. German National Academy of Sciences. ISBN 978-3-8047-3180-6.

(EC), 2009. European Parliament and Council Directive 2009/31/EC of 23 April 2009 on the Geological Storage of Carbon Dioxide and Amending Council Directive 85/337/EEC, European Parliament and Council Directives 2000/60/EC, 2001/80/EC, 2004/35/EC, 2006/12/EC, 2008/1/EC and Regulation (EC) No 1013/2006 ('CCS Directive') [2009] OJ L140/114.

(EC), 2015. Support the Review of Directive 2009/31/EC on the Geological Storage of Carbon Dioxide. Executive Summary. Final deliverable under Contract No 340201/2014/679421/SER/CLIMA.C1, pp. 1–192.

(EC), 2018. In-Depth Analysis in Support of the Commission Communication. COM (2018) 773, p. 198.

Elkerbout, M., Bryhn, J., 2019. An Enabling Framework for Carbon Capture and Storage (CCS) in Europe: An Overview of Key Issues. CEPS Policy Brief. No 2019/03, September 23, 2019.

(EPE), 2020. Estudos do Plano Decenal de Expansão de Energia 2030. Avaliação do Suprimento de Potência no Sistema Elétrico e impactos da Covid-19. Superintendência de Geração de Energia Elétrica. Agosto de 2020. https://www.epe.gov.br/pt/publicacoes-dados-abertos/publicacoes/plano-decenal-de-expansao-de-energia-2030.

(GCCSI), 2019. Global Status of CCS 2019: Targeting Climate Change. Global Status Report. pp. 1–46.

Fiorini, A., Pasimeni, F., Georgakaki, A., Tzimas, E., 2017. Analysis of the European CCS research and innovation landscape. Energy Procedia 114, 7651–7658. Available from: https://doi.org/10.1016/j.egypro.2017.03.1897.

Gerard, D., Wilson, E.J., 2009. Environmental bonds and the challenge of long-term carbon sequestration. Journal of Environmental Management 90 (2), 1097–1105. Available from: https://doi.org/10.1016/j.jenvman.2008.04.005.

GCCSI, GlobalInstitute, CCS, 2020. The Global Status of CCS. Global Status Report. GCCSI 1–44.

Graus, W., Roglieri, M., Jaworski, P., Alberio, L., 2011. The promise of carbon capture and storage: evaluating the capture readiness of new EU fossil fuel power plants. Climate Policy 11 (1), 789–812. Available from: https://doi.org/10.3763/cpol.2008.0615.

Haszeldine, S., 2019. Getting CO_2 storage right—arithmetically and politically. Carbon Capture and Storage. The Royal Society of Chemistry, pp. 563–567, Energy and Environment Series No. 26.

Hatimondi, S.A., Musse, A.P.S., Melo, C.L., Dino, R., De Castro Araujo Moreira, A., 2011. Initiatives in carbon capture and storage at PETROBRAS research and development centre. Energy Procedia 4, 6099–6103. Available from: https://doi.org/10.1016/j.egypro.2011.02.616.

Herzog, H., 2017. Financing CCS demonstration projects: lessons learned from two decades of experience. Energy Procedia 114, 5691–5700. Available from: https://doi.org/10.1016/j.egypro.2017.03.1708.

Herzog, H., Colomb, D., 2004. Carbon capture and storage from fossil fuel use. Encyclopedia of Energy, 19p.

(IEA), 2009. Carbon capture and storage: full-scale demonstration progress update. OCDE/IEA.

(IEA), 2015. Technology Roadmap: Carbon Capture and Storage. OECD/IEA 2013, 5.

(IEA), 2019. World Energy Outlook 2019. https://www.iea.org/reports/world-energy-outlook-2019 (accessed 18.04.20).

IEAGHG, 2019. The Shell Quest Carbon Capture and Storage Project. Executive Summary. International Energy Agency, GHG Technology Collaboration Programme, September 2019, pp. 1–94. http://documents.ieaghg.org/index.php/s/XWMnlSPwd2tmRDM/download (accessed 07.09.20).

IPCC, 2005. Special Report on Carbon Dioxide Capture and Storage. Prepared by Working Group III of the Intergovernmental Panel on Climate Change [Metz, B., Davidson, O., de Coninck, H.C., Loos, M., Meyer, L.A. (eds.)]. Cambridge University Press, Cambridge, United Kingdom and New York, NY, USA, 442 pp.

IPCC, 2014. Climate Change 2014: Synthesis Report of the Fifth Assessment Report of the Intergovernmental Panel on Climate Change. In: Core Writing Team, Pachauri, R.K., Meyer, L. (Eds.), Intergovernmental Panel on Climate Change (IPCC), Geneva, Switzerland.

IPCC, 2018. Global Warming of 1.5°C. Intergovernmental Panel on Climate Change (IPCC, Switzerland.

Ketzer, J.M., Machado, C.X., Rockett, G.C., Iglesias, R.S., 2015. Brazilian Atlas of CO_2 Capture and Geological Storage. EDIPUCRS, Porto Alegre, 2015.

Krevor, S., Blunt, M.J., Trusler, J.P.M., De Simone, S., 2019. An introduction to subsurface CO_2 storage. In: Carbon Capture and Storage. Energy and Environment Series No. 26. The Royal Society of Chemistry 2020, pp. 238–295.

Lako, P., van der Welle, A.J., Harmelink, M., van der Kuip, M.D.C., Haan-Kamminga, A., Blank, F., et al., 2011. Issues concerning the implementation of the CCS directive in the Netherlands. Energy Procedia 4, 5479–5486. Available from: https://doi.org/10.1016/j.egypro.2011.02.533.

Lal, R., 2005. Carbon sequestration. Philosophical Transaction of the Royal Society B 310, 1628–1632.

Lions, J., Devau, N., De Lary, L., Dupraz, S., Parmentier, M., Gombert, P., et al., 2014. Potential impacts of leakage from CO_2 geological storage on geochemical processes controlling fresh groundwater quality: a review. International Journal of Greenhouse Gas Control 22, 165–175. Available from: https://doi.org/10.1016/j.ijggc.2013.12.019.

(MME), 2019. Resenha Energética Brasileira. http://www.mme.gov.br/documents/36208/948169/Resenha + Energ%C3%A9tica + Brasileira + - + edi%C3%A7%C3%A3o + 2020/ab9143cc-b702-3700-d83a-65e76dc87a9e (accessed 20.04.20).

Maia da Costa, A.V.M., Costa, P.C.O., Miranda, A.B.R., Goulart, M.D., Udebhulu, O.F.F., Ebecken, N., et al., 2019. Experimental salt cavern in offshore ultra-deepwater and well design evaluation for CO_2 abatement. International Journal of Mining Science and Technology 29 (5), 641–656. Available from: https://doi.org/10.1016/j.ijmst.2019.05.002.

Morgan, W.T., Darbyshire, E., Spracklen, D.V., Artaxo, P., Coe, H., 2019. Non-deforestation drivers of fires are increasingly important sources of aerosol and carbon dioxide emissions across Amazonia. Scientific Reports 9, 16975. Available from: https://doi.org/10.1038/s41598-019-53112-6. accessed 23.04.20.

Musarra, R., Cupertino, S.A., Costa, H.K.de M., 2019. Liability in Civil and Environmental Subjects for Carbon Capture and Storage (CCS) Activities in Brazil. International Journal for Innovation Education and Research 7 (10), 501–524. Available from: https://doi.org/10.31686/ijier.vol7.iss10.1799.

Musarra, R.M.L.M., Costa, H.K.M., 2020. Normas Internas e Internacionais para Participação Pública em Projetos de CCS. Revista FSA (Faculdade Santo Agostinho) 17, 110–132.

NDC Brazil, 2015. Nationally Determined Contribution to United Nations Framework Convention on Climate Change (UNFCCC).

Nobre, C.A., et al., 2016. Land-use and climate change risks and the need of a novel sustainable development paradigm. Proceedings of the National Academy of Sciences 113 (39).

Nobre, C.A., Sampaio, G., Borma, L.S., Castilla-Rubio, J.C., Silva, J.S., Cardoso, M., 2016. Land-use and climate change risks in the Amazon and the need for a novel sustainable development paradigm. In: Proceedings of the National Academy of Sciences of the United States of America (PNAS), September 27, 2016, 113 (39) 10759−10768. Available from: https://doi.org/10.1073/pnas.1605516113 (accessed 23.04.20).

Oliveira, G.M., 2016. Roteiro Tecnológico (Roadmap) da Captação, Utilização e Armazenamento de Dióxido de Carbono (CCUS) em Portugal (Ph.D. thesis). Universidade Fernando Pessoa. Programa de Doutoramento em Ciências da Terra, Porto, Portugal, p. 1490.

Patil, R., 2012. Impacts of carbon dioxide gas leaks from geological storage sites on soil ecology and above-ground vegetation. Diversity of Ecosystems 2, 27−50.

PDE, 2020. Plano Decenal de Expansão de Energia 2029. Ministério de Minas e Energia e Empresa de Pesquisa Energética. MME/EPE, Brasília.

Petrobras. 2019. Tecnologias Pioneiras do Pré-Sal. webpage: https://presal.hotsitespetrobras.com.br/tecnologias-pioneiras/#0 (accessed 23.04.2020).

Pop, A., 2015. The EU legal liability framework for carbon capture and storage: managing the risk of leakage while encouraging investment. The university of Aberdeen. ASLR 6, 32−56.

Ramírez, A., 2020. Chapter 13: Carbon capture and utilisation. Carbon Capture and Storage. RSC Energy and Environment Series, pp. 426−44626. Available from: https://doi.org/10.1039/9781788012744-00426.

Rochedo, P.R.R., Costa, I.V.L., Império, M., Hoffmann, B.S., Merschmann, P.R.D.C., Oliveira, C.C.N., et al., 2016. Carbon capture potential and costs in Brazil. Journal of Cleaner Production 131, 280−295. Available from: https://doi.org/10.1016/j.jclepro.2016.05.033.

Román, M., 2011. Carbon capture and storage in developing countries: a comparison of Brazil, South Africa and India. Global Environmental Change 21 (2), 391−401. https://doi.org/10.1016/j.gloenvcha.2011.01.018.

Rodrigues, C.F.A., Dinis, M.A.P., Lemos de Sousa, M.J., 2015. Review of European energy policies regarding the recent "carbon capture, utilization and storage" technologies scenario and the role of coal seams. Environmental Earth Sciences 74 (3), 2553−2561. Available from: https://doi.org/10.1007/s12665-015-4275-0.

Román, M., 2011. Carbon capture and storage in developing countries: a comparison of Brazil, South Africa and India. Global Environmental Change 21 (2), 391−401. Available from: https://doi.org/10.1016/j.gloenvcha.2011.01.018.

Romeiro-Conturbia, V.R.D.S., 2014. Carbon Capture and Storage Legal and Regulatory Framework in Developing Countries: Proposals for Brazil (Ph.D. thesis). Institute of Energy and Environment, University of São Paulo.

Shogenova, A., Piessens, K., Holloway, S., Bentham, M., Martínez, R., Flornes, K.M., et al., 2014. Implementation of the EU CCS Directive in Europe: results and development in 2013. Energy Procedia 63, 6662−6670. Available from: https://doi.org/10.1016/j.egypro.2014.11.700.

van der Zwaan, B., Gerlagh, R., 2016. Offshore CCS and ocean acidification: a global long-term probabilistic cost−benefit analysis of climate change mitigation. Climatic Change 137, 157−170. Available from: https://doi.org/10.1007/s10584-016-1674-5.

Zahasky, C., Krevor, S., 2020. Global geologic carbon storage requirements of climate change mitigation scenarios. Energy and Environmental Science 13, 1561−1567. Available from: https://doi.org/10.1039/d0ee00674b.

Zhang, M., Bachu, S., 2011. Review of the integrity of existing wells in relation to CO_2 geological storage: what do we know? International Journal of Greenhouse Gas Control 5 (4), 826−840. Available from: https://doi.org/10.1016/j.ijggc.2010.11.006.

An overview of the existing carbon capture, utilization, and storage projects in Asia: Comparing policy choices and their consequences for sustainable development

Romario de Carvalho Nunes, Haline Rocha and Hirdan Katarina de Medeiros Costa

Institute of Energy and Environment, University of São Paulo, São Paulo, Brazil

15.1 Introduction

Carbon capture, utilization, and storage (CCUS) technologies are strategic for balancing economic and energy demand growth with the continued fossil fuel reliance in Asian Countries. CCUS technologies can contribute to reduce 32% of the global carbon dioxide (CO_2) emission by 2050 (IPCC, 2018). Accordingly, its commercial-scale deployment becomes imperative to meet net-zero climate goals in the short term, while nature-based solutions and renewable energy scale up in Asia and worldwide.

Asian countries can economically benefit from CCUS deployment, especially the developing industrial economies, such as China, India, Indonesia, Thailand, and Vietnam, which are characterized by their rapid industrialization and urbanization processes associated with very high and increasing greenhouse gas (GHG) emissions. Other similarities among Asian countries toward CCUS development include: (1) improved energy access and consequent increasing energy demand; (2) continued reliance on coal; (3) natural gas as a transitional fuel; (4) absence of a market mechanism for carbon reduction; and (5) absence of a favorable regulatory environment and of political incentives for CCUS. Although there are many specificities for each country and prospects for CCUS, this chapter addresses the general opportunities and barriers to large-scale CCUS development in Asia due to their relevance in the global perspective. The current status of CCUS deployment, future perspectives, opportunities, and barriers will be presented for China, Southeast Asia (including Indonesia, Philippines, Thailand, and Vietnam), Japan, and India. The energy scenario including energy mix, CO_2 emissions, and climate change mitigation goals will also be presented and discussed in this chapter.

CCUS commercial-scale deployment requires a favorable regulatory environment, which is not yet consolidated in Asia or in most developing countries worldwide. Moreover, most of the research on international CCUS policies and regulations to date cover mainly European and North American countries. Therefore this chapter contributes with a general overview of CCUS projects in Asian countries and their respective policies, laws, and regulations.

15.2 Energy

Energy is considered a key factor for economic growth, since it directly contributes to the manufacture of goods and represents a fundamental input for industry and a significant component of household consumption (Carfora, 2019).

Fossil fossils will continue to dominate energy systems for much of this century (Garg, 2009). In particular, coal is projected to remain the basis for electricity generation in many major economies, particularly in countries where coal is the main source of native and economically viable energy (Garg, 2009).

Economic growth plays a vital role in the economic and human development of any country. In recent years, developing countries have emphasized efforts to promote and to update their industrial activity, increasing their energy consumption to produce more goods and services very quickly (Hanif et al., 2019). Hence, the current challenge is to guarantee "sustainable economic growth" while encouraging "simple economic growth" (Hanif et al., 2019).

From an economic point of view, developing nations are facing several challenges. Poverty and low income per capita continue to be the most important challenges for the local population (Change to Hanif et al., 2019). In addition, efforts to increase per capita income in developing countries are causing new negative externalities in the form of the depletion of natural resources, environmental and land degradation, and global warming (Hanif et al., 2019).

Despite the growing recognition that CCUS are an important part of the portfolio of emission mitigation technologies, its deployment in the energy sector remains out of the way (Hanif et al., 2019). However, recent technological innovations and policy developments provide encouraging signs for the future of CCUS in power generation (IEA, 2020a).

In the Asian region, countries such as China and India (among others) are heavily involved in the race to achieve higher levels of production, but at the expense of their own environmental health, as national policies to safeguard the environment and control carbon emissions failed to meet the desired protection goals (Hanif et al., 2019).

The following are general aspects related to energy in each of the countries selected in this chapter.

15.2.1 China

China is a global leader in the energy sector and in domestic investments in renewable energy. During 2015 China invested around US$ 103 billion in the energy sector, two and a half times the amount budgeted by the United States (Sharvini, 2018).

In China the intensity of coal consumption was responsible for most of the variation in the growth rate of energy consumption. Its per capita gross domestic product (GDP) also played an important role in increasing energy use (Wang, 2020). The order of contribution to energy consumption is the intensity of coal consumption, GDP per capita, intensity of gas consumption, intensity of oil consumption, and population (Wang, 2020). In China all these factors are mutually potentiating to increase energy consumption (Wang, 2020). China should focus on limiting coal use and looking for renewable fuels to replace traditional coal consumption, especially in sectors such as electricity (Wang, 2020).

15.2.2 India

In India the main determining factor in energy consumption is the intensity of coal consumption (IEA, 2020a). Like any other large developing country, and similarly to China, the growth in coal consumption contributed most to the increase in total energy consumption (Wang, 2020). The Indian energy system has evolved around it and is designed to remain so in the future under a "usual business" scenario, in addition to mitigating national energy security risks. However, the use of coal exacerbates global climate change; even in the context of broad improvements in energy efficiency and the deployment of renewable energy and other fuel exchange measures, coal use in India will increase for many years (Garg, 2009).

However, India also stands out for being a country that has introduced policies in favor of renewable energy sources and intends to implement one of the largest renewable energy programs in the world in the next 10 years (Carfora, 2019). This program will also improve access to electricity in remote and rural areas, facilitating off-grid renewable energy, leading to increased demand and energy consumption also in homes (Carfora, 2019).

15.2.3 Indonesia

Indonesia's population reached about 266.79 million in 2018 and an annual growth rate of 5.0% is projected to reach 306 million in 2050 (Sharvini, 2018). In addition, domestic energy consumption was projected to triple during the 2010−30 period. Traditional production of energy from fossil fuels has failed to keep pace with development and growth (Sharvini, 2018). The use of coal, natural gas, and geothermal energy has increased together with the reduction in oil consumption (Sharvini, 2018). Thus energy policies in Indonesia still depend on the consumption of coal and natural gas, although it is depleting these resources (Sharvini, 2018). To ensure the continuous supply of energy, the government introduced policies to guarantee energy for the future, with an emphasis on reducing the consumption of nonrenewable energy resources, especially oil, and on increasing the consumption of renewable resources such as biomass, solar, wind, hydroelectric, and geothermal (Sharvini, 2018).

15.2.4 Japan

Japan is highly dependent on the import of nonfossil resources for much of its energy supply (Sharvini, 2018). Japan has enjoyed a steady supply of energy, implementing policies to diversify supply and to promote energy conservation since the 1970s oil crisis. In 2014 the Strategic Energy Plan was established after the Fukushima nuclear disaster (Sharvini, 2018).

On September 14, 2012, the Japanese government addressed the possible phasing out of nuclear energy by 2040 (Sharvini, 2018). Immediately after the shutdown of nuclear facilities across the country following the Fukushima accidents, political support to nuclear energy fell, thereby increasing the publicly declared emphasis on renewable energy sources (Hong, 2013).

Japan depends on imported energy resources for 95% of its domestic energy consumption (Hong, 2013). Therefore reducing dependence and spending on imports are always central issues in Japan. In 2013 the importation of natural gas represented more than half of the total expenditure on importing fuels for electricity generation (Hong, 2013).

15.2.5 The Philippines

The Philippines is an archipelagic country located in the Western Pacific Ocean, comprising three main islands: Luzon, Visayas, and Mindanao (Sumabat, 2016). With 7107 islands, with a total area of 300,000 square kilometers, the Philippines is highly vulnerable to the risks of climate change, specifically to rising sea levels (Sumabat, 2016).

The Philippine energy sector faces the double challenge of heavy dependence on fossil fuels and imported energy and high energy demand. The average annual growth of the Philippines' GDP over the past 10 years has been 5.4% and the country plans to increase its GDP growth to 7% by 2040 (Mondal et al., 2018). The planned and higher GDP growth will drive further growth in energy demand. The country's self-sufficiency in primary energy supply has declined in recent years and it has suffered power cuts or power outages, especially during the summer months, since the 1990s (Mondal et al., 2018).

The country has developed its geothermal resources and is currently the second largest producer of geothermal energy in the world (Sumabat, 2016). However, its reliance on hydropower exposes its vulnerabilities to droughts, more specifically in the south of the island of Mindanao (Sumabat, 2016). Therefore it is necessary to improve its performance in the use of other renewable sources, such as biomass, wind, solar, and oceanic, to reduce carbon emissions (Sumabat, 2016). The archipelagic geography of the Philippines offers an estimated potential of 170,000 MW of ocean energy that has not yet been developed (Sumabat, 2016).

15.2.6 Thailand

Thailand's energy policy consists of two main dimensions: the supply side and the demand side (Meangbua, 2019). The residential sector is mentioned prominently in Thailand's 20-year Energy Efficiency Development Plan (EEDP 2015) and in the Alternative Energy Development Plan (AEDP 2015) to achieve the energy efficiency and conservation goal (Meangbua, 2019).

Thailand's oil and electricity consumption in the residential sector accounted for approximately 20% of Thailand's total energy consumption in 1990, but in 2017 it increased to around 51% of its total energy consumption (Meangbua, 2019).

In addition to the challenge of mitigating GHG emissions, the security of energy supply has become a critical challenge for Thailand, especially in terms of natural gas. Thai power generation relies heavily on the raw material for domestic natural gas, representing 67% of the total fuel used in power generation in 2015. The increased demand for electricity has rapidly increased the extraction of natural gas from the Gulf of Thailand, with the country registering the lowest reserves for production (R/P) ratio among its neighboring countries, such as Malaysia, Indonesia, and Vietnam. In addition, in the current R/P ratio, it will be able to supply power to Thailand for only 5.5 years, which represents the lowest level of energy security in Thailand's history (Supasa, 2017). Energy consumption in residential sectors grew by 56.86% in 2010 from 1990. The final energy consumption per capita in Thailand increased from 0.8 toe per capita in 1990 to 1.1 toe per capita in 2010, or a 37.5% increase. Also significant, CO_2 emissions increased sharply by 101% from 1995 to 2010. From 1990 to 2010, CO_2 emissions from residential sectors increased by about 3.8% on average per year (Meangbua, 2019).

The rapid depletion of energy resources has become a critical challenge for Thailand, especially in terms of natural gas, the main source of power generation (Supasa, 2017). In 2015, 67% of the fuel used in power generation was natural gas, followed by coal (18%). Renewable energies for electricity production still represent a very low proportion, about 3% of its total energy generation (Supasa, 2017). Prolonged hot weather and higher temperatures—consequences of climate change—as well as urban sprawl have considerably increased electricity consumption in the domestic sector. Along with trade promotion policies and industrial growth, the demand for all types of energy, especially electricity and oil consumption, has also increased (Supasa, 2017).

15.2.7 Vietnam

In Vietnam in 2012 the share of emissions from the energy sector was 66% (Hiep, 2020). In recent years, with the rapid growth of new coal-fired plants, an increasing amount of CO_2 has been emitted into the atmosphere (Hiep, 2020). Among the technologies for power generation, coal plants have the largest emission factors between 740 and 910 g CO_2 equiv./kWh. Other fossil fuels, including oil and natural gas, also have significantly higher emission factors than hydro, solar, and wind energy (Hiep, 2020).

The current electricity production in Vietnam is largely generated from coal, gas, oil, and hydroelectric plants, while there is only a small contribution from wind and other energies (Nong, 2019). Of these, coal plants have recently been selected and developed quickly to meet substantially increasing demands, as investment and operating costs for coal plants are relatively low compared to other plants (Nong, 2019).

In Vietnam, coal burning was responsible for the largest share of electricity generation (37% in 2016, equivalent to around 70% of the CO_2 emissions from the electricity sector) (Nong, 2019). According to the revised National Energy Development Master Plan for 2011−20 with the Vision for 2030, the Vietnamese government intends to reach

25,620 MW (42.7%) in 2020 and 55,167 MW (42.6%) in 2030 from coal-fired power plants (Hiep, 2020).

15.2.8 Energy consumption brief

The authors, based on data from the British Petroleum's Statistical Review of World Energy, prepared Table 15.1, which shows the energy consumption by fuel type in each of the countries in 2019.[1]

We can thus highlight the high energy consumption from coal, which accounts for nearly 89% of the consumption of this fuel in Asia and more than 70% of the world consumption. China and India stand out at the global level in coal consumption while Indonesia at the regional level, as demonstrated in the next section. There is also a high consumption of natural gas, especially from CIS and the Middle East.

However, there is also the positive highlight that allows verifying that one-third of all the renewable energy produced in the world comes from these seven countries, with special emphasis on China.

These seven countries together consume more than a third of all the primary energy in the world, therefore confirming their relevance in the global energy scenario and, consequently, their importance both in relation to the implementation of measures to control CO_2 emissions and new technologies, including the CCS.

It is worth adding some aspects in relation to the Southeast Asia Block; note that in some of the Asian countries, including Brunei, Singapore, Indonesia, Malaysia, and Thailand, the CO_2 emission values per capita are comparable to those of developed countries (Lee, 2013). The dramatic increase in CO_2 emissions is related to the rapid industrial development of countries such as Indonesia, Malaysia, Thailand, and Vietnam (Lee, 2013). There is a demand for immediate attention to CO_2 emissions in this region, since the greatest challenge facing Southeast Asian countries now is to find a solution to keep global warming development and mitigation in balance (Lee, 2013).

Maintaining or reducing CO_2 emissions along with sustainable development has become a difficult task for most Southeast Asian countries for a variety of reasons, be it due to the lack of implementation of signed policies and agreements, divergence in country policies, restrictions to the penetration of renewable energy technologies as well as deficiencies in the clean development mechanism (CDM) (Lee, 2013). Southeast Asia is one of the most vulnerable regions to the threat of climate change and is in an unfavorable situation compared to other large developing countries, such as China, India, and Brazil, in obtaining CDM projects (Lee, 2013).

---

[1] Since Indonesia, the Philippines, Thailand, and Vietnam have a smaller contribution to the global energy scenario when compared to China, India, and Japan, and have largely similar energy aspects, the term "Southeast Asia Block" will be used, and appear in some graphs and tables to gather data from these countries. Thus the data will be arranged in a simpler way and will allow the reader to understand the relevance of this block in various aspects, both globally and continentally. For comparative purposes at the continental level, we add three more terms to the graphs and tables: CIS—Commonwealth of Independent States, which refers to the countries of the former Soviet bloc (in our chapter, only those Asian countries are present), the Middle East, and others.

TABLE 15.1   Primary energy: consumption by fuel (exajoules).

Country	Oil	Natural gas	Coal	Nuclear energy	Hydroelectric	Renewables	Total
China	27.91	11.06	81.67	3.11	11.32	6.63	141.70
India	10.24	2.15	18.62	0.40	1.44	1.21	34.06
Japan	7.53	3.89	4.91	0.59	0.66	1.10	18.67
Southeast Asia Block	8.09	3.91	6.92	0.00	0.88	0.87	20.67
Middle East	17.80	20.10	0.40	0.06	0.30	0.12	38.78
CIS	8.05	19.96	5.49	1.88	2.21	0.03	37.63
Others	15.28	8.20	8.25	1.67	1.25	0.49	35.13
Total	94.90	69.27	126.26	7.71	18.05	10.44	326.64
Participation Asia	56.65%	30.33%	88.79%	53.12%	79.20%	93.94%	65.85%
Participation World	27.85%	14.85%	71.02%	16.43%	37.97%	33.85%	36.83%

Adapted from British Petroleum, 2020. Statistical Review of World Energy (accessed 15.07.20).

Carbon dioxide, methane (CH_4), nitrous oxide (N_2O), hydrofluorocarbons, perfluoro-carbons, and sulfur hexafluoride (SF 6) were listed in the 1998 Kyoto Protocol as GHGs that contribute to global warming. Furthermore, there is a misconception that countries in Southeast Asia contribute little to global warming. Indonesia alone, with the fourth largest population in the world, contributed about 4.73% of the world's total GHG emissions.

CO_2 has a greater impact on global warming due to its relatively high emission to the atmosphere compared to other GHGs (Lee, 2013). The main contributors of CO_2 in Southeast Asia are power generation and heat production, manufacturing, construction, and transportation (Lee, 2013).

15.2.9 Energy mix and the role of fossil fuels in Asia

As already shown in Table 15.1, the consumption of fossil fuels represents a very significant share of the Asian energy mix and is globally representative.

In relation to the energy matrix shown in Table 15.2, which consolidates the data for 2019, we see in detail how dependent Asian countries are on fossil fuels, and the efforts needed to reduce both their dependence and the reduction of CO_2 emissions.

We can basically divide the Asian countries into three large blocks: (1) China and India—with great dependence on coal in their energy matrix; (2) Japan and Southeast Asia Block—with a significant share of oil as the main energy source; and (3) the Middle East and CIS—which depend on natural gas as much as China and India depend on coal.

Although nuclear, hydroelectric, and renewable energy have rapidly grown in some countries in this region, as seen further on, they still represent a very small portion in the context of energy matrix.

TABLE 15.2 Primary energy: participation in consumption.

Country	Oil	Natural gas	Coal	Nuclear energy	Hydroelectric	Renewables
China	19.7%	7.8%	57.6%	2.2%	8.0%	4.7%
India	30.1%	6.3%	54.7%	1.2%	4.2%	3.5%
Japan	40.3%	20.8%	26.3%	3.1%	3.5%	5.9%
Southeast Asia Block	39.1%	18.9%	33.5%	0.0%	4.2%	4.2%
Middle East	45.9%	51.8%	1.0%	0.1%	0.8%	0.3%
CIS	21.4%	53.1%	14.6%	5.0%	5.9%	0.1%
Others	43.5%	23.3%	23.5%	4.8%	3.5%	1.4%

Adapted from British Petroleum, 2020. Statistical Review of World Energy (accessed 15.07.20).

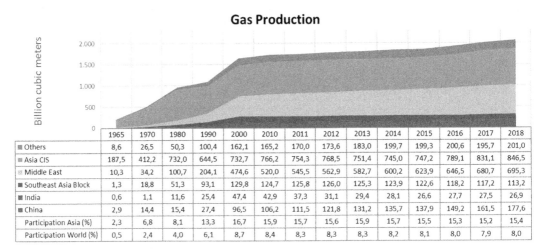

Gas Production

Billion cubic meters	1965	1970	1980	1990	2000	2010	2011	2012	2013	2014	2015	2016	2017	2018
Others	8,6	26,5	50,3	100,4	162,1	165,2	170,0	173,6	183,0	199,7	199,3	200,6	195,7	201,0
Asia CIS	187,5	412,2	732,0	644,5	732,7	766,2	754,3	768,5	751,4	745,0	747,2	789,1	831,1	846,5
Middle East	10,3	34,2	100,7	204,1	474,6	520,0	545,5	562,9	582,7	600,2	623,9	646,5	680,7	695,3
Southeast Asia Block	1,3	18,8	51,3	93,1	129,8	124,7	125,8	126,0	125,3	123,9	122,6	118,2	117,2	113,2
India	0,6	1,1	11,6	25,4	47,4	42,9	37,3	31,1	29,4	28,1	26,6	27,7	27,5	26,9
China	2,9	14,4	15,4	27,4	96,5	106,2	111,5	121,8	131,2	135,7	137,9	149,2	161,5	177,6
Participation Asia (%)	2,3	6,8	8,1	13,3	16,7	15,9	15,7	15,6	15,9	15,7	15,5	15,3	15,2	15,4
Participation World (%)	0,5	2,4	4,0	6,1	8,7	8,4	8,3	8,3	8,3	8,2	8,1	8,0	7,9	8,0

GRAPH 15.1 Historical survey of gas production. Source: *Adapted from British Petroleum, 2020. Statistical Review of World Energy (accessed 15.07.20).*

Note that most CO_2 emissions come from fossil fuels, such as coal, the main source of energy for the automotive industries, which is significantly linked to development and economic growth (Muhammad, 2019).

Seeing that oil, coal, and natural gas are the main fuels consumed in Asia, it is necessary to verify the production history of these fuels over the last decades in these countries. Therefore we prepared Graphs 15.1, 15.2, and 15.3 that respectively show the production of gas, oil, and coal in each of the countries selected in this chapter as well as in other Asian regions.

Graph 15.1 shows that gas production is mainly present in Commonwealth of Independent States (CIS) countries, especially in Russia, and countries in the Middle East, precisely the regions that have the highest consumption of this fuel within their energy

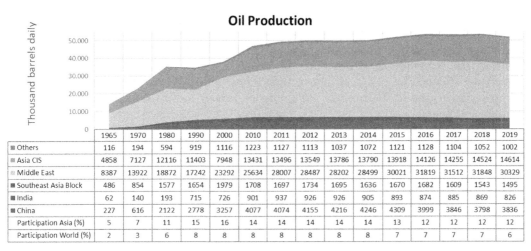

Oil Production

Thousand barrels daily

	1965	1970	1980	1990	2000	2010	2011	2012	2013	2014	2015	2016	2017	2018	2019
■ Others	116	194	594	919	1116	1223	1127	1113	1037	1072	1121	1128	1104	1052	1002
■ Asia CIS	4858	7127	12116	11403	7948	13431	13496	13549	13786	13790	13918	14126	14255	14524	14614
▦ Middle East	8387	13922	18872	17242	23292	25634	28007	28487	28202	28499	30021	31819	31512	31848	30329
■ Southeast Asia Block	486	854	1577	1654	1979	1708	1697	1734	1695	1636	1670	1682	1609	1543	1495
■ India	62	140	193	715	726	901	937	926	926	905	893	874	885	869	826
■ China	227	616	2122	2778	3257	4077	4074	4155	4216	4246	4309	3999	3846	3798	3836
Participation Asia (%)	5	7	11	15	16	14	14	14	14	14	13	12	12	12	12
Participation World (%)	2	3	6	8	8	8	8	8	8	8	7	7	7	7	6

GRAPH 15.2 Historical survey of oil production. Source: *Adapted from British Petroleum, 2020. Statistical Review of World Energy (accessed 15.07.20).*

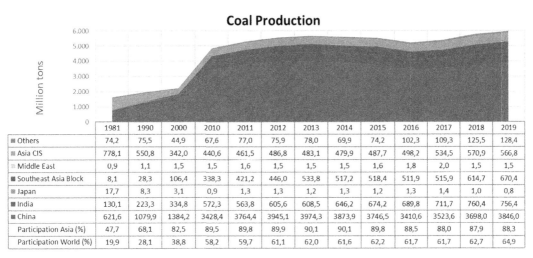

Coal Production

Million tons

	1981	1990	2000	2010	2011	2012	2013	2014	2015	2016	2017	2018	2019
■ Others	74,2	75,5	44,9	67,6	77,0	75,9	78,0	69,9	74,2	102,3	109,3	125,5	128,4
▦ Asia CIS	778,1	550,8	342,0	440,6	461,5	486,8	483,1	479,9	487,7	498,2	534,5	570,9	566,8
▦ Middle East	0,9	1,1	1,5	1,5	1,6	1,5	1,5	1,5	1,6	1,8	2,0	1,5	1,5
■ Southeast Asia Block	8,1	28,3	106,4	338,3	421,2	446,0	533,8	517,2	518,4	511,9	515,9	614,7	670,4
▦ Japan	17,7	8,3	3,1	0,9	1,3	1,3	1,2	1,3	1,2	1,3	1,4	1,0	0,8
■ India	130,1	223,3	334,8	572,3	563,8	605,6	608,5	646,2	674,2	689,8	711,7	760,4	756,4
■ China	621,6	1079,9	1384,2	3428,4	3764,4	3945,1	3974,3	3873,9	3746,5	3410,6	3523,6	3698,0	3846,0
Participation Asia (%)	47,7	68,1	82,5	89,5	89,8	89,9	90,1	90,1	89,8	88,5	88,0	87,9	88,3
Participation World (%)	19,9	28,1	38,8	58,2	59,7	61,1	62,0	61,6	62,2	61,7	61,7	62,7	64,9

GRAPH 15.3 Historical survey of coal production. Source: *Adapted from British Petroleum, 2020. Statistical Review of World Energy (accessed 15.07.20).*

matrices. In the 2010s the average participation of the countries selected in this chapter in the production of natural gas in relation to its production in Asia is around 15% and 8% in the world.

Note the growing production of natural gas in the Middle East since the 1980s; although the region had already produced an enormous amount of oil for decades, only after that decade did it begin to produce considerable volumes of natural gas.

Graph 15.2 shows the oil production time series in selected regions of Asia.

Although Japan and the Southeast Asia Block are countries where oil is predominant in the energy matrix, they do not emerge as major producers of this fuel. There is no oil production in Japan or the Philippines and oil production, decreasing even in Indonesia, Thailand, and Vietnam, does not meet all the energy consumption demanded. There is hence a dependence on this fuel in two directions, both in terms of energy and national energy security, due to the need to fully import part of this fuel, as is the case of Japan and the Philippines.

Also note that, although energy consumption is high in the selected countries (56.65% of all the Asian oil demand, as shown in Table 15.1), production historically corresponds to less than 10%.

Graph 15.3 shows the coal production in Asian countries and in other regions, with the emphasis on China and India.

As verified, China and India are the major consumers and producers of coal in the Asian continent and in the world. The dominance in relation is quite latent; China and India consumed 81.67 and 18.62 exajoules from coal (51.79% and 12% of world consumption) in 2019, respectively, and produced, respectively, 47.3% and 9.3% of all the coal in the world, showing that these two countries alone have more than half of the world consumption and production of this fuel (British Petroleum 2020).

Additionally, there is a resumption of coal production at CIS—the main coal producer in Asia in the early 1980s, which saw a sharp drop in production in the 1990s with the dissolution of the Soviet Union.

The Southeast Asia Block countries have an increasing dependence on coal and, as shown in the graph, coal production in these countries is growing sharply.

15.3 CO$_2$ emissions

The promotion of industrialization in developing economies, to improve the production of goods and to satisfy the demands of growing populations, is considered inevitable for the economic progress of these nations (Hanif et al., 2019). The corresponding technological advances have resulted in massive increases in production but have also created new challenges in the form of increased demand for energy, increased pressure on natural resources, and issues of solid waste management and carbon emissions (Hanif et al., 2019).

The link among energy consumption, economic growth, and environmental pollution has been the subject of considerable academic research over the years (Zeaei, 2015). The main reason for studying carbon emissions is that they play a central role in the current debate on protecting the environment, reducing climate change, energy security, sustainable use of available resources, and sustainable development (Zeaei, 2015). Economic growth is also closely linked to energy consumption, since a higher level of energy consumption leads to greater economic growth (Zeaei, 2015).

However, in the last few decades, the social benefits arising from the exploitation of biotic systems have decreased dramatically . Thus the need arose to face the depletion of resources seriously; if humans continue to consume natural resources at the current rate, the planet will no longer be able to meet the growing demand (Hanif et al., 2019).

Alarmingly, the East Asia and Pacific region as a whole has lost 70%–90% of its original wildlife habitat—particularly in Indonesia, Cambodia, Vietnam, and Thailand (Hanif et al., 2019)—in just a few decades, due to infrastructure and agricultural development, land degradation, deforestation, shrimp farming, and other practices. These habitat losses plus rising levels of carbon dioxide in the atmosphere have taken the region to a critical point (Hanif et al., 2019).

Carbon dioxide is considered the main global pollutant responsible for global warming and climate change. Globally, 75% of the total GHGs consist of carbon dioxide; these gases do not dissipate or disappear, but simply move from one area to another, remaining in the biosphere for thousands of years. According to (Hanif et al., 2019), the GHG emissions from coal alone are causing about 50,000 premature deaths and 0.4 million new cases of bronchitis each year. In the past few decades, due to these environmental changes, harmful smog from coal consumption has caused serious health problems, particularly in Indonesia, Thailand, Malaysia, Brunei, and Singapore (Hanif et al., 2019).

Data on energy consumption and CO$_2$ emissions show that Asian countries have been the major emitters in the world in the last decade (Zeaei, 2015). However, in recent years, the degree of CO$_2$ emissions and energy consumption in the European Union have been effectively reduced (Zeaei, 2015). European countries are trying to reduce pollution by employing various tools, such as the development of new technologies to reduce energy consumption, emphasizing environmental R&D and adopting new CO$_2$ emission rules (Zeaei, 2015).

Unlike the EU and *North American Free Trade Agreement* (NAFTA), where member countries share land borders, most Asian countries are island or peninsular countries and about 98% of the volumes of traded commodities are shipped by sea (Lee, 2016). This indicates that Asian economic integration may have a significant implication for regional maritime activities, consequently altering the associated CO$_2$ emissions, since about 3% of the global CO$_2$ emissions derive from this mode of transport (Lee, 2016).

Regarding CO$_2$ emissions in Asia, Graph 15.4 shows the decrease in the share of CIS—in early 1965, it accounted for more than half of the gas emissions in the atmosphere—and the rapid growth of China, the major CO$_2$ emitter on the planet. The countries selected in this chapter emit about 70% of all the CO$_2$ from Asia and more than 40% of the global CO$_2$.

Because CO$_2$ emissions are a key part of this chapter, we have stratified the Southeast Asia Block data, as shown in Graph 15.5.

As mentioned throughout this chapter, economic and population growth, increased income, and the use of coal as a fuel are the causes of the steady increase in carbon dioxide emissions into the atmosphere. In the last 10 years, there has been a more moderate growth in Thailand and the Philippines and aggressive growth in Indonesia and mainly Vietnam.

The rapid growth of CO$_2$ emissions in Southeast Asia is one of the consequences of population and economic growth that results in a dense energy intensity. In the transport sector, the focus is always on reducing excessive dependence on oil due to the unstable market price and limited oil reserves. Thus CO$_2$ emissions from transportation are also taking some credit from these issues. To mitigate CO$_2$ emissions, four main approaches were adopted, that is, alternative vehicles, alternative fuels, improved fuel efficiency, and intelligent transportation

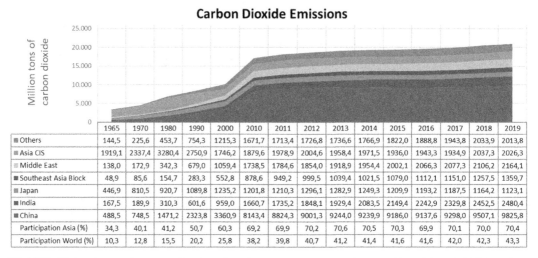

GRAPH 15.4 Historical survey of carbon dioxide emissions in Asia. Source: *Adapted from British Petroleum, 2020. Statistical Review of World Energy (accessed 15.07.20).*

	1965	1970	1980	1990	2000	2010	2011	2012	2013	2014	2015	2016	2017	2018	2019
▪ Others	144,5	225,6	453,7	754,3	1215,3	1671,7	1713,4	1726,8	1736,6	1766,9	1822,0	1888,8	1943,8	2033,9	2013,8
▪ Asia CIS	1919,1	2337,4	3280,4	2750,9	1746,2	1879,6	1978,9	2004,6	1958,4	1971,5	1936,0	1943,3	1934,9	2037,3	2026,3
▪ Middle East	138,0	172,9	342,3	679,0	1059,4	1738,5	1784,6	1854,0	1918,9	1954,4	2002,1	2066,3	2077,3	2106,2	2164,1
▪ Southeast Asia Block	48,9	85,6	154,7	283,3	552,8	878,6	949,2	999,5	1039,4	1021,5	1079,0	1112,1	1151,0	1257,5	1359,7
▪ Japan	446,9	810,5	920,7	1089,8	1235,2	1201,8	1210,3	1296,1	1282,9	1249,3	1209,9	1193,2	1187,5	1164,2	1123,1
▪ India	167,5	189,9	310,3	601,6	959,0	1660,7	1735,2	1848,1	1929,4	2083,5	2149,4	2242,9	2329,8	2452,5	2480,4
▪ China	488,5	748,5	1471,2	2323,8	3360,9	8143,4	8824,3	9001,3	9244,0	9239,9	9186,0	9137,6	9298,0	9507,1	9825,8
Participation Asia (%)	34,3	40,1	41,2	50,7	60,3	69,2	69,9	70,2	70,6	70,5	70,3	69,9	70,1	70,0	70,4
Participation World (%)	10,3	12,8	15,5	20,2	25,8	38,2	39,8	40,7	41,2	41,4	41,6	41,6	42,0	42,3	43,3

systems. The approaches were carried out through regulation, taxation, policies, and investment (Lee, 2013).

15.3.1 CO$_2$ emission projections

The materials produced by three heavy industries—chemicals, steel, and cement—play a critical role in our daily lives. In addition to their critical role today, heavy industry sectors will provide many of the key inputs needed for a sustainable energy sector transition. The production of these materials requires large amounts of energy, about 2300 Mtoe in 2019, roughly equivalent to the total primary energy demand in the United States. This fuel mixture has serious implications for emissions (IEA, 2020b). The steel and cement sectors each generate about 7% of the total CO$_2$ emissions from the energy system (including emissions from industrial processes), and the chemical sector, another 4%. Together, these heavy industries are directly responsible for an amount of emissions similar to those produced in all the road transport, including trucks, automobiles and two- or three-wheeled vehicles (IEA, 2020b).

CCUS and hydrogen are therefore two technology families critical to achieving substantial emission reductions; their applications to heavy industry, in most cases, are not yet commercially available (IEA, 2020b). They represent more than 50% of the annual emissions reductions in heavy industry in 2070 in our modeling of a zero net emissions energy system. Bringing all of these technologies into demonstration and prototype stages to markets quickly is critical if heavy industry sectors are to make their contribution to achieving a zero net emissions energy system (IEA, 2020b).

Renewable energy, energy savings, and increased energy efficiency can indirectly reduce CO$_2$ emissions (Lee, 2013). All the countries in Southeast Asia have their own national policies and legislations to increase energy efficiency and to promote renewable

Carbon Dioxide Emissions

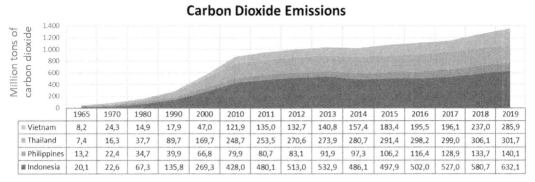

	1965	1970	1980	1990	2000	2010	2011	2012	2013	2014	2015	2016	2017	2018	2019
Vietnam	8,2	24,3	14,9	17,9	47,0	121,9	135,0	132,7	140,8	157,4	183,4	195,5	196,1	237,0	285,9
Thailand	7,4	16,3	37,7	89,7	169,7	248,7	253,5	270,6	273,9	280,7	291,4	298,2	299,0	306,1	301,7
Philippines	13,2	22,4	34,7	39,9	66,8	79,9	80,7	83,1	91,9	97,3	106,2	116,4	128,9	133,7	140,1
Indonesia	20,1	22,6	67,3	135,8	269,3	428,0	480,1	513,0	532,9	486,1	497,9	502,0	527,0	580,7	632,1

GRAPH 15.5 Historical survey of carbon dioxide emissions in Southeast Asia Block. Source: *Adapted from British Petroleum, 2020. Statistical Review of World Energy (accessed 15.07.20).*

energy. Being enriched with a variety of natural resources, renewable energy development is seen as having a bright future in Southeast Asia. Inspired by laws and policies, the potential of renewable energies, such as solar, wind, water, biomass, and geothermal, has been discovered in this region (Lee, 2013).

Thus Graph 15.5 shows the generation of energy from renewable sources. (Renewable power is based on the gross generation from renewable sources including wind, geothermal, solar, biomass and waste, and not accounting for cross-border electricity supply.)

Since the mid-2000s, China has stood out in relation to the generation of renewable energy, having a share of around 26% of this generation in 2019. There is a considerable growth in all countries. The rather small proportion of CIS stands out when compared to other countries.

Globally, 4% of the total anthropogenic CO$_2$ emissions are released by the oil refining sector. Carbon capture and storage (CCS) is a technological option with recognized potential to mitigate CO$_2$ emissions (GCCSI, 2019, 2020).

As shown in Graph 15.6, the selected countries currently refine almost 30% of the world's oil, making an important contribution to carbon dioxide emissions and excellent for the implementation of CCS projects, as well as coal plants in Graph 15.7.

Although there is no oil production in Japan, the country is observed to import a large amount of crude oil and refines it in its own territory. Regarding the Southeast Asia Block, even with the reduction in oil production, as shown in Graph 15.2, the amount of refined oil is growing, increasing the attention on meeting the conditions agreed in the Paris Agreement regarding the reduction of CO$_2$ emissions.

15.3.2 Paris Agreement and Asian countries

The Paris Agreement establishes a global framework to prevent dangerous climate change, limiting global warming to well below 2°C and endeavoring to limit it to 1.5°C. It also aims to strengthen the countries' capacity to deal with the impacts of climate change

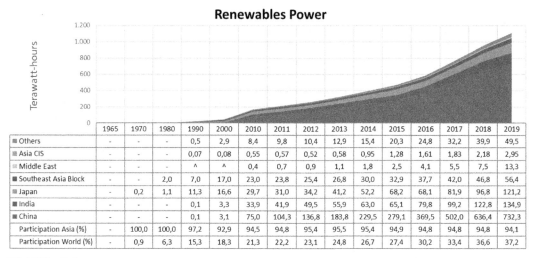

GRAPH 15.6 Historical survey of renewables power. Source: *Adapted from British Petroleum, 2020. Statistical Review of World Energy (accessed 15.07.20).*

	1965	1970	1980	1990	2000	2010	2011	2012	2013	2014	2015	2016	2017	2018	2019
■ Others	-	-	-	0,5	2,9	8,4	9,8	10,4	12,9	15,4	20,3	24,8	32,2	39,9	49,5
■ Asia CIS	-	-	-	0,07	0,08	0,55	0,57	0,52	0,58	0,95	1,28	1,61	1,83	2,18	2,95
■ Middle East	-	-	-	^	^	0,4	0,7	0,9	1,1	1,8	2,5	4,1	5,5	7,5	13,3
■ Southeast Asia Block	-	-	2,0	7,0	17,0	23,0	23,8	25,4	26,8	30,0	32,9	37,7	42,0	46,8	56,4
■ Japan	-	0,2	1,1	11,3	16,6	29,7	31,0	34,2	41,2	52,2	68,2	68,1	81,9	96,8	121,2
■ India	-	-	-	0,1	3,3	33,9	41,9	49,5	55,9	63,0	65,1	79,8	99,2	122,8	134,9
■ China	-	-	-	0,1	3,1	75,0	104,3	136,8	183,8	229,5	279,1	369,5	502,0	636,4	732,3
Participation Asia (%)	-	100,0	100,0	97,2	92,9	94,5	94,8	95,4	95,5	95,4	94,9	94,8	94,8	94,8	94,1
Participation World (%)	-	0,9	6,3	15,3	18,3	21,3	22,2	23,1	24,8	26,7	27,4	30,2	33,4	36,6	37,2

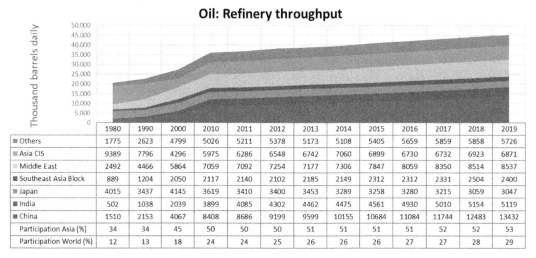

GRAPH 15.7 Historical survey of oil refinery throughput. Source: *Adapted from British Petroleum, 2020. Statistical Review of World Energy (accessed 15.07.20).*

	1980	1990	2000	2010	2011	2012	2013	2014	2015	2016	2017	2018	2019
■ Others	1775	2623	4799	5026	5211	5378	5173	5108	5405	5659	5859	5858	5726
■ Asia CIS	9389	7796	4296	5975	6286	6548	6742	7060	6899	6730	6732	6923	6871
■ Middle East	2492	4466	5864	7059	7092	7254	7177	7306	7847	8059	8350	8514	8537
■ Southeast Asia Block	889	1204	2050	2117	2140	2102	2185	2149	2312	2312	2331	2504	2400
■ Japan	4015	3437	4145	3619	3410	3400	3453	3289	3258	3280	3215	3059	3047
■ India	502	1038	2039	3899	4085	4302	4462	4475	4561	4930	5010	5154	5119
■ China	1510	2153	4067	8408	8686	9199	9599	10155	10684	11084	11744	12483	13432
Participation Asia (%)	34	34	45	50	50	50	51	51	51	51	52	52	53
Participation World (%)	12	13	18	24	24	25	26	26	26	27	27	28	29

and to support them in their efforts (Europe Union, 2020). The Paris Agreement is the first global and legally binding global agreement on climate change, adopted at the Paris Climate Conference (COP21) in December 2015. The EU and its Member States are among the approximately 190 Parties in the Paris Agreement. For the agreement to enter into force, at least 55 countries representing at least 55% of the global emissions have had to deposit their instruments of ratification (European Union).

As international discussions progressed, the requirement for additional mitigations was improved after the adoption of the Paris Agreement within the United Nations Framework Convention on Climate Change in December 2015. Each nation must increase its mitigation target to achieve the two-degree target as a long-term goal (Wakiyama, 2017). To meet the global two-degree target, countries need to reduce GHG emissions by 40%$-$70% compared to 2010 levels by 2050. The Intended Nationally Determined Contribution (INDC) targets presented by nations are not sufficient to achieve the two-degree target as a medium-term target, and additional mitigation actions are needed (Wakiyama, 2017).

The Paris Agreement is the second legally binding climate agreement after the Kyoto Protocol (Wei, 2019). This Agreement establishes the global layout of climate governance after 2020, and its entry into force reflects the determination of several countries to take action against climate change (Wei, 2019).

The proportion and growth rate of fossil energy will be further reduced with the help of the rapid development of renewable energies after the Paris Agreement. Since the determining power of energy intensity factors varies from country to country, the corresponding adjustments must be made accordingly. The intensity of coal consumption directly implies an energy consumption increase in China and India (Wang, 2020).

The effect of energy intensity must be treated as a fundamental issue, as it has a direct relationship with energy efficiency (Wang, 2020). Scientific research to improve the use and efficiency of energy conversion should be encouraged. In addition, for developing countries such as China and India, dependence on coal consumption is mainly due to the low price of traditional fossil fuels (Wang, 2020). However, in the long run, their governments may link the shift to a pattern of consumption development of more sustainable energy and an environment-friendly society with the offer of more job opportunities (Wang, 2020).

This reduction is made possible by a combination of reduction and recycling measures, such as increased low-carbon technologies (LCTs), clean and renewable coal technologies, and carbon capture and sequestration in our fossil fuel thermal plants (Wang, 2020). However, for CCS postcombustion plants in coal-fired units, capturing CO$_2$ by a 30% CO$_2$ capture amine system would mean an energy penalty of around 25%, including a minimum of 10% for compression and pumping to deep reserves, such as mineral rocks, gas hydrates, and the ocean (Wang, 2020). In any case, in the Indian context, when CO$_2$ sequestration is considered to be a much more appropriate option than CO$_2$ storage, the energy penalty still remains at the 15% level (Sethi, 2017).*

Under a strict global GHG mitigation regime, coal use would decline in the future. However, this would increase the dependence on imported natural gas, increasing the risks of energy security. CCS could therefore play a key role in mitigating energy security risks for India and the risks of global climate change (Garg, 2009). CCS is globally identified as a likely technology for large-scale CO$_2$ mitigation. Therefore CCS is an important technology to balance national and global risks (Garg, 2009). It can reduce the external cost of coal use by India in scenarios of greater sensitivity to the climate and, therefore, would keep domestic coal as a viable alternative (Garg, 2009). An indicative calculation suggests that the CCS potential could be in the order of 345 MtCO$_2$ nationally at the main coal deposits in India. In this context, CCS is of interest to reduce the carbon footprint of a fossil fuel-based energy system (Garg, 2009).

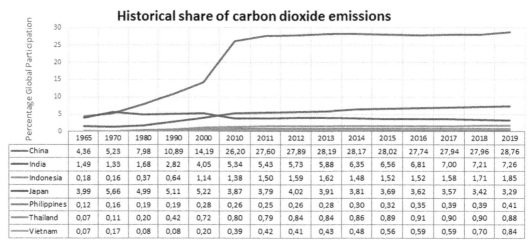

GRAPH 15.8 Historical share of carbon dioxide emissions. Source: *Adapted from British Petroleum, 2020. Statistical Review of World Energy (accessed 15.07.20).*

As discussed in Graph 15.4, there is an increase in CO_2 emissions in several Asian countries, but it is necessary to verify the increase or decrease in the global participation in CO_2 emissions by the countries selected in this chapter (Graph 15.8).

As can be seen, Japan is the only country whose global share of CO_2 emissions is decreasing. Even with the growth of renewable energy in China and India, efforts to meet the CO_2 emission reduction targets ratified in the Paris Agreement and energy policies are not enough to reverse this growth. This shows that CO_2 emissions are not decreasing, as can be seen in Graph 15.4, and/or the other countries are having more positive results in reducing CO_2 emissions, contributing even more to the share gain of China and India as the main emitters.

Table 15.3 shows the quantity of CO_2 storage facilities in the selected countries, the date of signature of the Paris Agreement by each country, as well as a summary of the agreed National Determined Contributions (NDCs).

Because China, India, and Japan are the selected countries that most emit CO_2, a brief aspect of the Paris Agreement will be discussed with regard to these countries.

15.3.3 China

In response to the call for international CO_2 emission reductions, China has pledged to reduce CO_2 emissions per unit of GDP by 60%–65% by 2030 compared to 2005 emissions levels (Wang, 2019).

Currently, the CCS operational capacity in China is no more than 2 million tons per year of CO_2 capture. This needs to increase by many magnitudes over the next 15 years. The Paris Agreement redirected attention to reducing emissions and CCS is becoming a more prominent part in this context in China. Carbon capture and storage are proven to be in use at various scales and in a variety of industries in China, highlighting their versatility (Global CCS Institute, 2018).

TABLE 15.3 Paris climate promise tracker and CCS facilities.

Country	CCS facilities	Signatory Paris agreement	Date	Summary of the INDC/NDC
China	19	Yes	30/06/2015	A peak in carbon dioxide emissions by 2030, with "best efforts" to peak earlier. China has also pledged to source 20% of its energy from low-carbon sources by 2030 and to cut emissions per unit of GDP by 60%−65% of 2005 levels by 2030, potentially putting it on course to peak by 2027.
India	2	Yes	01/10/2015	A 33%−35% reduction in emissions intensity by 2030, compared to 2005 levels. Also pledges to achieve 40% of cumulative electricity installed capacity from nonfossil fuel-based resources by 2030. Will also increase tree cover, creating an additional carbon sink of 2.5−3 billion tons of CO$_2$ equivalent by 2030. India intends to cover the $2.5 trillion cost of its pledge with both domestic and international funds.
Indonesia	1	Yes	23/09/2015	A 29% reduction in emissions by 2030, compared to business as usual. Indonesia says it will increase its reduction goal to 41%, conditional on "support from international cooperation". Includes a section on adaptation.
Japan	7	Yes	17/05/2015	A 26% reduction in emission on 2013 levels by 2030. Includes precise information on how it will generate its power by 2030.
Philippines	0	Yes	01/10/2015	A reduction in emissions of about 70% by 2030, relative to a business-as-usual scenario, on the condition of international support.
Thailand	0	Yes	01/10/2015	An unconditional 20% reduction in emissions by 2030, compared to business-as-usual levels. This could increase to 25%, conditional upon the provision of international support. Includes section on adaptation.
Vietnam	0	Yes	29/09/2015	An 8% reduction in emission by 2030, compared to a business-as-usual scenario. This could be increased to 25% conditional upon international support. Also pledges to increase forest cover to 45%.

INDC, Intended Nationally Determined Contribution; *NDC*, National Determined Contribution.
Adapted from Carbon Brief (2020).

China's 13th Five-Year Plan (FYP) for 2016−20 includes the goal of reducing energy intensity per unit of GDP by 15% and a goal of reducing CO$_2$ intensity by 18% by 2020 (Khanna, 2019). While energy consumption in China per unit of GDP decreased by 37% from 2005 to 2016, the total primary energy consumption increased by 167% in the same period and coal consumption still accounted for 62% of the primary energy consumption in 2016 (Khanna, 2019). Two short-term strategies being pursued in China, in addition to energy efficiency, include promoting the adoption of renewable sources, especially in the energy sector, and electrification (Khanna, 2019).

As in the rest of the world, for CCS to be widely deployed in China, a supporting business case must be made. At its core, this involves three interconnected factors: the establishment of national emission reduction targets consistent with the objectives of the Paris Agreement, the inclusion of CCS in a National Climate Action Plans and the establishment of policies that reward emission reductions by CCS (Global CCS Institute, 2018).

15.3.4 India

In the Paris Agreement in 2015, India committed to reducing the intensity of CO_2 emissions by about 30%−35% by 2030 compared to 2005 (Kumar, 2020). The coal-based electricity generation sector, with an installed capacity of 222 GW, is responsible for about three quarters of the total electricity generation and will continue to be the dominant source of power generation in India (Kumar, 2020). This sector contributes about half of the total CO_2 emissions generated in the country (Kumar, 2020). Therefore if India wishes to achieve the goals announced in the Paris Agreement, it is imperative to find economic measures to reduce CO_2 emissions in this sector (Kumar, 2020). Carbon pricing is economically the most efficient strategy for reducing emissions (Kumar, 2020).

India was not obliged to reduce carbon emissions under the Kyoto Protocol, but in the Paris Agreement, the country pledged to reduce the CO_2 intensity in terms of GDP. The Indian Government has undertaken some initiatives to discourage the generation of CO_2 emissions (Kumar, 2020). A kind of carbon tax, known as Clean Energy Cess, of Indian rupees (INR) 50 (about US $ 0.75) per ton on coal and lignite consumption was introduced in 2010−11. This tax was increased to INR 400 (more than $ 6) per ton in 2016 (Kumar, 2020). In addition, the Execute, Reach, and Negotiate program for energy efficiency and renewable energy certificates are market-based regulatory steps to price carbon emissions (Kumar, 2020).

15.3.5 Japan

In its Icf, Japan sets a 26% reduction target by 2030 compared to the 2013 level (equivalent to an 18% reduction from the 1990 level). The INDC target is based on the Long-Term Energy Supply and Demand Perspectives published in July 2015 by the Ministry of Economy, Trade and Industry (METI). This target considerably reduces the level of ambition in comparison to the emission reduction target announced in 2010 (25% reduction from the 1990 level in 2020) (Wakiyama, 2017).

In the Japanese INDC, the target energy mix in 2030 is 22%−24% renewable, 20%−22% nuclear, and 56% fossil fuel generation (Wakiyama, 2017).

According to INDC, the target for total CO_2 emissions in Japan in 2030 will be 927 $MtCO_2$ (308 $MtCO_2$ reduction in 2013) and the CO_2 emissions targeted in the residential sector will be 122 $MtCO_2$ (16% of total CO_2 emissions in 2030) (Wakiyama, 2017).

15.4 CCUS technologies in Asian countries

Prior to the Covid-19 crisis, CCUS was gaining new momentum. In 2020 eight new CCUS-equipped power plants were announced, spurred primarily by new investment incentives in the United States, bringing the total to 20 power plants in development globally. As with other clean-energy investments, CCUS projects now face considerable challenges. CCUS applied to power is at an early stage of commercialization; securing investments will thus require complementary and targeted policy measures, such as tax credits and grant funding (IEA, 2020a).

Most of the mentioned Asian countries are industrial economies that rely on fossil fuels for electricity generation and, accordingly, could largely benefit from CCUS (GCCSI, 2019). Additionally, due to technical similarities and overlapping technology between the power generation sector and the CCUS chain, the studied Asian countries are already familiar with some of the complexities and technologies of CCUS, such as geological studies for coal and petroleum exploration and production (E&P) and CO_2 geological storage research and development (R&D). This means that, with incentives, Asia could rapid deploy large-scale CCUS projects due to its technological background. The large-scale development of CCUS could enable Asian countries to meet the constantly increasing energy demand together with the transition to a more sustainable energy future.

Besides the regulatory and economic barriers, CCUS deployment is rising in Asia. Among the current 24 large-scale CCUS facilities worldwide, 15 are placed in Asian countries (GCCSI, 2019).

CCUS early deployment in Asian countries is mostly associated with CO_2 utilization, which can bring economic feasibility to the CCUS integration chain.

Considering its political and economic relevance, China's policy framework toward CCUS has an influencing role among other Asian countries. Therefore this section starts by analyzing China's existing CCUS policy system, at the national and provincial level, followed by an analysis of its effectiveness and identified gaps and barriers to CCUS large-scale deployment. Further, we present CCUS policies undertaken by other Asian countries that have initiatives with regard to CCUS, such as Japan, Indonesia, the Philippines, Thailand, Vietnam, and India. Finally, general conclusions about CCUS prospects in Asia will be addressed.

15.4.1 China

As the worldwide leader in carbon dioxide (CO_2) emissions and major industrial economy, China can largely benefit from CCUS technologies.

The power sector plays the most significant role in CO_2 emissions in China, due to its fossil-driven energy mix and consequent major carbon footprint. To date, electricity-related CO_2 emissions account for approximately 45% of the country's total carbon emissions (Ma et al., 2019; Wei et al., 2020a, 2020b). This applies especially to the Chinese thermal power generating sector, and its continued reliance on coal—approximately 80% of the electricity generation in China comes from coal (IEA, 2018). Additionally, China's installed generating capacity of power plants is constantly increasing. For instance, in 10 years, the installed capacity increased by about three times, from 517.2 GW in 2005 to 1525.3 GW in 2015 (Carbon Brief, 2020). According to the national 13th FYP from 2016 to 2020, 60 GW of new coal-fired power plants are built annually. Consequently, its CO_2 emission reduction strategy should focus on the power generation sector until an energy transition towards renewables is established.

The industrial sector is the second most significant contributor to CO_2 emissions in China. This is due to the country's high level of industrialization and the large-scale deployment of energy and carbon-intensive industries, such as iron, steel, and cement production (Ding et al., 2020). For instance, China's iron and steel sector contributed with more than 30% of the total industrial-related CO_2 emissions within the country (IEA, 2019; Ding et al., 2020).

In this scenario, China committed to reduce 60%−65% of its CO_2 emission per unit GDP until 2030, based on 2005 levels (IEA (International Energy Agency), 2019b). Additionally, in 2020 China announced its latest climate commitment, pledging to become carbon neutral by 2060 (GCCSI, 2020). Thus CCUS technologies are essential for achieving China's climate goals while meeting its fast-growing energy demand and should be incorporated into the country's climate policy portfolio. Establishing CCUS readiness in power plants and steel plants can be a low-cost technical approach toward CO_2 abatement in China's power and industrial sectors in the future (Wang, 2020). However, CCUS readiness relies on the CCUS segments integration, which requires access to CO_2 transport infrastructure and to geological storage sites. Additional challenges to CCUS readiness are related to CCUS cost path and to regulatory barriers.

Barriers to CCUS development include the high costs to retrofit a thermal power plant with CCUS technology. Understandably, Chinese climate policies have been more concerned with immediate and local environmental problems, such as sulfur dioxide, nitrogen oxide, and mercury pollution from coal-fired electricity generation and its health consequences, than with CO_2 emissions, which have long-term effects. Consequently, Beijing has been cautious in committing to an aggressive program for CCUS deployment (Jiang et al., 2020).

Despite its carbon footprint and the mentioned barriers, China is becoming a world leader in LCTs and a major contributor to the world's climate mitigation process (Ding et al., 2020). China has been rapidly pioneering net-zero technologies in solar, wind, and hydropower generation sectors, and has become the world's largest user of hydropower and solar thermal heating and one of the largest on wind power generation worldwide. In this sense, China's investment in net-zero technologies, such as CCUS, can enhance its global reputation as a responsible and climate-concerned rising industrial economy.

China currently leads CCUS activity in Asia, with more than 10 years of experience with CCUS research and development R&D, as well as demonstration projects. China has already 20 projects that include CCUS technologies, including eight large-scale CCUS facilities—one in operation, two in construction, and five in early development stages (GCCSI, 2019).

CCUS was first deployed in China with the Coalbed Methane Technology/CO_2 Sequestration Project in 2005. The project targeted CO_2 utilization to enhance coalbed methane production (CO_2-ECBM). Due to the high efficacy of CO_2 for ECBM and the high storage ratio of CO_2 in coals due to organic porosity,[2] China's government incorporated CCUS into its climate goals strategy. Accordingly, since 2005 China has invested and

[2] CO_2 geological storage in coal strata is a well-stabilized and efficient technology, due to coal's affinity for CO_2, which relies on the adsorption process. In coals and other organic-rich rocks, CO_2 storage occurs through adsorption into organic matter microporous structures (e.g., vitrinite). This adsorption process accounts for

conducted numerous studies on CCUS technical and economic aspects, especially regarding CO_2 utilization (Wang et al., 2019; Jiang and Ashworth, 2021).

Additional steps toward CCUS in China include changes to the existing regulatory regime from the government to support CCUS. Such changes include the Chinese FYP—blueprints of China's economic and social development over a five-year period, which started to emphasize CCUS relevance since the 12th FYP (Jiang et al., 2020). The FYPs are national guidelines that issue specific reports for different areas, such as environmental protection and energy development. These guidelines also require provinces and municipalities to promote local FYPs, the implementation of which is assisted by national and regional authorities.

In this sense, with the 12th FYP for National CCUS Technology Development, China officially established CCUS as part of national climate strategies. The 12th FYP focused on CCUS technical aspects and was accompanied by two governmental rules: (1) Notice on Promoting CCUS Test Demonstration, and (2) Notice on Strengthening the Environmental Protection of CCUS Demonstration Projects, both established in 2013. These governmental guidelines promoted scale expansion of regional CCUS project pilots to national demonstration by creating subsidy mechanisms, standardization of processes, international collaboration, and CCUS knowledge-sharing dissemination within China (Jiang et al., 2020).

In 2016, with the release of the 13th FYP and the release of *Catalogues of National Key Energy-saving and Low-carbon Technologies*, CCUS gained unprecedented relevance in China. These governmental guidelines have been promoting CCUS financially, for example, investment of 37.5 billion CNY since 2016 to scale up CCUS from 1% to 10% (Jiang et al., 2020), and technically, for example, through pipeline network expansion for CO_2 transport. Furthermore, also in 2016, a third notice on CCUS was released. The so-called *Technical Guidelines for Environmental Risk Assessment of CCUS* focused on CCUS' environmental risks, which are mostly associated with the geological risk of CO_2 leakage from storage sites. The technical guidelines contain provisions on CO_2 leakage addressing issues such as groundwater, soil and surface water pollution, and associated corrective measurements and liabilities.

To meet the Paris Agreement and to support CCUS deployment, China should focus on the power and industrial sectors. Regarding the power sector, the coal-fired thermal power generation model is continuously being implemented in China (IEA, 2018). Accordingly, China's CCUS project investment models must consider the retrofit of existing coal-fired power plants and of steel plants to install CO_2 capture technologies in the future. For instance, existing power plants could incorporate postcombustion CO_2 capture, and new power plants could be engineered with precombustion CO_2 capture technology together with integrated gasification combined-cycle technologies (Wang et al., 2018).

CCUS technologies are currently the only available strategy that can reduce CO_2 emissions from coal-fired power generation and large industrial processes by up to 90% (ADB, 2015). Additionally, CCUS is the most cost-effective alternative among climate change consequences and abatement strategies, for example, meeting climate goals without CCUS would represent a 25% cost increase (IEA, 2019). Therefore investing in the readiness of

95%–98% of the total gas storage, while 2%–5% are stored as free gas. (Bachu, 2007; Rodrigues et al., 2013; Rodrigues, 2002).

CCUS in these facilities could be fundamental for a large-scale promotion of CCUS and for transitioning from the current high-carbon to low-carbon thermal power generation in China (Jiang and Ashworth, 2021).

CO_2 capture readiness can also be applied to gasification units. China has installed more than 100 coal gasifiers that produce pure streams of CO_2 as a by-product, currently vented directly into the atmosphere. Emissions from these gasifier plants are more straightforward and less costly to capture than emissions from combustion plants and should therefore be the immediate focus of CO_2 capture projec (Li et al., 2012; Ding et al., 2020).

Additional opportunities for CCUS in China rely on CO_2 utilization technologies. CO_2-ECBM and CO_2-EOR technologies have the potential to create and to improve the initial commercial demonstration of CO_2 geological storage in China. Facing the scenario of absent carbon taxation mechanisms, CCUS deployment in China will most likely be related to these profitable technologies, for example, extra oil production and coal mining safety benefits may compensate the energy penalty concern in CCUS (Li et al., 2012).

Other commercial opportunities for CCUS in China include transfers of these cutting-edge technologies and technical expertise in an emerging global CCUS industry and future market, as well as external financial support and collaboration.

As shown, China has been progressively improving its CCUS policy system and, consequently, China's CCUS industry is experiencing rapid development. However, China's existing CCUS policy regime is still insufficient to promote CCUS deployment at the required rapid and large-scale. This is due to overlapped economic and regulatory barriers, which are characteristics of CCUS' early deployment. These barriers include the lack of a legal and regulatory regime that is supportive to CCUS, accompanied by the absence of CCUS market mechanisms, financial stimulus, or subsidies.

To promote CCUS deployment in China, it is essential to continue supporting CCUS R&D, engineering CCUS readiness, and acquiring geological knowledge for evaluating storage capacity and identifying potential storage sites. In addition, CCUS should be incorporated into Chinese national development plans to endorse stakeholders' investments (Li et al., 2012).

15.4.2 Southeast Asia

Indonesia, the Philippines, Thailand, and Vietnam are the southeast Asian countries with the fastest growing economies in the world. This economic growth is linked to an improved energy access and to a rapid industrialization process that led to an increasing energy demand (ADB, 2013).

The energy matrix of these countries historically relies on fossil fuels, which account for approximately 90% of their collective energy supply (ADB, 2013). Among the fossil fuel demand, there is a significant reliance on coal—with increasing domestic production and imports; transitioning to a growing demand for natural gas.

Southeast Asian countries commitments to CO_2 abatement include promoting energy efficiency, renewable energy, and LCTs (IPCC, 2018). Indonesia set the most ambitious goal and pledged to reduce its emissions by 26% by 2020, and recognized CCUS as a prospective technology to be incorporated into the country's climate strategy (ADB, 2013).

Philippines, Thailand, and Vietnam did not mention CCUS technologies in their climate actions, although they could largely benefit from it.

Coal and natural gas-powered plants emerge as an opportunity for CO_2 capture in Southeast Asian countries, especially in Indonesia and Thailand (Darmawan et al., 2017). Additionality, the natural gas growing demand and national production among these countries also led to opportunities for CCUS in natural gas-processing facilities and on power plants with natural gas combined cycle. Additionally, domestic coal occurrence can be associated with CO_2-enhanced coal methane technologies—a branch of the CCUS chain (CO_2 utilization and storage) that is heavily deployed in Asia.

The four countries combined have a geological storage capacity for about 54 Gt of CO_2. Most of this storage capacity is attributed to saline aquifers (ADB, 2013). However, oil and gas fields and coal strata could represent a more feasible starting point for CO_2 storage due to their potential for additional hydrocarbon/coalbed methane production through CO_2 utilization (CO_2-enhanced oil/gas/CBM recovery technologies), and the consequent revenue could abate CCUS costs. Additionally, the geological characteristics of these fields are well-known, and the capacity estimates are less uncertain.

Considering the economic aspects of CCUS deployment, economic analyses carried out by the Asian Development Bank attribute the lowest cost to natural gas-processing units for CO_2 capture with onshore storage into oil and gas fields (ADB, 2013). Other low-cost options include CO_2 capture from natural gas combined-cycle and supercritical pulverized coal power plants, all associated with onshore geological storage into oil and gas fields with CO_2-EOR.

Overall, considering the technical and economic aspects of CCUS, early-stage commercial deployment should target geological storage sites where CO_2-EOR is applicable. This is due to the significant cost and uncertainty reduction, which is required for promoting further large-scale CCUS deployment. Additionally, CO_2-EOR and CO_2-ECBM opportunities were already identified in Indonesia (Darmawan et al., 2017), Thailand, and Vietnam (ADB, 2013; Nguyen-Trinh and Ha-Duong, 2015), and could represent a feasible entry-point for CCUS among these countries.

Large-scale deployment of CCUS requires a supportive legal and regulatory regime. Despite the opportunities, none of the aforementioned Southeast Asian countries have specific regulations on CCUS. Although their existing legal and regulatory framework can be partially adapted for CCUS, since it covers common issues, for example, land and subsurface rights, environmental impact assessment requirements, and other relatable concerns, it is still insufficient to promote large-scale CCUS projects. The CCUS regulatory regime must address challenging issues that are specifically related to CCUS technologies, such as permitting, investment, ownership, storage site operation, closure, postclosure, and monitoring activities, CO_2 leakage risk, and environmental impacts and liabilities over a geologic timescale, which are not yet resolved by the exiting regulation.

Considering the complexity of regulating CCUS activities due to cross-chain integration and geological storage liabilities, as well as the many involved entities in such a process (e.g., component agencies, operators, and other stakeholders), is it unlikely that a comprehensive CCUS regulation will be established in the short term (ADB, 2013). Facing this challenge, the existing regulation that can be relatable to CCUS must be applied and adapted for promoting pilot and demonstration CCUS projects. In parallel to enabling a regulatory environment until pilot and demonstration projects are sufficiently deployed, a specific legal and regulatory regime should be developed to promote large-scale CCUS among these countries.

CCUS technologies have a key role in decarbonizing energy systems in Indonesia, Thailand, Vietnam, and the Philippines. In Southeast Asia, and worldwide, CCUS is strategically necessary to marry the rapidly growing energy demand and reliance on fossil fuels with climate goals. Consequently, it becomes essential to promote CCUS in these countries and therefore to establish a regulatory regime that is supportive to CCUS, as well as to promote financial incentives and climate policies (Nguyen-Trinh and Ha-Duong, 2015).

15.4.3 Japan

Japan already met its commitment of introducing CCUS technologies into the national climate change mitigation strategy until 2030 (IPCC, 2018), and already deployed a fully integrated CCUS value chain in a project entitled Tomakomai. The Tomakomai CCUS demonstration project is a full chain CCUS project applied to the industrial sector, with CO_2 capture from a hydrogen production unit and storage in deep saline formations. The project has been conducted by the METI, New Energy and Industrial Technology Development Organization, and Japan CCS Co., Ltd (JCCS)—a collaboration between the Japanese government, the private sector, and local communities, also supported by the Japanese Research Institute of Innovative Technology for the Earth (RITE, 2015; JCCS, 2020; GCCSI, 2020).

The CO_2 from the hydrogen production unit in the Idemitsu Kosan Co., Ltd. Hokkaido Refinery is pipelined for 1.4 km to a CO_2 capture facility. Later, the captured CO_2 is compressed and transported a further 4 km to the offshore storage unit, which includes two geological formations: Moebetsu and Takinoue. In 2016 the Tomakomai project CO_2 geological storage component stored 100,000 tons of CO_2 injection per annum and achieved 300,000 tons of cumulative CO_2 injection in November 2019 (METI, 2019).

The CO_2 geological storage in offshore sites in Japan abides by the Act on Prevention of Marine Pollution and Maritime Disaster, with a storage permit issued by the Minister of the Environment. The storage permit holder is required to conduct monitoring, as stated in the "Monitoring Plan" of the project. The project monitoring plan includes analysis of pressure and temperature of the reservoir, as well as seismic surveys to monitor the CO_2 plume. Additionally, marine environmental impact surveys are required quarterly (seasonal). The monitoring plan is submitted with the storage permit application and should guarantee that CCUS is being conducted safely as planned (Câmara et al., 2013; METI, 2019).

The Tomakomai project is one of the most advanced CCUS demonstration projects in Asia and has been an example of successful integration of the CCUS technologies to the global community. The Tomakomai project demonstrated that the CCUS chain deployment can be technically and economically feasible, as well as secure, as long as appropriate monitoring and environmental management are required. The project is also an example of public awareness toward CCUS, since its stakeholders have been enhancing the understanding of local communities about CCUS and its relevance. Communication efforts have been carried out continuously and have expanded to the international community and promoted public acceptance and collaborations (Câmara et al., 2013). For instance, in 2016 the Tomakomai project was recognized by the Carbon Sequestration Leadership Forum, a ministerial level initiative to advance CCUS technology, and JCCS Co., Ltd. was nominated the Asia—Pacific regional champion for stakeholder engagement (METI, 2019; JCCS, 2020).

Despite the significance of the Tomakomai project, Japan has no specific laws or regulations for CCUS. Consequently, the existing regulation was adapted and applied, such as the High-Pressure Gas Safety Act, Industrial Safety and Health Act and Gas Business Act for CO_2 capture; and the Mining Act and Mining Safety Act, as well as the Act on Prevention of Marine Pollution and Maritime Disaster were applied to CO_2 geological storage. Additional regulations were created for CO_2 injection management standards and monitoring, for instance the METI guideline "For safe operation of a CCS demonstration project" and "Reservoir Management Standards Manual During CO_2 Injection," which were based on international CCUS regulations, technical standards, and guidelines (METI, 2019).

Overall, the Japanese experience of CCUS has been very successful and its knowledge-sharing approach contributed to the global awareness toward CCUS feasibility. Although legal and regulatory frameworks for CCUS and conceptual design of business models are not very advanced in Japan yet, or in Asia in general, the Tomakomai CCUS demonstration project is an example of CCUS integration, effective monitoring, and public and private sector collaboration.

15.4.4 India

Opportunities for CCUS in India are mainly related to its significant coal reserves and exploration, and to associated "clean-coal" technologies, such as enhanced coalbed methane (CO_2-ECBM).

India continues to present rising CO_2 emissions. For instance, in 2019 CO_2 emissions in India rose by approximately 5%, driven by a rapid increase in energy demand and, consequently, rising demand for coal in the electricity generation (GCCSI, 2019). In accordance with the increasing share of coal in India's electricity mix, the International Energy Agency estimates that by 2060 India will account for 20% of the global CO_2 emissions capture and storage (IEA, 2019). In this scenario, India holds major opportunities for CCUS technologies, which are essential to meet climate goals and the energy demand within the country.

India also hosts opportunities for CCUS in its industrial sector, since the country deploys numerous carbon-intensive industries, such as steel and cement, of which India is the world's second largest global producer (Viebahn et al., 2011). In this context, the Indian company Dalmia Cement has committed to becoming carbon negative by 2040 by adopting CCUS technologies. In 2019 the cement company announced a large-scale carbon capture facility in Tamil Nadu, in a partnership with Carbon Clean Solutions, which will provide the plant's technology (GCCSI, 2019, 2020).

With regard to CO_2 geological storage, estimates suggest approximately 500 $GtCO_2$ of storage capacity, which is mainly attributed to saline aquifers, coal seams, and basalts (Singh et al., 2006). CO_2 geological storage in basalts could be an innovative technology in

India, due to the large occurrence of this lithology and its feasibility has been studied (e.g., mineral trapping in the Deccan basaltic formation accounts for 200 Gt of the total storage capacity in India).

Despite the opportunities, CCUS technologies establishment in India require further initiatives and investment. To date, India has few incentives to deploy large-scale CCUS projects due to the lack of financial support, business models, and regulatory regime (Kumar et al., 2020; Viebahn et al., 2011). Therefore CCUS in India, as in most developing and industrial economies, requires international collaboration.

15.5 Final remarks and conclusions

CCUS technologies are necessary for reducing the worst impacts of climate change, due to its potential for CO_2 abatement in the carbon-intensive power and industrial sectors worldwide (GCCSI, 2019, 2020). CCUS gained relevance with the Intergovernmental Panel on Climate Change 1.5-degree target scenario, where it accounts for 32% of global CO_2 emission reduction by 2050 (IPCC, 2018). However, to promote such a contribution to climate goals, CCUS must be deployed at the large scale, especially in developing industrial nations, such as most Asian countries.

The role of CCUS in Asia, as in most developing countries, is to be a transitional technology to a more sustainable energy mix and to provide CO_2 mitigation for its continued use of fossil fuels for electricity generation. Opportunities for CCUS in Asian countries are significant, due to the intensive industrialization process, the associated growing energy demand and CO_2 emissions, and due to geological storage capacity.

In this chapter, CCUS prospects for China, India, Indonesia, Thailand, Vietnam, and Japan were analyzed and discussed. The analysis identified similar opportunities to CCUS deployment among these countries, including the improved energy access and consequent increasing energy demand, with continued reliance on coal and natural gas for energy generation. This scenario leads to cost-effective opportunities for CO_2 capture from coal-fired power plants and natural gas-processing units. Further opportunities include CO_2 utilization, which can promote economic feasibility to CCUS early-stage development, especially through enhanced hydrocarbon technologies, such as CO_2-EOR, enhanced natural gas recovery, and ECBM. Note that most of the studied countries already deploy these CO_2 utilization technologies and have the required knowledge and technological background to CCUS large-scale deployment.

As regards climate goals, ambitious targets to reduce CO_2 emissions by mid-century, such as the net-zero targets recently announced by China, highlight the need of CCUS technologies for these targets to be achievable in the required time frame. Such relevance relies on CCUS being the only likely solution to decarbonize carbon-intensive industries, such as the power and industrial sectors. Additionally, CCUS deployment fosters a new low-emission economy that is rising globally, creating opportunities for sustaining the Asian rapid economic development.

Despite the opportunities for CCUS in Asia, legal and regulatory barriers, which overlap with economic uncertainty, are limiting its commercial-scale deployment. The barriers include the absence of a market mechanism for carbon reduction, absence of a favorable regulatory environment and lack of political incentives for CCUS. Although there are many specificities for each analyzed country, rapid CCUS deployment in Asia requires climate policies and a regulatory regime that is supportive to CCUS.

As shown in successful examples of CCUS implementation, such as Japan, national and regional governments have either enacted laws and/or adapted the existing legal regime for promoting CCUS. The successful regulatory framework includes jurisdictions similar to the international experience, such as a mechanism for permitting for CO_2 storage, management and monitoring of injection/storage during the postclosure period, reporting, verification, and transfer of liability, in addition to CCUS financing systems, such as tax credits and subsidies.

Overall, Asian countries could largely benefit from CCUS technologies. Large-scale deployment of CCUS technologies can bring revenue through CO_2 utilization and by exporting and improving this globally demanded strategy and emerging industry. Additionally, CCUS can mitigate CO_2 emissions from these carbon-intensive economies while supporting a sustainable energy transition.

Acknowledgments

We are grateful to the "Research Center for Gas Innovation—RCGI" (Fapesp Proc. 2014/50279-4), supported by FAPESP and Shell, organized by the University of São Paulo, and the strategic importance of the support granted by the ANP (National Agency of Petroleum, Natural Gas and Biofuels of Brazil) through the R&D clause. We also thank the support from the National Agency for Petroleum, Natural Gas and Biofuels Human Resources Program (PRH-ANP), funded by resources from the investment of oil companies qualified in the R,D&I clauses from ANP Resolution No. 50/2015 (PRH 33.1—Related to Call No. 1/2018/PRH-ANP; Grant FINEP/FUSP/USP Ref. 0443/19).

References

ADB (Asian Development Bank), 2013. Prospects for Carbon Capture and Storage in Southeast Asia. Asian Development Bank, Mandaluyong City, Philippines.

ADB (Asian Development Bank), 2015. Roadmap for carbon capture and storage demonstration and deployment in the People's Republic of China Mandaluyong City. Asian Development Bank, Philippines.

Bachu, S., 2007. Carbon dioxide storage capacity in uneconomic coal beds in Alberta, Canada: methodology, potential and site identification. International Journal of Greenhouse Gas Control 1 (3), 374–385. Available from: https://doi.org/10.1016/S1750-5836(07)00070-9.

British Petroleum, 2020. Statistical Review of World Energy (accessed 15.07.20).

Câmara, G.A.B., Andrade, J.C., Oliveira, J.O., Rocha, P.S., 2013. Status of CCS technology in Japan and Brazil: a comparative analysis. Journal of Environmental Science and Engineering B2 (2013), 155–176.

Carbon Brief, 2020. Retrieved from: <https://www.carbonbrief.org/paris-2015-tracking-country-climate-pledges> (accessed 11.09.20).

Carfora, A., Pansini, R.V., Scandurra, G., 2019. The causal relationship between energy consumption, energy prices and economic growth in Asian developing countries: a replication. Energy Strategy Reviews 23, 81–85. Available from: https://doi.org/10.1016/j.esr.2018.12.004.

Darmawan, A., Sugiyono, A., Liang, J., Tokimatsu, K., Murata, A., 2017. Analysis of potential for CCS in Indonesia. Energy Procedia 114 (2017), 7516–7520ISSN 1876-6102. Available from: https://doi.org/10.1016/j.egypro.2017.03.1884.

Ding, H., Zheng, H., Liang, X., Ren, L., 2020. Getting ready for carbon capture and storage in the iron and steel sector in China: assessing the value of capture readiness. Journal of Cleaner Production 244, 118953. Available from: https://doi.org/10.1016/j.jclepro.2019.118953.

Europe Union. Paris Agreement. Retrieved from: <https://ec.europa.eu/clima/policies/international/negotiations/paris_en> (accessed 11.09.2020).

Garg, A., 2009. Coal and energy security for India: role of carbon dioxide (CO$_2$) capture and storage (CCS). Energy 34 (8), 10321041. Available from: https://doi.org/10.1016/j.energy.2009.01.005 >.

Global CCS Institute, 2018. Carbon capture and storage in de-carbonising the Chinese economy. <https://www.globalccsinstitute.com/news-media/insights/carbon-capture-and-storage-in-de-carbonising-the-chinese-economy/> (accessed 10.20).

GCCSI, Global CCS Institute, 2019. Targeting Climate Change. Global Status Report. GCCSI, pp. 1−46.

GCCSI, Global CCS Institute, 2020. The Global Status of CCS. Global Status Report. GCCSI, pp. 1−44.

Hanif, I., et al., 2019. Fossil fuels, foreign direct investment, and economic growth have triggered CO$_2$ emissions in emerging Asian economies: some empirical evidence. Energy 171 (15), 493−501. Available from: https://doi.org/10.1016/j.energy.2019.01.011.

Hiep, D.T., Hoffmann, C., 2020. A power development planning for Vietnam under the CO$_2$ emission reduction targets. Energy Reports 6 (Supplement 2), 19−24. Available from: https://doi.org/10.1016/j.egyr.2019.11.036.

Hong, S., Bradshaw, C.J.A., Brook, B.W., 2013. Evaluating options for the future energy mix of Japan after the Fukushima nuclear crisis. Energy Policy 56, 418−424. Available from: https://doi.org/10.1016/j.enpol.2013.01.002.

IPCC (Intergovernmental Panel on Climate Change), 2018. Global warming of 1.5°C. An IPCC Special Report on the Impacts of Global Warming of 1.5°C Above Preindustrial Levels and Related Global Greenhouse Gas Emission Pathways, in the Context of Strengthening the Global Response to the Threat of Climate Change.

IEA (International Energy Agency), 2018. Coal 2018: analysis and forecasts to 2023, International Energy Agency, Paris. <https://webstore.InternationalEnergyAgency.org/marketreport-series-coal-2018>.

IEA (International Energy Agency), 2019a. Carbon capture, utilisation and storage. International Energy Agency.

IEA (International Energy Agency), 2019b. World Energy Outlook 2019. International Energy Agency.

IEA (International Energy Agency), 2020a. CCUS in power. International Energy Agency.

IEA (International Energy Agency), 2020b. The challenge of reaching zero emissions in heavy industry. International Energy Agency.

JCCS, 2020. Japan CCS Co., Ltd. (JCCS). New business development of CCS (Carbon Dioxide Capture and Storage). <https://www.japanccs.com/en/corporate/info.php> (accessed 09.20).

Jiang, K., Ashworth, P., 2021. The development of Carbon Capture Utilization and Storage (CCUS) research in China: A bibliometric perspective. Renewable and Sustainable Energy Reviews 138 (ISSN 1364-0321), https://doi.org/10.1016/j.rser.2020.110521.

Jiang, K., Ashworth, P., Zhang, S., Liang, X., Sun, Y., Angus, D., 2020. China's carbon capture, utilization and storage (CCUS) policy: a critical review. Renewable and Sustainable Energy Reviews 119 (April 2019), 109601. Available from: https://doi.org/10.1016/j.rser.2019.109601 >.

Khanna, N., et al., May 2019. Energy and CO$_2$ implications of decarbonization strategies for China beyond efficiency: modeling 2050 maximum renewable resources and accelerated electrification impacts. Applied Energy 242, 12−26. Available from: https://doi.org/10.1016/j.apenergy.2019.03.116.

Kumar, S., Managi, S., Jain, R.K., 2020. CO$_2$ mitigation policy for Indian thermal power sector: potential gains from emission trading. Energy Economics 86, 104653. Available from: https://doi.org/10.1016/j.eneco.2019.104653.

Lee, T.C., Lam, J.S.L., Lee, P.T.W., 2016. Asian economic integration and maritime CO$_2$ emissions. Transportation Research Part D: Transport and Environment 43, 226−237. Available from: https://doi.org/10.1016/j.trd.2015.12.015.

Lee, Z.H., et al., 2013. An overview on global warming in Southeast Asia: CO$_2$ emission status, efforts done, and barriers. Renewable and Sustainable Energy Reviews 28, 71−81. Available from: https://doi.org/10.1016/j.rser.2013.07.055.

Li, J., Liang, X., Cockerill, T., Gibbins, J., Reiner, D., 2012. Opportunities and barriers for implementing CO$_2$ capture ready designs: a case study of stakeholder perceptions in Guangdong, China. Energy Policy. 45, 243-251, ISSN 0301−4215, https://doi.org/10.1016/j.enpol.2012.02.025.

Ma, J.J., Du, G., Xie, B.C., 2019. CO$_2$ emission changes of China's power generation system: input−output subsystem analysis. Energy Policy 124, 1−12. Available from: https://doi.org/10.1016/j.enpol.2018.09.030.

Meangbua, O., Dhakal, S., Kuwornu, J.K.M., 2019. Factors influencing energy requirements and CO_2 emissions of households in Thailand: a panel data analysis. Energy Policy 129, 521−531. Available from: https://doi.org/10.1016/j.enpol.2019.02.050.

METI, 2019. Ministry of Economy, Trade and Industry (METI), Japan. Report on Large-Scale CCS Demonstration Project Compiled. <https://www.meti.go.jp/english/press/2020/0515_004.html> (accessed 08.20).

Mondal, M.A.H., et al., March 2018. The Philippines energy future and low-carbon development strategies. Energy 147, 142−154.

Nguyen-Trinh, H.A., Ha-Duong, M., 2015. Perspective of CO_2 capture and storage (CCS) development in Vietnam: results from expert interviews. International Journal of Greenhouse Gas Control 37 (2015), 220−227. ISSN 1750-5836. Available from: https://doi.org/10.1016/j.ijggc.2015.03.019.

Nong, D., et al., July 2019. Growth of low emission-intensive energy production and energy impacts in Vietnam under the new regulation. Journal of Cleaner Production 225, 90−103. Available from: https://doi.org/10.1016/j.jclepro.2019.03.299.

RITE, 2015. Research Institute of Innovative Technology for the Earth. Current status and issues of CCS development in Japan. <http://www.jcoal.or.jp/coaldb/shiryo/material/upload/2-7Speech%202_Tsuzuku%20presentation.pdf> (accessed 08.20).

Rodrigues, C., Dinis, M.A., Lemos De Sousa, M.J., 2013. Unconventional coal reservoir for CO_2 safe geological sequestration. International Journal of Global Warming 5 (1), 46−66. Available from: https://doi.org/10.1504/IJGW.2013.051481 >.

Rodrigues, C.F.A., 2002. The application of isotherm studies to evaluate the Coalbed Methane potential of the Waterberg Basin, South Africa, p. 287. Doctoral Thesis, Faculty of Sciences of Oporto University, Porto, Portugal, pp. 298.

Sethi, V.K., Vyas, S., 2017. An innovative approach for carbon capture and sequestration on a thermal power plant through conversion to multi-purpose fuels—a feasibility study in indian context. Energy Procedia 114, 1288−1296. Available from: https://doi.org/10.1016/j.egypro.2017.03.1240.

Sharvini, S.R., et al., 2018. Energy consumption trends and their linkages with renewable energy policies in East and Southeast Asian countries: challenges and opportunities. Sustainable Environment Research 28 (6), 257−266. Available from: https://doi.org/10.1016/j.serj.2018.08.006.

Singh, A.K., Mendhe, V., Garg, A., 2006. CO_2 sequestration potential of geological formations in India. In: Eighth International Conference on Greenhouse Gas Control Technologies, GHGT-8. Trondheim, Norway, June 19−22, 2006.

Sumabat, A.K., 2016. Decomposition analysis of Philippine CO_2 emissions from fuel combustion and electricity generation. Applied Energy 164, 795−804. Available from: https://doi.org/10.1016/j.apenergy.2015.12.023.

Supasa, T., et al., 2017. Sustainable energy and CO_2 reduction policy in Thailand: an input−output approach from production- and consumption-based perspectives. Energy for Sustainable Development 41, 36−48. Available from: https://doi.org/10.1016/j.esd.2017.08.006.

Viebahn, P., Höller, S., Vallentin, D., Liptow, H., Villar, A., 2011. Future CCS implementation in India: a systemic and long-term analysis. Energy Procedia 4 (2011), 2708−2715. ISSN 1876-6102. Available from: https://doi.org/10.1016/j.egypro.2011.02.172.

Wakiyama, T., Kuramochi, T., 2017. Scenario analysis of energy saving and CO_2 emissions reduction potentials to ratchet up Japanese mitigation target in 2030 in the residential sector. Energy Policy 103, 1−15. Available from: https://doi.org/10.1016/j.enpol.2016.12.059.

Wang, D., He, W., Shi, R., 2019. How to achieve the dual-control targets of China's CO_2 emission reduction in 2030? Future trends and prospective decomposition. Journal of Cleaner Production 213, 1251−1263. Available from: https://doi.org/10.1016/j.jclepro.2018.12.178.

Wang, N., Ren, Y., Zhu, T., Meng, F., Wen, Z., Liu, G., 2018. Life cycle carbon emission modelling of coal-fired power: Chinese case. Energy 162, 841−852. Available from: https://doi.org/10.1016/j.energy.2018.08.054.

Wang, P.-T., et al., 2020. Carbon capture and storage in China's power sector: Optimal planning under the 2 °C constraint. Applied Energy 263 (ISSN 0306-2619). https://doi.org/10.1016/j.apenergy.2020.114694. In press.

Wang, Q., et al., 2020. Comparative analysis of drivers of energy consumption in China, the USA and India—a perspective from stratified heterogeneity. Science of the Total Environment 698, 134117. Available from: https://doi.org/10.1016/j.scitotenv.2019.134117.

Wei, Y., et al., 2019. The decomposition of total-factor CO_2 emission efficiency of 97 contracting countries in Paris Agreement. Energy Economics 78, 365–378. Available from: https://doi.org/10.1016/j.eneco.2018.11.028.

Wei, W., Hao, S., Yao, M., Chen, W., Wang, S., Wang, Z., et al., 2020a. Unbalanced economic benefits and the electricity-related carbon emissions embodied in China's interprovincial trade. Journal of Environmental Management 263, 110390. Available from: https://doi.org/10.1016/j.jenvman.2020.110390.

Wei, W., Zhang, P., Yao, M., Xue, M., Miao, J., Liu, B., et al., 2020b. Multi-scope electricity-related carbon emissions accounting: a case study of Shanghai. Journal of Cleaner Production 252, 119789. Available from: https://doi.org/10.1016/j.jclepro.2019.119789.

Zeaei, S.M., 2015. Effects of financial development indicators on energy consumption and CO_2 emission of European, East Asian and Oceania countries. Renewable and Sustainable Energy Reviews 42, 752–759. Available from: https://doi.org/10.1016/j.rser.2014.10.085.

16

Relevant aspects of carbon storage activities' liability in paradigmatic countries: Australia, Brazil, Canada, European Union, Japan, Norway, United Kingdom, and United States

Silvia Andrea Cupertino, Romario de Carvalho Nunes and Hirdan Katarina de Medeiros Costa

Institute of Energy and Environment, University of São Paulo, São Paulo, Brazil

16.1 Introduction

Social pressure on environmental issues caused by anthropic actions is driving changes in the legal system to fulfill public and private demand on the insertion of Environmentally Sound Technologies addressing environmental problems (Câmara et al., 2011), such as carbon capture, utilization and storage (CCUS). Government actors willing to promote CCUS technology see legal liability as an important factor to ensure information about risks and appropriate measures to avoid unmeasurable consequences to the environment and society; as for private companies, unlimited liability associated with technological problems new to commercial scale deployment should be potentially known (Wilson et al., 2009).

CCUS has been slowly deployed and its investment represents less than 0.5% of the global investment in clean energy and efficiency technologies. Due to stronger climate goals, investment incentives are injecting a new impetus in CCUS projects. According to the International Energy Agency (IEA), in 2020 there were 19 large CCUS projects in industry and fuel transformation, recent project startups under development, indicating that CCUS is a technology option to reduce direct CO_2 emissions, including process emissions from the industrial sector, which represent one-quarter of the global CO_2 emissions. CCUS must play the central role as one of the four main pillars of global energy transitions alongside electrification based on renewable energy, bioenergy, and hydrogen (IEA, 2020a).

However, CCUS technology has an environmental risk bias, due to the criticality of the operation, leaving the operator with numerous measures in order to avoid environmental accidents, including the provision of a significant amount of information, detailing the project operation, result modeling, and a monitoring plan, and how the project will be completed, including decommissioning and rehabilitation plans (IEA, 2010). Potential liability and compensation issues need to be addressed due to long-term risks in a distant future and uncertainty, raising inter alia questions on the availability of financing if damage occurs and the operator is no longer in business, and potential damage of accidental release of CO_2 in the atmosphere, which poses the risk of instability (Faure, 2016).

Therefore financial responsibility and liability during postclosure care and long-term stewardship of carbon capture and storage (CCS) projects must balance global and local risks with the climate benefit of CCS deployment, requiring the development of institutional structures to manage CCS risks over the long term; moreover, resources must be guaranteed to cover public monitoring and potential remediation costs to make projects viable (Wilson et al., 2009). In this context, the implementation of a regulatory system for civil liability in CCUS activity must rely on policies that address the mechanisms of prevention, such as monitoring and verification protocols, mitigation and remediation, as well as the regulatory and legal framework (Wilson and Friedmann, 2007).

The chapter is divided into four sectors. The first part addresses prevention and environmental accidents management, social risk management, and security in the CO_2 storage activity. Then, it describes regulatory systems applied to civil liability in CCUS in the selected countries (Australia, Brazil, Canada, European Union, Japan, Norway, United Kingdom, and United States). All the countries chosen have specific legislation addressing liability and have projects in operation.

16.2 Prevention and control of environmental accidents

CCS literature recognizes potential damage categories, such as atmospheric release, impact on water resources, geological impacts, human health impacts, and ecological impacts, depending on the life cycle of the project. Potentials can be managed, monitored, and valuated, but there are uncertainties related to legal issues and technology (Faure, 2016), obliging the operator to take any corrective or remedial measures associated with the storage location and its respective costs (IEA, 2010). If the operator has been given CO_2 incentives for CCS operations, the operator may also be responsible for compensating for any leakage of CO_2 into the atmosphere in the context of the incentive regime (IEA, 2010).

Nardocci (2002) teaches that "the potential impact of technological development and changes in lifestyle, as well as the increased sensitivity to health and safety hazards, have placed environmental risks and quality among some of the greatest concerns of current society." Environmental accidents, in addition to affecting the environment, by corollary, also affect the human health, which raise the need for prevention and control. In this sense, Cunha (2005) states that the ecological disaster, which causes consequences for men, is taken as part of the damage to the environment.

There are three levels of compliance with the duty of environmental protection: (1) the duty to actively promote the improvement of the state of the environment, with the development of environmental improvement actions and investment in the rehabilitation of habitats and ecosystems; (2) the duty to avoid the progressive and gradual degradation of ecosystems, habitats and natural resources; and (3) the duty to prevent the occurrence of serious environmental accidents.

The use of new technologies for carbon sequestration and geological storage meet the duty to reduce environmental degradation in the face of a reduction in the emission of greenhouse gases (GHGs). However, carbon storage activities in long-term geological formations may cause accidents and/or environmental damage, resulting in civil liability. It is thus urgent to discuss the control and prevention of accidents in these activities, the social risk management and the safety of CO_2 storage activities.

Monitoring CCUS activities is essential to support several crucial elements of security and protection. The CO_2 monitoring practice involves several stakeholders, including the operator, the regulator, and other project stakeholders, including the public (IEA, 2010). The robust development of guidance for monitoring the design of the plan is recommended in secondary legal instruments (which are generally easier to update and to change than primary legislation), or in technical guidance documents (IEA, 2010).

A common element of many CCS-specific legal and regulatory frameworks is the inclusion of detailed requirements for site selection, monitoring, and verification. These requirements are an important aspect of the licensing process for the initial design, but in many cases remain obligations throughout the life of a storage operation. Taken in conjunction with financial security requirements, these obligations can be viewed as effective and anticipate the risks associated with the technology to ensure that they are minimized in the later stages of the project life cycle (Global CCS Institute, 2019).

Monitoring and verification requirements are therefore clearly a critical element of the liability regime in many jurisdictions. Effective monitoring and verification throughout the project life cycle will be critical to ensure that the CO_2 plume behavior is in line with the predicted models and that there is permanent containment of the injected CO_2. For those regimes that offer the opportunity to resign and/or to transfer an authority after the cessation of injection activities, the results of monitoring and verification are likely to be an important aspect in demonstrating compliance with regulatory standards (Global CCS Institute, 2019).

16.2.1 Social risk management

The need for social risk management arises from the paradoxical reflex of technological advances, as is the case with activities involving CCUS. According to Nardocci (2002), decision-making and definition of criteria regarding risks is a very complex process, which involves its definition, evaluation of subjective data and uncertainties, social acceptability, and management based on the distribution of risk and benefits, among other aspects.

Nardocci (2002) points out that risk management "(...) begins with the judgment on the acceptability of the calculated risk levels. More specifically, evaluation is the stage of understanding the problem and management is the stage of actions."

The authorization and/or implementation of new technologies and the resulting development must take into account economic and social issues, given that, as Cunha (2005) points out, in spite of all being affected, the less economically favored population will suffer the immediate consequences, as they are more susceptible to lack of sanitation, collection of waste of all kinds, and are randomly deposited in the poorest places.

Thus social risk management, in its analyses, must weigh social justice when defining the acceptability of risks and the prevention and control measures to be adopted. Public policies for risk prevention must associate preventive actions with social justice.

The leak or unintentional migration of CO_2 from storage sites can lead to a series of potential impacts such as risks associated with health, safety, and the environment (IEA, 2010), which can be categorized as:

- *Impact on the surface*: potential to cause asphyxiation and to damage ecosystems (effects of CO_2 leakage on neighboring populations, worker safety, and effects on the biosphere and hydrosphere, such as tree roots, terrestrial animals, and the quality of ground and surface water), as well as problems associated with impurities present in the injected material.
- *Impact on the subsurface*: contamination through the mobilization of metals or other contaminants that have an increased risk due to the presence of certain impurities. It also has physical effects, such as soil survey, induced seismicity, displacement of underground water resources, and damage to hydrocarbon production.

In this matter, the IEA brings the importance of providing a robust methodology and economic tools to identify, detect, and quantify the leakage of CO_2 from CO_2 storage reservoirs on the seabed, citing as an example the Project Strategies for Environmental Monitoring of Capture and Marine Carbon Storage (STEMM-CCS) (IEA, 2020b). This project made it possible to acquire previously unpublished data that made it possible to monitor, in real conditions, various aspects of CO_2 storage and possible leaks. STEMM-CCS provided best practices for selecting and operating CCS offshore sites, and the results were shared with industrial and regulatory stakeholders to help increase confidence in CCS physical security and support the progress of the European Union towards a carbon neutral society (STEMM-CCS, 2020).

Public opinion is always important and that is why governments and companies seek to minimize the negative impacts of their operations. In relation to CCS monitoring, unwanted advertising may result, for example, from observations of CO_2 bubbles emanating from the seabed near a storage location or a change in the local marine environment. In such cases, it is beneficial for the operator to minimize damage to reputation by documenting that it has a robust and efficient monitoring process able to locate, quantify, and characterize any leak still in the initial stage (Waarun et al., 2016).

16.2.2 Safety in carbon storage activity

The carbon storage activity, because it is potentially polluting due to the possibility of gas leakage from the geological reservoirs, must be structured in a stable manner and with containment parameters, in order to be an environmentally safe and acceptable undertaking, because, as already pointed out elsewhere, the storage of CO_2 must be less harmful to the environment than the continuous emission of the noncaptured gas (Ravagnani and Suslick, 2008).

In this context, the control and assessment of risks in carbon storage activities are guided by three specific criteria, namely: (1) the amount of CO_2 to be injected into the underground reservoir; (2) the density at which the gas remains when sequestered, which is less than the density of the upper containment layers; and (3) the storage capacity of the sealed tank after injection for hundreds of years. Variations in any of the aforementioned elements must be subject to constant analysis and restraint by the institutions that carry out CCUS activities and the competent state public authorities, in order to avoid the greatest of the damaging effects of the technique: carbon leakage (Carvalho, 2010). The necessary care with regard to safety can minimize the occurrence of damage, as well as the problems caused by an eventual accident (Cunha, 2005).

Corrective measures are needed to protect human health and the environment, and to maintain the effectiveness of a CCUS project as a method of reducing CO_2 emissions (IEA, 2010). Remediation is necessary to resolve any damage associated with significant leakage, unintended migration, or any other irregularity in the operation of a storage location. The best practice examples for such measures are those adopted in the oil field as well as clogging techniques using heavy mud, as applied in the case of blowouts, standard well repair techniques in case of well failure, and interception of leaking in a nearby well to intercept the leak (IEA, 2010).

16.2.2.1 *The key role of monitoring in ensuring the integrity and security of facilities*

Monitoring technologies are advancing and it would be ideal to use new technologies to process authorization requests by operators as well as to justify the reasons for the choices in planned monitoring including frequency (continuous, annual, etc.), which may be related to technical limitations and cost considerations, and a description of the locations in the storage complex and surrounding domains, where the technique will be applied (for passive in situ techniques), or broader descriptions for intermittent mobile techniques (IPCC, 2006).

The absence or failure in monitoring can lead to the occurrence of major socioenvironmental disasters, from which an immediate drop in the value of the shares of the responsible companies is common. In this case, many investors are expected to sell their shares because of the associated risk and because it may take years for the causes of the accident to be known (Varela and Milone, 2014).

Site-specific monitoring requirements [under the Intergovernmental Panel on Climate Change (IPCC) Guidelines for National Greenhouse Gas Inventories] have monitoring technologies that have been developed and refined primarily by the oil and gas sector. The suitability and effectiveness of these technologies can be strongly influenced by the geology paths and potential emissions at the storage sites; therefore the choice of monitoring technologies will need to be made site by site (IPCC, 2006).

The continuous and successful guarantee of CO_2 storage is based on monitoring activities with sufficient provision of information in the project for effectively calculating the tonnage of CO_2 storage and avoided CO_2 (IEA, 2010). The appropriate location for a monitoring program must be comparable to the risk of storing natural gas and oil extraction.

For Dean et al. (2017), the bowtie is one of the methods capable of previously identifying the degraded barriers for maintaining the integrity of the installation and proposing corrective barriers in case an unwanted event occurs, as it provides a structure for systematic evaluation of event risks with the potential to affect storage performance. The "bowtie" represents the relationship between the five key elements that form it, that is, between the (1) main event, (2) threats, (3) consequences, (4) preventive safeguards, and (5) safeguards of corrective measures, as shown in the table below:

1. Main event	The unwanted event, placed in the center of the tie. In this case, the main event is the movement of the CO_2 cloud outside the storage complex
2. Threats	Conditions can lead to the main event. For example, the presence of a permeable failure or fracture system, injection-related stresses (pressure/thermal) or poorly connected abandoned wells
3. Consequences	Possible adverse results due to the unexpected occurrence of the main event. For example, the emission to the marine environment locally impacting the flora
4. Preventive safeguards	Decrease the likelihood of a threat leading to the main event. For example, the effects of the injection pressure are likely to be small, as the injection is good and the storage location is under subhydrostatic pressure
5. Corrective safeguards	Decrease the likelihood of significant consequences due to a top event. For example, the presence of a permeable formation below the storage complex seal provides alternative secondary storage

Elaborated by the authors from Dean, M., et al., June 2017. A risk-based framework for measurement, monitoring and verification (MMV) of the goldeneye storage complex for the Peterhead CCS project, UK. International Journal of Greenhouse Gas Control 61, pp. 1–15.

Therefore the integrity of the CO_2 storage facility is ensured by the development of a structured risk management process based on the well-established barrier (safeguard) approach, whose objective is to identify the necessary monitoring tasks and their respective technologies to reduce storage risks to a minimum.

16.3 Chosen countries experience

Due to the need for specific regulation of the sector and the civil liability underlying the risk activity of geological carbon sequestration, we present regulatory experiences in the selected countries, namely, Australia, Brazil, Canada, countries belonging to the European Union, Japan, Norway, United Kingdom, and the United States.

The development of a permit or licensing model to regulate CCS activities, in part or throughout the project life cycle, has played a central role. To date, legislation has been developed at the national, regional, and/or state and provincial levels in the United States, Canada, Europe, and Australia. Within these jurisdictions, specifically, several parallels can be drawn regarding the handling of liability, as regulators have sought to clearly allocate a wide range of potential responsibilities between the operator and the regulator throughout the project life cycle. In some cases, this has been achieved through the design and implementation of new mechanisms; however, on many occasions, much broader obligations are likely to be borne by operators through the implied application of a broader body of legislation and jurisprudence (Global CCS Institute, 2019).

16.3.1 Australia

Australia has a constantly evolving CCUS regulation system. According to Cook (2009), a key factor in this situation is the fact that Australia is one of the largest emitters of GHG per capita in the world, despite being responsible for only 1.5% of the global emissions, due to its high dependence on energy from fossil fuels.

Therefore Australia was concerned with regulating the implementation of CCUS technology. The country has specific regulations on the mechanism of geological carbon sequestration, detailing several important criteria, such as competence, jurisdiction, emission limits, tax incentives, and accountability (Monteiro Júnior, 2015).

In Australia, there is legislation at the federal, state, and territorial levels on CCUS.[1] The Greenhouse Gas Storage Act 2009 foresees that the licenses for exploration and storage of carbon, to be granted and/or renewed, depend on an assessment by an environmental authority (40, 96, 118, and 130); and that in the activities to be developed, in which there is the implication of possible problems and/or water pollution, there must also be prior approval by the Minister responsible for managing the Water Act 2000 (Sections 54, 57, 144, 147, and 165) (Costa et al., 2018).

Under the provisions of Section 111 of the Petroleum and Geothermal Energy Act 2000, CCUS operators are obliged to pay environmental rehabilitation compensation to the state because of a potential threat or serious environmental damage. Those responsible for granting licenses can also adopt measures to prevent or to minimize environmental damage and land rehabilitation (Costa et al., 2018).

The Greenhouse Gas Geological Sequestration Act 2008 (Victorian Onshore Act) determines that CO_2 storage cannot be carried out when it represents a risk to human health or to the environment (Sections 40 and 96) (Costa et al., 2018). In turn, the Offshore Petroleum and Greenhouse Gas Storage Act 2010 (Victorian Offshore Act) brings environmental protection mechanisms linked mainly to the closure of the site and remediation after the end of the CO_2 injection (Section 371) (Victorian Legislation, 2008).

In Australia, the transfer of long-term responsibility for carbon storage activities takes place at least 15 years from the closure of the storage site, and provided that the stability of the site is demonstrated, there are no significant risks of gas leakage; there is a guarantee to cover monitoring expenses (Romeiro-Conturbia, 2014). Despite the regulation of the

---

[1] We highlight the following Australian rules that have implications for civil liability arising from carbon capture and storage activities: Greenhouse Gas Storage Act 2009 (Qld GHG Storage Act); Petroleum and Geothermal Energy Act 2000 (South Australia P&GE Act); Greenhouse Gas Geological Sequestration Act 2008 (Victorian Onshore Act); and Offshore Petroleum and Greenhouse Gas Storage Act 2010 (Victorian Offshore Act).

sector in Australia, there is still a lack of clear and detailed rules regarding long-term risks, and risk-taking and responsibility for the government (Cook, 2009).

In 2020 the IEA panel called CCUS in Clean Energy Transitions reinforced the item Risk Mitigation Measures as a key element in the formulation of a successful political structure for creating a sustainable and viable market for CCUS. In this item, the importance of transferring and/or sharing part of the responsibility for the stored CO_2 is emphasized, particularly after the project is closed. Australian legislation allowing the transfer of CO_2 responsibility to the state is cited as a world reference (IEA, 2020a).

16.3.2 Brazil

The Brazilian legislation, especially that regarding the environment, is very comprehensive, covering a wide range of topics (Costa et al., 2017), although there are gaps in specific activities (such as CCUS) (Costa et al., 2018). First, it is necessary to elucidate that the monitoring phase (including the issuance of reports and eventual inspections) may depend on the type of licensing issued to the operator. According to Article 225, § 1, IV, it is incumbent upon the Public Power to demand, in accordance with the law, a prior environmental impact study on the installation of a work or activity potentially causing significant environmental degradation. In addition, Article 23 defines the common competence of the Union, the States, the Federal District, and Municipalities to protect the environment and to combat pollution in any of its forms (Costa et al., 2017).

Therefore to perform this role, in its Article 10, § 4, Law no. 6938/81 (Law of the National Environment Policy) provides the competence of IBAMA for licensing activities and works with significant impact, national or regional, later regulated by Decree no. 99274/90. In the oil and gas industry, the execution of business activities is mostly monitored by the National Agency of Petroleum, Natural Gas, and Biofuels (ANP).

ANP already performs its functions based on the sector legislation and, in the case of criminal environmental liability, it follows Law no. 9605/98 (Environmental Crimes Law) and Decree no. 3179/99, which regulates it (Costa and Musarra, 2019). These rules deal with aspects related to criminal action, crimes against the environment, and administrative infractions. ANP inspection actions are carried out in the form of audits, through samples and analysis of data and evidence, which aim to verify the operator's compliance with the requirements of the technical documentation regulated by ANP Resolution 37/2015, which provides for the concession of deadlines for treating nonconformities and eventual elaboration of the infraction notice (Costa and Musarra, 2019). For this, Law 6938/81 (Law of the National Environment Policy) provides, in its Article 10, § 4, the competence of IBAMA for licensing activities and works with significant impact, national or regional, later regulated by Decree no. 99274/90 (Costa and Musarra, 2019).

The Environmental Crimes Law defines the—administrative, civil, and criminal—responsibility of the legal entity and also allows the individual that committed the offense to be prosecuted. Specifically, in relation to environmental licensing, sanctions may result in suspension or cancelation of registration, license, or authorization, loss or restriction of tax incentives and benefits, and loss or suspension of participation in financing to credit institutions.

Due to the robust Brazilian legislation on the environment, as well as the solid municipal structure to oversee numerous activities, in Brazil, there is an environment conducive to the creation of specific rules for CCUS activities (Costa et al., 2018). Regarding the inspection structure, the ANP's own regulation on operational safety already prescribes the exercise of its functions in maritime installations, onshore installations, pipelines, and submarine systems, to which the CCUS activities could be combined.

The development of a legal and regulatory apparatus must be carried out with full knowledge of the relevant existing laws, as the CCUS can be more easily regulated by modifying existing structures, rather than the elaboration of completely new ones. In general, this can occur in conjunction with jurisdictions in the oil and gas sector, as has been done in Canada, Norway, and Australia (Rathmann, 2017).

The regulatory framework should include a set of information and procedures (stages of a project, active agents, and supervisory bodies) for the safe and effective implementation of CCUS techniques in Brazil focusing mainly on the stages of transport and geological storage of CO_2 (Rathmann, 2017).

Note that the oil and gas sector dominate the techniques of capture, transport, and injection of gas in geological reservoirs. In other words, agents working in the oil and gas sector in Brazil have experience in using gas separation technologies in the production of natural gas that would be similar to the technologies used for capturing CO_2, for example. Therefore it makes sense that the regulatory body that should adjust and supervise CCUS projects in Brazil has expertise in regulation in the oil and natural gas sector. The ANP was considered a possible regulatory body in this proposal because it is responsible for regulating the activities of the oil sector in Brazil. The federal environmental agency (IBAMA) would also have a relevant role in the emissions of licenses (prior, installation, and operation) in the stages of a CCUS project when these are necessary (Rathmann, 2017).

16.3.3 Canada

At all levels of Canadian government there are laws regulating the impact of polluting activities on the environment. Environmental protection is not expressed in the Canadian Constitution; thus the competence to legislate is concurrent, presenting complex environmental regulations that, in many cases, become vague, which allows for a more flexible application of laws.

Canadian provinces have excelled in regulating carbon capture and geological storage activities; this position can be evidenced by the creation and/or adaptation of legislation promoted by the governments of Alberta, British Columbia, Nova Scotia, and Saskatchewan (Costa et al., 2018).

The province of Alberta has specific legislation regarding CO_2 capture and storage activities. The Mines and Minerals Act, R.S.A. 2000 (Alberta Queen's Printer, 2000), establishes a ban on the injection of CO_2 captured in a subsurface reservoir without authorization and/or license (Sections 54 and 117). The Carbon Sequestration Tenure Regulation lists specific rules for the CO_2 exploration and storage concession contracts, such as concession period and renewal of the CCUS activity contract (Regulations 10 and 11); the maximum area to be covered by a carbon sequestration contract (Regulation 12); permission to drill wells and inject

CO_2 into deep underground reservoirs (Regulation 9); prohibition on the right to earn, work, or recover any minerals found within the land covered by the carbon sequestration concession, among others (Alberta Queen's Printer, 2011).

The Canadian Carbon Sequestration Tenure Regulation complements the licensing issue as it determines the need for all Monitoring, Reporting, and Verification (MRV) plans to present an analysis of the likelihood that operations will interfere with mineral recovery in addition to linking contract renewal/lease to renew the MRV triennial (IEA, 2010). This law also establishes obligations to obtain contracts such as paying the application fee prescribed in the Regulation of Administration of Mines and Minerals, paying the rent applicable for the first year of the contract, presenting evidence that the area covered by the application is suitable for CO_2 sequestration, submit a monitoring, measurement and verification plan for approval, and submit a decommissioning plan (IEA, 2010).

As for long-term liability, Alberta's regulation on CCUS provides for the transfer of all liabilities to the government; however, there is no specific minimum term for the referred transfer. It is nevertheless recommended that a certificate of closure be granted only at least 10 years after the beginning of the period of removal of structures and cessation of injections, and provided that compliance with the performance criteria required for closure is demonstrated.

According to the "Mineral Act," which revises Chapter M-17 of the Alberta Statute of 2000, Part 9 deals with the sequestration of captured carbon dioxide, establishing the tenancy property rights for the geological or geophysical assessment of the well sustainability to assess the suitability of the location. A license and approval by the regulator for drilling the well and use for the purpose indicated in the standard may be granted, and the lessee must submit a monitoring, measurement, and verification plan for approval; comply with the approved plan, submit "compliance" reports of the plan, meeting the requirements of the agreement.

Rights related to the injection of carbon dioxide by the sequestration modality may form an integral part of the monitoring plan, in particular the granting of the right to inject carbon dioxide captured on the surface of the sequestration reservoir. In addition, one must obtain license and approval from the regulator under the rules of the "Oil and Gas Conservation Act" before the drilling or use of the well begins. According to the mentioned standard, there must be a monitoring, measurement, and verification plan for approval, presentation of "compliance" reports of the plan, and meeting of the requirements of the agreement for this injection and capture phase.

The transfer of drilling rights can only take place after the Minister's consent, being the lessee's responsibility to monitor all the wells and proceed with the closure of all of their related activities, according to the regulation. Afterwards, the Minister will issue a closing certificate if the lessee has performed the duties of cessation of the activity, or if it has abandoned all the wells and installations in accordance with the requirements of the "Oil and Gas Conservation Act," as well as with the recovery requirements contained in the "Environmental Protection and Enhancement Act." This indicates whether the closing period specified in the regulations has expired, there is compliance with the conditions specified in the regulation; the verification of stable and predictable behavior of carbon dioxide capture and the absence of significant risk of future leakage. Consequently, the issuance of the certificate must be notified to the regulator.

The certificate issuing has the effect of transferring the ownership of the captured-injected carbon dioxide to the Crown, which assumes the obligations as owner of the wells and facilities licensed under the Oil and Gas Conservation, Act, Environmental Protection and Enhancement Act, Environmental Protection and Enhancement Act, and Surface Rights Act, releasing the tenant from any obligation related to carbon dioxide injection. Therefore the Crown is responsible for compensating for damages arising from unlawful acts.

This rule provides for the creation of a postclosure management fund, which will be used to monitor the behavior of the captured carbon dioxide injected, as well as to pay the cost of suspending the well, abandonment costs, and related to claims and remediation in orphan facilities, as well as cost recovery incurred by the lessee, according to management under Section 40 of the "Financial Administration Act."

16.3.4 European Union

When analyzing Europe, the North Sea is observed to be at the center of the implementation of the CCUS. Two facilities already store 1.7 $MtCO_2$/year and at least 11 other projects with a combined capacity of almost 30 Mt/year are under development in Europe (IEA, 2020a). Almost 70% of the emissions from power generation and industry are located 100 km from a potential storage location and 50% are 50 km, although most of these locations are on land, where public opposition can prevent their development. The total storage capacity can reach 300 Gt, or almost 80 years of current emissions (IEA, 2020a).

The instrument that created the European Union—the 1957 Treaty of Rome—did not provide for the theme of the environment in the list of public policies in the regional bloc; nevertheless, a set of rules was developed involving the environment, mainly from the 1970s, in the form of directives and regulations, covering all areas of environmental protection (Queiroz, 2005). Currently, the protection of the environment occupies a central place in the discussion agenda of the European Union community process. Environmental issues fall within the field of shared competences, so both the European Union and the Member States can legislate on environmental policy (Cupertino, 2019).

This flexibility regarding the possibility of adopting unilateral environmental measures by member countries creates a new problem, namely, the need to harmonize environmental laws between domestic laws and the Community environmental policy (Queiroz, 2005). It is certain that in the search for compatibility of these regulations, as taught by Thorstensen (1998), the Court of Justice plays an important role in the environmental area through treaties, exercising a regulatory role at the discretion of member states, curbing measures of a protectionist character, which do not characterize trade.

Among the universe of environmental legislation in the EU, one states that member states must guarantee the compulsory license for carbon exploration (Article 5); meeting objective, published, and nondiscriminatory criteria, through procedures open to all entities with the necessary capacities, with limited area and volume; for the time necessary to carry out the relevant exploration activities; and conferring on the license holder the exclusive right to explore the relevant storage complex (European Union, 2009).

In the event of a leak of or significant irregularity, the operator must immediately notify the competent authority, as well as take the necessary corrective measures, at least according to the approved corrective measures plan (according to Article 16) (European Union, 2009).

According to the said Directive 2009/31/EC (European Union, 2009), the storage location will be closed in three situations: (1) meeting the relevant licensing conditions; (2) at the reasoned request of the operator, after authorization by the competent authority; and (3) decision of the competent authority as a result of the withdrawal of the storage license, due to the presence of leaks or significant irregularities, noncompliance with the terms of the license, due to technological changes or scientific findings; or 5 years after the license is issued and every 10 years thereafter as a result of license review.

After the storage site is closed, the operator is responsible for monitoring, communication, and corrective measures, in accordance with the requirements established in this directive, and must seal the site and remove the equipment intended for injection (European Union, 2009). When the authority decides to close after the storage license is withdrawn under the terms of paragraph 3 of Article 11, the responsibility for monitoring and corrective measures is transferred (European Union, 2009).

The transfer of long-term responsibility (Article 18), determined when all the available evidence indicates that the stored CO_2 is completely and permanently contained, occurs with the sealing of the site and the removal of the injection facilities; responsibility for the geological storage site will be transferred to the competent authority, provided that at least 20 years have passed (European Union, 2009). Still, it requires meeting financial obligations, and verification of the sealing of the place (European Union, 2009).

EU Directive 2009/29/EC (European Union, 2009) also includes the provision of a reserve fund that must be guaranteed when there is a transfer of long-term responsibility for future responsibilities and monitoring costs, in Article 20 (European Union, 2009).[2]

16.3.5 Norway

The Norwegian government attaches great importance to the CCUS. The geological storage of CO_2 in Norway has been carried out on a commercial scale, as is the case of the Sleipner Project, which started in 1996. From the implementation to the closure of the project, around 20 million tons of CO_2 are expected to be stored at a total cost of 80 million dollars; much less than what the company would have if CO_2 were being released into the atmosphere, close to 50 million dollars each year, taking into account emissions between 1996 and 1999 and the Norwegian government-imposed emission rates; and the project in the Snøhvit fields (in operation since 2007).

---

[2] Article 20.o.
Reserve Fund:
1. Before transferring responsibility in accordance with Article 18, Member States shall ensure that the operator makes a financial contribution available to the competent authority, in a manner to be decided by the Member States. The operator's contribution shall take into account the criteria referred to in Annex I and the elements relating to the history of CO_2 storage that are relevant for determining posttransfer obligations and covering at least the expected cost of monitoring over a 30-year period. This contribution can be used to cover the costs borne by the competent authority after the transfer of responsibility to ensure that CO_2 is completely and permanently confined to geological storage sites after the transfer of responsibility.

Although Norway does not have a specific CCS legal and regulatory framework, it has been working on new regulations that would cover the transport and storage of CO_2 in geological reservoirs on the Norwegian continental shelf, in accordance with the existing oil legislation, as well as ratifying international mechanisms for regulation of exploration and storage of CO_2. The central point of the CCUS activities assesses the impact not only from an economic but also from an environmental viewpoint, aiming at mitigating damages, according to Romeiro-Conturbia (2014, pp. 63−64).

Norway ratified the CO_2 export amendment for the 36th meeting of the London Convention and the 9th meeting of the London Protocol, November 3−7, 2015. Regulations related to the exploration of undersea reservoirs on the continental shelf for storage of CO_2 and related to the transport of CO_2 on the continental shelf were established by Royal Decree on December 5, 2014, in accordance with Section 3 of Law No. 12, of June 21, 1963, on scientific research and exploration of submarine natural resources other than petroleum resources. An important feature in these regulations is that they state that the plan for the development and operation of an underwater reservoir for CO_2 injection and storage must include preventive measures and mitigation of environmental impacts (Costa et al., 2018).

As for long-term liability, the country, following the European Union directive, indicated its intention to adhere to the minimum period of 20 years after the storage site is closed to transfer responsibility to the competent authority. It is also important to note that in Norway there is a law that imposes a tax on CO_2 emissions, which encourages the implementation of CCUS projects. According to Romeiro-Conturbia (2014), the main driver of this new CO_2 legislation in Norway is the mandatory implementation of the (European Union, 2009) in Norwegian law, covering the 20-year term proposed by the EU Directive (Global CCS Institute, 2019), adopting the European Union Emissions Trading Scheme with CCS projects.

16.3.6 Japan

Despite the lack of express regulation for the activity, Japan aims to be a carbon neutral country by 2050. The activities involving CCUS technology are also undergoing discussions and the implementation of regulatory frameworks. Among these, the following stand out: amendments to the 2007 Marine Pollution Prevention Law, to reflect the London Protocol 1996; participation in the International Convention for the Prevention of Pollution from Ships (MARPOL); establishment of Law Concerning the Promotion of Measures to Cope with Global Warming; and the development of the Kyoto Protocol Target Achievement Plan and the Action Plan for Building a Low Carbon Society (Câmara et al., 2011).

The country has committed to commercialize carbon capture and use technology by 2023 and CCS used in coal power generation by 2030. Japan has invested heavily in technology but has not shown any plans to tackle its dependence on coal. Thus the country has launched a long-term climate strategy (Sauer, 2019).

16.3.7 United States

The United States is a global leader in CCUS, accounting for more than 60% of the global CO_2 capture capacity and half of all the planned capacity, supported by new policy incentives and a favorable investment environment (IEA, 2020a). Most stationary emission sources in the United States are located close to potential geological storage sites: 85% of the emissions come from factories located 100 km from a site and 80% at 50 km. Total potential storage is estimated at 800 Gt, or 160 years from the current US energy sector emissions (IEA, 2020a).

In the United States, as Monteiro Júnior (2015) points out, the activities of injecting substances and waste underground and into geological formations have been carried out for more than 40 years through enhanced oil recovery (EOR). Due to the marked EOR activities, the United States has several regulations also applicable to specific geological CO_2 storage activities.

The regulation of activities involving the capture, transport, use, and storage of CO_2 (CCUS) reveals social and environmental criteria focused on safety and socioenvironmental responsibility (Nunes and Costa, 2020). The US Environmental Protection Agency (EPA) has established requirements (requirements for permits, site characterization, financial responsibility, etc.) at the federal level for operators who aim to store CO_2 in geological formations. Several federal states, in turn, have already created their sector regulations.

There is ample regulation for the sector in the United States, regarding civil liability for carbon storage and risk prevention mechanisms. As following, regulations and legislation on CCUS are highlighted (IEA, 2010):

1. *Chapter 43-05 Geologic Storage of Carbon Dioxide (North Dakota Storage Administration Statute)*: requires operators to implement the emergency and public safety response plan and the worker safety plan proposed in the respective permission request (Section 43). The Statute states that applications for licenses to operate storage facilities must include, among other things: (1) a map of the storage reservoir location and technical assessment of the storage facility; (2) a leak detection and monitoring plan for wells and surface installations; (3) a leak detection and monitoring plan using subsurface wells to monitor the movement of CO_2 outside the reservoir; (4) a performance guarantee to provide the North Dakota Industrial Commission with sufficient funds to satisfy any regulatory obligation that the operator fails to comply with; and (5) a closure plan (under Section 43-05-01-05).

2. *North Dakota Century Code Chapter 38-22 Carbon Dioxide Underground Storage (North Dakota CO_2 Storage Statute)*: the Charter requires that, before the North Dakota Industrial Commission issues a CO_2 storage permit, among other things (Section 38-22-08), the following is required: (1) the proposed storage facility be suitable for CO_2 injection and storage; (2) the proposed storage facility will not adversely affect surface water or fresh water; (3) CO_2 will not escape from the storage tank; (4) the storage reservoir will not be contaminated by substances that could compromise its integrity; (5) the storage facility will not endanger human health or the environment; and (6) that the storage facility be in the public interest.

3. *Chapter 3: Oklahoma Carbon Capture and Geologic Sequestration Act (Oklahoma CCS Act)*: the Oklahoma CCS standard requires that operators of CO_2 sequestration facilities obtain a license before operating the facility (§27A-3-5104(B)). Before obtaining such authorization: (1) the relevant body (as determined in accordance with 27A-3-5-103) must be requested to hold a hearing; and (2) the operator must provide hearing notice

to several interested parties in accordance with §27A-3-5-104(D). By the way, they have the Oklahoma federal underground injection control program (§27A-3-5-104(B)).

4. *Title 2: Water Administration, Subtitle D: Water Quality Control, Chapter 27: Injection Wells (Texas Injection Wells Act)*: the law says it can grant a request for permission to inject CO_2 for geological storage, in whole or in part, and issue the relevant license, if it finds that, among other things: (1) CO_2 injection and storage do not endanger or impair human health and safety; and (2) the reservoir into which CO_2 is injected is or may become suitable for protection against leakage or migration of CO_2 (under Section 27.051(b-1)). It must include terms and conditions reasonably necessary to protect fresh water from pollution, such as the necessary coating.

5. *Title 16: Economic Regulation, Part 1: Texas Railroad Commission, Chapter 5: Carbon Dioxide (Texas CO2 Code)*: the Texas CO_2 Code provides that a CO_2 installation license can be issued if the applicant has demonstrated, among other things, that: (1) with appropriate safeguards, ground drinking water and surface water sources can be adequately protected from the migration of CO_2 or displaced training fluids; (2) CO_2 injection will not jeopardize or harm human health and safety; and (3) the reservoir into which the CO_2 is to be injected is or may become suitable for protection against leakage or migration of CO_2 (§5,206(a)).

6. *Texas Health and Safety Code, Title 5: Sanitation And Environmental Quality, Subtitle C: Air Quality, Chapter 382: Clean Air Act, Subchapter K: Offshore Geologic Storage of Carbon Dioxide (Texas Offshore Storage Act)*: the law authorizes the Texas Natural Resources Conservation Commission to adopt standards for, among other things, the location of a CO_2 repository. It also provides that, if federal CO_2 capture requirements are issued, the Commission must ensure that the location of the CO_2 repository meets those requirements (Article 382.502).

7. *Title 54: Public Utilities, Chapter 17: Energy Resource Procurement Act, Section 701: Rules for Carbon Capture and Geological Storage (Utah CCS Rules Statute)—this standard requires under Section 54-17-701 (6) the following to be ensured*: (1) adequate health and safety standards are met; minimize the risk of unacceptable leakage of CO_2; and (2) adequate regulatory oversight and public information on carbon capture and sequestration.

8. *Title 35: Public Health and Safety, Chapter 11: Environmental Quality, Article 3: Water Quality (Wyoming Sequestration Permitting Statute)*: Wyoming Sequestration Permission Statute requires specific issues to be addressed by the recommendations, which include, among others: (1) the content required for permission applications, which should include an assessment of the expected impacts on fluid resources, underground structures and surfaces, measures to mitigate such impacts and plans for environmental surveillance and detection of excursions, prevention, and control programs; and (2) connection requirements and financial guarantee for CO_2 sequestration facilities and sites (Section 35-11-313).

9. *Code of Federal Regulations, Title 40: Protection of Environment, Parts 144 (Underground Injection Control Program), 145 (State UIC Program Requirements) and 146 (Underground Injection Control Program: Criteria and Standards) (USA Underground Injection Rules)*: as US Underground Injection Rules prohibit any injection activity, including those using Class VI wells in CO_2 storage operations, which allow the circulation of fluid containing any contaminants in underground drinking water sources, if the presence

of the contaminant could cause violation drinking water regulations or otherwise adversely affect human health (§144.12).

10. *Title 30: Minerals, Oil, and Gas and Environmental Quality, Chapter 11: Louisiana Geologic Sequestration of Carbon Dioxide Act (Louisiana CO$_2$ Sequestration Act)*: the Law says that the competent authority can edit rules, regulations, and orders for, among others things: preventing pollution of freshwater by oil, gas, salt water, or CO$_2$; and require the removal of equipment, structures, and waste, and otherwise require general cleaning of the site (§1104(A)). Paragraph 1104(C) says that before any reservoir is used for CO$_2$ storage, the Conservation Commission must have found, after a public hearing, that: (1) such use of the reservoir is adequate and feasible; and (2) that the proposed storage does not endanger human lives or cause risk to property.

11. *Title 30: Minerals, Oil, and Gas and Environmental Quality, Chapter 3: Exploration and Prospecting (Louisiana MOGEQ Act)*: the Law provides that the State Council for Minerals and Energy concludes operating contracts in which the state receives a share of the revenues CO$_2$ storage, among other things, and assumes all or part of the risk of the cost of this activity, in situations in which the board determines that this would be in the state best interest, for reasons of equity or promotion of conservation (§209).

12. *Title 82: Mineral, Oil and Gas, Chapter 11: Oil and Gas Conservation, Part 1: Regulation by the Board of Oil and Gas Regulation (Montana Oil and Gas Statute)*: the Statute says that the Oil and Gas Conservation Council is empowered, among other things, to require operators to take measures to prevent contamination or damage to surrounding land or underground strata caused by drilling operations. This power includes regulation of CO$_2$ injection (Section 82-11-111(2)). The rule prohibits, among other things: pollution of any state waters; and placing or causing any liquid, gaseous, solid or other substance to be placed in a location where it may cause pollution of any state waters (Section 82-11-127). It further says that the CO$_2$ injection permit system, under the terms of the Statute, must include, among other things, mitigation, the ability to stop and impacts of CO$_2$ leaks (Section 8211-123).

In addition, the North American Code of Federal Regulations, Title 40: Protection of the Environment, Parts 78 (Appeal Procedures) and 98 (Mandatory CO$_2$ Storage Reporting Rules), establishes important milestones and definitions as one of the few laws that spell out the difference between injector well for CO$_2$ storage and injector well for better hydrocarbon reserve performance and efficiency, besides determining the obligations and duties of both owner operators (IEA, 2010).

The North American Code of Federal Regulations also establishes well-defined technical and administrative guidelines, such as the need for owners and operators of such CO$_2$ sequestration facilities to follow procedures for monitoring and reporting, quality assurance, missing data estimation and maintenance of specified records, as well as carbon monitoring, reporting, for example, the amount of CO$_2$ received, injected, produced, emitted by superficial leak and emissions from equipment leaks and ventilated emissions from surface equipment (IEA, 2010).

The United States, in regulating activities involving CO$_2$ storage, in most of the aforementioned laws, limited the issuing of licenses and/or authorizations to the absence of risks to the environment, especially water contamination and safety for humans. It is also noted that the issue of civil liability for possible damages is widely addressed.

According to Romeiro-Conturbia (2014), the US EPA considers that CCS is an option within the mitigation strategies to reduce GHG emissions in the country; it established requirements, through legal and regulatory framework at the federal level for operators that aim to perform geological CO_2 storage (for permits, site characterization, financial responsibility, etc.), aiming to protect underground sources of drinking water that may be contaminated by eventual leaks of the injected carbon.

During the first 50 years after the closure of the CO_2 injection sites, operators are responsible for monitoring the storage locations and verifying the behavior and migration of carbon dioxide, ensuring that there are no leaks that cause risks to the environment. This period may be appropriate, reduced or extended, depending on the safety statements and potential risks presented by the operator after the end of the CO_2 injection (Romeiro-Conturbia, 2014).

16.3.8 United Kingdom

According to Câmara et al. (2011), in the United Kingdom, issues involving CCUS are linked to energy sources. This situation stems from the fact that the main source of GHG emissions in the United Kingdom is the production of energy from fossil fuels. Toledo Filho (2014) says that the climate policy in the United Kingdom is a complex and diversified set of instruments that internalize the costs of GHG emissions with the use of economic instruments, such as quota trading, carbon emissions, or rates.

Thus from this complex of instruments in force in the United Kingdom, the following stand out, with implications for CCUS technologies:

Energy Act 2008	Provides a regulatory framework for licensing offshore carbon dioxide storage
Energy Act 2010	Financial incentives law, in the form of fees for electricity providers, aiming at financing CCS projects in the United Kingdom
Energy Act 2008 (Consequential Modifications) (Offshore Environmental Protection) Order 2010	Provides a regulatory framework for licensing offshore carbon dioxide storage
The Storage of Carbon Dioxide (Licensing etc.) Regulations 2010	The regulations cover the conditions for granting licenses and operating authorizations; the obligations of the storage operator; the closure of the storage site; postclosure period; and financial security. They also amend the environmental damage regulations (prevention and remediation) that implement the environmental liability directive. The warehousing operation is added to the list of damage-causing activities

Elaborated by the authors after the United Kingdom Legislation; Global CCS Institute, 2019. Lessons and perceptions: adopting a commercial approach to CCS liability. Available from: https://www.globalccsinstitute.com/wp-content/uploads/2019/08/Adopting-a-Commercial-Appraoch-to-CCS-Liability_Thought-Leadership_August-2019.pdf (accessed 20.12.20).

Therefore the United Kingdom has maintained a leading role in the discussion and promotion of climate change mitigation policies. According to Toledo Filho (Toledo and Demétrio, 2014), for many years, the country was the first and only country that had mandatory long-term targets explicit in its national legislation aimed at reducing emissions. According to Romeiro-Conturbia (2014), the United Kingdom has instituted requirements for regulating long-term storage of CO_2 at sea, in its territory, through the EU CCS Directive on geological storage of carbon dioxide, introducing the license requirement by competent authority, which may contain requirements on the closure of the carbon storage site. The developer of the activity is responsible for the integrity of the storage facility for a period of 20 years, after which the responsibility is transferred to the relevant authority.

16.4 Conclusion

In carbon storage activities, the risks of environmental accidents are real and the range of negative results is great. Several factors can trigger them, for example, the lack of an emergency plan and inattention to prevention (Cunha, 2005). Thus it is essential that these activities undergo an environmental impact study before being installed.

Several developed countries have already created mechanisms and regulations for developing and using CO_2 sequestration and geological storage, recognizing the importance of this technology for protecting the environment and sustainable development. Much progress has been made in the regulation of CCUS technologies, detailing requirements related to licensing, the composition of the CO_2 flow, monitoring, inspections, preventive and corrective measures for damages, the closure of activities, obligations and/or responsibilities for damages, security of activities, etc. In the legislation of the selected countries, there are minimum requirements and deadlines for transferring the long-term responsibility of the explorer to the state in the activities of storage of carbon dioxide ranging from 10 to 50 years.

The model of the countries studied serves as a lesson for creating a long-term liability standard. In the cases studied, the responsibility is transferred to the state by proving the stability of the storage location, the adequacy of the safety of the location, the low risk forecast, and monetary compensation for this assumption of responsibility.

Therefore for carbon storage activities to spread on a large scale, it is necessary to address safety issues and social risk management, aiming to minimize damage and to rethink the environmental civil liability of polluting agents that operate CCUS activities in order to make the implementation of this technology more attractive, economically viable; the regulations already existing in other countries can be used as parameters for creating a specific national regulation system.

Acknowledgments

We are grateful to the "Research Center for Gas Innovation—RCGI" (Fapesp Proc. 2014/50279-4), supported by FAPESP and Shell, organized by the University of São Paulo, and the strategic importance of the support granted by the ANP (National Agency of Petroleum, Natural Gas, and Biofuels of Brazil) through the R&D clause. We also thank the support from the National Agency for Petroleum, Natural Gas and Biofuels Human Resources Program (PRH-ANP), funded by resources from the investment of oil companies qualified in the R,D&I clauses from ANP Resolution No. 50/2015 (PRH 33.1—Related to Call No. 1/2018/PRH-ANP; Grant FINEP/FUSP/USP Ref. 0443/19).

References

Alberta Queen's Printer, 2000. Mines and Minerals Act. Available from: https://www.qp.alberta.ca/documents/acts/m17.pdf (accessed 20.12.20).

Alberta Queen's Printer, 2011. Carbon Sequestration Tenure Regulation. Available from: https://qp.alberta.ca/documents/Regs/2011_068.pdf (accessed 20.12.20).

Câmara, G., Silva Júnior, A., Rochac, P., Andrade, C., 2011. Armazenamento de Dióxido de Carbono em Reservatórios Geológicos: Tecnologia Mais Limpa? "Cleaner Production Initiatives and Challenges for a Sustainable World." CSW, São Paulo.

Carvalho, L.L., 2010. The Risks of Carbon Capture for the Environment and the Civil Liability in Brazil. A Responsabilidade Civil do Estado Brasileiro perante os Riscos do Sequestro Geológico de Carbono para o Meio-Ambiente. Lawinter Review, vol. I, no. 02, April de 2010. New York, United States of America.

Cook, P.J., 2009. Demonstration and deployment of carbon dioxide capture and storage in Australia. Energy Procedia 1 (1), 3859–3866. Available from: https://www.sciencedirect.com/science/article/pii/S1876610209008315 (accessed 20.12.20).

Costa, H.K.M., Musarra, R., 2019. Princípios Gerais do Direito: Aplicabilidade nas Atividades CCS. In: Costa, K.K.M., Musarra, R. (Org.). Aspectos jurídicos da captura e armazenamento de carbono no Brasil. Lumen Juris, Rio de Janeiro, pp. 25–52.

Costa, H.K.M., Musarra, R., Miranda, M.F., 2018. The Main Environmental Permitting Requirements on CCS Activities in Brazil: Sustainability and Development Conference, 2018, Ann Arbor, MI, USA. Sustainability and Development Conference Proceedings. Michigan University, Ann Arbor, MI, pp. 01–23.

Costa, H.K.M., Musarra, R., Miranda, M.F., Moutinho Dos Santos, E., 2018. Environmental license for carbon capture and storage (CCS) projects in Brazil. Journal of Public Administration and Governance 8, 163–185.

Costa, H.K.M., Santos, M.M., Matai, P.H.L.S., 2017. Questões ambientais e Licenciamento ambiental—Cap. 17. In: José R. Simões Moreira (Org.). Energias Renováveis, Geração Distribuída e Eficiência Energética. LTC Editorial—Grupo GEN, Rio de Janeiro, pp. 354–367.

Cunha, B.P., 2005. Desenvolvimento sustentável e dignidade: considerações sobre os acidentes ambientais no Brasil. Verba Juris—Anuário da Pós-Graduação em Direito 4 (4). Available from: http://www.periodicos.ufpb.br/ojs2/index.php/vj/article/view/14821 (accessed 20.12.20).

Cupertino, S.A., 2019. A responsabilidade civil na estocagem de carbono no Brasil. Dissertação (Mestrado em) - Instituto de Energia e Ambiente, Universidade de São Paulo, São Paulo, 2019. Available from: https://doi.org/10.11606/D.106.2019.tde-29012020-173712.

Dean, M., et al., June 2017. A risk-based framework for measurement, monitoring and verification (MMV) of the goldeneye storage complex for the Peterhead CCS project, UK. International Journal of Greenhouse Gas Control 61, 1–15.

European Union, 2009. Directive 2009/31/EC.

Faure, M., 2016. Liability and compensation for damage resulting from CO_2 storage sites. William and Mary Environmental Law and Policy Review, 40 (2) (2015–2016). Symposium Issue: What's your water? A discussion of Threat of Virginia's Water quality, Article 3, February 2016.

Global CCS Institute, 2019. Lessons and perceptions: adopting a commercial approach to CCS liability. Available from: https://www.globalccsinstitute.com/wp-content/uploads/2019/08/Adopting-a-Commercial-Appraoch-to-CCS-Liability_Thought-Leadership_August-2019.pdf (accessed 20.12.20).

IEA, 2010. Carbon Capture and Storage—Model Regulatory Framework. OECD/IEA, Paris, p. 130.

IEA, 2020a. CCUS in Clean Energy Transitions. IEA, Paris. Available from: https://www.iea.org/reports/ccus-in-clean-energy-transitions.

IEA, 2020b. Monitoramento Ambiental de Captura e Armazenamento de Carbono Offshore. IEA, Paris. Available from: https://www.iea.org/articles/environmental-monitoring-of-offshore-carbon-capture-and-storage (accessed 20.12.20).

IPCC, 2006. Diretrizes para Inventários Nacionais de Gases de Efeito Estufa. Energia, 2 (Capítulo 5).

Monteiro Júnior, J.V., 2015. A regulação do sequestro geológico de carbono como instrumento de fomento ao desenvolvimento na indústria do petróleo brasileira. Natal, 2015. Available from: https://repositorio.ufrn.br/jspui/bitstream/123456789/21999/1/Regula%c3%a7%c3%a3oSequestroGeol%c3%b3gico_MonteiroJunior_2015.pdf (accessed 20.12.20).

Nardocci, A.C., 2002. Gerenciamento Social de Riscos. Revista de Direito Sanitário, 3, 1, março de 2002. Available from: periodicos.usp.br (accessed 20.12.20).

Nunes, R.C., Costa, H.K.M., 2020. An overview of international practices for authorization and monitoring CO_2 storage facilities. International Journal of Advanced Engineering Research and Science 7, 133−142.

Queiroz, F.A., December 2005. Meio ambiente e comércio na agenda internacional: a questão ambiental nas negociações da OMC e dos blocos econômicos regionais. Ambiente & Sociedade, Campinas 8 (2), 125−146. Available from: http://www.scielo.br/scielo.php?script=sci_arttext&pid=S1414753X2005000200007&lng=en&nrm=iso (accessed 20.12.20).

Rathmann, R., 2017. Opções transversais para mitigação de emissões de gases de efeito estufa captura, transporte e armazenamento de carbono. Ministério da Ciência, Tecnologia, Inovações e Comunicações, Brasília.

Ravagnani, A.T.F.S.G., Suslick, S.B., 2008. Modelo dinâmico de sequestro geológico de CO_2 em reservatórios de petróleo. Revista Brasileira de Geociências 38(1−Suplemento). Instituto de Geociências/USP, São Paulo. pp. 39−60.

Romeiro-Conturbia, V.R.S., 2014. Marco legal e regulatório da captura e armazenamento de carbono em países em desenvolvimento: propostas para o Brasil. 2014. Tese (Doutorado)−Programa de Pós Graduação em Energia, Universidade de São Paulo, São Paulo.

Sauer, N., June 2019. Japan sets carbon neutral goal with focus on capturing emissions. Climate Home News. Available from: https://www.climatechangenews.com/2019/06/12/japan-says-will-carbon-neutral-fails-settimeline (accessed 20.12.20).

STEMM-CCS, 2020. Strategies for Environmental Monitoring of Marine Carbon Capture and Storage. Available from: https://www.stemm-ccs.eu/ (accessed 20.12.20).

Thorstensen, V., 1998. A OMC—Organização Mundial do Comércio e as negociações sobre comércio, meio ambiente e padrões sociais. Revista Brasileira de Política Internacional 41 (2), 29−58. Available from: https://doi.org/10.1590/S0034-73291998000200003.

Toledo, F, Demétrio, F, 2014. Integração da política climática: segurança energética e proteção climática, lições das experiências da Alemanha e do Reino Unido. Tese de Doutorado − Universidade de Brasília. Centro de Desenvolvimento Sustentável.

Varela, C.A., Milone, D., 2014. A Resposta do mercado aos Acidentes Ambientais na Indústria Petrolífera: Estudo do Caso do Desastre no Golfo do México. Encontro Internacional sobre Gestão Empresarial e Meio Ambiente.

Victorian Legislation, 2008. Greenhouse Gas Geological Sequestration Act 2008. Available from: https://www.legislation.vic.gov.au/in-force/acts/greenhouse-gas-geological-sequestration-act-2008/013 (accessed 20.12.20).

Waarun, I.K., et al., 2016. CCS Leakage Detection Technology—Industry Needs, Government Regulations, and Sensor Performance. In: 13th International Conference on Greenhouse Gas Control Technologies. GHGT-13, November 14−18, 2016, Lausanne, Switzerland.

Wilson, E.J., Friedmann, S.J., 2007. Research for deployment: incorporating risk, regulation, and liability for carbon capture and sequestration. Environmental Science and Technology 41, 5945−5952. Available from: https://pubs.acs.org/doi/pdf/10.1021/es062272t (accessed 20.12.20).

Wilson, E.J., Klass. A. B., Bergan, S., 2009. Assessing a liability regime for carbon capture and storage. Energy Procedia 1 (1), 4575−4582. ISSN 1876-6102. Available from: https://doi.org/10.1016/j.egypro.2009.02.277.

PART IV

The current picture and future perspectives

A transitioning model: from oil companies to energy players

Alexandre Sales Cabral Arlota

Partner at Mattos Filho, Veiga Filho, Marrey Jr. e Quiroga Advogados, Rio de Janeiro, Brazil

17.1 Introduction

Petroleum and its derivatives have become the main energy source of the last century, through widespread use as fuels for thermal plants, automobiles, and machines. Consequently, the oil sector is one of the most relevant driving forces of the present economy (Yergin, 2009).

The oil industry's figures are still impressive: according to data presented by the US Energy Information Administration, the total world production for petroleum and other liquids amounted to 94.25 million barrels per day in 2020 and is expected to reach 97.42 million barrels per day in 2021 (Energy Information Administration (EIA), 2020). The Organization of the Petroleum Exporting Countries (OPEC),[1] leaded by Saudi Arabia, are expected to account for 32.4% of the global production in 2020, followed by the United States, the top producing country, with an average volume of 18.58 million barrels/day, representing approximately 19.7% of the international production, and then Russia (11.1%), Canada (5.5%), China (5.2%), and Brazil (4.1%). Until the crash caused by the coronavirus pandemic (Covid-19), global petroleum and other liquids consumption had risen systematically, reaching in 2019 the amount of 101.23 million barrels/day (Energy Information Administration (EIA), 2020).

Leading the international oil production are the major oil companies, defined as the group formed by transnational companies in the oil sector, based on metrics that consider

---

* The author would like to thank the assistance by Carlos Herculano Cubillas, Artur Avilla, and Giovanna Kaiser in the research of this chapter.

[1] Organization of Petroleum Exporting Countries: Algeria, Angola, Congo (Brazzaville), Equatorial Guinea, Gabon, Iran, Iraq, Kuwait, Libya, Nigeria, Saudi Arabia, the United Arab Emirates, and Venezuela.

economic dimension, years in activity, and market position. The majority of such companies are horizontally integrated and engage in the exploration, production, commercialization, refining, transportation, and distribution of petroleum and its products.

Major oil companies can be divided into: (1) those under private control and (2) those under governmental control. Among the first are companies such as British Petroleum (BP), Chevron, Royal Dutch Shell (Shell) and ExxonMobil (Exxon)—members of the Seven Sisters[2]—and European Total and Eni, which together form the "super-majors"—the six largest private oil exploration and production companies.

Among the State-owned companies are[3]: the NOCs (National Oil Companies), focused on local development and exploration, such as the Russian Rosneft, the Mexican PEMEX, the Venezuelan PDVSA, and the Brazilian Petrobras; as well as the INOCs (International National Oil Companies), which have large upstream investments in foreign countries, such as the Norwegian Equinor, the Chinese China National Petroleum Corporation (CNPC) and Sinopec, in addition to the Russian Gazprom (International Energy Agency (IEA), 2020).

Altogether—considering the private companies, NOCs, and INOCs—the major oil companies account for more than 70% of global oil production, according to a study conducted by the International Energy Agency in 2020 (International Energy Agency (IEA), 2020).

Despite the unparalleled numbers, it is not an exaggeration to say that the major oil companies are in a paradigmatic moment, as they are challenged to balance their short-term gains with the need to overcome the shortage of oil reserves, which may occur in the next 50 years (British Petroleum (BP), 2019). However, even more critical than the finitude of resources—which is relative, since the technological advances and cartelized oil prices expand the exploration frontiers to new horizons,[4] like the ultradeep Pre-salt layer and the shale gas—are the socioenvironmental challenges arising from a heavily polluting business, which, for this reason, is being pressured from all fronts to urgently implement a structural change.

Amid this changing reality, the major oil companies reorient their businesses, to adapt to the sustainability consensus that has been emerging gradually in the international community, having its most ambitious goals in the 2030 Agenda for Sustainable Development.

17.2 New paradigm

In the past, the business diversification in the oil industry represented only a repositioning of their asset portfolio, as a way of minimizing geopolitical risks, or a vertical integration, by controlling downstream distribution and investments. Such a dynamic reinforced

---

[2] Group formed by the following oil companies: Royal Dutch Shell, British Petroleum Amoco (BP), Standard Oil of New Jersey (Esso), Standard Oil of New York (Socony), Texaco, Standard Oil of California (Socal), and Gulf Oil.

[3] In fact, to state that those companies are State-owned is an imprecise simplification: although the country is the majority and/or controlling shareholder, there may be a sizable corporate stake in the hands of private parties and/or negotiated publicly on the stock market.

[4] The impacts of the recent dissent between OPEC and Russia are not entirely clear, especially when associated with the ongoing pandemic and the drop of oil demand, although the latest agreement by OPEC + members on January 2021 indicates that the cartel can still control oil prices and avert long-term slumps in oil prices.

a concentration on the oil business itself. In the last decade, however, there has been a real diversification, due to the increase in investments in other energy sources.

In addition to the challenges imposed by the finitude of the resources to be extracted and intrinsic geopolitical tensions, there are other externalities that directly affect the continuity of the traditional structure of oil companies, notably, stakeholders' actions. Accordingly, the conduct of such external agents ends up creating incentives for major oil companies to explore a broader energy market.[5]

Indeed, there are numerous initiatives of private equity fund managers, financiers, and governmental agents that push oil companies toward adopting ecologically sustainable measures. A relevant example was BlackRock's decision, the largest private equity manager in the world, to reduce its investment in high environmental risk assets, establishing sustainability as a standard for future investments (BlackRock). This movement, before being internalized by fund managers and companies, was largely driven by public demonstrators and consumers, who forced the decarbonization into the governmental and private entrepreneurs' business agenda.[6]

It is also worth noting that in early 2020, financial institutions controlling over US\$ \$90 trillion, among the world's largest public pension funds, endowments, foundations, and institutional investors, signed up to the United Nations' Principles for Responsible Investment, thus committing to incorporate sustainability into their investments and to abide to enhanced shale gas recovery requirements (Saa, 2020).

As another clear illustration of private investors' soft power to influence the behavior of other players, investors managing over 3 trillion dollars in assets issued an open letter to Brazilian embassies in Europe. According to a New York Times' publication in June 2020, the managers of these global investment funds indicated increasing restrictions, should the right-wing Brazilian government—marked by its denialism and by its support for the dismantling of environmental policies (Andreoni et al., 2020)—delay the implementation of effective measures to control the Amazon's deforestation or maintain its rhetoric of confronting indigenous peoples (Boadle, 2020). That reinforces a general perception that such private agents can modulate the activity of the entities traditionally considered the most powerful players in the international order: sovereign countries.

It is evermore clearer that the governments have recognized the role that capital markets play in promoting social and environmental targets, such as the EU's Action Plan on Sustainable Finance. This trend spills over into guiding, conforming, and conditioning sovereign wealth funds and central banks (Saa, 2020).

[5] For a historical overview of how externalities affect the behavior of economic agents in the oil sector, see: Parra, F., 2004. Oil Politics, A Modern History of Petroleum. IB Tauris, New York, p. 61.

[6] "Launched in late 2012 with the 'Do the Math' campaign on climate activist Bill McKibben's 350.org platform, the call for divestment aims to both delegitimize the fossil fuel industry and reduce its access to capital. From humble roadshow beginnings, the campaign has grown into a global movement, joined by a range of name-brand institutional investors, including pension funds, sovereign wealth funds, philanthropic foundations, educational institutions and faith-based organizations. As of Autumn 2020, more than 1200 institutional investors managing over \$14 trillion of assets around the world have committed to divestment of some, if not all, of their fossil fuel holdings." Mormann, F., 2020. Why the divestment movement is missing the mark. Nature Climate Change 3 (1), 1067–1068.

In yet another example, after 4 years using carbon dioxide emission levels as a criterion for vetoing the availability of resources, Norges—the world's largest sovereign wealth fund, with over US$ 1 trillion in assets, and part of the Norwegian Central Bank—decided to exclude certain oil and mining companies from its portfolio due to environmental damages, including Canadian Natural Resources, Imperial Oil, Anglo American, and Glencore (Arbex, 2020).

The Brazilian National Development Bank (*Banco Nacional de Desenvolvimento Econômico e Social*—BNDES), an export credit agency and the largest infrastructure financier in Brazil, developed special credit lines, with lower interest rates for renewable and energy-transitioning projects, such as *FINEM Geração de Energia*, exclusively for renewable energy generation projects and combined-cycle natural gas power plants (Brazilian National Development Bank (BNDES)).

Multilateral financial institutions, such as the Inter-American Development Bank (IDB), the International Finance Corporation (IFC), and the World Bank (IBRD), also provide preferential credit terms for the renewable sector. To that effect, the most recent changes to the Equator Principles, which are mandatory guidelines for projects eligible to IDB, IFC, and IBRD's credit lines, further enhanced the sustainability criteria and established specific mechanics to assess whether the projects are in line with the 2030 Agenda (Ahrens et al., 2020).

To expand stockholders' visibility, credit-rating agencies, such as standard and poors, recently created specific indexes to assess and weight publicly traded companies on how they adhere to a 1.5°C global warming climate scenario (S&P Dow Jones Indices, 2020).

Relying on public goals for social responsibility, some of the world's most influential companies are committing to sourcing all its energy from renewables, including diverse industries such as computer manufacturer Dell, investment bank Macquarie, and fashion brand Ralph Lauren. In 2019 alone American corporations entered into a record 13.6 GW of power purchase agreements for renewable energy, with Google as the largest offtaker, signing contracts for 1720 MW (Wong et al., 2020).

Therefore all these incentives are fostering energy transitioning and modulating the behavior of economic agents.

There are also inherent differences between the risk composition of oil projects and that of renewable energy projects, which can redirect financial flows, especially equity investments and loans: among others, the shorter development term for wind farms and solar parks reduces uncertainties regarding the contractor's ability to complete the projects on schedule and in compliance with the preestablished performance criteria, minimizing one of the main risks for financiers in project finance structures, namely the uncertainty of project completion.

In addition to being less capital intensive, renewable energy projects can be implemented without many complications in between 1 and 2 years; while oil and gas projects, especially offshore, often take up to 5 years to enter into operation.

The shorter term for the development of renewable energy assets also means that it will take a shorter amount of time for the asset to start generating revenue and, consequently, to repay the financing, which may allow more favorable conditions for raising funds and less debt service.

While the oil industry has the ever-present risk that prospecting will reveal a well with smaller reserves than those projected or whose exploitation is economically unfeasible; the evolution of metrics for calculating the incidence of sunlight and wind projections, as well as greater volume of available data, have contributed to the renewables sector having a substantially lower risk profile.

In this context, the major oil companies are pressured to communicate new business models, beyond the petroleum business, as well as to take opportunities and design investment diversification strategies—even if these companies maintain oil as their core business.

This tendency seems to suffer further deviations, as there is a progressive detachment between the gains obtained with oil exploration and the investments made by oil companies in other sources of energy. Due to the development of technologies—in addition to preferential fundraising conditions and tax incentives—the costs of implementing renewable energy projects have been reduced, making wind and solar assets highly profitable and consolidating them as independent from the oil businesses.

As an illustration, a survey conducted by Bloomberg indicated that, between 2018 and 2019, major oil companies carried out over 140 transactions for the acquisition of renewable energy assets and energy transition: BP, Chevron, Equinor, Repsol, Shell, Saudi Arabian Oil, and Total appeared as major investors and concentrated around 75% of the investments made by oil companies in that niche (Abington and Gilblom, 2019).

To reflect the widening of the company's scope of activities and to dilute the perception that it fundamentally exploits hydrocarbons, Statoil changed, in 2018, its name to Equinor. Among the consideration of continuing to operate under the cloak of an internationally recognized brand, with decades of success, but which included the word *"oil,"* the company preferred to use an entirely different semantics that would break with the previous lexicological allusion and reveal what it proposes be a new identity, adhering to the sustainable economy. It is not by chance that the Latin prefix *"Equi"* indicates the idea of balance, a synonym of sustainability (Statoil, 2018). A similar choice was made by Spanish Gas Natural Fenosa, which rebranded itself into Naturgy (Naturgy), opting for the word *"nature"* in its new name, pointing towards a deeper identity change.

BP is also presenting itself as an integrated energy company (British Petroleum (BP)); as oil service companies amend their business pitches and advertisement pieces to indicate that they are technology providers for the *energy industry*, and not only for the oil and gas segment, as exemplified by Baker.

The recently renamed Equinor says on its website that it is *"one of the world's most CO_2-efficient producers of oil and gas"* and will direct 15%−20% of its investments to new energy solutions (Equinor). As evidence of its commitment to the energy transition, Equinor increased its ownership in Scatec—a solar energy company, with assets located in Argentina, Brazil, Czech Republic, Honduras, Jordan, Malaysia, Mozambique, Rwanda, South Africa, and Ukraine (Scatec)—raising its shareholding to 15.2% of the votes (Adomaitis and Solsvik, 2019). Equinor also expects a production capacity from renewables to range between 4 and 6 GW by 2026, mainly based in the current project portfolio, focused on offshore wind projects in Norway, Poland, United Kingdom, and United States, where Equinor plans to build its first wind project outside of Europe (Equinor).

Equinor also indicates that the interplay between oil companies and renewables goes beyond the ramp-up of investments in sustainable assets. In fact, the engineering improvements and best practices of the oil and gas industry are being used to foster decisive advancements in renewables. Offshore wind, for example, will rely on controlling the force of waves, while producing and transmitting energy, something that the oil and gas industry has mastered for decades long. The use of contractual templates, standardized provisions, and specific liabilities regimes, which are typical of the oil and gas sector, will likely be replicated for renewables, as major oil companies become heavy players in renewable generation.[7]

In its turn, BP has pledged to neutralize its emissions by 2050, which would require taking measures to offset more than 360 million tons of carbon annually (Ambrose, 2020a). Its solar energy investment platform, Lightsource BP, is one of the companies with the largest installed capacity in the sector with a global capacity of 3 GW in projects developed to date and with 16 GW in its development pipeline, establishing a firm footprint in the international energy generation market (Lightsource).

Diversifying its investments even further, BP partnered with Bunge to operate in the bioenergy and sugarcane ethanol business, and formed BP Bunge Biofuels, with annual production capacity of 1.8 billion liters of ethanol and generation capacity of 1.2 GWh of energy (Gaudarde). To some extent, this structure has another success story: about 10 years ago, Anglo-Dutch Shell and the Brazilian group Cosan formed Raízen, with estimated market value of US$ 12 billion and approximately 40,000 employees, emerging as a global leader in sustainable energy market (Shell).

Shell's most recent investment plan foresees an annual budget of US$ 3 billion for renewables (Plano é investir US$ 3 bi em energias renováveis no mundo, diz Shell). There are also other examples: Total Eren is the renewable energy subsidiary of the French group Total, dedicated to the exploitation of renewable energy, generating more than 2700 GWh in 2019 and having more than 100 power plants either in operation or under construction in 18 countries (Total Eren).

In relation to investments for the expansion of the natural gas market, oil companies have entered as partners in thermoelectric plants and natural gas distribution hubs, which, at the same time, ensure a strategic position in energy transitioning and generates demand for gas, a by-product of oil, however, less polluting (David, 2019). It should be noted that the strategy of forming joint ventures, instead of investing alone all the amount necessary for the project, is common in the oil industry, which is capital intensive and high-risk and, for such reasons, has a special fondness for the pulverization of investments in multiple initiatives among partners (Ribeiro, 2014).

Solely in the Brazilian market, there are three relevant cases: (1) BP's partnership with Prumo (of which the private equity fund EIG Partners is a relevant shareholder) to develop the Gas Natural Açu thermoelectric park, with 3 GW capacity (Gás Natural Açu (GNA)); (2) Golar Power—a joint venture between a major Norwegian gas company and the private equity fund StonePeak—is shareholder of Celse and Celba, owners of gas-fueled thermal plants with 1.5 GW (Terminal Gás Sul) and 600 MW capacity (Projeto para térmica de R$ 1,5 bi a GNL no Pará prevê fornecer gás à indústria local); and (3) Shell—

---

[7] "However, the good news is that there will be no offshore wind industry that does not rely on the gold-standard engineering competence and experience of the offshore oil and gas industry. So, take a look at that business plan and look for possibilities through transformation, adaptation, or possibly cooperation". Equinor. We're determined to be a global offshore wind energy major. Here's how. https://www.equinor.com/en/what-we-do/wind.html (accessed 01.11.21).

partnered with a local private equity fund—has a corporate stake at the Marlim Azul Termelétrica, with 565 MW capacity (Brazilian National Development Bank (BNDES), 2019).

Regarding the oil companies' bet on the densification of the natural gas market and on its prevalence as an energy transition fuel, Qatar Petroleum reserved a major portion of the assembly line of the South Korean shipyards Daewoo, Hyundai, and Samsung, for the construction of a fleet of 100 LNG carriers by 2027, in an investment of US$ 19 billion (Chung and Yang, 2020).

17.3 New technologies

In addition to investing in less polluting sources of energy, oil companies have sought to develop techniques to mitigate the impact of their operations on the environment, especially through investments in research and development, directed at collaborations with academic centers of excellence.

This means that the reduction of the carbon footprint and the necessary measures to neutralize emissions depend not only on direct investments in renewable sources, but also on actions in the oil chain itself, through the development of new technologies and their use in oil and/or natural gas extraction, consequently lowering the concentration of carbon in its emissions. Among the measures in place, carbon capture and storage (CCS) stands out.

CCS consists of a chain of processes to capture, transport, and store carbon emissions, preventing it from entering the atmosphere (Carbon Capture and Storage Association). Currently, this technology represents one of the possible ways to minimize greenhouse gas (GHG) emissions, given its potential to capture up to 90% of the carbon dioxide produced from the burning of fossil fuels.

The CCS chain can be divided into three stages: (1) capturing the carbon dioxide, using separation/absorption technologies; (2) transporting the captured carbon dioxide; and (3) storing the carbon dioxide emissions underground.

Regarding the first stage in the CCS process, there are three common methods by which carbon dioxide is captured: precombustion, postcombustion, or using oxy-fuel combustion systems (Jha, 2008).

A precombustion system involves transforming the fuel, through gasification and reforming processes, into a mixture of carbon dioxide and hydrogen. These gases are subsequently separated in the process: carbon dioxide is transported and stored, while hydrogen can be used as a fuel for the generation of clean energy, with emission rates close to zero.

A postcombustion system concerns the absorption of carbon dioxide by means of solvents that act on the exhaust gases produced from the burning of fuels. Once captured, the carbon dioxide is released from the solvent, compressed, and transported to the storage location.

Oxy-fuel combustion is the burning of fuels in an atmosphere of pure oxygen. In this environment, the gases resulting from combustion will consist of carbon dioxide and water vapor, which are separated by means of condensation, allowing the compression and storage of pure carbon dioxide.

After the carbon dioxide is captured, it is then transported by ships, road tanker, or gas pipelines to storage sites—porous geological formations, located several kilometers under the Earth's surface, such as former gas and oil fields or deep saline formations.

Lastly, the storage of carbon dioxide, the final stage in the CCS process, is divided into four mechanisms. Firstly, it is injected under pressure into the geological formation until it reaches an impermeable layer of rock (caprock) that traps it. This is structural storage, the main storage mechanism for carbon dioxide used in CCS.

As the injected carbon dioxide moves through the storage site towards the caprock, a portion of the carbon dioxide is trapped in the microscopic pore spaces of the rock. This mechanism is known as residual storage.

Over time, the carbon dioxide stored will dissolve into the salty water in the surrounding rock. The water with dissolved carbon dioxide becomes denser, sinking down to the bottom of the storage site, through a mechanism known as dissolution storage.

Finally, when carbon dioxide binds chemically and irreversibly to the surrounding rock, mineral storage occurs, the last mechanism of the storage of carbon dioxide, which has been proved to be an effective and safe technology.[8]

Scientists have also developed CCUS (carbon capture, utilization and storage), which, in addition to carbon storage, provides other uses for the gas.[9] In this sense, it is worth mentioning the use of carbon dioxide in EOR (enhanced oil recovery) by which carbon gas is injected into primarily depleted oil reservoirs so that the oil is pumped into the exploration wells. This practice draws attention due to the possibility of, depending on the origin and volume of injected carbon gas, enabling more sustainable oil production—and, theoretically, even the possibility of a negative carbon footprint (International Energy Agency (IEA), 2020).

Thus both CCS and CCUS can play a crucial role in the transition to an ecologically sustainable matrix, with social and economic applications for the oil industry. Furthermore, by

---

[8] "CCS has been working safely and effectively for 45 years (since the Apollo 17 moon landing in 1972). Operations undertaken over almost half a century demonstrates that CO_2 can be safely stored deep below ground. Oil, gas and naturally occurring CO_2 reservoirs have proven that fluids can be safely sealed underground for millions of years. CCS facilities target the same geology. CCS technology is verifiably well tested. Seventeen largescale facilities are operating successfully around the world (with four more coming on stream shortly). These 17 facilities are currently capable of capturing more than 30 Mtpa of CO_2 per annum. [...] CCS is needed because the amount of fossil fuels we burn continues to rise. Last year, fossil fuels reached a record 83.6 billion barrels of oil equivalent (Bboe) compared to 73.3 Bboe 10 years ago. There are no signs of abatement. In 25 of the last 26 years, we burned more fossil fuels than the year before. The only year recording a decrease in the last 25 years was 2009 (caused by the global recession). CO_2 emissions have increased every year since 1960 and in the last 2 years, these hit all-time records. The renewables' (solar and wind's) share of gross electricity generation is currently less than 5%, rising to 17% by 2040.vi Fossil fuels' share of electricity generation will equate to 50% by 2040. This confirms the urgency at which CCS must be applied to power and wider industry." Global CCS Institute, 2017. The Global Status of CCS: 2017. Global CCS Institute, Melbourne. https://www.globalccsinstitute.com/wp-content/uploads/2018/12/2017-Global-Status-Report.pdf (accessed 01.10.20).

[9] "CCUS is central to the industry decarbonization portfolio. [...] CCUS is the third most-important lever for emissions reductions in these subsectors, contributing a cumulative 27% (21 $GtCO_2$) of emissions reductions by 2060." International Energy Agency (IEA), 2019. Transforming the World Through CCUS. IEA, Paris.

reducing the environmental impact of oil activities, these technologies would facilitate access to private investments and public incentives, including tax benefits, regulatory flexibility, and preferential lines of credit. On the other hand, the use of gas captured in the hydrocarbon exploration and production chain itself, via injection, would lead to direct economic gains for the oil industry, since it would dispense with the use of other high-priced inputs and chemicals.[10]

17.4 An example of energy transition in the developing world

Economic development relies on mandatory energy supply. It is therefore an essential investment for any country to guarantee the bases of its growth. That is to say, energy consumption is indispensable to modern activities, reinforcing the strategic profile of the sector. As a result, energy security is subject to political and regulatory factors that end up shaping its working framework.

In Latin America, due to various circumstances,[11] there are cyclical movements—referred to as *"backward movements of pendular effect"* (Ribeiro, 2009)—which sometimes guide countries to lesser discriminatory treatment of foreign investors, and other times restrict access to financial flows remitted from abroad (Ribeiro, 2010). In summary, the flexibilization of nationalist doctrines and the mitigation of barriers to the entry of foreign capital resulted from the acknowledgment of the local capital's insufficiency to promote national development, amid the internationalization promoted by globalization.

On the other hand, globalization is marked by contradictory tendencies (Santos, 2008) and, although it presents itself as a limited response to the aspirations of improving the living conditions of local populations, this same globalization gives rise to irascible movements regarding what comes from abroad, promoting the resurgence of nationalist rhetoric. Although this turn-away from cosmopolitanism is experienced all over the world, and with force in central economies, the frequent changes to the regulatory framework for the exploration of natural reserves form a peripheral, and most commonly Latin American, phenomenon (Ferreira and Borges, 2014). Especially considering that the exercise of sovereignty under an economic perspective and the State monopoly over the oil sector occurred at the same time in which the countries of Latin America were attempting an economic emancipation, toward their own development structure, thus creating a historical echo that reverberates intensely nowadays (Arlota, 2019).

From a broader perspective, and taking the entire energy sector into account, in Latin America there is also a direct link between the oscillations of governmental policies and the

---

[10] An indication of the accelerated pace of this technology's expansion and its commercial use is that the percentage of CCS installations operating today is only 11%, with the remainder being under construction. In this sense, see: study confirms the tendency of expanding the use of CCS technologies. Research Center for Gas Innovation (RCGI), June 14, 2020. https://www.rcgi.poli.usp.br/pt-br/study-confirms-trend-to-expand-the-use-of-ccs-technologies.

[11] For an accurate record of the development of nationalist doctrines in Latin American countries regarding diplomatic protection, international responsibility of the State and treatment of foreign people and goods, vide: Xavier Jr., E.C., 2015. Direito Internacional dos Investimentos: o tratamento justo e equitativo dos investidores estrangeiros e o Direito brasileiro. Gramma Editora, Rio de Janeiro, pp. 58–61.

strategies for expanding the energy supply. Accordingly, the national energy matrixes, especially the Brazilian, underwent significant changes due to the nonlinear economic guidelines enacted by different governments and the contradictions of, at sometimes, stimulating economic growth via direct intervention in the economy and, in other occasions and intermittently, rationalizing public resources and retracting the State's role in direct investments.

In a historical perspective, the Brazilian energy sector has been dependent on public activity. It is worth mentioning the huge investments in extraordinary large energy projects were perceived by the Brazilian government as a proper vector to economic development, generating thousands of jobs and producing what was seen as a virtuous circle of wealth.

That is, the national development model, despite deviations, predominantly permeated the Brazilian economy in the 20th century and infused the energy sector with a multifaceted functionality, simultaneously assuring that the energy supply would: (1) follow the growing demand; (2) create thousands of jobs in mega construction works, as well as the related development of an engineering industry in such a scale that it could be internationalized; and (3) lead to energy autonomy, which would be mistaken with the concept of national security. One of the most symptomatic expressions of the confluence of this multilayered perspective was the creation of Petrobras, a State-owned company designed to explore, produce, and refine hydrocarbons (Barros et al., 2012).

This model of an active and intervening public sector also included large hydroelectric plants, such as Itaipu (14 GW) and, most recently, Belo Monte (4.57 GW) and Jirau (3.75 GW). However, it is necessary to draw some distinctions: while Itaipu is part of the classic tradition of direct governmental investment, in accordance with the military thinking of the 1960s and 1970s; Belo Monte and Jirau are signs of a recent *aggiornamento*, and have relied on another model, similar to a public—private partnership, by which the contractors formed consortiums and financed the construction with their own funds and/or through BNDES' loans, to earn revenue from the asset exploration under concession. Thus the Brazilian government was released from the obligation to directly invest several billion dollars in these later hydroelectric plants.

The direct investment model would be overcome, initially because the Brazilian hydroelectric potential has reached a point of semisaturation. In this scenario, new large-scale hydroelectric projects would only be viable in regions distant from consumption centers and would depend on substantial investments in transmission lines. Also, the environmental and social impacts of the construction of dams has become evident to society, who could grasp the harms of flooding thousands of hectares of forests and the relocation of native and riverside communities.

On top of the increasing geographic and social difficulties, there was a dramatic change in the economic policy. It is important to note that the energy sector suffered severe turbulences from the oil shocks of the 1970s, which resulted in the imbalance of national payment accounts and the economic slowdown in the 1980s. This whole scenario seriously damaged Brazil's financial equilibrium (Goldenberg and Prado, 2003).

The public sector, until then the main investor in the energy park and the primary developer of the infrastructure necessary for the expansion of the matrix, retracted. In the 1990s and part of the 2000s, successive governments were characterized by fiscal rigidity, manifested by imperatives of austerity and budgetary control, which ultimately undermined the government's ability to finance large-scale projects. Meanwhile, the private sector, unprepared and hesitant to provide investments that would take decades to repay, was unable or unwilling to fill the gap, leading to a gradual mismatch between energy supply and demand.

The result could not had been any different: in 2001 the country suffered from a series of blackouts and widespread shutdowns, caused by long periods of drought, unveiling yet another problem with past choices: a matrix severely dependent on a seasonal factor, rain, affecting the country's energy security and autonomy, which had been the very driving force behind the choice of hydroelectric plants. Considering the rationing of energy, the Brazilian society was forced to respond quickly (Goldenberg and Prado, 2003), leading to the diversification of the Brazilian energy matrix.

To this end, the government started acting as a regulatory agent and stimulating investments by private agents, guaranteeing demand by entering into long-term energy supply contracts (Coelho, 2011). In parallel, the installation of new thermoelectric plants—mainly diesel and coal, which have a significantly shorter construction time than hydroelectric plants—was encouraged. The main reason was that these thermoelectric plants could be dispatched more rapidly in case of need, stabilizing the system (Goldenberg, 2015).

In a second investment cycle, Brazil continued contracting thermoelectric plants to expand its energy park, however, favoring natural gas this time, as to address concerns over the migration into a lesser polluting model.[12] In this context, six energy auctions were carried out between 2015 and 2019, which contracted 4.8 GW of thermal energy from natural gas (Brazilian National Agency of Electrical Energy (ANEEL)).

The recent choice of natural gas reveals that this fuel lies in the middle between polluting and renewable sources, making it the preferred fuel for the energy transition. It is noteworthy that, unlike solar and wind sources, natural gas is not subject to environmental factors for its availability and, in its liquefied state, it can be stored, traded, and transported on a larger scale. In the Brazilian scenario, as additional factors which contributed to the increasing relevance of natural gas, one can mention the need to stabilize the electrical system through thermal plants with "on and off" dispatch to counter the hydrological risk and the increase in supply resulting from the exploration of the Pre-salt (Costa et al., 2019). Globally, the increase in Australia's and the United States' production have also reduced the prices of this commodity over recent years (Santos, 2020).

However, due to pressures for a rapid migration towards sustainable bases, a commitment expressed in international agreements such as the Kyoto Protocol, Brazil cannot rely solely on natural gas. It is worth mentioning that Brazil also ratified the Paris Agreement, a document approved at the Conference of the Parties (COP 21), which establishes the signatories' commitment to limit the increase in the global average temperature to 2°C. Thus, Brazil undertook to reduce GHG emissions by 37% by 2025 and 43% by 2030, compared to 2005 levels.

To accelerate the pace of the transition of the Brazilian energy matrix, the Ministry of Mines and Energy has been promoting exclusive energy auctions for renewable sources (Brazilian National Agency of Electrical Energy (ANEEL), 2019). Solely in the wind sector, between 2008 and 2018, installed capacity in Brazil grew by an impressive 4.15%, now representing 9% of

---

[12] On average, coal-to-gas switching reduces emissions (including methane slip) by 50% when producing electricity.

the national energy matrix. Today, there are over 600 wind farms in operation, 72 under construction, and another 133 have been approved for short-term implementation.

In 2019 an aggregate capacity of 2.1 GW was installed in the solar sector, representing an increase of 88% over the previous year (Vialli, 2020) and totaling 5.76 GW, thus exceeding the combined installed capacity of coal and nuclear plants in Brazil (5.58 GW) (Casarin, 2020).

Because of a historical primacy of hydraulic power and the recent diversification into solar and wind, 45.3% of the Brazilian energy matrix comes from renewable sources, whereas 13.7% represents the global average and 9.7% for the Organization for Cooperation and Development countries.

However, it is worth mentioning that, in the most recent 10-year Energy Expansion Plan, the Brazilian government foresees that oil products will remain the main source of energy until 2029, with only a proportional reduction of its participation in the matrix.

Thus even though oil still plays a significant role in local energy production, natural gas, ethanol, and renewable sources are increasingly gaining importance. Consequently, the players in the oil sector are not only following this movement, but also, to some extent, inducing it.

17.5 Effects of Covid-19 pandemic

When Covid-19 spread worldwide, with devastating consequences, all economic growth forecasts were revised downwards, causing the perception that the world economy would have tough years ahead. One of the many short-term consequences was the unprecedented depression in future oil prices, as world demand for fuels was drastically reduced and the momentary cushioning capacity in the North American market had circumstantially reached its limit (Santos, 2020).

In a broader sense, there were two shocks: on the one hand, an oversupply caused by the misalignment between the OPEC + members' production goals and, on the other, an upheaval in demand (Energia, 2020). However, while immediate measures were being taken, such as the hiring of oil tankers to increase storage capacity and, in parallel, production units were suspended or demobilized, the dissent between Russia and Saudi Arabia lost strength, prompting a new agreement, in order to cut world production by almost 10 million barrels/day and try to stabilize the oil's price, albeit at substantially lower levels than those practiced in recent years (Energia, 2020).

In this scenario, the market's intuitive reaction seemed to indicate that investments by oil companies would redirect towards high production and profitable upstream assets, causing uncertainties regarding renewable energy investment commitments.[13]

[13] "A few years ago, the kind of double-digit drop in oil and gas prices the world is experiencing now because of the coronavirus pandemic might have increased the use of fossil fuels and hurt renewable energy sources like wind and solar farms. That is not happening. [...] And while work on some solar and wind projects has been delayed by the outbreak, industry executives and analysts expect the renewable business to continue growing in 2020 and next year even as oil, gas and coal companies struggle financially or seek bankruptcy protection." Penn, I., 2020. Oil companies are collapsing, but wind and solar energy keep growing. The New York Times, April 7. https://nyti.ms/39S48Ki.

Contrary to this rushed conclusion, all signs seem to point to perpetuity or, even, the strengthening of the direction toward a postcarbon economy, either through the enhancement of externalities—such as governmental action—or by the behavior of the oil companies themselves.[14]

With some surprise, not for the news itself, but for the moment of its announcement, the giants Shell[15] and Total, still during the pandemic peak, announced, between May and June 2020, their commitment to reduce emissions to net zero by 2050, joining other oil companies, such as BP and Eni (Paraskova, 2020). To achieve this ambitious goal, they will have to intensify their investments, both in the generation of energy from renewable sources and in carbon capture technologies (Ambrose, 2020b).

On the other hand, Exxon has come under increasing pressure from some of its largest shareholders, including Legal and General Investment Management, Church Commissioners, and New York State Common Retirement Fund, to announce a similar commitment to reduce its carbon footprint (Financial Times).

Financial analysis indicates, that—in addition to the volatility of oil prices, new local urban mobility patterns, and long-lasting restrictions on international travel—an aggravation of cost-capital conditions for the oil industry will follow, causing fundraisings to be more expensive and penalizing long-term returns (UBS Switzerland AG, 2020).

Given the increase in the scale of solar and wind industries, renewable generation has become largely competitive with oil, especially when the latter is at extraordinarily low levels. Furthermore, as many components of renewable energy generation are highly dependent on petroleum products, such as the blades of wind turbines, low oil prices reduce further the costs of installing new wind farms.[16]

[14] "Big Oil faces a future where it may not be so big, and may have less to do with oil. The industry has faced an increasingly uncertain future as climate change has moved to the forefront of the public's consciousness. Investors, too, have stepped up their calls for action. Now the coronavirus pandemic, by radically cutting demand for oil and gas and giving governments the whip hand in directing the revival of their economies, looks likely to accelerate the long-term shift away from fossil fuels in many nations." Big Oil faces up to a future beyond petroleum: the pandemic is set to accelerate the shift away from fossil fuels. Financial Times. June 15, 2020. https://www.ft.com/content/590b1fec-af0d-11ea-a4b6-31f1eedf762e?desktop = true&segmentId = d8d3e364-5197-20eb-17cf-2437841d178a&_lrsc = %E2%80%A6.

[15] In an interview for the Financial Times in early 2021, Shell's officer stated that, should Shell meet its aspiring schedule to cut carbon emissions, Shell would become the largest energy player in the world by 2035. Crooks. E., Raval, A., 2019. Shell aims to become world's largest electricity company. Financial Review, March 13. https://www.afr.com/companies/energy/shell-aims-to-become-world-s-largest-electricity-company-20190313-p513zb.

[16] "Overall, we don't currently foresee a significant long- term impact on the pace of renewables infrastructure investment as a result of low oil prices or the COVID-19 pandemic. From a longer-term perspective, we do not anticipate low crude oil prices will have much impact on renewables growth globally, particularly in the developed world where the use of crude oil for power generation is more limited. Ironically, with fossil fuel products used to manufacture parts for wind turbine blades, for example, low oil prices could actually help to further reduce the installed cost of new renewable (wind) generation. Beyond these direct impacts, it is important to keep in mind that renewables generation costs—both wind and solar—have continued to decline and are generally competitive with the lowest cost fossil fuel generation, which is currently natural gas." UBS Switzerland AG, 2020. After COVID-19: How to Invest in a More Indebted, Less Global, and More Digital Word. UBS.

Preliminary assessments have also indicated smaller average losses and greater resilience in the share value of renewable energy companies, when compared to those of oil companies. This may enhance the tendency of segmenting these assets in different legal entities, to avoid that the depreciation of equity value of oil companies undermine the shares of other business divisions, which are environmentally sustainable (Denning, 2020).

If, before the Covid-19 pandemic, the debate on governmental stimulus was emphasized through economic packages for a greener economy, such as the European Green Deal launched in 2019; the sanitary emergency and the consequent global economic crisis have promoted cleaner investments to new heights.

Not by chance, in March 2020, South Korea announced its Green New Deal, putting an end in funding coal projects. Almost simultaneously, China launched its 14th Five-Year Plan for the National Economy, which takes sustainable development as a fundamental assumption. Indeed, China cannot afford anymore to completely ignore that economic growth and environmental goals must be balanced.[17]

In its turn, the European Union is seeking to build consensus for its Green Deal (International Renewable Energy Agency (IRENA), 2020). Therefore, economic agents, including several governments, appear to be articulating to combine short-term measures

---

[17] "China's priority is to build a modern energy industry that is secure, stable, economically efficient and based on clean and sustainable energy resources (China's Energy Policy White Paper, 2012). As the Chinese economy grew, so did the energy demand. China recognizes that it cannot continue on the path of economic growth regardless of the environmental consequences. And that is why it has committed itself to fulfilling the commitments on energy efficiency and reducing the burden on the environment set out in the 13th Five-Year Plan and the Paris Agreement. [...] Efforts to limit the impact of climate change have not always met with understanding in China. From the perspective of China's leaders in the 1990s, environmental degradation was a necessary price to pay to alleviate large-scale poverty. [...] In 1973 the government created a national environmental protection organ and environmental planning became included in national plans. Chinese government promulgated the Environmental Protection Law in 1979 for 'trial implementation' and the concept of harmonious development (Xietiao Fazhari), similar to the idea of sustainable development. Both of these policies, however, fell short of their enactment. The full Environmental Protection Law was adopted in 1989. [...] The beginning of the 21st century was marked by tensions between China's position on its global environmental obligations and the international community's expectations. China's position changed at the Paris Conference, at which the country committed itself to the following action by 2030: peaking of carbon dioxide emissions around 2030 and making best efforts to peak early; lowering carbon dioxide intensity (carbon dioxide emissions per unit of gross domestic product—GDP) by 60%–65% from the 2005 level; increasing the share of nonfossil fuels in primary energy consumption to around 20%; and increasing the forest stock volume by around 4.5 billion cubic meters from the 2005 level. [...] The first goal—to achieve peak emissions around 2030 and make best efforts to peak early—was announced by President Xi Jinping in November 2014 at a summit with US President Barack Obama in Beijing. The second goal—to lower carbon dioxide emissions per unit of GDP ("carbon intensity") by 60%–65% from the 2005 level by 2030—builds on a similar pledge for 2020 announced by Premier Wen Jiabao just before the Copenhagen climate conference in 2009. The third goal—to increase the share of nonfossil fuels in primary energy to around 20% by 2030—was also announced at the November 2014 Beijing summit with President Obama. In light of China's size and projected economic growth, this goal implies a very substantial increase in renewable and nuclear power generation in the next decade. [...] To meet the criteria, China must pursue a wide range of policies to decarbonize and peak its emissions. Notably, China aims to peak its coal consumption by 2020 and limit its oil consumption growth. It intends to grow its share of natural gas, nuclear, hydro and other renewables." Cibul'a, A., 2020. China and its way to a low carbon economy: current issues and challenges. Geopolitics of Energy 42 (8).

with the medium- and long-term goals of the 2030 Agenda for Sustainable Development and the Paris Agreement (International Renewable Energy Agency (IRENA), 2020).

Although no longer a European Union member, the United Kingdom committed to a green recovery, fighting climate change, by investing £12 billion in creating 250 thousand climate-friendly jobs (Johnson, 2020) and reinforcing its legal undertaking to reaching carbon net zero emissions by 2050.[18] Moreover, the Boris Johnson Administration decided to end the sales of new petrol and diesel vehicles by 2030, putting the United Kingdom ahead of France, which has designed a 2040 ban, and in line with Germany, Ireland, and the Netherlands, just falling behind Norway, which has banned fossil fueled cars from as early as 2025 (Ambrose, 2020c).

The United States' general elections held on November 2020, among other structural issues for North Americans as a society, can also be viewed as a referendum between a president, who refused to acknowledge the urgency of the climate crisis, seeking reelection[19]; and a contender who the committed to replace the United States in the international framework and to take an active role in establishing a way forward in meeting the goals of the 2030 Agenda, thus putting pressure in other governments.[20]

For that reason, during his campaign president-elect Biden pledged to rejoin the Paris Agreement[21] and presented the boldest climate plan of any presidential candidate in history, proposing a US$ 2 trillion investment in clean energy over four years and called for

[18] "The United Kingdom is one of the first major economies to set a legally binding target to reduce emissions to net zero by 2050. The target came into force on June 27, 2019. The Committee on Climate Change advised in a report dated May 2, 2019 that the United Kingdom could require up to four times the amount of renewable generation available at today's levels, complemented by firm low-carbon power options such as nuclear power and carbon capture and storage (applied to biomass or gas-fired plants) to achieve its net zero emissions target." Dewar, J., Cintré, K., 2020. Renewable energy law review: United Kingdom. Renewable Energy Law Review 3 (1), 220—228.

[19] About the bias of the Trump Administration and its neglect to consider basic evidences that the benefits outweighed largely the costs in remaining the Paris Agreement, vide: Arlota, C. Does the United States' withdrawal from the paris agreement on climate change pass the cost—benefit analysis test? University of Pennsylvania Journal of International Law 41 (4), 881—938. https://scholarship.law.upenn.edu/jil/vol41/iss4/1.

[20] "The United States will hold a climate summit of the world's major economies early next year, within 100 days of Joe Biden taking office, and seek to rejoin the Paris agreement on the first day of his presidency, in a boost to international climate action. [. . .] Donald Trump, whose withdrawal of the United States from the Paris Agreement took effect on the day after the US election in November, shunned the Climate Ambition Summit. Countries including Russia, Saudi Arabia and Mexico were excluded as they had failed to commit to climate targets in line with the Paris accord. Australia's prime minister, Scott Morrison, had sought to join the summit but his commitments were judged inadequate, and an announcement from Brazil's president, Jair Bolsonaro, of a net zero target just before the summit was derided as lacking credibility." Harvey, F., 2020. US to hold climate summit early next year and seek to rejoin Paris accord. The Guardian, December 14. https://www.theguardian.com/environment/2020/dec/14/us-to-hold-world-climate-summit-early-next-year-and-seek-to-rejoin-paris-accord.

[21] "[. . .] President-elect Joe Biden blasted out a statement, vowing to rejoin the Paris Agreement "on day one" and to restore the United States as a world leader in climate action: 'I'll immediately start working with my counterparts around the world to do all that we possibly can, including by convening the leaders of major economies for a climate summit within my first 100 days in office,' he said. [. . .] Biden promised to put the nation on a path to achieve net-zero emissions by 2050, to ensure that the shift toward cleaner energy brings new US jobs." Dennis, B., 2020. The US will soon rejoin Paris climate accord. Then comes the hard part. The Washington Post, December 22. https://www.washingtonpost.com/politics/2020/12/22/biden-paris-climate-accord/.

100% clean electricity by 2035 (Boyle, 2020). Although the volume of investment is impressive, many critics dispute that it still falls short of what will be effectively needed. Doubts were also raised, as Biden refrained from openly criticizing fracking, which could have costed decisive votes in the swing state of Pennsylvania.[22]

In Brazil, the CEOs of industry leaders, including BP, Chevron, CNPC, Eni, Equinor, Exxon, Petrobras, Shell, and Total, among others, released a joint letter, which reiterated the continuous efforts to reduce carbon emissions, develop solutions for low carbon content and support the local government in efficient energy transitioning policies (Oil and Gas Climate Initiative (OGCI)). At the governmental level, Decree No. 10,387 (Federal Decree 10387, 2020) was enacted on June 5, 2020, extending other industries' tax benefits for renewable projects.

From a social perspective, by forcing a pause for reflection about the future of international society, the Covid-19 pandemic created a scenario in which world leaders are pressured to promote a climate-friendly economic recovery, with a choice of words that reveal some degree of optimism: *build back better*.

Whether concrete actions will be consistent with the public reports and other pitches to the press, as to set aside the gap indicated by the International Renewable Energy Agency's 2020 report, between rhetoric and effective measures, it can only be confirmed in the medium and long term, when the pandemic's enduring economic repercussions become clearer (Denning, 2020). In any case, the mere fact that there is no indication of a change in direction evidences that economic transformation on a sustainable basis is an irreversible vector and that an overall structural change is greater solidified than previously thought.[23]

17.6 Conclusion

There is no doubt that the global energy matrix will change significantly in the coming decades. Faced with this scenario, oil companies send clear signals that they intend to participate in this transition and are taking measures to add new sources of energy to their portfolios.

[22] "[...] Biden's climate plan would lay out $2 trillion over 4 years towards clean energy and infrastructure, which he says will create 'millions' of jobs and move the US closer to a carbon-free future. For comparison, the cost, while expensive, it is still short of the 1-year, $2.2 trillion price tag for US coronavirus stimulus measures to date). [...] However, fracking is a substantial source of jobs and revenue in the crucial swing state of Pennsylvania, where some 32,000 workers are employed in the fracking and natural gas industry. Biden told voters there in July that fracking 'is not on the chopping block,' though his campaign says he supports no new fracking on federal land." Berardelli, J., 2020. How Joe Biden's climate plan compares to the Green New Deal. CBS News. October 5. https://www.cbsnews.com/news/green-new-deal-joe-biden-climate-change-plan.

[23] BP's CEO, Bernard Looney, is revealing: "How does Covid change things? I think it has only emphasized, reemphasized, recommitted me to the need to take BP on the energy transition. Why? First of all, I think oil has its challenges and I can only see Covid adding to the challenges of oil in the years ahead. We don't know how it's all going to play out. But it's gotten more likely to have oil be less in demand. Second, I think people are more aware of the fragility, the frailty of the ecosystem that we're living in—that things can change overnight. People are looking up at the skies and seeing clean skies and things are quieter on their roads—people will emerge from this potentially more conscious of the quality of air and the environment." Tett, G., Nauman, B., Raval, A., 2020. Moral Money special edition: in depth with BP's Bernard Looney. Financial Times, May 13. https://www.ft.com/content/635aa603-f17b-4795-9d7a-4f935ed59c8c.

The question that arises is whether the financial pressures resulting from the Covid-19 pandemic will force an influx in this trajectory, reducing the flow of investments by oil companies in renewable energy or, on the contrary, will be a new driving force to even deeper structural changes and, consequently, will end up contributing to reinforce a multi-faceted action of agents pertaining to the sector and those external to it.

References

Abington, T., Gilblom, K., 2019. Shell leads Big Oil in the race to invest in clean energy. Bloomberg September 4. Available from: https://www.bloomberg.com/news/articles/2019-09-04/shell-leads-big-oil-in-the-race-to-invest-in-clean-energy-tech.

Adomaitis, N., Solsvik, T., 2019. Petroleira Equinor amplia participação na empresa de energia solar Scatec. Reuters December 20. Available from: https://economia.uol.com.br/noticias/reuters/2019/12/20/petroleira-equinor-amplia-participacao-na-empresa-de-energia-solar-scatec.htm.

Ahrens, M.H., Hussain, M., Harman, R., 2020. The new equator principles—climate change, US application and what EP4 means for you. Milbank September 29.

Ambrose, J., 2020a. BP sets net zero carbon target for 2050. The Guardian February 12. Available from: https://www.theguardian.com/business/2020/feb/12/bp-sets-net-zero-carbon-target-for-2050.

Ambrose, J., 2020b. Shell unveils plans to become net-zero carbon company by 2050. The Guardian April 16. Available from: https://www.theguardian.com/business/2020/apr/16/shell-unveils-plans-to-become-net-zero-carbon-company-by-050#:~:text=Royal%20Dutch%20Shell%20plans%20to,carbon%20intensity%20of%20its%20business.&text=The%20plan%20includes%20an%20interim,%2C%20up%20from%2020%25%20previously.

Ambrose, J., 2020c. UK plans to bring forward ban on fossil fuel vehicles to 2030. The Guardian September 21. Available from: https://www.theguardian.com/environment/2020/sep/21/uk-plans-to-bring-forward-ban-on-fossil-fuel-vehicles-to-2030.

Andreoni, M., Casado, L., Londoño, E., 2020. Amazon deforestation soars as pandemic hobbles enforcement. The New York Times June 6. Available from: https://www.nytimes.com/2020/06/06/world/americas/amazon-deforestation-brazil.html.

Arbex, P., 2020. Blacklisted: aumenta a pressão sobre petroleiras e mineradoras. Brazil Journal March 13. Available from: https://braziljournal.com/blacklisted-aumenta-a-pressao-sobre-petroleiras-e-mineradoras.

Arlota, A.S.C., 2019. A Cláusula Knock-for-knock: admissibilidade à luz do direito brasileiro. Rio de Janeiro, Lumen Juris.

Baker, H., We are Baker Hughes, an energy technology company. https://www.bakerhughes.com/ (accessed 13.12.20).

Barros, P.S., Schutte, G.R., Pinto, L.F.S., 2012. Além da Autossuficiência: O Brasil como Protagonista do Setor Energético. IPEA, Brasilia.

BlackRock. Sustainability as BlackRock's new standard for investing. https://www.blackrock.com/br/blackrock-client-letter (accessed 14.01.21).

Boadle, A., 2020. Brazil's Bolsonaro blames indigenous people for Amazon fires in U.N. speech. Reuters September 22. Available from: https://www.reuters.com/article/un-assembly-brazil/brazils-bolsonaro-blames-indigenous-people-for-amazon-fires-in-u-n-speech-idUSKCN26E0AM.

Boyle, L., 2020. What are Joe Biden's plans to fight climate change? The Independent December 16. Available from: https://www.independent.co.uk/environment/biden-climate-change-plan-paris-agreement-b1775187.html.

Brazilian National Agency of Electrical Energy (ANEEL). Editais de Geração. https://www.aneel.gov.br/geracao4 (accessed: 25.03.20).

Brazilian National Agency of Electrical Energy (ANEEL), 2019. LEILÃO de energia renovável tem deságio de 45% e gera investimentos de R$ 1,9 bi. Agência Nacional de Energia Elétrica. September 2. www.aneel.gov.br/sala-de-imprensa-exibicao-2/-/asset_publisher/zXQREz8EVlZ6/content/leilao-de-energia-renovavel-tem-desagio-de-45-e-gera-investimentos-de-r-1-9-bi/656877?inheritRedirect = false.

Brazilian National Development Bank (BNDES), 2019. BNDES destina R$ 2 bilhões para implantação de usina termelétrica a gás natural proveniente do Pré-sal. December 20. https://www.bndes.gov.br/wps/portal/site/

home/imprensa/noticias/conteudo/bndes-destina-r2-bilhoes-para-implantacao-de-usina-termeletrica-a-gas-natural-proveniente-do-pre-sal.

Brazilian National Development Bank (BNDES). BNDES Finem—Geração de Energia. https://www.bndes.gov.br/wps/portal/site/home/financiamento/produto/bndes-finem-energia (accessed 23.03.20).

British Petroleum (BP), 2019. Statistical Review of World Energy 2019. British Petroleum, London.

British Petroleum (BP). What we do. https://www.bp.com/en/global/corporate/what-we-do.html (accessed 13.12.20).

Carbon Capture & Storage Association. What is CCS? http://www.ccsassociation.org/what-is-ccs (accessed 25.03.20).

Casarin, R., 2020. Potência instalada de energia solar ultrapassa carvão e nuclear somadas no Brasil, informa ABSOLAR. Portal Solar June 4. Available from: https://www.portalsolar.com.br/blog-solar/energia-solar/potencia-instalada-de-energia-solar-ultrapassa-carvao-e-nuclear-somadas-no-brasil-informa-absolar.html.

Chung, J., Yang, H., 2020. Qatar Petroleum's $19 billion LNG vessel order boon for South Korean shipbuilders. Reuters June 1. Available from: https://www.reuters.com/article/us-south-korea-shipping-qatar/qatar-petroleums-19-billion-lng-vessel-order-boon-for-south-korean-shipbuilders-idUSKBN23908K.

Coelho, C.R., 2011. Energia Elétrica: Contratos e Gestão de Riscos. Insper, São Paulo.

Costa, H.K.M., Tomé, F.M.C., e Silva, I.M.M., 2019. A Regulação do gás natural canalizado, Substitutivo ao Projeto de Lei no 6.407/2013 e Novo Mercado de Gás no Brasil. In: Costa, H.K.M. (Ed.), A Regulação do Gás Natural no Brasil. Lumen Juris, pp. 1—18.

David, S., 2019. A Relevância do Gás Natural no Setor Elétrico Brasileiro. In: Costa, H.K.M. (Ed.), A Regulação do Gás Natural no Brasil. Lumen Juris, pp. 237—262.

Denning, L., 2020. Big Oil can help renewables by spinning them off: spinning energy-transition assets into new companies would attract new investors and mollify old ones. Bloomberg June 3. Available from: https://www.bloomberg.com/opinion/articles/2020-06-03/big-oil-can-help-renewable-energy-with-spinoff-stocks.

Energy Information Administration (EIA), 2020. December 2020 Short-Term Energy Outlook. EIA, Washington, DC.

Equinor. Energias renováveis para um futuro mais sustentável. https://www.equinor.com.br/pt/energias-renovaveis.html (accessed 19.03.20).

Equinor. We're determined to be a global offshore wind energy major. Here's how. https://www.equinor.com/en/what-we-do/wind.html#us-offshore-wind (accessed 01.08.21).

Energia, F.G.V., 2020. Informe Petropolítica. Fundação Getúlio Vargas;, Rio de Janeiro.

Federal Decree 10387, 2020. http://www.planalto.gov.br/ccivil_03/_ato2019-2022/2020/decreto/D10387.htm (accessed 11.10.20).

Ferreira, L.P., Borges, M.A.S., 2014. Empresas Estrangeiras e o Direito ao Desenvolvimento: uma análise jurídica da política de conteúdo local no setor petrolífero nacional. In: Rosado, M.R.S. (Ed.), Direito Internacional dos Investimentos. Renovar;, Rio de Janeiro.

Gás Natural Açu (GNA). Quem somos. https://www.gna.com.br/a-gna/quem_somos (accessed 23.03.20).

Gas Natural Acu Financial Times. https://www.ft.com/content/590b1fec-af0d-11ea-a4b6-31f1eedf762e (accessed 12.11.2020).https://www.ft.com/content/590b1fec-af0d-11ea-a4b6-31f1eedf762e (accessed 13.12.20).

Gaudarde, G., 2019. BP e Bunge criam empresa para atuar nos mercados de etanol e energia no Brasil. EPBR July 22. Available from: https://epbr.com.br/bp-e-bunge-criam-empresa-para-atuar-nos-mercados-de-etanol-e-energia-no-brasil.

Goldenberg, J., 2015. Energia e Sustentabilidade. Revista de Cultura e Extensão da USP 1 (14), 33–43.

Goldenberg, J., Prado, L.T.S., 2003. Reforma e Crise do Setor Elétrico no Período FHC. Tempo Social (USP) 15 (2), 219–235.

International Energy Agency (IEA), 2020. The Oil and Gas Industry in Energy Transitions: Insights From IEA Analysis. IEA;, Paris.

International Renewable Energy Agency (IRENA), 2020. Global Renewables Outlook: Energy Transformation 2050, 2020th edition Irena, Abu Dhabi.

Jha, A., 2008. Explainer: how carbon is captured and stored. The three main techniques preventing carbon dioxide from coal-fired power stations contributing to global warming. The Guardian. September 5. Available from: https://www.theguardian.com/environment/2008/sep/05/carboncapturestorage.carbonemissions1#:~:text=Carbon%20capture%20and%20storage%20(CCS)%20is%20a%20range%20of%20technologies,and%20contribute%20to%20climate%20change.

Johnson, B., 2020. Boris Johnson: now is the time to plan our green recovery: we will use Britain's powers to invention to repair the pandemic's damage and fight climate change. Financial Times November 17. Available from: https://amp.ft.com/content/6c112691-fa2f-491a-85b2-b03fc2e38a30.

Lightsource, B.P., Solar power for our world: a global leader in solar development. https://www.lightsourcebp.com (accessed 01.08.21).

Naturgy. Presence in Spain. https://www.naturgy.com/en/get_to_know_us/international_presence/spain (accessed 12.12.20).

Oil and Gas Climate Initiative (OGCI), 2020. Delivering on a low carbon future: a progress report from the oil and gas climate initiative, December 2020. oilandgasclimateinitiative.com.

Paraskova, T., 2020. Shell joins other oil majors in energy transition push. Oil Price June 23. Available from: https://oilprice.com/Latest-Energy-News/World-News/Shell-Joins-Other-Oil-Majors-In-Energy-Transition-Push.html.

Pedroso, J., Ribeiro, M.R.S., 2020. Implicações da Pandemia do COVID-19 para o E&P e a Indústria do Petróleo e Gás no Brasil. Revista Energia, Ambiente e Regulação (RCGILex) 1 (4), 13−21.

Plano é investir US$ 3 bi em energias renováveis no mundo, diz Shell. Exame. September 9, 2020. https://exame.abril.com.br/negocios/plano-e-investir-us-3-bi-em-energias-renovaveis-no-mundo-diz-shell.

Projeto para térmica de R$ 1,5 bi a GNL no Pará prevê fornecer gás à indústria local. Reuters. October 30, 2019. https://epocanegocios.globo.com/Brasil/noticia/2019/10/epoca-negocios-projeto-para-termica-de-r15-bi-a-gnl-no-para-preve-fornecer-gas-a-industria-local.html.

Ribeiro, M.R.S., 2014. Joint Ventures na Indústria do Petróleo, second ed. Renovar;, Rio de Janeiro.

Ribeiro, M.R.S., 2009. Sovereignty over natural resources, Investment Law and expropriation: the case of Bolivia and Brazil. Journal of World Energy Law & Business 2 (2), 129−148.

Ribeiro, M.R.S., 2010. Direito dos Investimentos e o petróleo. Revista da Faculdade de Direito da UERJ 1 (18), 1−37.

S&P Dow Jones Indices, 2020. Transition to a 1.5°C World with the S&P PACT Indices (S&P Paris-Aligned & Climate Transition Indices). S&P Global June 02. Available from: https://www.spglobal.com/spdji/en/education/article/transition-to-a-15-c-world-with-the-sp-paris-aligned-climate-transition-indices/.

Saa, L., 2020. PRI welcomes 500th asset owner signatory. Principles for Responsible Investment January 27. Available from: https://www.unpri.org/pri-blogs/pri-welcomes-500th-asset-owner-signatory/5367.article.

Santos, M., 2008. Por Uma Outra Globalização—do Pensamento Único à Consciência Universal, fifteenth ed. Record, São Paulo.

Santos, E.M., 2020. Os Impactos na Mobilidade e o Petróleo a Preço Negativo. Revista Energia, Ambiente e Regulação (RCGILEX) 1 (4), 5−12.

Scatec. Asset portfolio overview. https://scatec.com/asset-portfolio-overview (accessed 01.07.21).

Shell, 2020. Shell e Cosan anunciam Raízen. February 14. https://www.shell.com.br/imprensa/comunicados-para-a-imprensa-2011/shell-cosan-jointventure-100211.html.

Statoil, 2018. Statoil to change name to Equinor. March 15. https://www.equinor.com/en/news/15mar2018-statoil.html.

Terminal Gás Sul. Quem somos. https://www.terminalgassul.com.br/quem-somos (accessed 24.03. 20).

Total Eren. Our global footprint. https://www.total-eren.com/en/renewable-energy/our-global-footprint (accessed 01.09.21).

UBS Switzerland AG, 2020. After COVID-19: how to invest in a more indebted, less global, and more digital word. UBS.

Vialli, A., 2020. Energia solar puxa avanço global das renováveis e Brasil lidera na América Latina. Portal Solar June 16. Available from: https://www.portalsolar.com.br/blog-solar/energia-solar/energia-solar-puxa-avanco-global-das-renovaveis-e-brasil-lidera-na-america-latina.html.

Wong, K.B., Scott, H.T., Bloom, C.S., 2020. Renewable energy law review: United States. Renewable Energy Law Review 3 (1), 229−242.

Yergin, D., 2009. The Prize, The Epic Quest For Oil, Money and Power. Free Press, New York.

Sustainable development and its link to Carbon Capture and Storage (CCS) technology: toward an equitable energy transition

Hirdan Katarina de Medeiros Costa[1],
Paulo Negrais Seabra[1]*, Carolina Arlota*[2,3] *and*
Edmilson Moutinho dos Santos[1]

[1]Institute of Energy and Environment, University of São Paulo, São Paulo, Brazil
[2]Visiting Assistant Professor of Law at the University of Oklahoma, College of Law, Norman, OK, United States [3]Law and Economics Research Fellow, Capitalism and Rule of Law Project (Cap Law) at Antonin Scalia Law School, George Mason University, Fairfax, VA, United States

18.1 Introduction

Sustainable development, as a concept, first appeared in the United Nations Conference on Human Environment in 1972 (Patel and Nagar, 2018). Nonetheless, this concept gained significant traction in the 1980s, when the Brundtland Commission Report defined it as "development that meets the needs of the present without compromising the ability of future generations to meet their own needs."

Later, the Rio Declaration on Environment and Development established it as a quite anthropocentric dimension in its principle 1, bringing back human beings to the nuclei of sustainable development concerns, in which they have the right to claim for a life healthy and in harmony with nature (Rio Declaration on Environment and Development adopted by the United Nations Conference on Environment and Development, June 14, 1992).

The concept of sustainable development has been significantly debated, with many questioning its legal force, and whether it has some other normative value, even a nonlegal or quasilegal character (Weiss, 1993; Weiss et al., 2016). Nowadays, there appears to be some consensus on understanding sustainable development through a triangular framework that addresses economic growth, social development, and environmental protection (Magraw and Hawke, 2006).

This chapter proceeds as follows: Section 18.2 analyzes the triangular proposition that sustainable development encompasses economic development, environmental protection, and social progress. Then, it discusses the implications of this conceptual framework. This chapter further considers this conceptualization of sustainable development and its impact for global climate debate, including the dialogue with Sustainable Development Goals (SDGs) of the Agenda 2030. In addition, this chapter connects sustainability to climate change issues, whereby carbon capture and storage (CCS) technology appears as a crucial tool to mitigate climate change. It concludes that this technology may foster a fair energy transition, which encompasses the best interests of the most vulnerable population as to the consequences of climate change and its disparate impact.

18.2 Sustainable development: concept overview

What is sustainability? Regarding this concept, Marquardt (2006) points out that its origin lies in the word "Nachhaltigkeit," developed by Hans Carlowitz, who in 1713 developed a theory about the optimal use of forests, which were the main source of energy of the time.

The term "Sustainability" has been associated to a dynamic condition or state that can be sustained for an indeterminate, but long period (Daly, 1996; Moutinho dos Santos, 2004; Costa and Moutinho dos Santos, 2011; Costa et al., 2017; Arodudu, 2017; Costa, 2018; Paglia, 2018). Achieving sustainable conditions involves a simultaneous understanding of three main aspects: environmental, economic, and social needs of communities (and through their generations) (Gomis et al., 2011; Jeppesen, 2011; Gatti and Seele, 2014; Opp, 2017; Nabavi et al., 2017; Tapia-Fonllem et al., 2017).

These dimensions can be detailed, for example, in the environmental dimension aspects; one may look at natural resource use and its economic consequences (Soini and Dessein, 2016). For instance, Hardin (1968) reflected on the intrinsic limits of the planet, and Auty (2007) discussed natural resources economics and its problems toward achieving development.

Moreover, sustainability dimensions can be expanded at the confluence with other issues, such as those connected to energy matters or to values orientations. For example, Georgescu-Rogen (1971, 1976) warned that the economy depends on energy to produce goods and services, and the second law of thermodynamics limits energy; thus economic growth is also limited by it; and, Egmond and Vries (2011) drew on the integral sustainability aspect associated to human dignity.

Regarding the economic dimension of sustainability, the World Commission on Environment and Development (1987) expressed in "Our Common Future Report" that sustainable development shall embrace the needs of present and future generations (Pietrzak and Balcerzak, 2016; Rogge and Reichardt, 2016; Espinosa and Walker, 2017). This approach was incorporated worldwide by governments, companies, nongovernmental organizations, international forums, etc. However, Sen (2013, p. 6) directs his analysis to address that although there is a conceptual improvement, it is still incomplete because "there are important grounds for favoring a freedom-oriented view, focusing on crucial freedoms that people have reason to value." Therefore, in his opinion, the sustainability concept should "aim at sustaining human freedoms, rather than only at our ability to fulfil our felt needs" (Sen, 2013, p. 6).

Currently, the sustainability concept, as well as the three dimensions referred to above, are connected to climate change-related issues. In this regard, Nath and Roberts (2021, p. 3) pointed out that "climate change and global warming have now occupied the center stage in the discussion on growth and development issues for the current and future generations."

It is thus necessary to look at the emergence of the climate crisis and consider how the sustainability concept may evolve and be incorporated as a new dimension (Hák et al., 2016). From this point, the next section will examine the international treaties on sustainable development related to the climate change subject, including the technological aspects discussed to reduce carbon emissions.

18.3 The international agenda and its relationship with sustainable development and climate change

In 1987 the Brundtland Report characterized a new type of development, called sustainable development, in which environmental issues were a clear concern, and developed the idea that progress should encompass the needs of the present generations without compromising those of future generations (Birnie and Boyle, 2002).

The international agenda related to the environmental cause, specifically climate change, became robust in 1988, when the Intergovernmental Panel on Climate Change (IPCC) was founded as the international body that reviews and assesses the latest science on climate change (Paglia and Parker, 2021). Clearly, IPCC reports have had influence on combatting climate change and have proposed international policies to avoid or to minimize climate change crises (IPCC, 1988, 1990, 2007, 2018, 2020).

Another important landmark was the United Nations Conference on Environment and Development, also called ECO-92 (or Earth Summit or Rio 92), in which the concept of sustainable development was consolidated with the publication of Agenda XXI (Fontanillas, 2020), which is inseparable from the problem of using natural resources (Pequim, 2020) as well as from the issue related to emissions, specifically carbon, which implies concerns with climate justice (Moss, 2009; Humphreys, 2009; Althor et al., 2016).

In the 1990s the United Nations Framework Convention on Climate Change (UNFCCC) implementing process, and the institution of the Kyoto Protocol (1997) were key advancements in climate and environmental global policy. They provided a new perspective for the future of international climate policy, with the purpose of reducing greenhouse gas (GHG) emissions, reversing the increase in carbon dioxide concentration in the atmosphere that has occurred since 1850, mainly from using fossil fuels.

In 2015 the Paris Agreement on Climate Change was created within the umbrella of the UNFCCC (Arlota, 2020). The Paris Agreement specifically determines: "This Agreement, in enhancing the implementation of the Convention, including its objective, aims to strengthen the global response to the threat of climate change, in the context of sustainable development and efforts to eradicate poverty, including by: (a) Holding the increase in the global average temperature to well below 2°C above preindustrial levels and pursuing efforts to limit the temperature increase to 1.5°C above preindustrial levels, recognizing that this would significantly reduce the risks and impacts of climate change" (Paris Agreement, Article 2). Many countries have submitted their iNDCs (intended nationally determined contributions) (Costa and Musarra, 2020) and are committing to new NDCS under the legal framework of the Paris Agreement (Articles 3 and 4 of the Paris Agreement).

In addition, it is important to emphasize that the Conference of the Parties (COP) held the global and annual commitment to discuss sustainability and climate issues. The first COP meeting was held in Berlin, Germany in March 1995 (UNFCC, 2020). COP became a relevant arena to debate climate change effects and its decisions work toward the implementation of adaptation and mitigation measures (UNFCC, 2020).

To connect sustainability and development and to achieve the purpose to collaborate with humankind's evolution, the United Nations launched the Millennium Development Goals in 2000, with eight challenges: Goal 1, eradicate extreme poverty and hunger; Goal 2, achieve universal primary education; Goal 3, promote gender equality and empower women; Goal 4, reduce child mortality; Goal 5, improve maternal health; Goal 6, combat HIV/AIDS, malaria and other diseases; Goal 7, ensure environmental sustainability; and Goal 8, develop a global partnership for development (United Nations, 2015b).

Under Agenda 21, a supplementary UN initiative refers to the National Sustainable Development Strategy (NSDS), and currently defined as a key issue is the definition of how each country would deal with NSDS, which means they need to define the best approach for implementing it, being dependent on the dominant political, social, and environmental circumstances. A standard prior prepared model, as an approach to NSDS is not achievable, and a particular label is not important, as long as the fundamental principles of NSDS observe the economic, social, and environmental objectives equitably and unified (United Nations, 2020c). These international initiatives preceded the SDGs.

In 2012, 20 years after Eco-92, at the Rio + 20 conference, the focus of the discussions became the green economy in the context of sustainable development and poverty eradication and the institutional framework for sustainable development (Oliveira, 2012). At that conference, no specific objectives were defined, but it was emphasized that future SDGs should be restricted in terms of how many will be proposed, they should be feasible and easily understandable. Moreover, the objectives should address all the three dimensions of sustainable development in a balanced manner and should be coherent and integrated into the UN development agenda beyond 2015 (United Nations, 2015a). The conference generated the document "The Future We Want," recognizing that the formulation of goals could be useful for launching a coherent global action focused on sustainable development.

After Rio + 20, an Open Working Group was established for preparing the SDGs. The result was the proposal of the 17 SDGs, and of the 169 targets associated with the assessment of the UN General Assembly in 2015. The next section details some SDGs and their connections with the climate change crisis.

18.4 Agenda 2030 and its role to sustainable development and the climate change crisis

As mentioned, the sustainable development concept was put in place with the United Nations 2030 Agenda for Sustainable Development by its 17 SDGs. Achieving those targets will certainly improve the sustainability of the planet. Environmental protection is one of the three dimensions of sustainable development addressed by Agenda 2030 and most of its SDGs are linked to global warming and the climate change crisis. Therefore to achieve the majority of these goals a global coordinated action will be necessary to prevent the increase of GHG emissions.

The two first goals—SDG1 (no poverty) and SDG2 (zero hunger)—establish that the key targets are eradicating poverty and hunger, ensuring for all a possibility to achieve equality and dignity on a healthy environment (United Nations, 2015a). The Food and Agriculture Organization estimated that 2 billion people worldwide (around 25.9% of the world's population) had moderate or severe levels of food insecurity in 2019, and the number of people affected by hunger in the world continues to rise. The COVID-19 pandemic evolution will increase the portion of the world's population affected by moderate or severe food insecurity (FAO, 2020). To reach these first two goals, it is essential to increase the food supply, secure clean drinking water and sanitation, as well as other actions respecting the environment.

Nowadays, food production is predominantly dependent on the use of fossil energy (oil, coal, and natural gas) and nitrogen fertilizer, such as urea that still comes mostly from fossil sources. Another issue related to agricultural production is anthropogenic methane emission to the atmosphere; methane is a GHG with stronger heat absorption than CO_2. The world's largest anthropogenic sources of CH_4 are ruminant livestock, responsible for 31% of the current total CH_4 emissions, and rice production, due to the anaerobic decomposition of organic matter, responsible for 10% of these global emissions (Olivier and Peters, 2020). Therefore, to meet these challenges, it is essential to encourage the use of renewables, efficient agricultural practices, and low-carbon transportation besides other measures.

The SDG6 (clean water and sanitation) is a huge target because 2.2 billion people in the world lack safely managed drinking water and 4.2 billion lack safely managed sanitation (United Nations, 2020a). To overcome this deficit of good quality water, it will be necessary to guarantee the supply in a climate crisis environment, such as droughts and floods, which is a huge challenge.

Three billion people worldwide still rely on wood, coal, charcoal, or animal waste for cooking and heating (UNEP, 2020a). This puts a lot of pressure on SDG7 (affordable and clean energy). In the world, 789 million people lack electricity (United Nations, 2020a; Costa, 2019). Every year, indoor air pollution kills 4.3 million people (UNEP, 2020a). The energy sector is the main GHG emitter worldwide, accounting for over 70% of the total (Climate Watch, 2017). Hence, in order to achieve this objective, it is important, for instance, to invest in renewable energy, to disseminate its use and to put in place energy saving policies.

One of the most challenging SDGs is number 11 (sustainable cities and communities). Cities occupy just 3% of the Earth's land but account for up to 80% of the energy consumption as well as 75% of the global waste and GHG emissions, and concentrate people, infrastructures, housing, and economic activities. They are especially vulnerable to climate change and natural disaster impacts, such as floods and hurricanes. SDG11 seeks to promote fair urban planning, sustainable construction, low-carbon transport, green spaces, and sustainable lifestyles (UNEP, 2020a).

One of the SDGs directly related to the warming climate system—SDG13 (climate action)—aims to "take urgent action to combat climate change and its impacts" (United Nations, 2015a,b), as climate change is increasing the rise in sea level, droughts, and the likelihood of natural disasters, such as wildfires, floods earthquakes, tornados, winter storms, and hurricanes (IPCC, 2018). SDG13 mainly targets the following actions: fostering resilience and adaptive capacity to deal with hazards and natural catastrophes caused by climate change globally; integrating actions against global warming into governmental guides, public policies, and planning; refining education, awareness-raising, and human and institutional capacity on climate change mitigation, adaptation, impact reduction and early warning; bringing effectiveness to the commitment undertaken by developed country parties to the UNFCC to a goal of jointly mobilizing $100 billion annually by 2020 from all sources to address the needs of developing countries in the context of meaningful mitigation actions and transparency in implementation, and to fully operationalize the Green Climate Fund through its capitalization as soon as possible; and fostering mechanisms for raising capacity for effective climate change-related planning and management in less developed countries and small island developing states, including focusing on women, youth and local and marginalized communities (UNEP, 2020b).

More specifically to protect the environment there is SDG14 (life below water) that aims to "conserve and sustainably use the oceans, seas and marine resources for sustainable development" (United Nations, 2015a; Costa and Ladeira, 2019). One of the main threats to ocean life is the continuous process of acidification due to the increase of CO_2 content in the water. The rise of acidity is projected to be 100%−150% by 2100, affecting half of all marine life (United Nations, 2020a).

Another planet-related SDG is number 15 (life on land) that deals with the surge in rate of forest destruction, mainly driven by agricultural expansion. This poses a real danger to the biodiversity in the world, where over 31,000 land species are threatened with extinction (United Nations, 2020a). Other problems of deforestation include the carbon stored in the vegetation being released to the atmosphere and the disturbances in the rainfall regime.

The pre-COVID-19 benchmark scenario presented by an independent group of scientists in the 2019 classified the SDGs goals into three groups: (1) SDG targets on track; (2) those which are considered within reach with extra efforts; and (3) SDGs whereby the process for implementing them are going toward the opposite side. Most of the SDGs associated to human development are in the second group, with a few being in the first. However, all the planet-related SDGs fall into the third group (Independent Group of Scientists appointed by the Secretary-General, 2019). Hence, the planet-related SDGs are far from being achieved, which is a major concern for global climate action. This situation also exposes island countries, indigenous populations, and developing countries more significantly to the adverse impact of climate change (IPCC, 2018). Accordingly, there are significant inequalities regarding the disproportionate impact of climate change at the local and global levels.

Moreover, the difficulties in implementing the planet-related goals have been aggravated by the COVID-19 pandemic. Despite CO_2 emissions falling an appraised 17% in early April 2020 in relation to the 2019 average levels (mainly due to the shutdown of many industrial and commercial enterprises and the dramatic reduction in travel), emissions are expected to quickly rebound (Le Quéré et al., 2020). Therefore overcoming crisis could cause a post-COVID-19 negative scenario, in which the world will quickly get back to the business-as-usual path pre-COVID-19, diverting from the Paris Agreement goals.

In such a scenario, it is possible that the COVID-19 crisis will impact the SDGs achievements in the short and medium term by 2030. The extent of this impact relies on how civilizations reply to the pandemic and its repercussions. In an optimistic scenario about the outcome of the COVID-19 crisis, it is possible to reenergize the global effort toward sustainable development. For this, however, it is important to quickly recover from the damage that will be different for each country and region. It is also important to persist and advance positive changes of healthcare, social protection, and governance systems established during the pandemic in most countries. Another essential action is to maintain and to reinforce the improvements observed concerning the planet-related SDGs during the COVID-19 crisis, such as the decline in CO_2 emissions and the improvement of air and water quality (United Nations, 2020b).

Sustainable recovery and planet protection need concrete efforts in which to invest: power generation using renewable energy sources; building smart grids; carbon capture and storage (CCS) during the energy transition; development of modern low carbon emission clean fuels; development and use of power-saving appliances; and switching to electric vehicles supplied by clean energy. Another very important action is to end fossil fuel subsidies and to impose gasoline taxes taking advantage of the currently prevailing historically low oil prices. Investments are also needed for adaptation purposes, particularly in developing low-lying countries that are hard hit by the effects of climate change.

There is a final lesson provided by the COVID-19 crisis: the neglect of the planet, particularly in the context of preserving its biomes, increases the danger of frequent epidemics and pandemics that can put people and the economy at risk.

18.5 The place of CCS in the fair energy transition to sustainable development

The climate crisis needs a global action from national governments, policymakers, industry, financial sector, academia, nongovernmental organizations, and the public that can lead to net zero carbon emission. Renewable energies, such as solar and wind, play an important role to achieve the neutral carbon ambitions. Yet without carbon capture and storage (CCS), the GHG emissions reduction goals for 2050 will be very difficult to attain (IEA, 2016). CCS is an additional avenue for reducing GHG emissions, as it enables the use of fossil fuels with significant reduction in such emissions (Dessler and Parson, 2020). Accordingly, the use of CCS would greatly assist the energy transition to cleaner and greener energy sources, such as renewables.

Furthermore, it can also assist in implementing fairness considerations domestically as well as globally. Environmental and climate justice considers the disproportionate impact of environmental pollution and regulations on minorities as well as those with less economic resources. These justice-based considerations include a multitude of actions, such as critical enforcement of environmental regulation, public participation in decision-making, transportation equity, workplace and school exposures to toxic materials, while acknowledging that even global climate policies will be grounded on their local level effects (Outka, 2012). Therefore an equitable energy transition considers fairness and equality when designing polices on energy access and security aiming at the displacement of fossil fuels. As low-carbon technologies themselves can lead to injustice (such as the takeover by corporations from community ownership, poor work conditions in the biofuel industry, among others), policymakers and key stakeholders are urged to include such considerations when developing and implementing energy policies (Gambhir et al., 2018).

Besides equitable considerations, which are also present in the concept of sustainable development, as previously outlined in Section 18.4, the principle of sustainable development is intimately connected to the development and access to technologies. This is the case not only because of specific provisions in the UNFCCC (its preamble and Articles 3 and 4, for example) and in the Paris Agreement (such as in Articles 2 and 7, for instance), but also due to related efforts regarding the transfer of technology and financial resources for climate action globally (Arlota, 2021). In this context, sustainable development policies that foster technology transfer need to consider the local needs and priorities of the country receiving such a technology, including its stage of development and social context (such as educational skills, institutions, and civil society) for this technology transfer to be meaningful over time. The IPCC has long recognized the relevance of technology transfer as a crucial component to promote adaptation and to foster climate resilience. (IPCC, 2000).

Globally, the primary sources of GHG emissions in 2016 were energy production of all types (73%), agriculture (11%), industrial processes (5.5%), land-use change and forestry (6.5%), and wastes (3%) of a total of 49.6 $GtCO_2e$ (Climate Watch, 2017). The world total primary energy consumption by fuel is mainly fossil, around 33% from oil, 27% from coal, and 24% from natural gas, and most of the future scenarios suggest that fossil fuels will continue to be predominant (BP, 2020). China and India, the first and fourth highest GHG emitters in the world, are still very reliant on coal, for instance.

The CCS technologies, alongside electrification, hydrogen, and sustainable bioenergy, can contribute to clean energy transitions in several ways. CCS concerns retrofitting existing fossil fuel-based power and industrial plants. It can also enable clean hydrogen production from fossil fuels, predominantly from coal and natural gas, the main current sources of hydrogen production (IEA, 2019). This will open a new market for low-carbon hydrogen.

There are currently 28 commercial CCS facilities in operation in the world, two suspended temporarily, closely connected to the oil and gas industry and, in particular, to enhanced oil recovery (23 facilities), with the capacity to capture up to 40 $MtCO_2$ each year (GCCSI, 2020). The deployment of CCS to date has been concentrated in the United States, which is home to 16 operating facilities. But to achieve the neutral carbon goal in 2050, CCS needs to be deployed in a huge number of facilities. The International Energy Agency Sustainable Development Scenario aims at a global decline in CO_2 emissions from the energy sector to net zero by 2070; CCS shall account for nearly 15% of the cumulative reduction in emissions compared with the Stated Policies Scenario (government policies and pledges that have already been adopted or declared related to energy and the environment) (IEA, 2020; Magraw and Hawke, 2006).

There are many reasons for CCS to be deployed in the world at a slow pace. Among these reasons are (1) lack of legal and regulatory framework that curbs CO_2 emissions in most jurisdictions (Costa et al., 2018; Morbach and Costa, 2020; IEA, 2020); (2) no attractiveness of CCS as a business, especially where CO_2 has no important value as raw material in the production of fuels, chemicals or building materials (Musarra and Costa, 2018; IEA, 2020); (3) high costs of installing the infrastructure (Almeida et al., 2017; IEA, 2020); (4) difficulties in establishing an integrated CO_2 supply chain (Costa, 2020; IEA, 2020; Morbach and Costa, 2020) (5) public opinion against storage, mainly onshore (IEA, 2020; Musarra and Costa, 2019, 2020; Musarra et al., 2019).

Another issue that hinders CCS' wider acceptance is the view that it is a fossil fuel technology that competes with renewables for investments. In reality, the investment in CCS has also fallen well behind that of other clean energy technologies. Annual investment in CCS has consistently accounted for less than 0.5% of the global investment in renewables and efficiency technologies (IEA, 2020).

A new challenge for a fair energy transition is the COVID-19 crisis. Several new CCS projects and investment incentives have been announced in 2020, but the economic recession due to the COVID-19 pandemic can threaten the implementation of these projects. The low oil price because of the pandemic affected the enhanced oil recovery (EOR) projects in place and in the planning phase. In May 2020 Petra Nova Carbon Capture System, in Texas, United States, suspended the CO_2 capture in an existing coal-fired generating station due to the low oil price that made the use of CO_2 for EOR uneconomic (NRG, 2020).

In spite of that, 2020 forecasts for CCS have been enhanced by a number of new funding announcements, involving CCS infrastructure, and project developments. In March, the United Kingdom government confirmed its pledge to invest at least £800 million to establish CCS clusters on at least two UK sites by 2030 (GCCSI, 2020). The United States government announced US$ 203 million in grants for CCS development and deployment (USDOE, 2020a,b). In May, the Australian government announced plans to allow CCS to qualify for existing funding programs for clean technologies. The Norwegian government announced it would provide US$ 1.8 billion in funding for the Longship CCS project (IEA, 2020).

So far it has been unclear whether the deployment of CCS and, consequently, the energy transition to a low carbon economy will be affected by the COVID-19 crisis. However, CCS is the only group of technologies that reduces GHG emissions in key sectors directly, including some types of industry (especially steel, chemicals, and cement), aviation, trucking, and shipping, and has an important role in the energy transition to net zero carbon emission in 2050 (IEA, 2020).

18.6 Conclusion

In this chapter, we intended to link sustainable development, the climate change debate, and CCS technology. From sustainable development, we highlighted its main features, which maintain its continuous meaningful expansion, including the climate change movement approach. Moreover, it was significant to discuss how the international agenda encompasses a series of landmarks that show the evolution of humankind's perception of reality toward a climate change crisis that impacts everyone and that not only needs to be dealt by each nation, but also jointly, as an international community.

The discussion of the SDGs shows the challenges ahead. Sustainable development and its dimensions, such as economic development, environmental protection, social progress, and climate change measures, need to be implemented, which is very complex and difficult for each nation. Moreover, we highlighted that many SDGs are connected to global warming and the climate change crisis.

In fact, GHG emissions are a great concern to humankind; thus measures should be implemented to contain their advance: enhancing renewables, efficient agricultural practices, low carbon transportation, and CCS technologies. As mentioned, SDG7 is directed mainly at investment in renewable energy; SDG11 intends to change lifestyles toward a sustainable approach; SDG13 is specifically related to climate action, whereby sustainability and CCS technologies find a place to interact; and SDG14 concerns life below water; thus it is really correlated to the energy–sustainability–climate change nexus.

As already analyzed, COVID-19 may impact the SDGs achievements by 2030, which depend on a complex scenario after the pandemic. Hence, predictions at this point are quite complex since they are conditioned upon the domestic and international policies being considered at the time of this writing. From our point of view, considering an optimistic scenario, the international agenda may be organized toward sustainable development achievements, and the climate change crisis may be contained with the increasing CCS technologies uses, as well as by controlling CO_2 emissions. Yet, in a pessimistic scenario, the world may give up the Paris Agreement goals and end up rapidly increasing those emissions.

It is possible to affirm that a mix of solutions and creativity will be necessary, such as renewable energy sources continually increasing; widescale smart grid adoption; CCS technologies intensification by many projects; development of modern low-carbon emission clean fuels; and fomenting the production and use of new power-saving applications. The list is not exhaustive and includes phasing out fossil fuel subsidies as well as investments or adaptation.

One may say that the COVID-19 crisis shows human beings how the neglect of the planet causes epidemics and pandemics and the price to be paid is high. In this line, we have argued that CCS technologies may be an effective tool for reducing GHG emissions; therefore, its use would benefit the energy transition to cleaner and greener energy sources. Moreover, CCS technologies may serve for retrofitting existing fossil fuel-based power and industrial plants.

Barriers exist to CCS deployment in the world, despite its benefits. Due to the COVID-19 pandemic, the implementation of CCS projects and, consequently, the energy transition to a low carbon economy may have been affected. Nevertheless, we must highlight that CCS has an important role in net zero carbon emission in 2050.

Acknowledgments

We are grateful to the "Research Center for Gas Innovation–RCGI" (Fapesp Proc. 2014/50279-4), supported by FAPESP and Shell, organized by the University of São Paulo, and the strategic importance of the support granted by the ANP (National Agency of Petroleum, Natural Gas and Biofuels of Brazil) through the R&D clause. We also thank the support from the National Agency for Petroleum, Natural Gas and Biofuels Human Resources Program

(PRH-ANP), funded by resources from the investment of oil companies qualified in the R,D&I clauses from ANP Resolution No. 50/2015 (PRH 33.1—Related to Call No. 1/2018/PRH-ANP; Grant FINEP/FUSP/USP Ref. 0443/19).

References

Almeida, J.R.L., Rocha, H.V., Costa, H.M., Santos, E.M., Rodrigues, C.F., Desousa, M.J.L., 2017. Analysis of civil liability regarding CCS: the Brazilian case. Modern Environmental Science and Engineering 03, 382—395.

Althor, G., Watson, J., Fuller, R., 2016. Global mismatch between greenhouse gas emissions and the burden of climate change, vol. 6 (20281). Scientific Reports, London. < https://www.nature.com/articles/srep20281#-citeas > (accessed 03.11.20), Available from: https://doi.org/10.1038/srep20281.

Arlota, C., 2020. Does the United States withdrawal from the Paris Agreement on climate change pass the cost—benefit analysis test. University of Pennsylvania Journal of International Law 41, 881—938.

Arlota, C., 2021. International Energy Law and the Paris Agreement in the aftermath of the COVID-19 pandemic: challenges and possibilities. ILSA Journal of International & Comparative Law 27.

Arodudu, O., et al., 2017. Towards a more holistic sustainability assessment framework for agrobioenergy systems—a review. Environmental Impact Assessment Review 62, 61—75.

Auty, R.M., 2007. Natural resources, capital accumulation and the resource curse. Ecological Economics 61, 627—634.

Birnie, P., Boyle, A., 2002. International Law and the Environment, second ed. Oxford University Press Inc., New York.

BP, 2020. BP Statistical Review of World Energy 2020. < https://www.bp.com/content/dam/bp/business-sites/en/global/corporate/pdfs/energy-economics/statistical-review/bp-stats-review-2020-full-report.pdf > (accessed 30.11.20).

Climate Watch, 2017. Data Explorer. < https://www.climatewatchdata.org/data-explorer/historical-emissions?historical-emissions-data-sources=cait&historical-emissions-gases=all-ghg&historical-emissions-regions=All%20Selected&historical-emissions-sectors=total-including-lucf&page=1 > (accessed 30.11.20).

Costa, H.K.M., Moutinho dos Santos, E., 2011. Sustainability and the Allocation of oil royalties: a theoretical contribution. In Giannetti, B.F., Almeida, C.M.V.B., Bonilla, S.H. (Eds.), International Workshop Advances in Cleaner Production, 3, 2011, São Paulo. Proceedings. UNIP, São Paulo.

Costa, H.K.M., Brasil, N.W., Moutinho Dos Santos, E., 2017. Reflections on the concept of sustainability, its adjectives and human unity. In: VI International Workshop Advances in Clean Production, 2017, São Paulo. VI International Workshop Advances in Cleaner Production. UNIP, São Paulo, p. 110.

Costa, H.K.M., 2018. Royalties, Justice and Sustainability. Rio de Janeiro, Lumen Juris.

Costa, H.K.M., Musarra, R.M.L.M., Miranda, M.F., Moutinho Dos Santos, E., 2018. Environmental license for carbon capture and storage (CCS) projects in Brazil. Journal of Public Administration and Governance 8, 163—185.

Costa, H.K.M., 2019. ODS 7. Direito à Energia e o caso do Programa Luz Para Todos. In: de Medeiros Costa, H.K. (Ed.), Direitos Humanos e Objetivos de Desenvolvimento Sustentável. Lumen Juris, Rio de Janeiro, pp. 155—170.

Costa, H.K.M., Ladeira, N.L., 2019. ODS 14: Diálogo entre Direitos Humanos de Terceira Geração, Direito do Mar e os Impactos Ambientais. In: de Medeiros Costa, H.K. (Ed.), Direitos Humanos e Objetivos de Desenvolvimento Sustentável. Lumen Juris, Rio de Janeiro, pp. 295—310.

Costa, H.K.M., Musarra, R.M.L.M., 2020. Law sources and CCS (carbon capture and storage) regulation in Brazil. International Journal of Advanced Engineering Research and Science 7, 195—201.

Costa, H.K.M. (Ed.), 2020. Transição energética, justiça geracional e mudanças climáticas: o papel dos fósseis e a economia de baixo carbono. Lumen Juris, Rio de Janeiro, p. 364.

Daly, H.E., 1996. Beyond Growth: The Economics of Sustainable Development. Beacon Press, Boston, MA.

Dessler, A.D., Parson, E., 2020. The Science and Politics of Global Climate change: A Guide to the Debate. Cambridge University Press, New York.

Egmond, N.D., Vries, H.J.M., 2011. Sustainability: the search for the integral worldview. Futures 43, 853—867.

Espinosa, A., Walker, J., 2017. A Complexity Approach to Sustainability: Theory and Application. World Scientific.

FAO (The Food and Agriculture Organization), 2020. The State of Food Security and Nutrition in the World 2020. < http://www.fao.org/documents/card/en/c/ca9692en > (accessed 30.11.20).

Fontanillas, C.N., 2020. Sustainable development and competitiveness: a study focused on the doctrinal environmental aspect. International Journal of Advanced Engineering Research and Science 7 (4). Retrieved from: http://journal-repository.com/index.php/ijaers/article/view/1894.

Gatti, L., Seele, P., 2014. Evidence for the prevalence of the sustainability concept in European corporate responsibility reporting (Report). Sustainability Science 9 (1), 89 (14).

Gambhir, A., Green, F., Pearson, P.J., 2018. Towards a Just and Equitable Low-Carbon Energy Transition. Grantham Institute: Brief, p. 26.

GCCSI (Global CCS Institute), 2020. Global Status of CCS Report 2020.

Georgescu-Rogen, N., 1971. The Entropy Law and The Economic Process. Harvard University Press, Cambridge, MA.

Georgescu-Rogen, N., 1976. Energy and Economic Myths. Institutional and Analytical Economic Essays. Pergamon Press, New York.

Gomis, A.J.B., Parra, M.G., Hoffman, W.M., Mcnulty, R.E., 2011. Rethinking the Concept of Sustainability (Report). Business and Society Review, Summer 116 (2), 171(21).

Hák, T., Janoušková, S., Moldan, B., 2016. Sustainable development goals: a need for relevant indicators. Ecological Indicators 60, 565–573.

Hardin, G., 1968. The Tragedy of Commons. Science 162, 1244–1245.

Humphreys, S., 2009. Human Rights and Climate Change. Cambridge University Press, New York. < https://www.cambridge.org/core/books/human-rights-and-climate-change/introduction-human-rights-and-climate-change/B89D34682C9C05FF50914706A342A275 > (accessed 03.11.20).

IEA (International Energy Agency), 2016. 20 Years of Carbon Capture and Storage—Accelerating Future Deployment. OECD, Paris. < https://www.oecd-ilibrary.org/energy/20-years-of-carbon-capture-and-storage_9789264267800-en > (accessed 03.11.20).

IEA (International Energy Agency), 2019. The Future of Hydrogen. < https://www.iea.org/reports/the-future-of-hydrogen > (accessed 30.11.20).

IEA (International Energy Agency), 2020. Energy Technology Perspectives 2020: Special Report on Carbon Capture. Utilisation and Storage, Paris, France.

Independent Group of Scientists appointed by the Secretary—General, 2019. Global Sustainable Development Report 2019: The Future is Now—Science for Achieving Sustainable Development. New York, United Nations.

IPCC, 1988. Report of the First Session of the WMO/UNEP Intergovernmental Panel on Climate Change. World Meteorological Organization, Geneva.

IPCC, 1990. IPCC First Assessment Report. World Meteorological Organization, Geneva.

IPCC, 2000. IPCC Methodological and Technological Issues in Technology Transfer: Summary for Policy Makers.

IPCC, 2007. Climate Change 2007: Synthesis Report. Contribution of working groups I, II and III to the Fourth Assessment Report of the IPCC. IPCC, Geneva.

IPCC, 2018. Global Warming of 1.5°C (Summary for Policymakers). World Meteorological Organization, Geneva.

IPCC, 2020. IPCC Website. < https://www.ipcc.ch > (accessed 04.12.20).

Jeppesen, S., 2011. Exploring an explicit use of the concept of sustainability in transport planning. Systemic Pract. Action. Res. 24 (2), 133–146.

Le Quéré, C., Jackson, R.B., Jones, M.W., Smith, A.J.P., Abernethy, S., Andrew, R.M., et al., 2020. Temporary reduction in daily global CO_2 emissions during the COVID-19 forced confinement. Nature Climate Change 10, 647–653. Available from: https://doi.org/10.1038/s41558-020-0797-x.

Magraw, D.B., Hawke, L.D., 2006. Sustainable development. In: Bodanski, D., et al. (Ed.), The Oxford Handbook of International Environmental Law: Oxford.

Marquardt, B., 2006. History of sustainability. An environmental concept in the history of Central Europe (1000-2006). Critical History, 32, 172–197.

Morbach, I., Costa, H.K.M., 2020. CO_2 capture and storage: property rights overview in Brazil. International Journal of Advanced Engineering Research and Science 7, 191–198.

Moutinho Dos Santos, E., 2004. Energia, gás natural & sustentabilidade. Associate Professor in Energy. Post-Graduation in Energy. Environment and Energy Institute (IEE), USP, São Paulo.

Moss, J., 2009. Climate justice. In: Jeremy, M. (Ed.), Climate Change and Social Justice. Melbourne University Press, Victoria, pp. 51–66.

Musarra, R.M.L.M., Costa, H.K.M., 2018. Elements of public action and governance in capture, stocking and carbon transportation activities. International Journal of Humanities and Social Science Invention (IJHSSI) 7, 46–53.

Musarra, R.M.L.M., Costa, H.K.M., 2019. Comparative International Law: the scope and management of public participation rights related to CCS activities. Journal of Public Administration and Governance 9, 93–109.

Musarra, R.M.L.M., Cupertino, S.A., Costa, H.K.M., 2019. Liability in civil and environmental subjects for carbon capture and storage (CCS) activities in Brazil. International Journal for Innovation Education and Research 7, 501–524.

Nabavi, E., Daniell, K.A., Najafi, H., 2017. Boundary matters: the potential of system dynamics to support sustainability? Journal of Cleaner Production 140, 312–323.

Nath, S., Roberts, J.L., 2021. Transition from economic progress to sustainable development: missing links. In: Roberts, J.L., Nath, S., Paul, S., Madhoo, Y.N. (Eds.), Shaping the Future of Small Islands. Palgrave Macmillan, Singapore, pp. 3–19. Available from: https://doi.org/10.1007/978-981-15-4883-3_1.

NRG, 2020. Petra Nova Carbon Capture System (CCS) placed in reserve shutdown. < https://www.nrg.com/about/newsroom/2020/petra-nova-status-update.html > (accessed 30.11.20).

Olivier, J.G.J., Peters, J.A.H.W., 2020. Trends in Global CO_2 and Total Greenhouse Gas Emissions: 2019 Report. PBL Netherlands Environmental Assessment Agency, The Hague.

Oliveira, J.A.P., 2012. Rio + 20: what we can learn from the process and what is missing. Cadernos EBAPE.BR 10 (3), 492–507. Available from: https://doi.org/10.1590/S1679-39512012000300003.

Opp, S.M., 2017. The forgotten pillar: a definition for the measurement of social sustainability in American cities. Local Environment 22 (3), 286–305.

Outka, U., 2012. Environmental justice issues in sustainable development: Environmental justice in the renewable energy transition. Journal of Environmental and Sustainability Law 19 (1), 60–122.

Paglia, E., 2018. The socio-scientific construction of global climate crisis. Geopolitics 23 (1), 96–123.

Paglia, E., Parker, C., 2021. The intergovernmental panel on climate change: guardian of climate science. In: Boin, A., et al., (Eds.), Guardians of Public Value. Available from: https://doi.org/10.1007/978-3-030-51701-4_12.

Patel, B.N., Nagar, R., 2018. Introduction. In: Patel, B.N., Nagar, R. (Eds.), Sustainable Development and India: Convergence of Law, Economics, Science, and Politics, Oxford, pp. 1 – 10.

Pequim, C.N., 2020. State, Sustainable Development and Governance in Brazil: Public Policies for Energy and Water Post Rio-92. Esboços, Florianópolis, 27 (44), pp. 78–93. Available from: https://doi.org/10.5007/2175-7976.2020.e63220 (jan./abr. 2020) ISSN 2175-7976, pp. 78–156.

Pietrzak, M.B., Balcerzak, A.P., 2016. Assessment of Socio-Economic Sustainability in New European Union Members States in the years 2004–2012. < http://www.badania-gospodarcze.pl/images/Working_Papers/2016_No_5.pdf > (accessed 30.11.20).

Rogge, K.S., Reichardt, K., 2016. Policy mixes for sustainability transitions: an extended concept and framework for analysis. Research Policy 45 (8), 1620–1635.

Sen, A., 2013. The ends and means of sustainability. Journal of Human Development and Capabilities, Taylor & Francis Journals 14 (1), 6–20.

Soini, K., Dessein, J., 2016. Culture-sustainability relation: towards a conceptual framework. Sustainability 8 (2), 167.

Tapia-Fonllem, C., Corral-Verdugo, V., Fraijo-Sing, B., 2017. Sustainable behavior and quality of life. Handbook of Environmental Psychology and Quality of Life Research. Springer International Publishing, pp. 173–184.

UNEP (United Nation Environment Programme), 2020a. Goal 7: Affordable and clean energy. < https://www.unep.org/explore-topics/sustainable-development-goals/why-do-sustainable-development-goals-matter/goal-7 > (accessed 15.12.20).

UNEP (United Nation Environment Programme), 2020b. Goal 13: Climate Action. < https://www.unenvironment.org/explore-topics/sustainable-development-goals/why-do-sustainable-development-goals-matter/goal-13 > (accessed 03.01.21).

United Nations, 2015a. Transforming Our World: the 2030 Agenda for Sustainable Development (A/RES/70/1). Resolution Adopted by the General Assembly on September 25, 2015, New York.

United Nations, 2015b. The Millennium Development Goals Report 2015. < https://www.un.org/millenniumgoals/2015_MDG_Report/pdf/MDG%202015%20rev%20(July%201).pdf > (accessed 30.11.20).

United Nations, 2020a. Sustainable Development Goals. < https://sdgs.un.org/ > (accessed 30.11.20).

United Nations, 2020b. Sustainable Development Outlook 2020—Achieving SDGs in the Wake of COVID-19: Scenarios for Policymakers. < https://sdgs.un.org/sites/default/files/2020-07/SDO2020_Book.pdf (accessed 30.11.20).

United Nations, 2020c. National Sustainable Development Strategies. https://sdgs.un.org/topics/national-sustainable-development-strategies > (accessed 04.01.21).

UNFCC, 2020. Website: < https://unfccc.int/process/bodies/supreme-bodies/conference-of-the-parties-cop > (accessed 30.11.20).

USDOE (United States Department of Energy), 2020a. U.S. Department of Energy Announces $131 Million for CCUS Technologies. < https://www.energy.gov/articles/us-department-energy-announces-131-million-ccus-technologies > (accessed 30.11. 20).

USDOE (United States Department of Energy), 2020b. DOE Dedicated to Innovating Technology That Keeps our Air, Water, and Land Clean While Providing American Jobs. < https://www.energy.gov/articles/doe-dedicated-innovating-technology-keeps-our-air-water-and-land-clean-while-providing > (accessed 30.11.20).

WCED—The World Commission on Environment and Development (Gro Harlem Brundtland, Chair), 1987. Our Common Future. Oxford University Press, Oxford.

Weiss, E.B., 1993. Justice pour les générations futures. Editions Sang de la Terre, Paris.

Weiss, E.B., et al., 2016. International Law for the Environment. West Publishing, Minnesota.

Why is social acceptance important for capture, storage, and transport of carbon (CCS) projects?

Yane Marcelle Pereira Silva and Hirdan Katarina de Medeiros Costa

Institute of Energy and Environment, University of São Paulo, São Paulo, Brazil

19.1 Introduction

The main global production, circulation, and distribution of goods and services must be done under premises that support environmental recovery and economic growth over time, and the presence of normative and social validation gives an optimistic value to them (Brundtland, 1987).

The activities of capture, storage, and transport of carbon (CCS) emerge among actions that aim to attain the sustainable development of ecosystems recovery criteria side by side with economic growth and mitigating undesirable effects of anthropic origin in the environment, such as climate change and acidification of the oceans (Musarra et al., 2019; Craig and Porter, 2006).

However, the more expansive are social, economic, and environmental impacts of a project, the more difficult it becomes to get the social license. The energy sector has such sensibility and needs to incorporate all impacts inside its projects, including CCS technologies.

The social license may currently be characterized when a project has ongoing approval within the local community and other stakeholders (SL, 2020). On the other hand, legal and regulatory constraints are specifically related to a governmental law or regulation applied to a project (Craig and Porter, 2006).

Thus to understand the social license and legal/regulatory constraints, popular participation is paramount in creating the rules of democracies. These will indicate the needs and level of quality of life of people that are appropriate to improve. Beyond that, public

371

perception and public acceptance are crucial for support or for opposition to a project. Environmental rules might also be a constraint to large-scale energy systems, such as natural gas production or carbon sequestration projects.

Regarding those environmental issues on CCS, the regulations and the legal framework can follow national rules (Costa et al., 2018). Investors who have access to the rules and know which rules should be followed may feel more confident to expend capital in CCS projects. Thus it is important to know and to be prepared for legal/regulatory limitations, which are mainly environmental guidelines. For instance, how will a community be involved under a CCS project? Is public participation a regulatory limitation? Which is meant by social license, public acceptance, and public perception? Are they also constraints?

This chapter aims to present a theoretical approach to the social license, public acceptance, public perception, and public participation considering their key features. Focusing on CCS technologies, projects, or facilities, regarding broadly social acceptance, this chapter intends to explore a literature review on how communities have seen these technologies as a path to decarbonize hard to reduce industries and the social license of operating on them.

19.2 Social license to operate, public acceptance, public perception, and public participation features and goals

The energy sector and its large-scale projects have shown for decades many social and environmental impacts, modifying natural landscapes along transmission lines path or even changing the way isolated native communities organize themselves and their relation to nature. Therefore it is important to understand social participation besides the legal and regulatory constraints.

For that, a few questions emerge to help to clarify it:

1. How can the concept of public participation be defined?
2. What are the differences between it, public perception, and public acceptance?
3. What are the links among them?
4. How can social license be affected to operate a specific project?
5. What can those definitions imply with regard to social license to a Carbon Capture Utilization and Storage project from a society and community?
6. Are there different forms of social participation?

19.2.1 Social license to operate

Related to the social license to operate, International Council on Mining and Metals (2012, p. 5) pointed out the significance of discourse and perspective and their influence over the corporate decision: the industry has widely accepted the concept of a social license to operate as an essential attribute of success. It has prompted companies to look well beyond only operational issues related to technical and economic assessment of their projects. The social license may also surpass community approval when there is a connection, mainly emotional, between the project and collective identity. In this line, the community becomes defenders of the project since they consider themselves coowners (SL, 2020).

More frequently, social license norms may be found in common law countries. On the other hand, in civil law countries, the social license concept runs into difficulties because an official authority can grant a license to operate independently from social permission. For example, in countries in Latin America, such as Brazil, environmental rules state a license to operate is granted by an official environmental agency; during this process, there is an audience with public participation (Nelsen, 2006; Slack, 2009; Musarra et al., 2019; SL, 2020).

Even though the term license suggests parallelism to legal approval, social license to operate (SLO) is not a formal or legal license and fundamentally differs from its legal counterparts. The idea of SLO, under the normative view, covers a kind of authorization to exist and to remain for a project known and determined for an activity in a given location or region, and also the building system of acceptance and validation before the community. Therefore it is possible to have legal authorization to operate without integral support from the local community, which could, after the community disapproval or opposition, affect or compromise the CCS project's viability (Gough et al., 2018; Syn, 2014).

Within the mining industry context, Thomson and Boutilier (2011) defined SLO as an informal permission given by the local community and broader society to industry to pursue technical work. It can be inferred, from the late 1990s, when the term was first used, until nowadays, that it is related to relationships between community and industry, regardless of governmental interventions. The concept has been enhanced to allow applications other than mining businesses, incorporating notions like acceptance, desire, beneficial relationships or mutual benefit, and lack of opposition (Dowd and James, 2014; Gough et al., 2018).

Although it presented itself as a versatile and sensitive concept in diverse contexts, specifically in CCS projects, the importance of community and social validation was evidenced through studies performed by CO2CRC Otway Project in Australia in 2006 and 2011 (Steeper, 2013). With regard to on-field surveys, an increase in acceptance was observed to the project by the public where it was located. Based on Steeper (2013), some findings and insights into CCS projects communications and direction provided for refinement of community engagement can be emphasized:

1. awareness of CCS and the project;
2. major improvement in relationships with stakeholders (landowners) due to the efforts of a mediator (diligent community liaison officer);
3. a strong profile among governments, industry, and CCS researchers;
4. through intensive local media, improve the project's profile; and
5. establish an active link to local community organizations to improve relationships and make them more informed and involved with the project.

In this line, Gough et al. (2018) has framed their research around the SLO concept, some of the key elements relevant to understanding the social license: scale, stakeholders' relationships, dynamism, trust risk, and confidence. These elements can ideally conceptualize SLO as legal and social dynamic approval of a given project resulting from a long-term relationship and trust-based partnerships between operators, government, and communities (host communities and communities of interest).

Also, SLO consists of elements from social legitimacy when there is an alignment with local social rules, formal and informal, such as legal, social, and cultural; credibility when there is true and clear information and by complying with all obligations made to the local community; and trust (Bastida et al., 2005; O'Faircheallaigh, 2010; SL, 2020).

SLO for a CCS project encompassing capture, transport, and storage requires consideration at different scales for different elements of the chain to be included. SLO has different levels that vary in strength and quality, which may range from an absence or removal of an SLO up to a sense of "coownership" between community and industry (Dowd and James, 2014; Gough et al., 2018; Hall and Carr-Cornish, 2015).

In addition to the idea of appropriate scales for SLO, Gough et al. (2018) noted that social license could not be treated as a unique and static decision procedure, but as an ongoing process, such that over time the strength and quality of a project's social license may change as attitudes and opinions evolve. It depends on maintaining functional community relationships and dialogue through which "sociopolitical legitimacy" is established.

Following it, in her major study, Raufflet et al. (2013) considers that SLO has a constant cyclical process of learning, doing, and assessing, under the conditions of the capability to dialogue among the whole parties (local community and project managers involved), which happen repeatedly through. Then, the ability to adapt the dialogue and their approach to the locals is the key to success during the project life cycle. Consequently, SLO is always under the process to be fully conceded.

There are other principles that may help support a social license for CCS projects, which means, basically, that to build the conditions and strategies for acceptance it is important to understand the area's social context and apply the key arguments folded into the context. Offline communication should be grown with key stakeholders and the press to invest in a long-term trust-based process relationship with stakeholders' networks. Online communication networks should be established to build an identity on social media that extends into wider online communities associated with the "CCS" acronym.

Conclusively, building trust and confidence in institutions and procedures is fundamental to assure long-term social pacts and reduce the perception of risks on CCS activities, since it takes time to reveal to people the integrity of new technology. However, one single failed procedure can damage trust in other future technological options (also in this sense, see Liu et al. (2019)).

19.2.2 Public acceptance

Acceptance, or what some authors would call attitude, has emerged as a concept used in CCS public perception research (Anderson et al., 2009; Ashworth et al., 2013; Brunsting et al., 2011; Capstick et al., 2015; Karimi et al., 2016), and also a key finding of a result (positive) from processual established relations between communities and stakeholders related to one given CCS project (Reiner et al., 2006; Makki, 2012; Markusson et al., 2012; Oltra et al., 2012).

Seigo et al. (2014) distinguish expressed acceptance as an agreement with specific statements, such as "I would accept a CCS project in my community," and revealed acceptance as an act of engaging, or not, in activities to promote or prevent activities related to CCS projects. In this line, public acceptance is associated with the nonnormative dimension of the SLO; that is, it is situated in relationships with communities (and with the subjects that integrate them), who have the power to accept, remain neutral, or oppose a given project in your area of interest (physical or symbolic location) (Gough et al., 2018).

Beyond the mere pursuit of acceptance or support of certain technology, it is important to recognize complexities in the relationships between communities and technologies, for instance, considering a variety of communities around a project, that is, host communities and communities of interest (Gough et al., 2018; Syn, 2014).

Gough et al. (2018) highlighted that since carbon dioxide sequestration a technology in which elements can present a geographical variation, the SLO scale is key, and when storage is developed in the offshore site, there may not be an obvious host community, making it harder to establish long-standing relationships. Thus, it is inferred, with regard to public acceptance, that offshore carbon dioxide storage might be more feasible than inland projects, since it goes together with gains in scale or already established economic activities, such as enhanced oil recovery on mature fields, using essential facilities available and already remunerated.

Their findings demonstrated that a problem remained regarding procedural justice and the governance process, particularly about the broader democratic process (Krause et al., 2014; Gough et al., 2018).

19.2.3 Public perception

Public perception comprises studies and surveys carried out to assess the way the public has seen CCS technology as a role in climate change or a new long-term business in its own community (Fazey, 2018; Seigo et al., 2014; Thomas et al., 2018; Upham and Roberts, 2011a, 2011b; Ter Mors et al., 2006). The idea of perception is associated with how stakeholders are positioned and how they take sides, within or outside a social group, regarding a certain subject, theme, or issue based on their own existence and personal history, not necessarily based on a rational decision process (Itaoka et al., 2009, 2011; Seigo et al., 2014).

Regarding CCS technology, this idea involves surveys of the public's intuitive reaction to the technology, the public's mental models and the subsurface, and their awareness of climate change (Terwel et al., 2009; Feenstra et al., 2011; Carley et al., 2012). In part, it includes the mechanics of forming public opinion and monitoring the communication's process and the spread of information about risks and uncertainties of technological applications of CCS (Malone, 2005; Pietzner et al., 2011).

Seigo et al. (2014) have reviewed previous research on CCS public perception and presented a general picture of the variables that involve its formation and complexities: trust, fairness, affect, acceptance, knowledge, experience, perceived costs, perceived risks, perceived benefits, outcome efficacy, problem perception, energy context, and interference with nature.

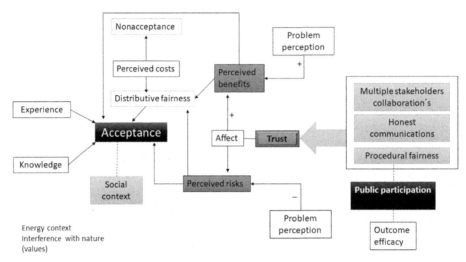

FIGURE 19.1   Fig. 19.1 A conceptual framework for public perception. *Source: Elaborated by authors from a review of the literature Seigo, S.L.O., Dohle, S., Siegrist, M., 2014. Public perception of carbon capture and storage (CCS): a review. Renewable and Sustainable Energy Reviews 38, 848–863.*

The concepts are articulated in different ways, showing interconnections that input public perception through acceptance or positive or negatives tendencies (Seigo et al., 2014), as shown in Fig. 19.1. This summarizes the idea that acceptance, although being presented as a concept and the whole assembly, assumes a nuclear function due to the possibility of being influenced by other concepts. The most expressive aspects for benchmarking acceptance would be (1) perceived risks, (2) perceived benefits, and (3) trust, as well as (4) the social context of the project site (Raven et al., 2009; Seigo et al., 2014; Shackley et al., 2009).

Lastly, the awareness of CCS also influences acceptance, but with limited impact, in which preexisting knowledge and information about CCS and eventual direct experiences with it or related technologies, such as fossil fuel supply chain and gas storage on depleted fields, reduce prior resistance about carbon storage projects (Van Alphen et al., 2007; Tokushige et al., 2007; Shackley et al., 2009; Seigo et al., 2014).

Also, influence acceptance, but with limited impact, the awareness of CCS, for instance, preexisting knowledge and information about CCS as well as eventual direct experience with CCS or with related technologies (e.g., oil production and underground gas storage) (Tokushige et al., 2007; Shackley et al., 2009; Seigo et al., 2014).

Trust is recognized as a key variable for technology acceptance. Studies have found that the effect of trust on acceptance is indirect, interceded through perceived benefits and perceived risks (Huijts et al., 2012; Karimi et al., 2016; Karimi and Toikka, 2014; Raven et al., 2009; Seigo et al., 2014). Furthermore, the relationship between trust and perceived risks operates through effect. Trust can be heightened through fair procedures, honest communications, and collaboration of multiple stakeholders. Three critical issues for acceptance were reported: trust in the developer, the quality of public engagement activities, and the public's and stakeholders' perceptions of the need for the facility (Oltra et al., 2012; Seigo et al., 2014).

Both procedural fairness and distributive justice are important to the formation of perception (Gough and Boucher, 2013; Seigo et al., 2014). First, procedural justice relates to the equality of processes through which decisions are made. Procedures that are perceived as fair lead to more trust and, finally, higher acceptance (Gough and Boucher, 2013; Seigo et al., 2014).

Additionally, distributive fairness is related to the distribution of effective costs, risks, and benefits, which implies the extended concept of environmental justice, which covers, in addition to the distribution of environmental harm, the search for equity and the protection of minority or marginalized individuals or communities (human and nonhuman) against environmental disadvantage (Gough and Boucher, 2013; Schlosberg, 2013). Positive and negative effects tend to be influenced by trust and influence perceived benefits and risks. Studies showed that pro- or anti-CCS messages are most persuasive if they activate a sensitive response or "emotionally self-referent" (Seigo et al., 2014; Wong-Parodi et al., 2011).

Perceptions about the high costs of CCS, specifically financial and psychological costs for individuals and society, are a major disadvantage of the technology, and have been a key factor against its acceptance (Seigo et al., 2014; Wong-Parodi et al., 2011).

Perceived benefits of CCS technology reflect how it is seen for oneself, society, and the environment. Adequate information, trust, and positive affect can build benefit perceptions at the broader, societal level (Seigo et al., 2014; Wong-Parodi et al., 2011). Outcome efficacy means people's belief that personal behavior modifies the implementation path of CCS technology. Besides, the connections and networks among individuals and their trust are linked to social capital (Seigo et al., 2014). These concepts are linked to public participation as well as to procedural fairness.

Researchers have warned that establishing trust among stakeholders is a long-term process that depends on the interactions of a variety of factors, including current and past associations that guide knowledge and confidence between stakeholders, their reliability and predictability, and expectations of how the industry will operate (Gough et al., 2018; Moffat et al., 2016). Thus part of this complex arrangement is the role of scientists, scientific evidence, and technical expertise and how this information is translated and communicated.

Problem perception relates to the awareness of the CCS technology and the climate change threat are real, even having a minority oppositions. Findings point to a negative influence of climate change awareness on risk perception and a positive influence on benefit perceptions (Seigo et al., 2014).

19.2.4 Public participation

Public participation is an idea broadly understood together with others, for instance, concepts of democracy, citizenship, and its rights or even collective decision theory (Bobbio et al., 2004). It is connected to legal procedures to run projects according to environmental law (Musarra et al., 2019) .

The concept of participation is filled with inconsistencies that have emerged from different terms of public involvement (social participation, community participation, or popular participation). In general, it is understood that participation presupposes that government institutions consider citizens' interests and opinions before making decisions or executing them (Furriela, 2002; Jacobi, 2002).

It can be affirmed that public participation occurs at the boundary of community members' public spirit in their relationship with governmental institutions and into the public arena. It involves decision-making processes on projects that affect their interests directly or indirectly, predominantly through formal consultation instruments. In its turn, positive public acceptance may correspond to the result of a well-conducted public participation process, despite the potential to become the main obstacles of business projects.

Public participation should be an integral part of any CCS project. A reasonable public participation process needs to certify that local citizens involved have suitable financial and technical resources beyond social capital to manifest their needs clearly and have their concerns addressed (Anderson et al., 2009; Seigo et al., 2014).

Inequitable processes and outcomes in terms of siting decisions and social exchange can potentially impact the SLO formation within local host communities negatively and beyond (Gough et al. 2018).

The capability of the society to participate in the SLO procedure depends on abstract and subjective aspects such as interest in participating and the sense of empowerment, that is, the community's ability to exercise voice and seek redress. Thus it tends to ensure some defense for local communities and citizens against the downside risk of government or corporate neglect.

Wong-Parodi and Isha (2009) argued that empowerment was rooted in the community's history and its material and social asset base, comprising the capability to mitigate for the community the well-defined risks of the CCS technology and ensure just procedures would be followed during technological development. Therefore the sense of empowerment brings the notion of procedural and distributive fairness.

The apparent silence or passive resignation on the opportunities of public participation should not be confused with acceptance. The passivity or reaction can be affected by socioeconomic standing and local history. It has been observed that the poorer communities felt less capable of acting, and they did not have confidence they would have much to do with participating in the siting process or that the entrepreneur would compensate them in the case that anything went wrong (Wong-Parodi and Isha, 2009).

Despite relatively different governance frameworks, the absence of demonstrable social justice, procedural fairness, and the limited opportunity for the public to influence decisions limited the potential to establish the SLO (Whitton et al., 2017). These circumstances lead to societal resistance and opposition, a political critique.

Consultative participation does not interfere directly in the decision-making process; already, the resolutive participation and the control participation imply the intervention in the course of the public activity (Jacobi, 2002).

As local legislation, juridical participation opens the possibility of collective procedural implementation through public civil action at the court system.

Public hearings are formal procedures used during the environmental license phase of work or activity, where citizens involved can take directly to the project details of impacts by presenting the report of environmental impacts.

Regarding CCS projects, public participation is a moment when citizens, as stakeholders, adopt actions to have their opinion and thoughts considered during life cycle projects, from the conceptual phase and SLO process to the postclosure monitoring stage. Also, public participation can emerge in the political arena as far as public hearings are performed to discuss climate change tools.

19.3 Current trends on CCS projects and social acceptance

After the conceptual approach from topic 2, where social acceptance was treated from a broad perspective, we intend to embrace cases and articulate their impressions to CCS projects by literature research mainly from the last years. As we already emphasized, social license to operate, public acceptance, and public perception are perspectives on broader social acceptance. On the other hand, public participation implies rules connected to a procedure that could be inside the social license to operate, and it is correlated to social acceptance as well. All concepts are important to CCS projects worldwide.

As mentioned, CCS projects may face barriers to be deployed, such as public opinion against storage (IEA, 2020; Musarra et al., 2019). Because of this reality, as studied in Section 19.2, social acceptance is widely analyzed on CCS projects worldwide, so the literature review has become vast (Chalmers et al., 2013; Huijts et al., 2012; Markusson et al., 2012; Raven et al., 2009; Tokushige et al., 2007).

As an example, Boyd et al. (2017) studied Canadian cases. During their survey, they found that there is low support for CCS in Canada. Although, they realized that results varied when considering participants' proximity to projects. In fact, they conclude that the publics' perceptions of the risk and benefits of CCS influenced support for or opposition to the technology.

Dütschke et al. (2016), looking at the German case, highlight that the public's negative perception is a key issue with regard to CCS projects. They ran an online survey and used a methodology approach to design different CCS scenarios separated by carbon dioxide source categories, transport possibilities, and sequestration phase options. A key result was a low rate of coal-based CCS projects compared to others, for example, a coal-fired power plant versus biomass power plant or industrial sources of CO_2.

There is a trend to see CCS projects as beneficial when linked to the net-zero carbon emission perspective instead of the negative view of competing with renewables for investments (Upham and Roberts, 2011a) and keeping the coal or oil industry alive.

Van Egmond and Hekkert (2015) analyzed an emblematic Dutch case. They studied the unsuccessful Barendrecht CO_2 storage project. Using a methodology based on debates' analyses, interviews, and a literature review, they found that the local opposition intensified over time. First, there was no solid outside support; second, the opponents were strongly represented and credible. Thus, in their opinion, the novelty and relatively unknown quantity were relevant points in this unsuccessful case.

Ashworth et al. (2013) examined Australia, the Netherlands, Canada, and Scotland, and they realized that there is a low level of public awareness about CCS technology. They found perceived risks and uncertainties, and concerns on leakages and their consequences to health, property value, and land use. These authors observed little tolerance in trading off between CCS and renewables. Thus it is relevant that the CCS projects and renewables go on in parallel.

As a solution, we have already drawn comments on Section 19.2 of this book. However, we highlight that operators should diversify their social acceptance suggestions, mainly on social license to operate, public acceptance, and public perception. For instance, we may list: an early engagement process with stakeholders, dissemination of knowledge from different energy sources and technologies, open to listening to the community, develop a trustful relationship among stakeholders (Ashworth et al., 2013), and create communicative tools effectively (Brunsting et al., 2011).

Considering public participation, a comparative approach developed by Musarra et al., (2019) embraces many jurisdictions, such as the United States, Canada, EU members, Brazil, and Australia, to analyze the main rules.

Musarra et al., (2019) have pointed out that scarce transparency and limited participation are the main issues in those jurisdictions. In general, rules state the right to information as a category of public participation. Moreover, they found the existence of the right to consult records and licenses. Nevertheless, there are charges for obtaining copies in some jurisdictions, and some information is classified due to public safety matters remaining confidential by the state authority's choice.

Furthermore, in some jurisdictions (Australia, some US, EU, and Brazil), public hearings and consultations are a formal procedure. However, it is unclear whether popular opinion is considered, answered, and has some priority. There is no description of how a state authority's decision is influenced by popular participation (Musarra et al., 2019).

As suggested by Musarra et al., (2019), public participation is an important procedure to guarantee democratic behavior along with the whole licensing and regulatory processes, discussing risk prevention, environmental management, as well as social and political implications of CCS projects. For instance, some rules are created after a public consultation following by a hearing. In this regard, a broad view of tools to participate should be considered, such as online citizenship portals with easy interface and dissemination for all audiences, as well as free access to technical information, and wide announcements of hearings, public calls, contracts, licenses, and renewals. The creation of rules for implementing CCS projects must consider the demands of climate justice and fair energy transition beyond the national level, reaching other countries.

Therefore CCS legislation to be implemented, considering public participation, should encompass landowners and occupants involvement and rules; should mantain documents and records available for public inspection as a way of input transparency about the public consultation and hearing procedures, the necessary public advertisement about CCS projects, the participation of citizens, making clear the right to be informed about carbon storage risks; the consultation with specific stakeholder groups; and, finally, a prediction that recommendations may only be made after the social license (Musarra et al., 2019).

19.4 Conclusion

We have started this chapter with the title "Why is social acceptance important for CCS projects?" Thus, inside our question, one may infer that we will defend social acceptance. However, our intention was broadly conceptual.

CCS projects may be hampered, for example, by public opinion against storage (IEA, 2020; Musarra et al., 2019). For this reason, a better understanding of how CCS's public perception works has become strategic and is of concern for project managers, academics, and policymakers.

Social acceptance may be seen as a generic term that covers SLO, public acceptance, public perception, and public participation. The social acceptance related to SLO and public participation refers to a specific project, such as a factory or an industrial plant.

Public participation can be viewed as SLO's relevant part, which concerns the community's decision-making process involvement for a local CCS project implementation, and it is relevant for informing public policy that enables legitimacy and strengthening of instruments. However, it is not restricted to SLO; it may cover wider guidelines on public policies.

On the other hand, when public acceptance appears in the context of public perception, it involves a generic theme, a category with multiple study variables about CCS technology and climate change.

Collecting key elements pointed out by Gough et al. (2018), the SLO is conceptualized as being legal and social dynamic approval of a given project resulting from a long-term relationship and trust-based partnerships between operators, government, and communities (host communities and communities of interest).

Cases illustrated that SLO's nonnormative dimension could interfere with the CCS's project viability once communities have the power to accept, remain neutral, or oppose a given project in a location (Gough et al., 2018; Syn, 2014).

Furthermore, public acceptance into SLO is a dynamic condition that admits different levels, may change over time and can vary in strength and quality, which varies from the limit of the absence of an SLO up to a sense of ownership jointly between citizens and agents involved in the project (Dowd and James, 2014; Gough et al., 2018; Hall and Carr-Cornish, 2015).

Besides SLO, we have found extensive literature on public acceptance and public perception as important to effectively reach CCS projects worldwide.

Social acceptance is difficult to reach and easy to lose. As has been stated, "engagement around CCS does not necessarily guarantee acceptance and deployment of the technology, it certainly assists in evaluating the public's understanding and what may constitute it being accepted" (Ashworth et al., 2013, p. 7418).

Acknowledgments

We are grateful to the "Research Center for Gas Innovation—RCGI" (Fapesp Proc. 2014/50279-4), supported by FAPESP and Shell, organized by the University of São Paulo, and the strategic importance of the support granted by the ANP (National Agency of Petroleum, Natural Gas and Biofuels of Brazil) through the R&D clause. We also thank the support from the National Agency for Petroleum, Natural Gas and Biofuels Human Resources Program (PRH-ANP), funded by resources from the investment of oil companies qualified in the R,D&I Clauses from ANP Resolution No. 50/2015 (PRH 33.1—Related to Call No. 1/2018/PRH-ANP; Grant FINEP/FUSP/USP Ref. 0443/19).

References

Anderson, J., Chiavari, H., de Coninck, S., Shackley, G., Sigurthorsson, T., Flach, D., et al., 2009. Results from the project' acceptance of CO2 capture and storage: economics, policy and technology (accept)'. In: Gale, I.J., Herzog, H., Braitsch, J. (Eds.), Greenhouse Gas Control Technologies, vol. 9. Elsevier Science Bv, Amsterdam, pp. 4649–4653.

Ashworth, P., Einsiedel, E., Howell, R., Brunsting, S., Boughen, N., Boyd, A., et al., 2013. Public preferences to CCS: how does it change across countries? Energy Procedia 37, 7410—7418.

Bastida, E., Irarra' zabal, R., Labo, R., 2005. Mining investment and policy developments: Argentina, Chile and Peru Centre for Energy. Petroleum and Mineral Law and Policy Internet Journal, 16.

Bobbio, N., Mateucci, N., Pasquino, G., 2004. Dicionário de política (2 volumes). Trad. Carmen C. Varrialle, Gaetano Loiai Mônaco, João Ferreira, Luis Guerreiro Pinto Cacais, Renzo Dini. UnB, Brasília.

Boyd, A.D., Hmielowski, J.D., David, P., 2017. Public perceptions of carbon capture and storage in Canada: results of a national survey. International Journal of Greenhouse Gas Control 67, 1—9. ISSN 1750-5836.

Brundtland, G.H., 1987. World Commission on Environment and Development. Our Common Future, pp. 8—9.

Brunsting, S., Upham, P., Dütschke, E., De Best Waldhober, M., Oltra, C., Desbarats, J., et al., 2011. Communicating CCS: applying communications theory to public perceptions of carbon capture and storage. International Journal of Greenhouse Gas Control 5 (6), 1651—1662. Available from: https://doi.org/10.1016/j.ijggc.2011.09.012.

Capstick, S., Whitmarsh, L., Poortinga, W., Pidgeon, N., Upham, P., 2015. International trends in public perceptions of climate change over the past quarter century. Wiley Interdisciplinary Reviews Climate Change 6 (1), 35—61. Available from: https://doi.org/10.1002/wcc.32.

Carley, S.R., Krause, R.M., Warren, D.C., Rupp, J.A., Graham, J.D., 2012. Early public impressions of terrestrial carbon capture and storage in a coal-intensive state. Environmental Science and Technology 46 (13), 7086—7093.

Chalmers, H., Gibbins, J., Gross, R., Haszeldine, S., Heptonstall, P., Kern, F., Markusson, N., Pearson, P., Watson, J., Wiskel, M., 2013. Analysing uncertainties for CCS: From historical analogues to future deployment pathways in the UK. Energy Procedia 37, 7668—7679.

Costa, H.K.D.M., Miranda, M.F., Musarra, R.M.L.M.M., Santos, E.M., 2018. Environmental license for carbon capture and storage (CCS) projects in Brazil. Journal of Public Administration and Governance 8 (3).

Craig, D., Porter, D., 2006. Development Beyond Neoliberalism? Governance, Open Process Framework (OPF). Available from: http://www.opfro.org/index.html?Components/WorkProducts/RequirementsSet/Requirements/LegalAndRegulatoryConstraints.html ~ Contents (accessed 23.11.20).

Dowd, A.-M., James, M., 2014. A social licence for carbon dioxide capture and storage: How engineers and managers describe community relations. Social Epistemology 28 (3—4), 364—384.

Dütschke, E., Wohlfarth, K., Höller, S., Viebahn, P., Schumann, D., Pietzner, K., 2016. Differences in the public perception of CCS in Germany depending on CO2 source, transport option and storage location. International Journal of Greenhouse Gas Control 53, 149—159. Available from: https://doi.org/10.1016/j.ijggc.2016.07.043.

Fazey, I., et al., 2018. Ten essentials for action-oriented and second order energy transitions, transformations and climate change research. Energy Research and Social Science 40, 54—70.

Feenstra, C.F.J., Mikunda, T., Brunsting, S., 2011. What Happened in Barendrecht? Case Study on the Planned Onshore Carbon Dioxide Storage in Barendrecht. ECN, Amsterdam, The Netherlands.

Furriela, R.B., 2002. Democracia, cidadania e proteção do meio ambiente. A. Fapesp.

Gough, C., Boucher, P., 2013. Ethical attitudes to underground CO_2 storage: Points of convergence and potential faultlines. International Journal of Greenhouse Gas Control 13, 156—167.

Gough, C., Cunningham, R., Mander, S., 2018. Understanding key elements in establishing a social license for CCS: an empirical approach. International Journal of Greenhouse Gas Control . Available from: https://doi.org/10.1016/j.ijggc.2017.11.003.

International Council on Mining and Metals (ICMM), 2012. In Brief: Mining's Contribution to Sustainable Development—An Overview London. International Council on Mining and Metals; Poverty Reduction and Political Economy. Routledge, London.

International Energy Agency (IEA), 2020. Energy Technology Perspectives 2020: Special Report on Carbon Capture, Utilisation and Storage. Paris, France.

Hall, L., Carr-Cornish, D., 2015. Social licence to operate: understanding how a concept has beentranslated into practice in energy industries. Journal of Cleaner Production 86, 301—310.

Huijts, N., Molin, E., Chorus, C., van Wee, B., 2012. Public acceptance of hydrogen technologies in transport: A review of and reflection on empirical studies. In: Geerlings, H., Shiftan, Y., Stead, D. (Eds.), Transition towards Sustainable Mobility: The Role of Instruments, Individuals and Institutions. Ashgate Publishing Ltd, pp. 137—164.

Itaoka, K., Okuda, Y., Saito, A., Akai, M., 2009. Influential information and factors for social acceptance of CCS: the 2nd round survey of public opinion in Japan. Energy Procedia 1 (1), 4803–4810. ISSN 1876-6102.

Itaoka, K., Saito, A., Akai, M., 2011. A study on roles of public survey and focus groups to assess public opinions for CCS implementation. Energy Procedia 4, 6330–6337. ISSN 1876-6102.

Jacobi, P.R., 2002. Políticas sociais locais e os desafios da participação citadina. Cien. Saude Colet. 7, 443–454. Available from: https://doi.org/10.1590/s1413-81232002000300005.

Karimi, F., Toikka, A., 2014. The relation between cultural structures and risk perception: how does social acceptance of carbon capture and storage emerge? Energy Procedia 63, 7087–7095. ISSN 1876-6102.

Karimi, F., Toikka, A., Hukkinen, J., 2016. Comparative socio-cultural analysis of risk perception of Carbon Capture and Storage in the European Union. Energy Research and Social Science 21, 114–122.

Krause, R.M., Carley, S.R., Warren, D.C., Rupp, J.A., Graham, J.D., 2014. "Not in (or under) my backyard": geographic proximity and public acceptance of carbon capture and storage facilities. Risk Analysis 34 (3), 529–540.

Liu, L., Bouman, T., Perlaviciute, G., Steg, L., 2019. Effects of trust and public participation on acceptability of renewable energy projects in the Netherlands and China. Energy Research and Social Science 53, 137–144.

Makki, M., 2012. Evaluating arctic dialogue: a case study of stakeholder relations for sustainable oil and gas development. Journal of Sustainable Development 5 (3), 34–45.

Malone, E.L., 2005. Finding a way: the potential for adoption and diffusion of carbon dioxide capture and sequestration technologies, Greenhouse Gas Control Technologies, vol. 7. Elsevier Science Ltd, Oxford, pp. 1531–1536, ISBN 9780080447049, 2005.

Markusson, N., Kern, F., Watson, J., Arapostathis, S., Chalmers, H., Ghaleigh, N., et al., 2012. A socio-technical framework for assessing the viability of carbon capture and storage technology. Technological Forecasting and Social Change 79 (5), 903–918.

Moffat, K., Lacey, J., Zhang, A., Leipold, S., 2016. The social licence to operate: A critical review. Forestry: An International Journal of Forest Research 89 (5), 477–488.

Musarra, R.M.L.M., Costa, H.K., de, M., 2019. Comparative International Law: the scope and management of public participation rights related to CCS activities. Journal of Public Administration and Governance 9 (2), 93. Available from: https://doi.org/10.5296/jpag.v9i2.14559.

Nelsen, J.L., 2006. Social license to operate. International Journal of Mining, Reclamation and Environment 20, 161–162.

O'Faircheallaigh, C., 2010. Aboriginal-mining company contractual agreements in Australia and Canada: implications for political autonomy and community development. Canadian Journal of Development Studies 30, 69–86.

Oltra, P., Upham, H., Riesch, A., Boso, S., Brunsting, E., Dutschke, A.L., 2012. Public responses to CO_2 storage sites: lessons from five European cases. Energy and Environment 23 (2–3), 227–248.

Pietzner, K., Schumann, D., Tvedt, S.D., Torvatn, H.Y., Næss, R., Reiner, D.M., et al., 2011. Public awareness and perceptions of carbon dioxide capture and storage (CCS): insights from surveys administered to representative samples in six European countries. Energy Procedia 4, 6300–6306. ISSN 1876-6102.

Raufflet, E., Baba, S., Perras, C., Delannon, N., 2013. Social License. In: Idowu, S.O., Capaldi, N., Gupta, A.D. (Eds.), Encyclopedia of Corporate Social Responsibility. Springer, Berlin.

Raven, R.P.J.M., Mourik, R.M., Feenstra, C.F.J., Heiskanen, E., 2009. Modulating societal acceptance in new energy projects: towards a toolkit methodology for project managers. Energy 34 (5), 564–574. ISSN 0360-5442.

Reiner, D., Curry, T., De Figueiredo, M., Herzog, H., Ansolabehere, S., Itaoka, K., et al., 2006. An International Comparison of Public Attitudes Towards Carbon Capture and Storage Technologies. Eighth International Conference on Greenhouse Gas Control Technologies (GHGT-8), Trondheim, Norway.

Schlosberg, D., 2013. Theorising environmental justice: The expanding sphere of a discourse. Environmental Politics 22 (1), 37–55.

Seigo, S.L.O., Dohle, S., Siegrist, M., 2014. Public perception of carbon capture and storage (CCS): a review. Renewable and Sustainable Energy Reviews 38, 848–863.

Shackley, S., Reiner, D., Upham, P., Coninck, H., Sigurthorsson, G., Anderson, J., 2009. The acceptability of CO_2 capture and storage (CCS) in Europe: an assessment of the key determining factors: part 2. The social acceptability of CCS and the wider impacts and repercussions of its implementation. International Journal of Greenhouse Gas Control 3 (3), 344–356. ISSN 1750-5836, 2009.

Slack, K., 2009. Mining Conflicts in Peru: Condition Critical. Oxfam America, Boston, MA.

Social License.com (SL), 2020. What is social license? Available from: http://socialicense.com/definition.html (accessed 23.11.20).

Steeper, T., 2013. CO$_2$CRC Otway project social research: assessing CCS community consultation. Energy Procedia 37, 7454–7461. Available from: https://doi.org/10.1016/j.egypro.2013.06.688.

Syn, J., 2014. The social license: Empowering communities and a better way forward. Social Epistemology 28, 318–339.

Ter Mors, E., Weenig, M., Ellemers, N., 2006. The influence of (in)congruence of communicator expertise and trustworthiness on acceptance of CCS technologies. In: International Conference on Greenhouse Gas Control Technologies, Trondheim, Norway.

Terwel, B.W., Harinck, F., Ellemers, N., Daamen, D.D.L., 2009. Competence-based and integrity-based trust as predictors of acceptance of carbon dioxide capture and storage (CCS). Risk Analysis 29 (8), 1129–1140.

Thomas, G., Pidgeon, N., Robert, E., 2018. Ambivalence, naturalness and normality in public perceptions of carbon capture and storage in biomass, fossil energy, and industrial applications in the United Kingdom. Energy Research and Social Science 46, 1–9.

Thomson, I., Boutilier, R., 2011. The Social Licence to Operate. SME Mining Engineering Handbook. Society for Mining, Metallurgy, and Exploration, Colorado.

Tokushige, K., Akimoto, K., Tomoda, T., 2007. Public perceptions on the acceptance of geological storage of carbon dioxide and information influencing the acceptance. International Journal of Greenhouse Gas Control 1 (1), 101–112. ISSN 1750-5836.

Upham, P., Roberts, T., 2011a. Public perceptions of CCS in context: results of NearCO$_2$ focus groups in the UK, Belgium, the Netherlands, Germany, Spain and Poland. Energy Procedia 4, 6338–6344.

Upham, P., Roberts, T., 2011b. Public perceptions of CCS: emergent themes in pan-European focus groups and implications for communications. International Journal of Greenhouse Gas Control 5 (5), 1359–1367.

Van Alphen, K., Voorst, Q.V., Hekkert, M.P., Smits, R.E.H.M., 2007. Societal acceptance of carbon capture and storage technologies. Energy Policy 35 (8), 4368–4380. ISSN 0301-4215.

Van Egmond, S., Hekkert, M.P., 2015. Analysis of a prominent carbon storage project failure—the role of the national government as initiator and decision-maker in the Barendrecht case. International Journal of Greenhouse Gas Control 34, 1–11.

Whitton, J., Brasier, K., Charnley-Parry, I., Cotton, M., 2017. Shale gas governance in the United Kingdom and the United States: Opportunities for public participation and the implications for social justice. Energy Research & Social Science 26, 11–22.

Wong-Parodi, G., Dowlatabadi, H., MacDaniels, T., Ray, I., 2011. Influencing attitudes toward carbon capture and sequestration: A social marketing approach. Environmental Science & Technology 45 (16), 6743–6751.

Wong-Parodi, G., Isha, R., 2009. Community perceptions of carbon sequestration: Insights from California. Environmental Research Letters 4 (3).

Climate change, Carbon Capture and Storage (CCS), energy transition, and justice: where we are now, and where are (should be) we headed?

Carolina Arlota[1,2] and Hirdan Katarina de Medeiros Costa[3]

[1]Visiting Assistant Professor of Law at the University of Oklahoma, College of Law, Norman, OK, United States [2]Law and Economics Research Fellow, Capitalism and Rule of Law Project (Cap Law) at Antonin Scalia Law School, George Mason University, Fairfax, VA, United States [3]Institute of Energy and Environment, University of São Paulo, São Paulo, Brazil

20.1 Introduction

Energy transition addresses how humankind uses energy for their needs and reconciles it with social, environmental, and economic interests. As the awareness about the consequences of climate change increases worldwide, so does the pressure on different society segments and critical stakeholders. In such a context, the role assigned to each target group and, ultimately, to every one of us, including businesses, consumers, scholars, academics, lawyers, public opinion, and policymakers, has changed significantly. Nowadays, immediate calls for action on climate and environmental justice, access to energy, and a fair energy transition are recurrent topics in media outlets. These points of view are part of the same movement in joining forces to combat climate change. Exhibit 1 for this argument would be the Swedish activist Greta Thunberg. All leading institutions, such as the United Nations and its organs, highlight the urgency and need for an equitable and fair energy transition. Therefore, defining who may be heard when determining energy transition policies is controversial domestically as well as internationally. As it concludes our book, this article advances a modern approach discussing the interplay of climate change, energy transition, and its new perspectives on justice.

This chapter starts with an historical overview of environmental justice and how it informs and shapes the current conceptualizations of climate justice. The chapter proceeds to discuss significant pressure points on climate change, namely, migration and justice. Also, this chapter summarizes significant findings of previous chapters, articulating these findings concerning current trends on international energy law, carbon capture and sequestration (CCS), and climate justice.

Ultimately, this chapter shows how policymakers and scholars may be better equipped with the analysis and policies discussed throughout this work. And, to the extent that some of the challenges coming in their direction may be anticipated in the book, readers may behave (or change their behavior) accordingly. As for the unforeseeable challenges, the regulatory framework presented should also be determinative of new policies in the complex, unique, and interesting times to come. One thing is certain: justice and fairness are no longer abstract propositions. They now can be evaluated and incorporated in the domestic and international spheres. Accordingly, there are reasons to be cautiously optimistic about the future.

20.2 A historical overview: from environmental justice to climate justice

From the initial experience of social movements in the United States, in addition to social and economic inequalities, environmental ones began to be the target of the demand of poor citizens and socially discriminated and vulnerable ethnic groups (Herculano, 2002). The existence of principles of climate justice is perceived in the document published in Bali (2002).

The discussion on environmental justice advocates the uniform scope of environmental goods and the benefits of the concrete application of sustainable development for all members of the current society (Acselrad et al., 2009; Costa and Santos, 2013), as well as the sharing of the burden of progress, is borne by the entire community, without discrimination on racial, ethnic, or economic issues (Ferraresi, 2012). In this vein, Fagundez et al. (2020) correlate extreme events, such as Hurricane Katrina, as one of the events that the scientific community observes as the intersection between environmental and climate justice.

The problem of climate change from the perspective of relational perception understands the human being as a subject of rights, which are intrinsically linked to climate justice. Climate security is a common necessity; nonetheless, it affects different people in distinct ways. Therefore the climate vulnerabilities are not the same (Seck, 2020).

Schlosberg and Collins (2014) address climate justice as concerned with local impacts and experiences, vulnerabilities, inequalities, the importance of the community's active participation movement, and demands for community sovereignty and functioning. Even in California, the creation of California's Global Warming Solutions Act 2006 is cited because of all this discussion of community empowerment.

Yildirim (2020) points to climate justice as the action against the harms of excess greenhouse gas (GHG) emissions in a city, in which it must be analyzed from a cross-sectional perspective, considering gender, housing, and socioeconomic structures. This is because the negative externalities of climate problems fall sharply on social minorities in vulnerable places. Low-income social groups are more vulnerable to the climate change process due to their cities' lack of structure.

Gonzalez (2020) addresses climate justice as originating from the concept of Environmental Justice, exploring it on four aspects: (1) distributive injustice since the northern countries are the largest emitters of GHGs and those that profit the most from this activity while those primarily affected by climate change are the countries of the South, which contribute less to the problem; (2) procedural injustice, northern countries dominate global economic and environmental governance structures ignoring the perspectives and priorities of southern countries; (3) corrective injustice, the countries of the South, most affected by climate change, are unable to obtain compensation for the damage caused by large emitters of GHGs; and (4) social injustice, considering that we live in an economic order that stimulates poverty and inequality, ignoring the finitude of the natural resources of the terrestrial ecosystem. In addition to these four aspects, it is possible to affirm, according to the author, that the concept of climate justice is upheld by human rights since the treaties of the last decade have sought to include recommendations to ensure that state policies for climate change ensure human rights for populations.

As a study by Althor et al. (2016) found out, vulnerability to the effects of climate change is uneven in the global sphere. Despite having high carbon emissions, countries such as China and the United States do not suffer the adverse consequences in the same way as African countries, which do not have negative emissions but are more vulnerable to climate change's damaging effects.

The United Nations Framework Convention on Climate Change (UNFCCC) recognized the climate debt of northern countries, establishing the principle of "common but differentiated responsibilities and respective capabilities (CBDR–RC)," which recognizes that all countries have shared obligations in the destruction of the environment, but denies equitable responsibility among them (Gonzalez, 2020). This same principle was reaffirmed by the Paris Agreement, which references climate justice and immigrant rights. Second, Intended Nationally Determined Contributions GHG emissions have already resulted in a 3°C increase in global temperature (1.5°C above the Paris Agreement target), enough to cause flooding in the most diverse regions of the planet. Gonzalez (2020) points out that without measures to reduce emissions of these gases, the globe may face its most significant migratory wave. The first document to explore ways to address climate change harm was the Warsaw International Mechanism on Loss and Damage, which included provisions for loss and damage. However, for the mitigation and adaptation of climate migrants, the document does not address relocation and resettlement measures, an idea advocated by small island developing states.

20.3 Climate change and its significant pressure points: migration and justice

As Humphreys (2009) notes, several protected human rights are threatened by the effects of climate change, namely, rights to health and life; rights to water, food, shelter, and property; rights related to livelihood and culture; with personal security in the event of conflict; migration and resettlement. Moss (2009) argues that, in order to deal with the potential for inequalities and vulnerabilities in the responsibilities of adaptation or mitigation of climate change, one must build a justice system that is capable of bringing answers with respect for human rights within a theory of social justice encompassing collective solution (Moss, 2009).

For Schlosberg and Collins (2014), adaptation is an interaction between environmental justice, climate justice, and social justice for the most vulnerable. According to Peel and Lin (2019), litigation and the development of climate policies are linked to adaptation because, in the absence of mitigation measures, the expansion of risks and extreme climatic events is visualized. As a previous policy to cope with the effects of climate change, adaptation helps in the best scenario of preventing vulnerabilities from being extreme (Barnett, 2009).

According to Gonzalez (2020), for climate justice to be achieved, it is necessary to develop legal approaches to climate displacement through which social movements and climate-vulnerable states can come together to combat these approaches. From this, the international right can be used in an antihegemonic way through racialized communities' social movements. An example of a policy originating in southern countries is the right of people relocated to self-determination recognized in article one of the International Covenant on Civil and Political Rights and other treaties (Gonzalez, 2020). This alternative approach recognizes climate refugees as political subjects capable of collectively deciding their destinies, reserving peoples, and states the right to preserve their cultural and community integrity and migrate with dignity. The approach to self-determination is a process that aims to create a bridge between the abyssal line that divides North and South by granting communities vulnerable to climate change the right to migrate collectively to preserve their culture, language, customs, and political organization.

However, the heterogeneity of climate-relocated communities, the need for conflict resolution mechanisms, and adaptation to migration present important obstacles to applying this approach. Another possible challenge is the fact that the right to collective migration must be complemented by the responsibility of northern countries to receive immigrants who decide to migrate individually. To this end, Gonzalez (2020) suggests the creation of passports for deterritorialized individuals, allowing climate migrants access to different states with the possibility of naturalization, as well as a liability-based legal framework that would emphasize the duty of Northern countries to reduce GHG emissions to prevent displacement. Gonzalez (2020) also contends that financial resources for climate adaptation and disaster risk reduction should be considered.

20.4 Where we are on CCS, energy transition, and climate justice

The CCS technology is crucial to attaining GHG emissions reduction goals by 2050 (IEA, 2016). CCS enables the use of fossil fuels to significantly reduce GHG emissions (Dessler and Parson, 2020). Accordingly, CCS would greatly assist in the energy transition to cleaner and greener energy sources, such as renewables.

Furthermore, it can also assist in implementing fairness considerations domestically as well as globally. Indigenous peoples, coastal and island populations, and inhabitants of the developing world are disproportionally affected by the consequences of climate change (IPCC, 2018). Environmental and climate justice conceptualizations encompass the disproportionate impact of environmental pollution and regulations on minorities and those with less economic resources. These justice-based considerations include a multitude of actions, such as critical enforcement of environmental regulation, public participation in decision-making, transportation equity, workplace and school exposure to toxic materials, while acknowledging that even global climate policies will be grounded on their local level effects (Outka, 2012). Therefore an equitable energy transition considers fairness and equality when designing policies on energy access and security to displace fossil fuels. It is noteworthy that low-carbon technologies themselves can lead to injustice (such as the takeover by corporations over community ownership, poor work conditions in the biofuel industry, for instance), so policymakers and key stakeholders are urged to include such considerations when developing and implementing energy policies (Gambhir et al., 2018).

In light of the above and considering the main findings presented in previous chapters throughout this book, CCS is not a panacea. Despite being a new technology deemed essential to significantly mitigate carbon emissions, CCS offers an intricate set of particular challenges. Major hurdles for CCS's widespread implementation include its lack of afford-ability and concerns over leakage, and the long-term stabilization required for carbon cap-ture benefits to keep accruing. A related challenge is a need for technological expertise. Besides, CCS may be perceived as an enabler of a carbon-based economy instead of foster-ing the transition to a fossil-fuel-free economic model. The market itself appears reluctant to support the widespread use of CCS, as the annual investment in CCS has consistently accounted for less than 0.5% of global investment in renewables and efficiency technolo-gies (IEA, 2020a). Climate justice, information and education training, and expertise may help to overcome such reluctance. As key chapters discussed throughout this book, open participation of all sectors of industrial activities and energy production as well as of all segments of civil society may advance the development of policies more aligned with car-bon neutrality by 2050 while not adversely impacting those less well-off.

Institutional action is needed to enable the use of CCS globally and regardless of the level of development of a given country. Under international law, there are currently two significant principles to support the implementation of CCS technology. The first principle refers to sustainable development, as defined in the UNFCCC (1992) and its implementing treaty, namely, the Paris Agreement on Climate Change (Paris Agreement, 2015). Despite a multitude of definitions and the imprecise scope of its application, sustainable develop-ment nowadays is interpreted as reconciling social development, environmental protec-tion, and economic growth (Arlota, 2021, and references therein). Hence, CCS would advance the energy transition by making the current energy production significantly less polluting, enabling recently industrialized nations, in particular, to secure access to energy sources and economic growth in a sustainable manner.

The second fundamental principle is designated as the principle of CBDR–RC, which refers specifically to different responsibilities allocated among countries, as defined under Article 3(1) of the UNFCCC, namely: "The Parties should protect the climate system for the benefit of present and future generations of humankind, on the basis of equity and in accordance with their CBDR–RC" (UNFCCC, 1992). This principle also informed the Paris Agreement on Climate Change (Paris Agreement, 2015), whose Article 2(2) determines the following: "This Agreement will be implemented to reflect equity and the principle of CBDR–RC, in the light of different national circumstances." In this context, it is worth praising the Intergovernmental Panel on Climate Change (IPCC)'s acknowledgment, decades ago, of the importance of the transfer of technology as an essential feature to foster adaptation and enhance climate resilience (IPCC, 2000). As the chapters in this book have demonstrated, intellectual property rights can be a powerful tool in fostering such transfer of technology in the CCS context. Nonetheless, as of today, they can be viewed as less than supportive of such a role.

In such a scenario, institutional action, under the terms of the UNFCCC and the Paris Agreement on Climate Change, requires developed countries to take the lead in reducing emissions. Notably, the increasing accessibility of renewable energy sources and technological advancements have rebuked the myth that economic development and increasing GHG emissions must coexist (Arlota, 2020b, and references therein). In this vein, it is noteworthy that the Paris Agreement is routinely cited as the only effective institutional solution to climate change (Arlota, 2020a). According to the Paris Agreement, the most outstanding international obligations of the member states who are signatories are their financial contributions to the Green Climate Fund (Paris Agreement, Article 9), which determines that: "Developed country Parties shall provide financial resources to assist developing country Parties concerning both mitigation and adaptation in continuation of their existing obligations under the Convention. Other Parties are encouraged to provide or continue to provide such support voluntarily." Another essential international obligation established under the Paris Agreement is fulfilling each country's National Determined Contributions (NDCs, under Article 4(2). As the treaty's language conveys, financial contributions and NDCs have voluntarily established commitments determined by each country exclusively.

20.5 Current trends

As discussed previously in this book, to the extent that International Energy Law gains momentum as an autonomous legal field, climate financing is expected to increase academic interest. Likewise, it is likely to occur with international investment arbitration in the energy sector, specifically. These appealing features of International Energy Law are reflective of particular proposals reducing or eliminating the financing of fossil fuel projects and general implementation of more stringent carbon taxes worldwide (see, e.g., the European Union and the Canadian general initiatives). The increasing interest in International Energy Law also evidenced the critical role of the Paris Agreement on Climate Change and its dual approach reconciling bottom-up measures (such as the NDC, for instance, which are established from each country's level to the transnational sphere) with top–bottom goals (as each NDC of a given country is progressively more ambitious than the country's previously submitted NDC) (Arlota, 2020a). Therefore the participation of international organizations supporting national institutions that are willing to build consensus around climate change is of paramount importance to fostering domestic policies aligned with the Paris Agreement's main objectives.

Amid the current COVID-19 pandemic, as declared by the WHO (2020), an economic crisis started to develop based on the drop in oil prices due to a price war between Russia and Saudi Arabia. Global oil and gas markets were facing unprecedented difficulties because of collapsing demand coupled with an already abundant supply that continues to increase in light of the reduced consumption caused by the pandemic (IEA, 2020b).

This can be an opportunity to rebuild the economy on a greener basis. For this to occur, however, significant changes need to be implemented on several fronts. A key strategy will be to reduce (aiming at elimination) financing and subsidies for fossil fuels. A report by authors affiliated with the International Monetary Fund found that coal, diesel, and natural gas subsidies do not reflect environmental costs (Coady et al., 2019). This is very concerning because the disregard of environmental costs is equivalent to, in practice, subsidizing environmental damage. This financing scheme (and the absence of stringent regulatory actions steering economic actions in a different direction) significantly imperils the energy transition to a carbon-neutral world by 2050.

Despite such a challenging scenario, there are reasons for some cautious optimism. On December 12, 2020, the United Nations, France, and the United Kingdom, in partnership with Chile and Italy, jointly hosted the Climate Ambition Summit in preparation for the next Conference of the Parties—COP 26, which was postponed to November 2021, in the aftermath of the pandemic. The Summit's principal objective was to discuss more ambitious NDCs (Climate Ambition Summit, 2020). During this extraordinary Summit, the European Union reiterated its earlier commitment to reduce its GHG emissions by at least 55%, compared to the net carbon emissions measured in 1990. The highest GHGs emitters, such as the European Union, Japan, and China (committing to achieve net-neutrality by 2060), have already explicitly embraced significant reductions (Arlota, 2021). The Biden—Harris administration has manifested its intention to act in a similar direction (Climate Ambition Summit, 2020). Notably, the increasing number of countries committing to net-zero emissions goals by 2050 has been deemed "the most significant and encouraging climate policy development of 2020" (U.N. Emissions Gap Report, 2020, at V).

20.6 Conclusion

This chapter provided a historical overview of environmental justice and how it informs and shapes the current conceptualizations of climate justice. The chapter proceeds to discuss significant pressure points on climate change, namely, migration and justice. In addition, this chapter summarized the main findings of previous chapters of this book, articulating these findings in relation to current trends on international energy law, CCS, and climate justice.

Finally, this chapter showed how policymakers and scholars might be better equipped with the analysis and policies discussed throughout this work. And, to the extent that some of the challenges coming in their direction may have been anticipated in the book, readers may behave (or change their behavior) accordingly. As for the unforeseeable challenges, the regulatory framework presented should also be determinative of new policies in the complex, unique, and interesting times to come. One thing is certain: justice and fairness are no longer abstract propositions. They now can be evaluated and incorporated in the domestic and international spheres. The rejoining of the United States to the Paris Agreement increases the momentum on meaningful global action to achieve the treaty's main goals regarding GHG emissions. Moreover, as many countries pledge to achieve carbon neutrality by 2050, CCS and more stringent regulations aiming to reduce GHG emissions are expected to face less opposition. Accordingly, there are reasons to be cautiously optimistic about the future.

Acknowledgments

Hirdan Katarina de Medeiros Costa is grateful to the "Research Center for Gas Innovation—RCGI" (Fapesp Proc. 2014/50279-4), supported by FAPESP and Shell, organized by the University of São Paulo, and the strategic importance of the support granted by the ANP (National Agency of Petroleum, Natural Gas and Biofuels of Brazil) through the R&D clause. She also thanks to the support from the National Agency for Petroleum, Natural Gas and Biofuels Human Resources Program (PRH-ANP), funded by resources from the investment of oil companies qualified in the R,D&I clauses from ANP Resolution No. 50/2015 (PRH 33.1—Related to Call No. 1/2018/ PRH-ANP; Grant FINEP/FUSP/USP Ref. 0443/19).

References

Acselrad, H., Mello, C.C.A., Bezerra, G. das N., 2009. O que é Justiça Ambiental. Garamond, Rio de Janeiro.

Althor, G., Watson, J., Fuller, R., 2016. Global Mismatch Between Greenhouse Gas Emissions and the Burden of Climate Change. Scientifc Reports, London, 6, 20281, 2016. Disponível em: https://www.nature.com/articles/ srep20281#-citeas (accessed 03.11.20). Available from: https://doi.org/10.1038/srep20281.

Arlota, C., 2020a. Does the United States withdrawal from the Paris Agreement on climate change pass the cost- −benefit analysis test. University of Pennsylvania Journal of International Law 41, 881−938.

Arlota, C., 2020b. The Amazon is burning—is Paris, too? A comparative analysis between the United States and Brazil based on the Paris Agreement on climate change. Georgetown Journal of International Law 52 (1), 161−214.

Arlota, C., 2021. International energy law and the Paris Agreement in the aftermath of the COVID-19 pandemic: challenges and possibilities. ILSA Journal of International and Comparative Law 27, 275−292.

Bali, 2002. Principles of climate justice. Available from: http://www.ejnet.org/ej/bali.pdf (accessed 03.11.20).

Barnett, J., 2009. Human rights and vulnerability to climate change. In: Humphreys, S. (Ed.), Human Rights and Climate Change. Cambridge University Press, Nova Iorque, pp. 257−271.

Climate Ambition Summit, 2020. https://www.climateambitionsummit2020.org/ (last visited 28.12.20).

Coady, D., et al., 2019. Global Fossil Fuel Subsidies Remain Large: An Update Based on Country-Level Estimates, pp. 17 − 19. IMF Working Paper WP/19/89.

Costa, H.K.M., Santos, E.M., 2013. Justiça e sustentabilidade: a destinação dos royalties de petróleo. Estud. av., São Paulo, 27 (77), 143−160. Available from: http://www.scielo.br/scielo.php?script = sci_arttext&pid = S0103- 40142013000100011&lng = en&nrm = iso; https://doi.org/10.1590/S0103-40142013000100011 (accessed 04.11.20).

Dessler, A., Parson, E., 2020. The Science and Politics of Global Climate Change. Cambridge University Press, Cambridge.

Fagundez, G.T., Albuquerque, L., Filpi, H.F.F.C.M., 2020. Violação de direitos humanos e esforços de adaptação e mitigação: uma análise sob a perspectiva da justiça climática. RIDH I Bauru 8 (1), 227−240.

Ferraresi, P., 2020. Racismo Ambiental e justiça social. Available from: https://escola.mpu.mp.br/publicacoes/ boletim-cientifico/edicoes-do-boletim/boletim-cientifico-n-37-edicao-especial-2012-direito-a-nao-discriminacao/ racismo-ambiental-e-justica-social (accessed 25.06.21).

Gambhir, A., Green, F., Pearson, P.J., 2018. Towards a Just and Equitable Low-Carbon Energy Transition. Grantham Institute, Briefing Paper 26.

Gonzalez, C.G., 2020. Racial Capitalism, Climate Justice, and Climate Displacement. Oñati Socio-Legal Series, USA.

Herculano, S., 2002. Riscos e desigualdades social: a temática da justiça ambiental no Brasil. I Encontro da ANPPAS— Indaiatuba, São Paulo, GT Teoria e Ambiente. Available from: www.anppas.org.br (accessed 25.06.21).

Humphreys, S., 2009. Human Rights and Climate Change. Cambridge University Press, New York. Available from: https://www.cambridge.org/core/books/human-rights-and-climate-change/introduction- human-rights-and-climate-change/B89D34682C9C05FF50914706A342A275 (accessed 03.11.20).

IEA (International Energy Agency), 2016. 20 Years of Carbon Capture and Storage—Accelerating Future Deployment. OECD, Paris. Available from: https://www.oecd-ilibrary.org/energy/20-years-of-carbon-cap- ture-and-storage_9789264267800-en (accessed 03.11.20).

IEA (International Energy Agency), 2020a. Energy Technology Perspectives 2020: Special Report on Carbon Capture, Utilisation and Storage. Paris, France.

IEA (International Energy Agency), 2020b. An Unprecedented Global Health and Economic Crisis. https://www. iea.org/topics/covid-19 (accessed 14.01.20).

IPCC, 2018. Global Warming of 1.5°C (Summary for Policymakers). World Meteorological Organization, Geneva.

IPCC (Intergovernmental Panel on Climate Change), 2000. Special Report on Emissions Scenarios. Cambridge University Press, Cambridge. Available from: https://www.ipcc.ch/site/assets/uploads/2018/03/emissions_scenarios-1.pdf (accessed 01.09.21).

Moss, J. (Ed.), 2009. Climate justice. In: Climate Change and Social Justice. Melbourne University Press, Victoria, pp. 51–66.

Outka, U., 2012. Environmental justice issues in sustainable development: environmental justice in the renewable energy transition. Journal of Environmental and Sustainability Law 19 (1), 60–122.

Peel, J., Lin, J., 2019. Transnational climate litigation: the contribution of the global south. American Journal of International Law, Cambridge 113 (4), 679–726. Available from: https://www.cambridge.org/core/journals/american-journal-of-international-law/article/transnational-climate-litigation-the-contribution-of-the-global-south/ABE6CC59AB7BC276A3550B9935E7145A (accessed 04.11.20).

Schlosberg, D., Collins, L.B., 2014. From environmental to climate justice: climate change and the discourse of environmental justice. WIREs Climate Change, Hoboken, NJ (EUA), 5, p. 363.

Seck, S.L., 2020. A relational analysis of enterprise obligations and carbon majors for climate justice. Oñati Socio-Legal Series: Climate Justice in the Anthropocene.

UNFCC, 1992. The United Nations Framework Convention on Climate Change art. 23, September 5, 1771 U.N.T.S. 107.

United Nations Environmental Programme, 2020. Emissions Gap Report 2020: Executive Summary. https://wedocs.unep.org/bitstream/handle/20.500.11822/34438/EGR20ESE.pdf?sequence = 8.

United Nations Paris Agreement on Climate Change, 2015. December 12, 2015, 54113 U.N.R.N. 88.

World Health Organization, 2020. WHO Director-General's Opening Remarks at the Media Briefing on COVID-19. https://www.who.int/dg/speeches/detail/who-director-general-s-opening-remarks-at-the-media-briefing-on-covid-19—11-march-2020 (accessed 14.01.20).

Yildirim, B.S., 2020. Climate justice at the local level: the case of Turkey. Journal of Political Science 45, 7–30.

Further reading

First Global Report (UNEP). Available from: https://wedocs.unep.org/bitstream/handle/20.500.11822/27279/Environmental_ru-le_of_l-aw.pdf-?sequence = 1&isAllowed = y (accessed 29.12.20).

Schlosberg, D., 2007. Defining Environmental Justice: Theories, Movements, and Nature. Oxford University Press, United Kingdom.

Index